Dynamo and Dynamics,
a Mathematical Challenge

# NATO Science Series

*A Series presenting the results of scientific meetings supported under the NATO Science Programme.*

The Series is published by IOS Press, Amsterdam, and Kluwer Academic Publishers in conjunction with the NATO Scientific Affairs Division

*Sub-Series*

| | |
|---|---|
| **I. Life and Behavioural Sciences** | IOS Press |
| **II. Mathematics, Physics and Chemistry** | Kluwer Academic Publishers |
| **III. Computer and Systems Science** | IOS Press |
| **IV. Earth and Environmental Sciences** | Kluwer Academic Publishers |

The NATO Science Series continues the series of books published formerly as the NATO ASI Series.

The NATO Science Programme offers support for collaboration in civil science between scientists of countries of the Euro-Atlantic Partnership Council. The types of scientific meeting generally supported are "Advanced Study Institutes" and "Advanced Research Workshops", and the NATO Science Series collects together the results of these meetings. The meetings are co-organized bij scientists from NATO countries and scientists from NATO's Partner countries – countries of the CIS and Central and Eastern Europe.

**Advanced Study Institutes** are high-level tutorial courses offering in-depth study of latest advances in a field.
**Advanced Research Workshops** are expert meetings aimed at critical assessment of a field, and identification of directions for future action.

As a consequence of the restructuring of the NATO Science Programme in 1999, the NATO Science Series was re-organized to the four sub-series noted above. Please consult the following web sites for information on previous volumes published in the Series.

http://www.nato.int/science
http://www.wkap.nl
http://www.iospress.nl
http://www.wtv-books.de/nato-pco.htm

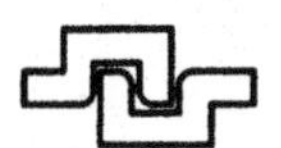

**Series II: Mathematics, Physics and Chemistry – Vol. 26**

# Dynamo and Dynamics, a Mathematical Challenge

edited by


## P. Chossat

I.N.L.N. (CNRS, UMR 6618),
Valbonne, France

## D. Ambruster

Department of Mathematics,
Arizona State University,
Tempe, Arizona, U.S.A.

and

## I. Oprea

Faculty of Mathematics,
University of Bucarest, Romania

**Kluwer Academic Publishers**

Dordrecht / Boston / London

Published in cooperation with NATO Scientific Affairs Division

Proceedings of the NATO Advanced Research Workshop on
Dynamo and Dynamics, a Mathematical Challenge
Cargèse, France
21–26 August, 2000

A C.I.P. Catalogue record for this book is available from the Library of Congress.

ISBN 0-7923-7069-4 (HB)
ISBN 0-7923-7070-8 (PB)

---

Published by Kluwer Academic Publishers,
P.O. Box 17, 3300 AA Dordrecht, The Netherlands.

Sold and distributed in North, Central and South America
by Kluwer Academic Publishers,
101 Philip Drive, Norwell, MA 02061, U.S.A.

In all other countries, sold and distributed
by Kluwer Academic Publishers,
P.O. Box 322, 3300 AH Dordrecht, The Netherlands.

*Printed on acid-free paper*

---



Printed in the Netherlands.

**TABLE OF CONTENTS**

## PREFACE

This book contains the lectures given at the workshop "Dynamo and dynamics, a mathematical challenge" held in Cargèse from August 21 to 26, 2000. The workshop differed from most previous conferences on the dynamo effect in two important respects. First, it was at this international conference that the experimental observation of homogeneous fluid dynamos was first reported. Second, the conference gathered scientists from very different fields, thus showing that the dynamo problem has become an interdisciplinary subject involving not only astrophysicists and geophysicists, but also scientists working in dynamical systems theory, hydrodynamics, and numerical simulation, as well as several groups in experimental physics.

This book thus reports important results on various dynamo studies in these different contexts:

- Decades after the discovery of the first analytic examples of laminar fluid dynamos, the self-generation of a magnetic field by a flow of liquid sodium has been reported by the Karlsruhe and Riga groups. Although there were no doubts concerning the self generation by the laminar Roberts-type or Ponomarenko-type flows that were used, these experiments have raised interesting questions about the influence of the turbulent fluctuations on the dynamo threshold and on the saturation level of the magnetic field.
- The increase in computing power now allows the simulation of planetary dynamos in almost realistic geometries and provides some results in agreement with the observations. Moreover, direct numerical simulations of the MHD equations yield insights into the different possible equilibria between magnetic and kinetic degrees of freedom, leading to different saturation levels of the magnetic field generated.
- Recent observations of stellar and galactic magnetic fields are discussed mostly in the context of mean field magnetohydrodynamics. These studies are likely to improve our understanding of the internal structure and the dynamics of the sun, and to clarify the origin and the role of galactic magnetic fields and their connection with various dynamical processes occuring in the interstellar medium.

- The tools of dynamical systems theory are used to further a qualitative understanding of dynamo dynamics. Using symmetry arguments, the full MHD equations can be reduced in the vicinity of bifurcation points, leading to low-dimensional dynamical systems providing possible models for the solar cycle or for the reversals of the Earth's magnetic field.

However, there are still large gaps to be filled between these different levels of description in order to obtain a full understanding of dynamo processes:
- It is still difficult to connect low-dimensional models with numerical simulations of the full problem and to determine how robust are the observed phenomena in the presence of many turbulent modes.
- Direct numerical simulations are still restricted to values of the fluid parameters which are very unrealistic when compared to laboratory experiments. Turbulence may in some cases have little influence on the dynamo threshold but is likely to strongly modify the saturation level of the magnetic field.
- An experimental demonstration of an unconstrained turbulent dynamo is still lacking and a turbulent dynamo with a negligible mean flow appears even more difficult to achieve, although phenomenological models with these properties have been known for a long time.
- A large gap also exists between laboratory experiments and ``natural" dynamos. Although the effect of the Coriolis force can be easily studied, many astrophysical plasmas strongly differ from liquid metals and may involve effects that are not described by the simple MHD equations valid for liquid metals. There are still many unknowns, both for stellar and planetary dynamos, that complicate the simulation of the relevant mechanisms in laboratory experiments. It is also unlikely that laboratory experiments will succeed in attaining magnetic Reynolds numbers that significantly exceed the dynamo threshold.
- Finally, at a more mathematical level, the problem of describing a bifurcation towards a dynamo from a strongly turbulent velocity field involves both technical difficulties and probable new qualitative effects.

Although it is impossible to know if one of the above open questions will lead to successful research in the future, we can certainly conclude that, despite considerable progress, the dynamo problem still merits a great deal of study.

Paris, March 5th, 2001,

Stéphan Fauve

## ACKNOWLEDGMENTS

This conference was made possible thanks to a NATO Advanced Research Workshop grant. We are also grateful to the Centre National de la Recherche Scientifique and to the National Science Foundation for their financial support. Many thanks to the staff of the Institut d'Etudes Scientifiques de Cargèse and to Isabelle Larochette for their efficient help to the organizers.

**PARTICIPANTS**

Dieter ARMBRUSTER dieter@source.la.asu.edu
Department of Mathematics, ASU, Tempe, Arizona 85287-1804, USA

Elena BENEVOLENSKAYA elena@quake.stanford.edu
Pulkovo Astronomical Observatory, St.Petersburg, Russia

Alberto BIGAZZI alberto.bigazzi@mate.polimi.it
Dept. of Mathematics, Politecnico di Milano, Piazza Leonardo da Vinci 32
20133 Milano, Italy

Marc-Etienne BRACHET Marc-Etienne.Brachet@lps.ens.fr
Ecole Normale Supérieure, LPS, 24 rue Lhomond, 75231 Paris cedex 05, France

Axel BRANDENBURG brandenb@nordita.dk
Nordita, Blegdamsvej 17, 2100 Copenhagen 0, Danmark

Werner BRASCH egbers@zarm.uni-bremen.de
ZARM, University of Bremen, Am Fallturm, 28359 Bremen, Germany

Jozef BRESTENSKY Jozef.Brestensky@fmph.uniba.sk
Dept of Geophysics, Faculty of Maths and Physics, Comenius University
Mlynska dolina, 842 48 Bratislava, Slovakia

Peter BROOKE j.m.brooke@mcc.ac.uk
CSAR Applications Support Group, Manchester Computing, Manchester University
Oxford Road, Manchester M13 9PL, U.K.

Nic BRUMMELL brummel@solarz.colorado.edu
JILA, University of Colorado, Boulder, CO 80309-0440 USA

Fritz BUSSE Busse@uni-bayreuth.de
University of Bayreuth, Physikalisches Institut, 95440 Bayreuth, Germany

Fausto CATTANEO cattaneo@mhd2.uchicago.edu
ASCI/FLASH Center, U. of Chicago, 5640 S. Ellis Ave., Chicago, IL 60615, USA

Pascal CHOSSAT chossat@inln.cnrs ;fr
I.N.L.N., 1361, route des lucioles, 06560 Valbonne, France

Kelly CLINE clinek@solarz.colorado.edu
JILA, campus box 440, University of Colorado
Boulder, CO 80309-0440, USA

Eurico COVAS E.O.Covas@qmw.ac.uk
Astronomy Unit, School of Math Sci. , Queen Mary and Westfield College
Mile End Road , London E1 4NS, U.K.

Ivan CUPAL ic@ig.cas.cz
Geophysical Institute ACSR, Bocni II 1401, 141 31 Prague 4, Czech Republic

François DAVIAUD daviaud@drecam.saclay.cea.fr
SPEC, CEA SACLAY, 91191 Gif sur Yvette Cedex, France

John DONALD jdonald@math.ucla.edu
University of California at Los Angeles
USA

Emmanuel DORMY dormy@ipgp.jussieu.fr
Institut de physique du Globe de Paris
4, place Jussieu
75252 Paris Cedex 05 FRANCE

Christoph EGBERS egbers@las.tu-cottbus.de
Lehrstuhl für Aerodynamik und Strömungslehre, Brandenburg Technical University
Karl-Liebknecht-Str.102, 03046 Cottbus, Germany

Thierry EMONET emonet@flash.uchicago.edu
The University of Chicago, Astronomy and Astrophysics
5640 S. Ellis ave, Chicago IL 60637 USA

Stephan FAUVE Stephan.Fauve@lps.ens.fr
Ecole Normale Superieure, LPS, 24, rue Lhomond, 75005 Paris, France

Katia FERRIERE ferriere@ast.obs-mip.fr
Observatoire Midi-Pyrénées, 14, av Ed Belin, 31400 Toulouse, France

Cary FOREST cbforest@facstaff.wisc.edu
Dept of Physics, 3277 Chamberlin Hall, University of Wisconsin
1150 University Ave, Madison, WI 53706 USA

Peter FRICK frick@icmm.ru
Institute of Continuous Media Mechanics, Korolev 1, Perm, 614061, Russia

Helmut FUCHS hfuchs@aip.de
Astrophysikalisches Institut, An der Sternwarte 16, 14482 Potsdam, Germany

A. GABOV alex@gabov.srcc.msu.su
Dept. of Physics, Moscow State University, Moscow 119899, Russia

Agris GAILITIS gailitis@sal.lv
Institute of Physics, 2169 Salaspils-1, Lettonia

Pavel HEJDA ph@ig.cas.cz
Geophysical Institute ACSR, Bocni II 1401, 141 31 Prague 4, Czech republic

Rainer HOLLERBACH rainer@maths.gla.ac.uk
Dept of Mathematics, University of Glasgow, Glasgow, G12 8QW, U.K.

David W. HUGHES dwh@amsta.leeds.ac.uk
Dept of Applied Mathematics, University of Leeds, Leeds LS2 9JT, U.K.

David IVERS david@maths.usyd.edu.au
School of Maths and Statistics F07, U. of Sydney, NSW 2006, Australia

Dominique JAULT Dominique.Jault@ujf-grenoble.fr
LGIT, Bat IRIGM BP 53
38041 Grenoble Cedex 09 FRANCE

Christopher A. JONES CAJones@maths.ex.ac.uk
School of Math Sciences, U. of Exeter, Exeter, EX4 4QE, U.K.

Markus JUNK junk@zarm.uni-bremen.de
ZARM, University of Bremen, Am Fallturm, 28359 Bremen, Germany

Ralf KAISER btm601@uni-bayreuth.de
U. of Bayreuth, Facultät für Mathematik und Physik, 95440 Bayreuth, Germany

Nathan KLEEORIN nat@menix.bgu.ac.il
Dept of Mechanical Engineering , The Ben-Gurion University of the Negev
84105 Beer-Sheva, P. O. Box 653, Israel

Edgar KNOBLOCH knobloch@physics.berkeley.edu
Dept of Physics, University of California, Berkeley, CA 94720, USA

Alexander KOSOVICHEV sasha@khors.Stanford.edu
W.W.Hansen Experimental Physics Lab, Stanford U., 455 via Palou, CA 94305, USA

Jurgen KURTHS juergen@agnld.uni-potsdam.de
Institut für Physik, University of Potsdam
PF 601553 Potsdam GERMANY

Kirill KUZANYAN kuzanyan@dnttm.ru
IZMIRAN, Lab. Heliophysics, Troitsk, Moscow region, 142092 Russia

Daniel P. LATHROP dpl@complex.umd.edu
Institute of Plasma Research, Dept. of Physics, University of Maryland
3319A A.V. Williams Bldg, College Park, MD 20742 , USA

Patrice LAURE laure@inln.cnrs.fr
I.N.L.N., 1361, route des lucioles, 06560 Valbonne, France

Reiner LAUTERBACH lauterbach@math.uni-hamburg.de
Fachbereich Mathematik , Universitaet Hamburg, Bundesstr. 55
20146 Hamburg, Germany

Jacques LEORAT leorat@obspm.fr
DAEC, Observatoire de Paris-Meudon, 92195 Meudon, France

Paul MATTHEWS paul.matthews@notthingham.ac.uk
School of Math Sciences, University of Nottingham, University Park,
Nottingham NG7 2RD, U.K.

Ian MELBOURNE ism@math.UH.EDU
Dept of Mathematics, University of Houston, Houston TX 77204-3476, USA

Ulrich MUELLER ulrich.mueller@ike.fzk.de
Forschungszentrum Karlsruhe I K E
Postfach 36 40
76021 Karlsruhe, GERMANY

Caroline NORE nore@themis.limsi.fr
LIMSI, batiment 508/BP 133, 91403 ORSAY Cedex, France

Christiane NORMAND normand@spht.saclay.cea.fr
CEA SACLAY
91191 Gif sur Yvette Cedex FRANCE

Marc NORNBERG mdnornberg@studenbts.wisc.edu
Dept of Physics, 3277 Chamberlin Hall, U. of Wisconsin, Madison, WI 53706, USA

Rob O'CONNELL roconnell@facstaff.wisc.edu
Plasma Physics, 3277 Chamberlin Hall, U. of Winsconsin
1150 University Avenue, Madison WI 53706, USA

Philippe ODIER podier@physique.ens-lyon.fr
Laboratoire de Physique, ENS lyon, 46, allée d'Italie, 69367 Lyon Cedex 07, France

Iuliana OPREA juliana@math.colostate.edu
Faculty of Mathematics, University of Bucarest, Str Academiei 14, sector 1
70109 Bucarest, Romania

Katarzyna OTMIANOWSKA-MAZUR otmian@oa.uj.edu.pl
Astronomical Observatory, Jagiellonian University, ul. Orla 171
30-244 Krakow, Poland

François PETRELIS petrelis@lps.ens.fr
Ecole Normale Supérieure, LPS, 24, rue Lhomond, 75231 PARIS cedex 05, France

Jean-François PINTON pinton@physique.ens-lyon.fr
Laboratoire de Physique, ENS Lyon, 46, allée d'Italie, 69367 Lyon Cedex 07, France

Yannick PONTY ponty@maths.ex.ac.uk
Observatoire de la Côte d'Azur, labo Cassini, BP 4229, O6304 Nice Cedex 4, France

Michael PROCTOR M.R.E.Proctor@damtp.cam.ac.uk
DAMTP, University of Cambridge, Silver St., Cambridge, CB3 9EW , U.K.

Karl-Heinz RÄDLER khraedler@aip.de
Astrophysical Institute Potsdam, An der Sternwarte 16, 14482 Potsdam, Germany

Michel RIEUTORD rieutord@ast.obs-mip.fr
Observatoire Midi-Pyrénées, 14, avenue Edouard Belin, 31400 Toulouse, France

Paul ROBERTS roberts@math.ucla.edu
UCLA Institute of Geophysics and Planetary Physics
405 Hilgard Avenue
Los Angeles, CA 90095 USA

Igor ROGACHEVSKII gary@menix.bgu.ac.il
Dept of Mechanical Eng. The Ben-Gurion University of the Negev
84105 Beer-Sheva, P. O. Box 653, Israel

Alastair RUCKLIDGE A.M.Rucklidge@damtp.cam.ac.uk
DAMTP, University of Cambridge, Silver St., Cambridge, CB3 9EW U.K.

Guenther RUEDIGER gruediger@aip.de
Astrophysical Institute Potsdam
An der Sternwarte 16, 14482 Potsdam GERMANY

Alexander RUZMAIKIN aruzmaikin@jplsp.jpl.nasa.gov
Jet Propulsion Laboratory, CalTech, mail stop 169-506
4800 Oak Grove Drive, Pasadena, CA 91109, USA

Denys SCHMITT Denys.Schmitt@polycnrs-gre.fr
Labo Magnétisme Louis Neel – CNRS
BP 166, 38042 Grenoble cedex 9, FRANCE

Norbert SEEHAFER seehafer@agnld.uni-potsdam.de
Institut für Physik, University of Potsdam, PF 601553 Potsdam, Germany

Woodrow SHEW wshew@Glue.umd.edu
Energy research, Bldg 223, U. of Maryland, College Park, MD 20472, USA

Mary SILBER silber@nimbus.esam.nwu.edu
Dept of Eng. Sciences and Applied Maths Northwestern University
Evanston, IL 60208 USA

Jan SIMKANIN simkanin@fmph.uniba.sk
Dept of Geophysics, Faculty of Maths and Physics, Comenius University
Mlynska dolina, 842 48 Bratislava, Slovakia

Daniel SISAN sisan@complex.umd.edu
Energy research, Bldg 223, U. of Maryland, College Park, MD 20472, USA

Andrew SOWARD A.M.Soward@exeter.ac.uk
School of Math Sciences, University of Exeter, Laver Building , North Park Road
Exeter, EX4 4QE, U.K.

Serguei V. STARCHENKO starche@borok.adm.yar.ru
GFO, Borok 42 - 4, Nekouzskiy, Yaroslavskaya obl.152742, Russia

Rodion STEPANOV rodion@icmm.ru
Institute of Continuous Media Mechanics, Korolev 1,, Perm, 614061, Russia

Robert STIEGLITZ robert.stieglitz@ike.fzk.de
Forschungszentrum Karlsruhe I K E
Postfach 36 40, 76021 Karlsruhe GERMANY

Reza TAVAKOL reza@maths.qmw.ac.uk
Astronomy Unit, School of Math Sciences, Queen Mary & Westfield College,
University of London, Mile End Road, London E1 4NS, U.K.

Jean-Claude THELEN jct@mhd11.uchicago.edu
ASCI/FLASH Center, U. of Chicago, 5640 S. Ellis Ave., Chicago, IL 60615, USA

Andreas TILGNER andreas.tilgner@uni-bayreuth.de
Physikalisches Institut, University of Bayreuth, 95440 Bayreuth, Germany

Steve TOBIAS smt@amsta.leeds.ac.uk
Dept of Applied Mathematics, University of Leeds, Leeds LS2 9JT, U.K.

Lorenzo VALDETTARO valde@pesto.mate.polimi.it
Dipartimento di Matematica, Politecnico di Milano, 20133 Milano, ITALY

Brigitta Von REKOWSKI B.M.Von-Rekowski@newcastle.ac.uk
Dept of Mathematics, U. of Newcastle, Newcastle upon Tyne NE1 7 RU, U.K.

Nigel O. WEISS N.O.Weiss@damtp.cam.ac.uk
D.A.M.T.P. , Silver Street, Cambridge CB3 9EW, U.K.

# A NONSTATIONARY DYNAMO EXPERIMENT IN A BRAKED TORUS

*Current State Of Perm Project*

P. FRICK, S. DENISOV, S. KHRIPCHENKO, V. NOSKOV, D. SOKOLOFF AND R. STEPANOV
*Institute of Continuous Media Mechanics,*
*Korolev 1, 614061, Perm, Russia*

**Abstract.**
An experiment for study magnetic field evolution in nonstationary turbulent flow of liquid sodium is discussed. Kinetic energy is accumulating during relatively long acceleration of the toroidal vessel (diameter about 1 m) with liquid sodium and the large power screw flow will be obtained only during the abrupt brake. Internal diverters will enable the required profile of velocity. The expected flow can act as a dynamo provided its magnetic Reynolds $R_m$ number is expected to exceed several tens. The experiment is developing in the Institute of Continuous Media Mechanics of Ural Branch of Russian Academy of Sciences, Perm, Russia.

## 1. Introduction

A fundamental result of Latvian and German teams obtained experimental verification for dynamo action [1, 2] stimulates further experimental efforts to realize various dynamo mechanisms specific for magnetic field generation in celestial bodies. Here we present a new project, which is intended to study the magnetic field evolution in nonstationary free decaying turbulent flow of liquid sodium [3]. A toroidal vessel (of about 1 m in diameter) with sodium will be slowly accelerated by a relatively low-power engine up to the velocities of about 100 $ms^{-1}$ and will be stopped abruptly by an external force. The internal diverters will provide the required screw profile of the sodium velocity. The expected flow can act as a screw dynamo provided its magnetic Reynolds $R_m$ number exceeds several dozens.

Free decaying turbulent flows are specific, say, for astrophysical jets accelerated by an active galactic nucleus or for turbulent flow in a galactic



*P. Chossat et al. (eds.), Dynamo and Dynamics, a Mathematical Challenge,* 1–8.

cluster between merges provided the flow is driven by merges. Resent studies for a given type of dynamo in forced [4] and free decaying turbulence [5] demonstrate that they can result is different magnetic field configurations.

## 2. Hydrodynamic experiment: Apparatus

The evolution of swirling flow in the braked channel was studied using a water prototype. A toroidal channel was cut through a plexiglass cylinder, made up of two halves and installed on a hub of a car wheel incorporating the braking system. The hub was rotated by the electromotor with frequency up to 50 R.P.S. The braking system allowed us to vary the braking force, i.e. the amplitude and the time of channel acceleration.

Two channels (A and B) were used. Their parameters as well as the expected parameters of the MHD apparatus are given in the Table. The screw flow in the channel is created by 6-blades (channel A) or 8-blades (channel B) diverters. The number of diverters varied from 1 to 4. The

| Parameters | Model A | Model B | MHD |
|---|---|---|---|
| Median radius of the torus $R_0$, m | 0.103 | 0.154 | 0.5 |
| Radius of the cross-section $r_0$, m | 0.027 | 0.04 | 0.1 |
| Mass of empty model, kg | 5.6 | 24.5 | 300 |
| Mass of fluid, kg | 1.25 | 4.86 | 100 |
| Inertia moment (channel), kg $m^2$ | 0.072 | 0.58 | 80 |
| Inertia moment (fluid), kg $m^2$ | 0.018 | 0.15 | 25 |
| Frequency of rotation, R.P.S. | 50 | 30 | 50 |
| Effective Re | $10^5$ | $5 \cdot 10^5$ | $3 \cdot 10^6$ |
| Effective Rm | - | - | 40 |
| Energy of rotation, J | $4.4 \cdot 10^3$ | $17.3 \cdot 10^3$ | $5 \cdot 10^6$ |
| Dissipated power, Wt | $4.4 \cdot 10^4$ | $8.7 \cdot 10^4$ | $2 \cdot 10^7$ |

braking time was varied in the range 0.2 – 1.0 sec. The flow was visualized with the help of polystyrene particles suspended in a weak NaCl solution. The size of particles was about 2mm because the smaller particles become invisible at the early stage of the flow evolution, when a strong dispersion arises due to small scale gaseous bubbles generated by the diverter. In Fig.1 the general structure of the screw flow is illustrated by a photo made with a relatively long exposure (0.1 sec) at late stage of evolution when the light dispersion does not hinder visualization of the tracks.

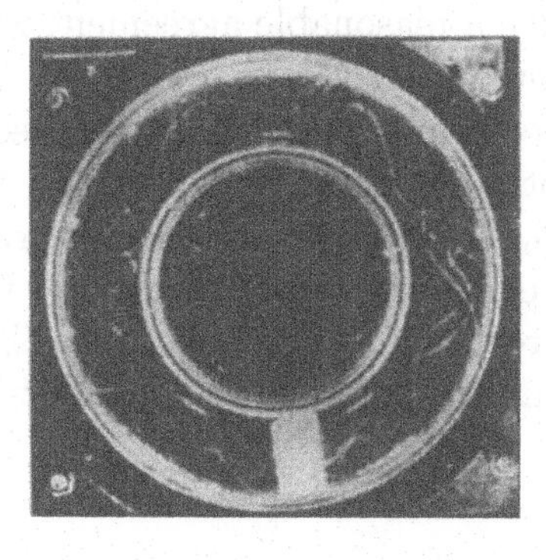

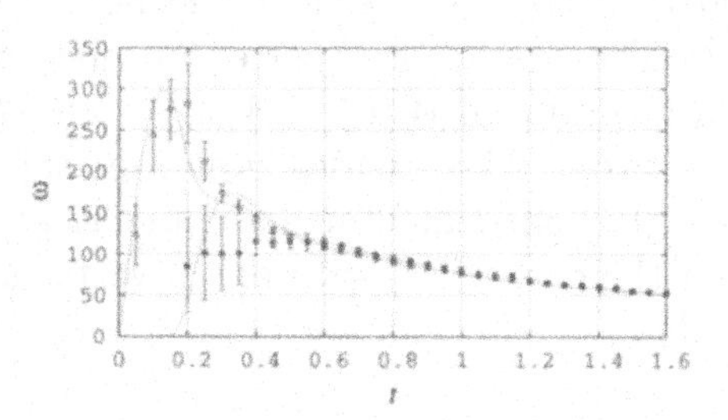


*Figure 1.* Screw flow in the channel A, at the late stage of evolution (1.5 sec after the full stop). The diverter is observable in the lower part of the channel as a light body.

*Figure 2.* Evolution of the azimuthal velocity near the diverter. $T_b = 0.19\,s$. White circles – angular velocity of the downstream blade, black circles – angular velocity of the upstream blade. Lines show the estimations calculated by the model (1)-(3).

## 3. Evolution of swirling flow

We introduce a simple model based on the mean longitudinal velocity $U(t)$, which is the same for any channel cross-section, and on the angular momentum $W(z,t)$ averaged over a given channel cross-section $z$ ( $z=0$ is the diverter position). $W$ propagates due to the mean velocity $U$ and decays due to the friction at the wall. We accept that the friction losses for the rotating mode are proportional to that at the mean velocity $U$. An additional decay of rotation occurs due to the action of the Coriolis force during the braking period and is proportional to $W\Omega$, where $\Omega$ is the rotation rate of the channel itself. As a result, we arrive at the following transport equation

$$\frac{\partial W}{\partial t} + U\frac{\partial W}{\partial z} = -\mu\left(\frac{v_* W(z,t)}{W(0)}\right)^2 - CW\Omega, \tag{1}$$

$$U = v_*(2.5\ln\frac{r_0 v_*}{\nu} + 1.75), \tag{2}$$

where $\mu$ and $C$ are dimensionless empirical coefficients and the so-called dynamical velocity $v_*$ is defined by the semiempirical law for turbulent flow in a tube[6].

The evolution equation for $U$ includes the frontal resistance and the losses caused by the action of diverter, which transfers energy to the rotating mode:

$$\frac{dU}{dt} = -R_0\frac{d\Omega}{dt} - \beta\frac{2v_*^2}{r_0} - \eta\frac{NU^2}{L} - \zeta\frac{N}{r_0^2 L}\left(W^2(0) - W^2(L)\right). \tag{3}$$

Here $N$ is the number of diverters, $L = 2\pi R_0$ is the channel circumference, $\zeta$ and $\eta$ are dimensionless empirical coefficients. We have found that for our model $\zeta = 5$, $\eta = 0.5$, $\mu = 2$, $C = 0.5$ give a reasonable agreement with experimental data, and the second term in the right part of Eq.(3) can be neglected ($\beta = 0$) because the next two terms dominate in the decay process. The action of diverter on the rotation mode is defined by the boundary condition $W(0) = W(L) + \sigma\,[UR_0/2 - W(L)]$ with the empirical coefficient $\sigma = 0.75$ , which describes the efficiency of the diverter. Fig.2 presents the experimental and numerical results for braking times $T_b = 0.19\,s$ in the channel B with one diverter and supports the applicability of the model.

## 4. Estimations of the MHD apparatus parameters

Fig.2 shows that after a full stop of the torus a regime with relatively stable value of velocity is established during a time, indicated as $T_{ef}$. We assume that the effective magnetic Reynolds number defined by $Rm_{\mathrm{ef}} = \sqrt{r_0^2U^2 + 4W(L)^2}/\nu_m$ is the characteristic of this regime. The model equations (1)-(3) were used to estimate the characteristics of nonstationary MHD flow in a MHD apparatus at various braking time $T_b$. The apparatus parameters are listed in third column of the Table. The results of these simulations for $N = 2$ (two diverters in the channel) are presented in Fig.3. The graph is done in log-log scale and shows that both data sets ($Rm_{\mathrm{ef}}$ and $T_{\mathrm{ef}}$) follow the power laws. The corresponding fitting gives $Rm_{\mathrm{ef}} = 7.5T_b^{-2/3}$, $T_{\mathrm{ef}} = 0.6T_b^{2/3}$ (time is measured in seconds). These fittings are shown in Fig.3 by solid lines and lead to a conclusion that the production $Rm_{\mathrm{ef}}T_{\mathrm{ef}}$ for a given channel and initial energy (frequency of torus revolution) is constant.

To verify this prediction we have investigated the dependence of the effective Reynolds number on the braking time in the water model. The corresponding results are also shown in Fig.3: crosses indicate the "fictitious" $Rm$ (defined in terms of experimental water velocity and magnetic diffusivity of sodium) obtained for the model B at different braking time $T_b$. The dotted line traces the slope "-2/3" and supports the validity of this power law.

The simulations were also done for the MHD channel with the same initial angular velocity, but one diverter. The system shows the same power law time behavior with different prefactors. Now $Rm_{\mathrm{ef}} = 8.6T_b^{-2/3}$, $T_{\mathrm{ef}} = 0.9T_b^{2/3}$. Thus, the use of one diverter slightly improves the effective Reynolds number and essentially alters the duration of expected regime. The production $Rm_{\mathrm{ef}}T_{ef}$ for $N = 1$ is 72% larger than for $N = 2$, which is a strong argument in favor of using only one diverter in the MHD apparatus.

## 5. Screw-dynamo in real flows in conducting channel

### 5.1. LINEAR DYNAMO PROBLEM IN CILINDRICAL GEOMETRY

We start by analyzing the linear dynamo problem for the screw flow in a cylindrical conducting tube with the finite wall thickness $d = r_1 - r_0$ (the inner radius of the tube is $r_0$, the outer - $r_1$), surrounded by air. It is close to the experimental conditions for thin torus.

Introducing the magnetic field in form $\mathbf{H}(r, \phi, z, t) = \mathbf{h}(r)e^{\gamma t + i(kz+m\phi)}$ described in cylindrical coordinates $\{r, \phi, z\}$ into the induction equation one gets

$$\begin{aligned}
\gamma h_r &= -i(m\omega + kv_z)h_r + \\
&Rm^{-1}\left[\frac{d^2h_r}{dr^2} + \frac{1}{r}\frac{dh_r}{dr} - \left(\frac{m^2+1}{r^2} + k^2\right)h_r - \frac{2im}{r^2}h_\phi\right] \\
\gamma h_\phi &= -i(m\omega + kv_z)h_\phi + r\frac{d\omega}{dr}h_r + \\
&Rm^{-1}\left[\frac{d^2h_\phi}{dr^2} + \frac{1}{r}\frac{dh_\phi}{dr} - \left(\frac{m^2+1}{r^2} + k^2\right)h_\phi + \frac{2im}{r^2}h_r\right],
\end{aligned} \tag{4}$$

where an axysimmetrical screw velocity field is $\mathbf{V}(r) = \{0, r\omega(r), v_z(r)\}$. Here distances are measured in units of the jet radius $r_0$, velocity is measured in units of longitudinal velocity on the axis of the jet $U$, so $Rm = \sigma U r_0$.

In the outer domain (air, which is dielectric and $\nabla \times \mathbf{H} = 0$) a potential $P$, can be introduced so, that at $r > r_1$

$$P(r, \phi, z, t) = CH_m^{(1)}(i|k|r)e^{\gamma t + i(m\phi + kz)}. \tag{5}$$

where $H_m^{(1)}(w)$- is the Hankel function of order $m$, $C$ - is a constante. The continuity of $\mathbf{h}$ on the outer border of the conductive tube leads to

$$\begin{aligned}
\frac{h_r(r_1)}{h_\phi(r_1)} &= \frac{|k|r_1}{mH_m^{(1)}(w)}\frac{dH_m^{(1)}(w)}{dw}\bigg|_{w=i|k|r_1}, \\
h_r(r_1) + r_1h_r'(r_1) &= -i\left(\frac{k^2r_1^2}{m} + m\right)h_\phi(r_1).
\end{aligned} \tag{6}$$

The axial symmetry of $\mathbf{h}$ in the center of the cylinder results in relations:

$$h_r'(0) = h_\phi'(0) = 0 \text{ for } m = \pm 1 \;\text{ or }\; h_r(0) = h_\phi(0) = 0 \text{ for } m \neq \pm 1. \tag{7}$$

The system (4) with boundary conditions (6) and (7) results in the eigenvalue problem for linear differential functional. The problem provides the search for eigenvalues of the matrix, which appears if the equations are replaced by their finite-difference analogies. The field generation implies the existence of eigenvalues with $\Re\gamma > 0$. The corresponding simulations were performed using the QR-algorithm and 200-800 grid points.

The conductivity of the tube can be different from the conductivity of the fluid. Let $\sigma_0$ be the conductivity of the fluid, which is used in definition of magnetic Reynolds number. $\sigma_1$ is the conductivity of the tube and is measured in units of $\sigma_0$ (thus the case $\sigma_1 = 1$ corresponds to identical conductivity of the fluid and of the tube. On the boundary fluid-tube one has to demand the continuity of magnetic field and the continuity of tangential component of electrical field.

## 5.2. CHANNEL OPTIMIZATION

Two kinds of velocity profiles were used in our simulations. Firstly, a $\xi$-family of velocity field for $r \leq 1$ described by

$$v_z(r) = \frac{\cosh(\xi) - \cosh(r\xi)}{\cosh(\xi) - \cosh(0)}, \tag{8}$$

where $\xi \geq 1$, $\chi$ - is the pitch parameter of the flow, (in the limit of solid rotation it is the screw pitch). Formula (8) describes the whole spectrum of profiles from the Poiseille solution ($\xi = 1$) up to rigid-body motion in the limit of high $\xi$ (in reality, for $\xi = 100$, $d = 5$ and $\sigma_1 = 1$ the critical Reynolds number differs from Ponomarenko solution less as $0,1\%$).

Secondly, we used the logarithmic velocity profile, which is widely used to describe the mean velocity profile of turbulent flow in tubes under high Reynolds numbers [6] for $r \leq 1$

$$v(r) = 1 - 5.75 k_1 \ln \frac{k_2}{k_2 - r}, \tag{9}$$

where $k_1$ and $k_2$ can be defined using the relations $k_1(2.5 \ln k_1 Re + 5.5) = 1$, $v_z(1) = 0$. For the azimuthal component one implies the "local solid rotation" $\omega(r) = \chi v_z(r)$.

The experimental data concerning the profile do not allow to definitely confirm the structure of the stream, thus we prefer to consider the whole family of profiles described by (8). It should be said that the logarithmic profile describes the velocity excluding the vicinity of the wall, where it should be corrected to satisfy the boundary condition. In Fig. 4 we show some typical velocity profiles used in the simulations.

The first crucial for the experimental apparatus question concerns the thickness of the tube wall. Fig. 5 shows the critical value of $Rm$ versus the flow parameter $\xi$ for three different wall thickness. The thickness $d$ is given in units of radius $r_0$, thus the value $d = 1$ corresponds to a tube with a wall

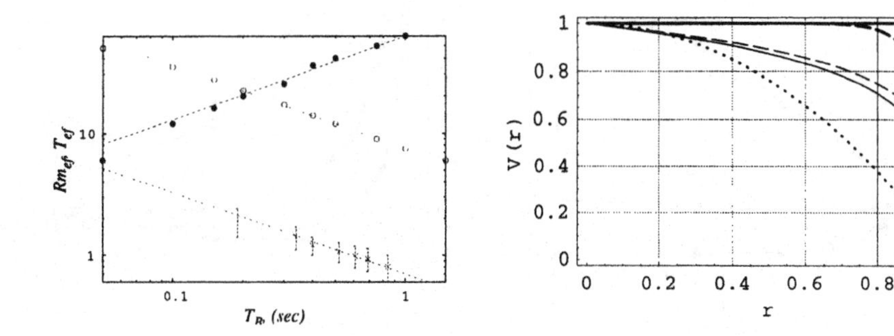


*Figure 3.* Characteristics of the nonstationary MHD flow in a braked torus with 2 diverters. Expected effective magnetic Reynolds number $Rm_{\text{ef}}$ and duration of the regime with this characteristics $T_{\text{ef}}$ versus the braking time $T_b$. Solid lines show the fits $Rm_{\text{ef}} = 7.5 T_b^{-2/3}$, and $T_{\text{ef}} = 0.6 T_b^{2/3}$.

*Figure 4.* Velocity profiles used in simulations. Thick curves show the $\xi$-profiles (8): $\xi = 1$ (dotted), $\xi = 18$ (dashed), $\xi = 100$ (solid). Thin curves show the logarithmic profiles (9): $Re = 10^6$ (solid), $Re = 10^7$ (dashed).

$d = r_0$. All the curves on Fig. 5 are given for the wall conductivity, which is equal to the conductivity of the fluid ($\sigma_1 = 1$).

One can see that the threshold depends on the profile parameter in qualitatively different way for different wall thickness. In the limit of thin conducting wall ($d = 0$ or $d = 0.1$) the increase of parameter $\xi$ leads to monotonous increase of $Rm_*$. For thick conducting wall (e.g. the case $d = 1$ in Fig. 5) the opposite dependence is true: as larger is $\xi$ (quasi solid motion) as lower is the threshold. At intermediate values of $d$ the neutral curve gets a minimum. For $d = 0.3$ the minimal $Rm_*$ can be seen at $\xi \sim 5$.

Next figure 6 shows how the threshold depends on the conductivity of the wall. All the curves are given for same velocity profile ($\xi = 18$, solid lines) and different wall thickness. It is known, that the Ponomarenko dynamo becomes impossible in the limit of ideal conducting media and the neutral curve $Rm_*(\sigma_1)$ has a stretched minimum with pronounced increase of critical $Rm$ for $\sigma_1 > 1000$. Our calculations shows that in the case of smooth velocity profiles the position of the minimum is shifted in the domain of moderate conductivity. Thus, for $d = 0.3$ the minimum arises at $\sigma_1 = 3.5$. From the point of view of experimental apparatus one should try to get a low threshold under a relatively thin wall. As optimal solution we suggest $d = 0.15$. Then the minimum of the neutral curve is got at $\sigma_1 = 5.5$, which corresponds to the conductivity of copper (if liquid sodium is considered as the fluid). Then $Rm_* = 27$.

Similar simulations were performed for the logarithmic profile (9). Firstly we got that in the range of Reynolds numbers, which can be interesting in the frame of our problem ($10^6 < Re < 10^7$) the neutral curve is practically

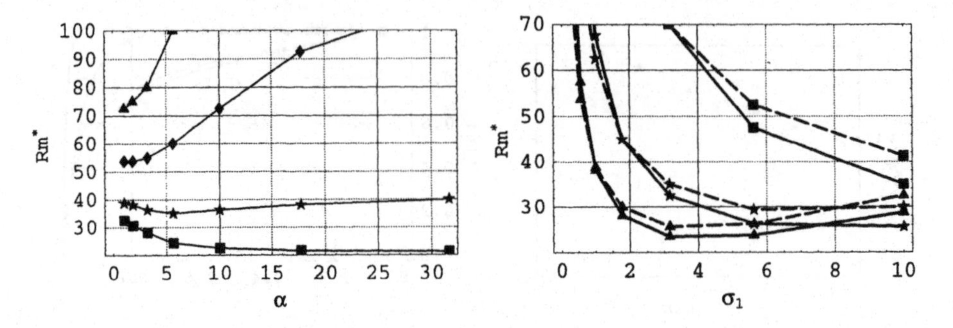


*Figure 5.* $Rm_*$ versus the velocity parameter $\xi$. The neutral curves for $k = -1$, $m = 1$, $\sigma_1 = 1$ are calculated for different shell thickness: $d = 0$ (triangle), $d = 0.1$ (romb), $d = 0.3$ (star), $d = 1$(box).

*Figure 6.* $Rm_*$ versus the shell conductivity. The neutral curves for different shell thickness using $\xi$-profiles for $\xi = 18$ (solid curves) and logarithmic profile for $Re = 5 \cdot 10^6$ (dashed curves): $d = 0.3$ (triangle), $d = 0.15$ (stars), $d = 0.05$ (boxs).

independent on the exact value of $Re$. Thus we fixed the Reynolds number as $Re = 5 \cdot 10^6$. Secondly, the general behavior of neutral curves $Rm_*(\sigma_1)$ for the logarithmic profiles and $\xi$-profiles is similar, but for the latter case the minimum of $Rm_*$ is slightly higher ($Rm_* = 30$, see Fig. 6).

This work is supported by Russian Foundation of Basic Researches (grant 99-01-00362).

## References


1. Gailitis, A., Lielausis, O., Dement'ev, S., Platacis, E., Cifersons, A., Gerberth, G., Gundrum, T., Stefani, F., Christen, M., Hänel, H. and Will, G. (2000) Detection of a flow induced magnetic field eigenmode in the Riga dynamo facility, *Phys.Rev.Lett.* **Vol. no. 84**, pp. 4365–4368
2. Stielitz R., Muler U. Experemental demonstartion of a homogeneous two-scale dynamo, *proceedings of IV PAMIR conference*, **Vol. no. 1**, pp. 175–182
3. Denisov, S.A., Noskov, V.I., Sokoloff, D.D., Frick, P.G. and Khripchenko, S.Yu. (1999) On a possibility of laboratory realisation of nonstationary MHD dynamo, *Doklady Mechanics* **Vol. no. 365**, pp. 478–481
4. Frick, P., Boffetta, G., Giuliani, P., Lozhkin, S. and Sokoloff, D. (2000) Long-time behaviour of MHD shell models, CD/0009053, p. 6.
5. Antonov, T., Frick, P. and Sokoloff, D. (2000) Alignement in free decaying MHD turbulence, *Numerical Methods and Advanced Computing,* **Vol. no. 1**, pp. 14 – 19, www.srcc.msu.su/num-meth/english/index.html.
6. Schlichting, H., *Grenzschicht-Theorie* G.Braun, Karlsruhe, 1964.

# RIGA DYNAMO EXPERIMENT

A. GAILITIS, O. LIELAUSIS AND E. PLATACIS
*Institute of Physics, Latvian University*
*LV-2169 Salaspils 1, Riga, Latvia*

AND

G. GERBETH AND F. STEFANI
*Forschungszentrum Rossendorf*
*P.O. Box 510119, D-01314 Dresden, Germany*

## 1. Introduction

It is widely believed that almost all magnetic fields in a natural environment are a result of the dynamo process – field generation in a moving nearly homogeneous electroconducting fluid of celestial bodies. Such are fields of the Earth, most of the planets, Sun, another stars and even galaxies.

The believe is based on the theory and numerical simulation. Until very recently no direct laboratory experiment was supporting this important point.

We are not going to model in the laboratory any particular celestial body. Our aim is to demonstrate the very idea – by intense stirring in a large volume of good electroconducting liquid one can generate a magnetic field. As the working fluid we are using 2 qm of molten sodium – the best electroconducting liquid available.

The fluid part of celestial bodies is stirred by thermal and other types of natural convection. In laboratory circumstances such stirring is much too slow. Hence we are stirring sodium by an outside forced propeller. Even using an outside force for reproduction of a natural phenomenon the scale of laboratory experiment needs to be larger as usual in hydraulic experiments.

The first magnetic field produced by stirring a liquid electroconductor was observed at the facility shown in Fig. 1 (Gailitis *et al.*, 2000a, and closer explored by Gailitis *et al.*, 2000b).



*P. Chossat et al. (eds.), Dynamo and Dynamics, a Mathematical Challenge,* 9–16.

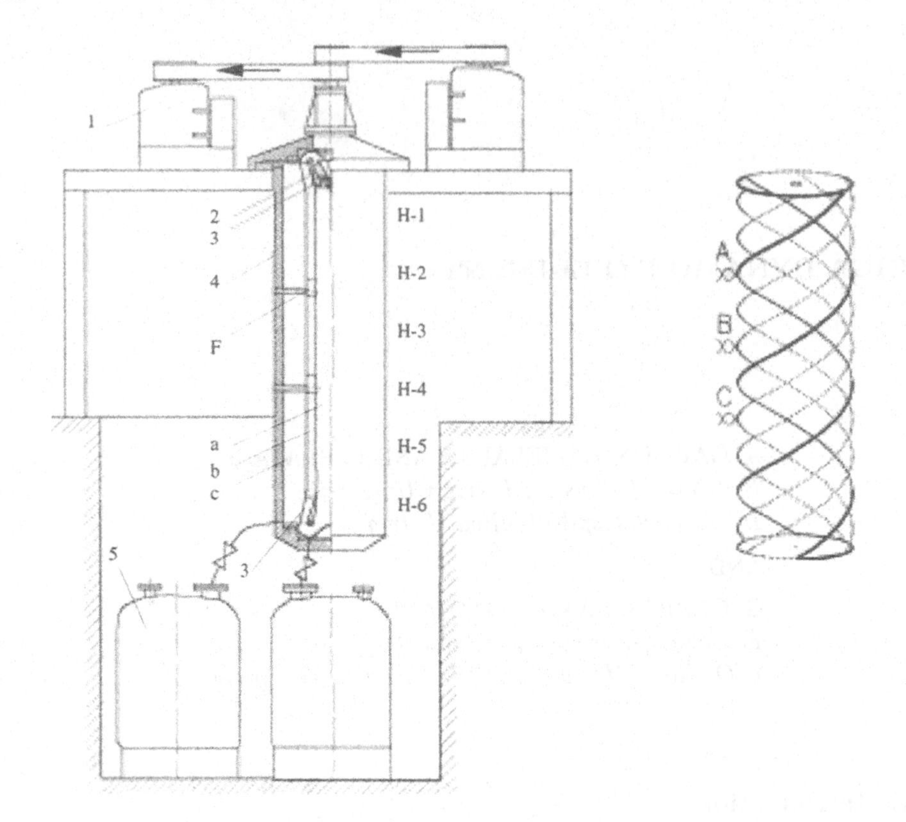


*Figure 1.* Riga dynamo facility: 1 – Motors, 2 – Propeller, 3 – Vanes, 4 – Thermal insulation, 5 – Na storage, a – Helical flow region, b – Back-flow region, c – Sodium at rest, F – Position of the flux-gate sensor and the inner induction coil, H1...H6 – Positions of Hall sensors. *Right* – The helical 3 - phase coil

## 2. The experimental setup

Two 55 kW motors 1 (Fig. 1) are rotating a propeller 2 which forces liquid sodium to circulate inside an annular vessel called dynamo module, part of which is located in the basement of sodium lab. The sodium flow is directed by two thin electroconducting cylindrical partition walls. In the central channel *a* sodium is flowing down from the propeller. In the coaxial counterflow channel *b* the flow is upwards to the propeller. In an outer part *c* of the vessel the sodium is moveless, it serves for electrical connection.

Vanes 3 located both before and after the propeller are designed to swirl the central flow as needed for field generation. The straight parts of channels are without any flowguides. Hence the swirl is maintained by inertia only until the downstream end of the central channel and stopped by other vanes when the flow is being reversed. According to the numerical simulation at a sodium temperature as low as $170^oC$ the device should start to generate a magnetic field at propeller rotation rate of appr. 2000 rpm. The critical flowrate is about 0.6 qm/s.

## 3. Mathematical background

Ponomarenko (1973) was the first who considered an endless helical stream moving as a solid cylinder and maintaining in full electrical contact with its immobile surrounding. Because of symmetry the generated field depends on axial distance $z$, azimuth $\phi$ and time $t$ exponentially:

$$\sim exp(pt + ikz + im\phi)$$

Hence the only nontrivial field dependence is a radial one. So the problem is an 1D problem. In general the $k$ and $p$ are complex constants to be evaluated when the problem is solved. In our case $m = 1$.

As a consequence of the induction equation the radial field dependence results in Bessel functions of complex argument. Continuity of electric and magnetic fields on contact surface and field vanishing on radial infinity leads to a closed complex and transcendent (Bessel functions!) secular equation in a form

$$F(Rm, k, p, geom.param.) = 0$$

having no analytic solution.

For high $Rm = \mu_0 \sigma v_{max} R >> 1$ (electroconductivity $\times$ maximum velocity $\times$ radius of the stream) an asymptotic solution gives some real $k$ interval where field is growing ($\Re p > 0$). Tracing back to smaller $Rm$ Gailitis & Freibergs (1976) found out that field can grow if $Rm > 17.7$ (numerical solution at optimum pitch). But field growing on infinite length does not mean growing in a limited space. The situation is called convective instability as all growing field perturbations are flowing downwards and with some group velocity $i\partial p/\partial k$ out of the module giving no feedback for the field generation. The group velocity can be modified and made close to zero by adjusting the size of the counterflow (Gailitis & Freibergs, 1980). The zero group velocity as a saddle point on complex $p(k)$ plane is a branching point on the inverse $k(p)$ plane. Hence close to it for any $p$ there are two different $k$ values $k_1$ and $k_2$. Two solutions moving in opposite directions can give the necessary feedback.

For the first instance the dynamo module on Fig. 1 is a composition of three coaxial solid like conductors (inner helical flow, counter flow and immobile volume of sodium). In each of them and in the outer insulating space as well the field radial dependence gives Bessel functions. Resistances of the internal walls can be included in the boundary conditions at contact surfaces. The secular equation remaines exact but much more complicate as in Ponomarenko case.

The mathematical model is endless while the dynamo module has finite length $L$. The both ends of the module are really complicate – propeller and vanes on top, flow reversals at bottom, thick walls, etc. No prospect to formulate exact boundary conditions there. Instead we apply an asymptotic constraint $(k_1 - k_2)L = 2\pi$. It means the generated field is a superposition of two solutions with the same time behaviour, the same spatial rate $-\Im k$ and having the node separation equal to $L$. It looks like as a usual standing wave deformed by factor $exp(-\Im kz)$. Solving together two secular equations (one with $k = k_1$, another with $k = k_2$) and the constraint one can get $p$, $k_1$ and $k_2$ for any $Rm$ and geometry.

By those 1D calculations the design of dynamo module was optimized (Gailitis, 1996) to get the phenomenon at costs as low as possible.

Without any care the real helical stream has radial velocity profiles far from solid ones. When these profiles are approximated by polynomials the inner solution is no more a Bessel function. We computed it either as power series evaluated directly from the induction equation or by means of finite difference methods. Comparing different profiles we found that helicity maximizing Bessel function profiles are optimal (Stefani *et al.*, 1999). A lot of effort was then spent modifying pre- and post-propeller vanes in water experiments to optimize the real profiles (Christen *et al.*, 1998).

## 4. Instrumentation

When generated the magnetic field is asymmetric with respect to the devices axis and slowly rotates round it. Hence each magnetic field sensor feels an AC field with the field rotation frequency. The sensors are located:

1) high sensitive flux-gate for vertical field - at the inner end of the measuring channel inside one of the upper supports F,

2) less sensitive induction coil for the radial field - at the very position inside another upper support,

3) array H1...H6 of 3D Hall sensors outside the thermoinsulation and

4) 3-phase helical coil (Fig. 1 on right) wound over the whole dynamo module just outside the thermoinsulation.

Originally the coil was designed to give an external excitation at subcritical flowrate. In the below reported experiment external excitation is not used and the voltage induced in the coil is measured, recorded and interpreted as an integral signal from the generated magnetic field over the whole outer surface of the dynamo module.

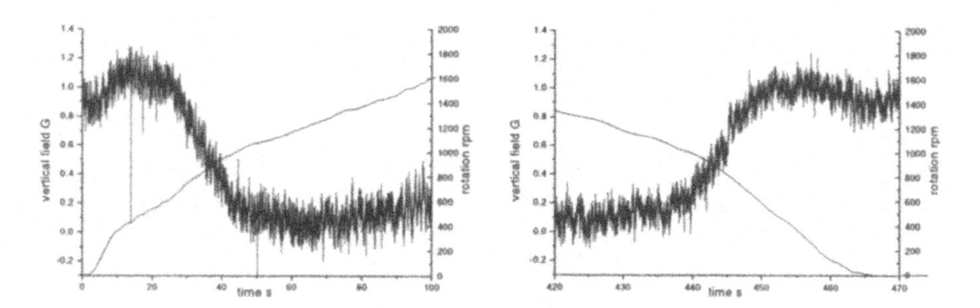


*Figure 2.* Deformation of the Earth magnetic field: left – at the start and right – at the end of the run

Together with magnetic field we record the propeller rotation rate, sodium temperature and motor power. There were no measurements for total flow rate and sodium velocities. The velocities were measured when our device was tested by water. With sodium they can be assumed the same and proportional to the propeller rotation rate but only before the field is generated. We have observed some indirect evidences for the field back-reaction on the flow. The direct measurements are left for future.

## 5. Operations

After the device is completely assembled and checked it is vacuumed and filled with argon. Each experiment series start with melting sodium by heat from electrical heaters. Then the module and all support parts are heated up to $200^0C$ or so. Using argon pressure sodium is filled from storage into dynamo module. When module is full and valves closed, the dynamo module is heated up to $300^0C$ and kept such for 24h. It is done to assure complete electrical contact between sodium and the thin inner walls. After this the dynamo module is cooled down to $150^0C$ and measurements can start. As the running propeller is heating sodium by hundred kW power and we have no sodium cooler we can not provide an experiment at fixed temperature. Hence the experiment series is divided into 5-7 min long runs by much longer silent periods for cooling in a natural way.

## 6. Field records

On Figs. 2-5 are field records for one typical run which started with $153^0C$ and ended with $175^0C$. Signals from different sensors are presented together with the propellers speed versus time. For the field the y axis is on left, for the propeller speed on the right. The time count is the same for all Figs. 2-5.

The beginning of the run is on Fig. 2. When nothing moves the flux-

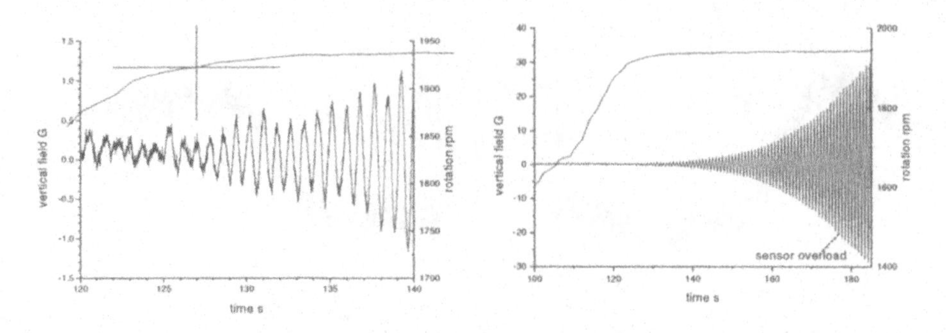


*Figure 3.* Field starts (left) and rises almost exponentially (right)

gate feels Earth magnetic field deformed by our steel-concrete building and the steel frame of our experiment. After the propeller starts the Earth field is at first slightly enhanced but then completely suppressed (Fig. 2 left). Until the propeller rotation rate about 1920 rpm we see only noise, mainly instrumental (Fig. 3). Close to 1920 rpm the noise starts to organize itself into a periodic structure. At 1920 rpm the pattern suddenly changes - a regular signal is rising. Seeing the event on a screen we stopped to accelerate propeller further and tried to keep the rotation rate as constant as possible for 80 s. During this time the field was almost exponentially rising and overloading the flux-gate over its normal range of $\pm 20$ G. The further field development is recorded by induction coils (Fig. 4).

As the kept rotation rate was only slightly over the critical one the field relaxation was too long to get the field saturated in these 80 s. Hence rotation rate was risen step wise and the saturation observed.

At the end of the run the propeller rotation was continuously lowered and finally stopped. During the braking time the generated field disappeared and after some time the Earth field was coming back into the dynamo in a sequence opposite as it was suppressed (Fig. 2 right).

The record from the helical outer coil (Fig. 4 right) is very similar to the record from the inner one (Fig. 4 left). With some delay the both amplitudes are following the propeller rate.

## 7. The frequencies and field values

Measuring on the fluxgate record the distances between maxima, minima and even or odd zeros we have determined the local field frequencies. All such points are plotted together on Fig. 6. For the most of the run time the flux-gate was overloaded and hence was giving wrong amplitudes of the field. Nevertheless the frequency remains right. It remarkably follows the propeller rotation. At the beginning and at the end of the field gener-

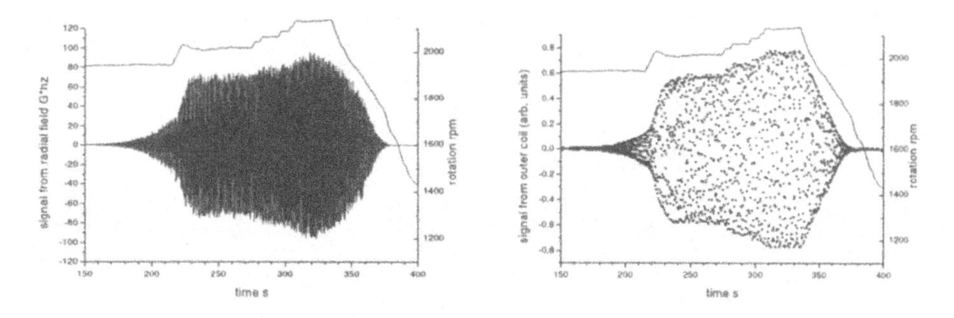


*Figure 4.* Signal from the inner induction coil (left) and outer helical coil (right)

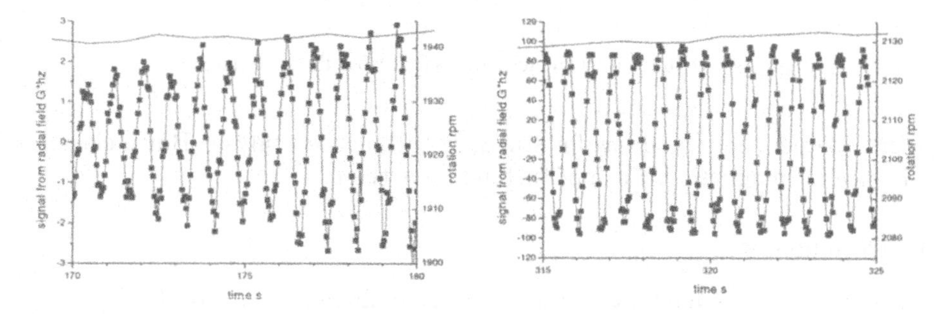


*Figure 5.* Two fragments from the left field record on Fig. 4

ation the points are dispersed because a weak signal can not give a precise frequency.

On Figs. 4 and 5 the signal from the inner coil is in units G*Hz. Taking frequencies from Fig. 6 we get the radial field equal to 2 G for the 177 s and 63 G for the 310 s of the run. The vertical component is larger. As for 177 s it was 20 G (Fig. 3) we can estimate the maximum value for 310 s at about 600 G.

## 8. The turbulence

On Fig. 6 some disperse remains even when the signal is high enough. An attempt to make a similar plot from the inner induction coil record results in much higher disperse. It is because the irregularity in an oscillating field record is far beyond the instrumental noise (Fig. 5). The magnetic field pattern is by no means rotating as a solid magnet. The overall rotation is superimposed by higher frequency fluctuations. They are more enhanced in induction coil signals as in flux-gate ones as the first is a time derivative from the field. Obviously the base for all this is the hydrodynamic turbulence in the sodium flow deforming magnetic field. The used rate for induction coil

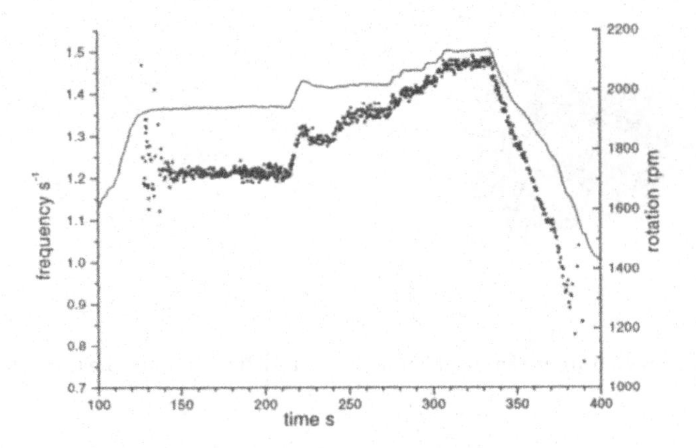


*Figure 6.* The local frequencies extracted from flux-gate signal

record (25 per sec) is too low for standard turbulence analysis. But a clear evidence for magnetic branch of turbulence is there.

## 9. Conclusion

Our experiment demonstrates stable and reproducible magnetic field produced by stirring the fluid electroconductor.

## References


1. Christen, M., Hänel, H.,Will, G. (1998): Entwicklung der Pumpe für den hydrodynamischen Kreislauf des Rigaer "Zylinderexperimentes". in *Beiträge zu Fluidenergiemaschienen* **4**, edited by W. H. Faragallah, G. Grabow, (Faragallah-Verlag und Bildarchiv, Sulzbach/Ts.), pp. 111–119
2. Gailitis, A., Freibergs, J (1976): To the theory of a helical MHD–dynamo. Magnetohydrodynamics **12**, 127–129
3. Gailitis, A., Freibergs, J (1980): Instability type of screw dynamo. Magnetohydrodynamics **16**, 116–121
4. Gailitis, A. (1996): Project of a liquid sodium MHD dynamo experiment. Magnetohydrodynamics **32**, 58–62
5. Gailitis, A. et al. (2000a): Detection of a flow induced magnetic field eigenmode in the Riga dynamo facility. Phys. Rev. Lett. **84**, 4365–4368
6. Gailitis, A. et al. (2000b): Magnetic field saturation in the Riga dynamo experiment. Phys. Rev. Lett., submitted
7. Ponomarenko, Yu. B. (1973): On the theory of hydromagnetic dynamo. J. Appl. Mech. Tech. Phys. **14**, 775–779
8. Stefani, F., Gerbeth, G., Gailitis, A. (1999); Velocity profile optimization for the Riga dynamo experiment. in *Transfer Phenomena in Magnetohydrodynamic and Electroconducting Flows*, edited by A. Alemany, Ph. Marty, J. P. Thibault, (Kluwer Academic Publishers, Dordrecht), pp. 31–44

# GENERATION OF MAGNETIC FIELD IN THE COUETTE-TAYLOR SYSTEM

P. LAURE AND P. CHOSSAT
*Institut Non-Linéaire de Nice, UMR 6618 - CNRS & UNSA*
*1361 route des Lucioles, 06560 Valbonne, France*

AND

F. DAVIAUD
*SPEC, CEA Saclay*
*91191 Gif sur Yvette cedex, France*

## 1. Introduction

The governing equation for the magnetic field $\mathbf{B}$ in an electrically conducting fluid with conductivity $\sigma$ and velocity $\mathbf{v}$ is the so-called induction equation

$$\frac{\partial \mathbf{B}}{\partial t} = \mathrm{curl}(\mathbf{v} \times \mathbf{B}) + \frac{1}{\mu_0 \sigma} \Delta \mathbf{B} \tag{1}$$

which follows from Maxwell equations and Ohm's law. The solution $\mathbf{B} = \mathbf{0}$ may become unstable for some critical value $\mathrm{Re}_{\mathrm{mc}}$ of the magnetic Reynolds number,

$$\mathrm{Re}_{\mathrm{m}} = \mu_o \sigma \mathrm{LV}, \tag{2}$$

$L$ and $V$ being respectively typical length and velocity scales.

The induction equation can be coupled with the full Navier-Stokes equation, but the the problem becomes numerically very arduous (see [5]). Another approach consists to impose an arbitrary velocity field $\mathbf{v}$ and looking at if the solution $\mathbf{B}$ is amplified or not.

In the sequel, we assume that the velocity field is the Taylor vortex flow coming from the bifurcation of the Couette flow.

## 2. The Taylor vortex flow

The Couette-Taylor apparatus consists of two rotating cylinders moving with different angular velocities $\Omega_i$ and $\Omega_o$ [4].



*P. Chossat et al. (eds.), Dynamo and Dynamics, a Mathematical Challenge,* 17–24.

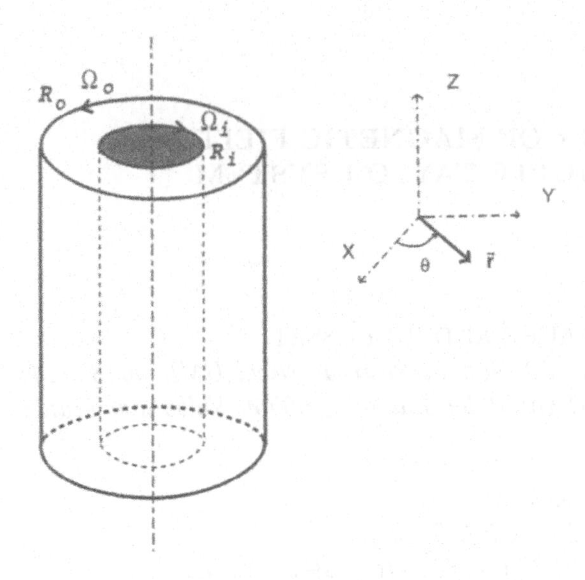


*Figure 1.* Geometry of the Couette-Taylor problem.

The parameters of this problem are the ratio of the inner and outer cylinder $\eta = R_i/R_o$, the inner and outer Reynolds numbers

$$Re_{i,o} = R_{i,o}\, \Omega_{i,o}(R_o - R_i)/\nu.$$

For small $Re_i$, the flow motion is purely azimuthal $V^c = (0, v_\theta^c(r), 0)$ (namely the Couette flow). A transition to axisymmetric motion (Taylor vortex) is observed for a value of the inner Reynolds number denoted $Re_c$ if the outer Reynolds number $Re_o$ is not too negative. This flow is stationary and periodic along the $z$-axis ($\alpha_c$ is its wavenumber).

In the following, we assume that the outer cylinder is at rest ($Re_o = 0$) and that the radii ratio is equal to .5. Indeed for theses parameter values [6], the second transition towards wavy vortices occurs at a large inner Reynolds number ($\epsilon_R = (Re_i - Re_c)/Re_c > 3$).

The asymptotic expansion of the Taylor flow in the neighborhood of the bifurcation point has the form

$$U = A\; v^t(r)\mathrm{e}^{\mathrm{i}\alpha_c \mathrm{z}} + \bar{A}\; \bar{v}^t(\mathrm{r})\mathrm{e}^{-\mathrm{i}\alpha_c \mathrm{z}} + \; .... \tag{3}$$

where $A$ is the amplitude of the bifurcated solution. $A$ can be evaluated thanks to amplitude equation [7]. Finally, the flow motion for $Re_i > Re_c$ is of the form

$$V = V^c + \rho(Re_i)V^t + ... \tag{4}$$

assuming that $V^t = 2\ \mathrm{Real}(v^t(\mathrm{r})\ \mathrm{e}^{\mathrm{i}\alpha_c \mathrm{z}})$ with the scaling $||v^t((R_i + R_o)/2)|| = 1$.

## 3. Equation for the Dynamo problem

We assume that the dynamo effect comes in as a secondary bifurcation after the Taylor vortex and the problem is now to determine the Reynolds number $Re_i$ such that the linear operator in (5) has eigenvalues belonging to the imaginary axis. The dynamo problem reads for a velocity field $V$ defined by relation (4) :

$$\frac{\partial \mathbf{B}}{\partial t} = \mathrm{Re_m}\ \mathrm{curl}(\mathbf{V} \times \mathbf{B}) + \mathbf{\Delta B} \tag{5}$$

$$\mathrm{div}(\mathbf{B}) = \mathbf{0} \tag{6}$$

with conductor boundary condition for $r = R_i, R_o$

$$\mathbf{B}.\mathbf{n} = 0 \tag{7}$$

$$\mathrm{curl}(\mathbf{B}) \wedge \mathbf{n} = 0 \tag{8}$$

and periodic boundary conditions in the $z$ direction. Moreover, we also assume conductor boundary condition on the top and the bottom of cylinders. In this way, we fixe the representation of the symmetry, $z \to -z$, which is now "natural" [2, 1] :

$$(r, \theta, z) \to (r, \theta, -z)\ ,\ (v_r, v_\theta, v_z) \to (v_r, v_\theta, -v_z). \tag{9}$$

Therefore, a Fourier decomposition will be done in this direction and $\mathrm{Re_m}$ is defined with the same scaling as the Reynolds number and then

$$\mathrm{Re_m} = \mathrm{P_m}\ Re_\mathrm{i} \tag{10}$$

where $\mathrm{P_m} = \mu_o \sigma \nu$ is the magnetic Prandtl number. We change the scale in axial direction so that $z$ is defined between 0 and 2 $\pi$ and $n_V$ is now the Fourier mode of $V^t$ in the $z$-direction (namely the number of Taylor vortices).

At this stage, the number of parameters is rather important and it is very tedious to determine directly the critical $Re_i$. First, we seek if the first terms in the expansion (4) can give rise to dynamo effect. In fact, with the two first terms $V^c$ and $V^t$, the velocity field has non-zero helicity density, $h = \mathbf{V}.\mathrm{curl}\mathbf{V}$ and therefore could exhibit dynamo action. More precisely, the helicity density is split in two terms

$$h(r, z) = \rho \sin(n_V\ z)\ h^c(r) + \rho^2 \sin(n_V\ z)\ \cos(n_V\ z)\ h^t(r)$$

$$h^c = v^c_\theta \left[ D v^t_z - \alpha_c\ v^t_r \right] + D v^c_\theta\ v_z \text{ and } h^t = v^t_z\ D v_\theta\ -\ v^t_\theta\ D v^t_z. \tag{11}$$

The first one, $h^c$, comes from the coupling between the Couette and the Taylor flows and the other one, $h^t$, is due to the Taylor vortex. As shown in Fig. 2, the maximum of the first term is located close to the moving wall whereas the maximum of second term is close to the middle of the cylinder.

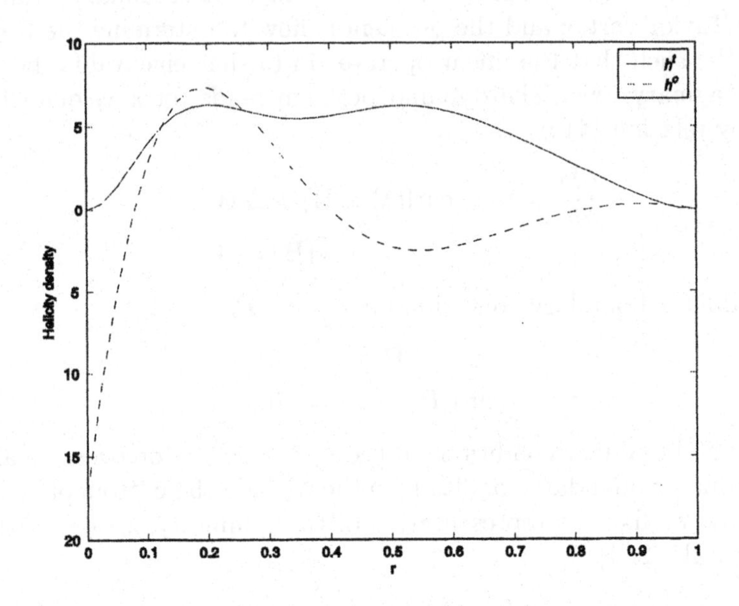


*Figure 2.* Helicity density of the basic flow versus the radial component : $Re_o = 0$ and $\eta = .5$

Finally, the methodology applied is the following : one fixes $\rho$ and $n_V$ and determines if there exists a critical magnetic Reynolds number.

### 3.1. NUMERICAL PROCEDURE

A Fourier analysis is made in the axial and azimuthal directions :

$$B(r, \theta, z) = \sum_{m,n} b^{m,n}(r)\ \mathrm{e}^{\mathrm{i}\ (\mathrm{m}\theta+\mathrm{nz})} \text{ with } \mathrm{b}^{-\mathrm{n},-\mathrm{m}} = \bar{\mathrm{b}}^{\mathrm{n},\mathrm{m}} \tag{12}$$

Each component $b^{m,n}(r)$ is discretized on $N$ collocation points [3]. The condition of zero divergence is used to eliminate the components $b_z^{m,n}$ or $b_\theta^{m,n}$ according to the value of $(m, n)$ :

$$\frac{1}{r} b_r^{m,n} + D b_r^{m,n} + \frac{\mathrm{i\ m}}{r}\ b_\theta^{m,n} + \mathrm{in\ b_z^{n,m}} = 0 \tag{13}$$

Due to reflection symmetry, $z \to -z$, we restrict our expansion to positive Fourier modes. The other modes can be deduced by the relation

$$b_r^{-n,m} = b_r^{n,m} \; ; \; b_\theta^{-n,m} = b_\theta^{n,m} \; ; \; b_z^{-n,m} = -b_z^{n,m} \tag{14}$$

Finally, one gets for each azimuthal mode $m$ a generalized eigenvalue problem in a matrix form,

$$M(m, \mathrm{Re_m}, \rho, \mathrm{n_V})[\,\mathrm{b}\,] = \mathrm{s}\;\mathrm{N}[\,\mathrm{b}\,] \tag{15}$$

where $M$ is a real sparse matrix, $N$ is a noninvertible diagonal matrix and the real part of the eigenvalue $s$ is the dynamo growth rate. The aim is to compute the eigenvalue $s$ with the largest real part and look at if this real part becomes positive as the magnetic Reynolds number increases. This algebraic problem is solved by an Arnoldi method coupled with an inverse iterative method [10, 11]. As the size of matrix $M$ becomes rapidly important ($\sim 4\; N * (N_z + 1)$ for $N_z$ Fourier modes), the GMRES algorithm with a suitable preconditioner is used to make the Arnoldi decomposition (instead of the usual LU factorization).

## 4. Numerical results : influence of $n_V$

The Fig. 3 shows the velocity field used in the computation. It consists of two counter-rotating vortices in the $(r, z)$-plane. It is displayed on one axial period and this pattern is repeated $n_V$ times if the number of Taylor vortices is $n_V$.

The computations are made for $n_V = 1, 2, 3, 4, 6, 7$; $\rho = 1.$ and various azimuthal modes (see Table 1). First for $n_V > 1$ and $m = 1$, there always exists a critical Magnetic Reynolds number from which the dynamo growth rate, Real($s$), is positive. Moreover, the amplified magnetic field is time-dependent (the imaginary part of $s$ is always non zero). For higher values of the azimuthal modes $m$, additional computations show that the real part of eigenvalues $s$ are always negative for the critical magnetic Reynolds numbers obtained for the mode 1. If $n_V = 1$, the perturbation are never amplified for magnetic Reynolds number up to 800. We have stopped our computation to this value as higher resolution is required [8] for large values of $\mathrm{Re_m}$. This is due to the concentration of magnetic field in sheets of thickness $O(\mathrm{Re_m}^{1/2})$ [9].

For $n_V = 2$, the two sets of two contra-rotating vortex (cf. Fig. 3) are located respectively between 0-$\pi$ and $\pi$-$2\pi$ (there are a total amount of four rolls). Fig 4a shows that the magnetic field is concentrated along the inner cylinder and along the outflow boundaries ($z = 0, \pi$ and $2\pi$). As usual the magnetic field is more important near the stagnation points [9].

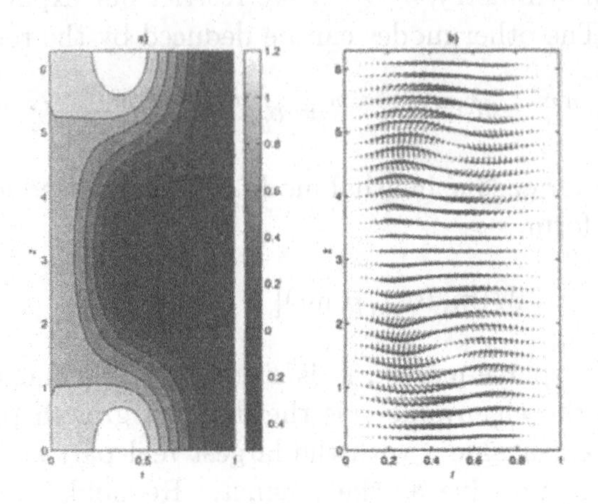

*Figure 3.* Velocity field $V^c + \rho V^t$ for $\rho = 1.$ and $Re_o = 0$ (the flow does not depend on $Re_i$ if $Re_o = 0$ : (a) isoline of $V_\theta$ in the $(r, z)$- plane; (b) $(V_r, V_\theta)$ in the $(r, z)$-plane.

| $n_V$ | $\mathrm{Re}_{mc}$ | Im($s$) | $N$ | $N_Z$ | First non null Fourier modes |
|---|---|---|---|---|---|
| 2 | 134.9 | -72.56 | 31 | 20 | 1, 3, 5, 7, ... |
| 3 | 180. | -24.51 | 41 | 50 | 1, 2, 4, 5, 7 .... |
| 4 | 135.7 | -76.10 | 41 | 50 | 2, 6, 10 ... |
| 5 | 142.1 | -80.06 | 41 | 70 | 2, 3, 7 .... |
| 6 | 134.9 | -75.57 | 41 | 90 | 3, 9, 15 .. |
| 7 | 137.9 | -77.40 | 41 | 110 | 3, 4, 10 ... |

TABLE 1. Results : $\eta = .5$, $Re_o = 0$, $\rho = 1$.

For $n_V = 3$, the critical Reynolds number is higher than for $n_V = 2$ whereas the magnetic fields is rather concentrated along the inflow boundaries.

For $n_V > 3$, the characteristic length of the magnetic field seems to be always twice the height occupied by the basic Taylor vortex (c.f. Fig. 3). For odd $n_V$, this is only approximated and the system tends to neglect the contribution due to the rolls located around $z = \pi$, in order to get the magnetic field generated for an even $n_V$. This point is depicted in Figs. 4c-d and Figs. 4e-f for the case $n_V = 5$ and $n_V = 7$.

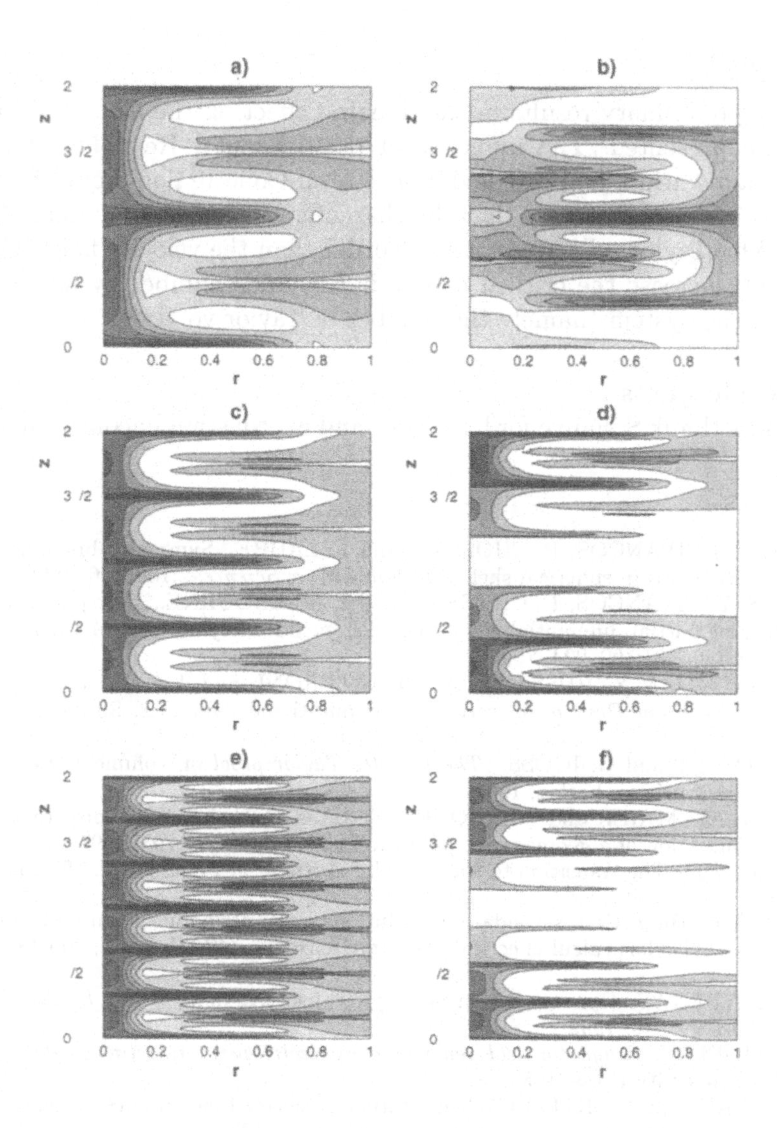


*Figure 4.* Isolines of modulus $|B|$ of the critical magnetic field in the $(r,z)$- plane for $\theta = 0$ and $t = 0$. The other parameters are $m = 1$, $\rho = 1.$, $Re_o = 0$ : (a) $n_V = 2$, (b) $n_V = 3$, (c) $n_V = 4$, (d) $n_V = 5$, (e) $n_V = 6$, (f) $n_V = 7$. The maximum value corresponds to the black color.

## 5. Conclusion

Only very preliminary results of the dynamo effect on the Couette-Taylor system are presented. The influence of positive outer Reynolds number $Re_o$, the radii ratio $\eta$ and the deviation $\rho$ from Couette flow would be still studied. However, on can see that the characteristic length of dynamo field generated by Taylor cells are in a same order than the velocity field. Then, one cannot decrease the critical magnetic Reynolds number by increasing the size of the system (namely the number of Taylor vortices).

**Acknowledgments**
The authors thank S. Fauve for his advice and his constant encouragements.

## References

1. I. BOSCH-VIVANCOS, P. CHOSSAT, and P. LAURE. Symmetry-breaking convective dynamos in spherical shells. *J. Non Linear Science*, 2:169–196, 1999.
2. I. BOSCH-VIVANCOS, P. CHOSSAT, and I. MELBOURNE. New planforms in systems of partial differential equations with euclidean symmetry. *Arch. Rational Mech. Anal.*, 131:199–224, 1995.
3. C. CANUTO, M.Y. HUSSAINI, A. QUARTERONI, and T.A. ZANG. *Spectral methods in Fluid Dynamics.* series in computational dynamics. Springer-Verlag, 1988.
4. P. CHOSSAT and G. IOOSS. *The Couette-Taylor problem*, volume 102 of *Appl. Math. Sci.* Springer-Verlag, 1994.
5. G.A. GLATZMAIER and P.H. ROBERTS. three-dimensional self-consistent computer simulation of a geomagnetic field reversal. *Nature*, 377:203, 1995.
6. C.A. JONES. The transition to wavy Taylor vortices. *J. Fluid Mech.*, 157:135–162, 1985.
7. P. LAURE. Bifurcation secondaire de solutions quasi-périodiques pour le problème de Couette-Taylor, calcul effectif de la forme normale. *CRAS série I*, 305:493–496, 1987.
8. P.C. MATTEWS. Dynamo action in simple convective flows. *Proc. R. Soc. Lond.* **A**, 455:1829–1840, 1999.
9. H.K MOFFAT. *Magnetic field generation in electrically conducting fluids.* Cambridge University Press, 1988.
10. N. NAYAR and J. ORTEGA. Computation of selected eigenvalues of generalized eigenvalue problems. *J. Comput. Physics*, 108:8–14, 1993.
11. Y. SAAD. Variations on Arnoldi's method for computing eigenelements of large unsymmetric matrices. *Linear Algebra and its Applications*, 34:269–295, 1980.

# DYNAMO ACTION, BETWEEN NUMERICAL EXPERIMENTS AND LIQUID SODIUM DEVICES

JACQUES LÉORAT
*Observatoire de Paris-Meudon 92195 Meudon*

P. LALLEMAND
*ASCI, bat.506, Université Paris-Sud,91405 Orsay*

J.L. GUERMOND
*LIMSI, Université Paris-Sud,91405 Orsay*

AND

F. PLUNIAN
*LEGI, Institut de Mécanique, Université J. Fourier, Grenoble*

## 1. Introduction

Conversion of kinetic energy into magnetic energy within a conducting fluid is but an example of various physical phenomena occuring in MHD turbulence at magnetic Reynolds numbers $R_m$ greater than a few tens. While the first numerical evidence of this phenomenon has been obtained since almost 45 years, it was only at the end of the year 1999 that it could be experimentally verified in liquid sodium flows. There is still a great gap between such experimental approaches and the understanding of natural dynamos, which are responsible for example of solar magnetic activity or of the geomagnetic field. To get closer to the real MHD problems, new designs of experimental devices and new numerical codes are needed in order to achieve greater Rm, study the turbulent characteristics, the nonlinear saturation regime and the influence of large-scale flow configuration and boundaries.

Most numerical dynamos are based either on laminar or on low kinetic Reynolds number flows. Moreover, the magnetic Prandtl number, which is of the the order of $10^{-5}$ for liquid metals (Earth's case) is often chosen closer to unity in numerical simulations. Experimental fluid dynamos are needed because the spatial resolutions achieved in direct numerical simulations are unable to describe high kinetic Reynolds number flows which are the



*P. Chossat et al. (eds.), Dynamo and Dynamics, a Mathematical Challenge,* 25–33.

rule in astrophysical situations. While large eddy simulations or statistical models (such as EDQNM), leading to turbulent transport coefficients, could be usefull to describe fully developped turbulence, they must be validated using real MHD flow experiments. Finally, the exploration of the nonlinear saturation regime may be so expensive in computation time, as to prevent for instance the self consistent numerical study of an Earthlike chaotic dynamo or long magnetic cycles.

After presenting the energetic considerations of the experimental dynamo problem in the next Section, and comparing some driving configurations, we will examine the specific case of the precession driving using moderate Reynolds number simulations (Section 3). To prepare the design of a large scale precession experiment which will involve a fully turbulent flow, a water experiment has been designed and is planned to be in operation at the beginning of year 2001 (Section 4). The occurrence of the dynamo action and the non linear regime will be investigated with a numerical code convenient for cylindrical geometry (Section 5).

## 2. Turbulence and energetics of a fluid dynamo

There is no general sufficient condition for dynamo action which could be used as a guide for the selection of an efficient driving. Empirical computational experience shows that, in order to occur, dynamo action asks that the magnetic number of a given flow configuration be greater that a critical number $R_m^*$, which depends on the flow geometry. Positive numerical results are known either for "laminar dynamos", where the flow is at large scales and steady, or for "turbulent dynamos", where the velocity is fluctuating in space and time, with a zero mean. A real fluid dynamo cannot be based on a purely laminar (steady) flow nor on homogenous turbulence. However, to begin the feasibility study of an experimental fluid dynamo, it is usefull to ask first which dynamo model, laminar or turbulent, couldbe easier to realize, and which main parameters control the energy dissipation.

### 2.1. TURBULENT DYNAMOS VERSUS LAMINAR ONES

Different kinds of laminar dynamos are known in spheres, which is the simplest geometry for the numerical treatment of magnetic boundary conditions. In many cases, the corresponding $R_m^*$ lies below 100, which may be taken as a rough technical objective. Practical conditions for dynamo action in turbulent flows have not been investigated systematically, as they are generally obtained from non linear MHD computations which are rather demanding in numerical power. To the author's knowledge, they seem not to contradict the predictions of the EDQNM model for homogeneous MHD turbulence (vanishing mean velocity and helicity), where turbulent dynamo action occurs for *turbulent* magnetic Reynolds numbers $R_{mt}$ (based on rms speed and integral scale) greater than about 30 (Léorat et al 1981).

Injection of kinetic energy with zero mean is possible in principle using for example a sufficient number of turbines acting at a fraction of the container scale,. However, the maximal mean flow speed generated by a turbine will exceed the turbulent rms speed, so that one expects that, in practice, $R_{mt} < 10R_m$ (say), where the *laminar* magnetic Reynolds $R_m$ is based on the largest scale of the container and maximal speed of the mean flow: accordingly, when the turbulent dynamo could be excited,the laminar $R_m$ would be above 300. This last number, whatever the crudeness of its estimation, is well above the critical threshhold $R_m^*$ of many laminar dynamos.

Another look on this question may be obtained from turbulence phenomenology if non-local interactions in wavenumber space may be neglected (leading possibly to an inverse helicity cascade). In the Kolmogorov inertial range for the kinetic energy spectrum which extends for wavenumbers $k > k_0$, the local stretching rate $ku(k)$ scales as $k^{2/3}$, while the ohmic dissipation rate scales as $k^2$. The ratio between the two (which is a kind of local magnetic Reynolds number) scales as $k^{-4/3}$, and increases towards larger scales, starting from unity at the ohmic dissipation scale. For homogenous turbulence, it is maximal around the integral scale, while if a mean flow is present, it may continue to increase at larger scales. This argue again for the importance of the largest scales to achieve large magnetic Reynolds number flows.

Note also that in presence of a mean velocity flow, turbulence at smaller scales may favor dynamo action , as it happens that the ad-hoc introduction of small scales or a finite correlation time in laminar dynamos seems to reduce $R_m^*$ compared to the purely laminar case.

Turbulent dynamo action with a vanishing mean flow is certainly realized in some natural dynamos as the Sun, and is indeed also obtained in numerical experiments at large magnetic Prandtl numbers. It seems irrelevant for liquid sodium experiments (very small magnetic Prandtl number), where the main obstacle is the power constraint required to overcome the turbulent dissipation of kinetic energy.

### 2.2. DRIVING POWER

Liquid sodium, with its high electrical conductivity, seems the best fluid to be used in an experimental fluid dynamo. Many efficient numerical laminar dynamos are known to have a critical $R_m^*$ between 30 and 100, and thus the kinetic Reynolds number of the flow will be around $10^7$. The flow which must be driven will be fully turbulent and a basic question of any experimental design involving such flows is to evaluate the amount of kinetic power needed to support the flow against viscous and ohmic losses. Assuming that the Kolmogorov cascade controls the non linear energy flux in the inertial range, and merging for simplicity turbulent and mean quantities (contrary to the preceeding sub- section), it is easy to obtain an expression for the power dissipated $P = kR_m^3/L$ where k is a coefficient characteristics of the geometry of the experiment (container, driving, etc...). The scaling of P with the Reyn olds number is experimentally verified, which supports the role of the Kolmogorov cascade in the flows driven by various means in a finite tank.

For a given available power, the scaling with L is in favor of large scale experiments, while the cubic scaling with $R_m$ would forbid to increase this number substantially. We want to stress here that the value of the power coefficient k should not be put aside in such a discussion. Its value is known (see these proceedings) in the following cases, say, $k_R$ (Riga), $k_K$ (Karlsruhe), $k_C$ (Cadarache) , which may be compared with the case of pipe flow, $k_P$, where the turbulence level is minimal and which is not a candidate for dynamo action. For studies of some hydrodynamical features of the Cadarache experiment, see for example, Fauve et al 1993, Pinton and Labbé 1994 and Mordant et al 1997. From the published values of these experiments, one gets the following ordering, corresponding also to an ordering in turbulence levels:

$$k_R/k_P = 5, k_K/k_P = 6.4, k_C/k_P = 8.6$$

This suggests that the path to larger magnetic Reynolds number flows has to be considered, using low k and large L. While everybody knows how to increase L and P, the search for lower k opens a new field : how store useful kinetic energy in a flow with the minimal turbulent dissipation ? "useful" is related to the occurence of dynamo action, which implies for example that any mean solid body rotation has to be substracted from the flow. As no sufficient condition for dynamo action is available, once a driving is selected on the basis of a low k, the flow must be numerically tested with a MHD code. We will consider such a proposal below.

## 3. Precession driving

The interest towards flow generation at large scale without internal walls leads to consider driving the flow by the container wall itself, as occurs in the case of a precessing container. Upon the fast rotation of the container is superimposed a slow rotation (precession) at an angle with the rotation axis. For an arbitrary container shape, conservation of angular momentum of the fluid leads to a large scale circulation driven by pressure forces transmitted by the walls. In the case of the sphere, there is a peculiar inviscid solution due to Poincaré, with a transition to the wall through a viscous boundary layer. The case of the ellipsoidal container is interesting for the Earth dynamo: in this latter context, it has been first studied by Malkus (1968) and is still under active study (for France, cf studies at LGIT, Grenoble). An obvious technical advantage of precession driving for liquid sodium flows is the absence of seals on the container.

We propose to examine further the precession driving in the framework of experimental dynamos, using a cylindrical tank. This question has already been studied by Gans (1970), using a liquid sodium experimental device which gave torque measurements (and magnetic field), but without information on the actual flow configuration and was not developped further. To our knowledge, no related numerical simulations have been published on this topic.

The first issue which can be numerically tackled, at least for moderate Reynolds numbers flows, concerns the relative amount of toroidal and poloidal speeds which may be generated. Indeed, one of the few practical recipes derived from studies of linear numerical dynamos is that the ratio of these speeds should be close to unity to get optimized dynamos (it leads to efficient coupling between the magnetic field components). Simulations have been performed up to $R_e = 5000$, using two different hydrodynamical codes, one based on lattice Boltzmann method (P. Lallemand, see d'Humières 1992) and the other on finite elements (Guermond and Quartapelle, 1997).

We consider a cylinder rotating around its symmetry axis and precessing around an axis orthogonal to the rotation one. The parameters of the problems are the aspect ratio (we choose heigth equal to diameter), the kinetic Reynolds number $R_e$ (defined as the inverse of the Taylor number ) and the precession rate, $pr$, i.e. the ratio of the precession to the rotation angular velocity. When $pr = 0$, the steady flow is a solid body rotation and increasing $pr$ leads to an axial (poloidal) flow, accompanied by a decrease of the azimuthal velocity (see figure 1). At Reynolds numbers below about 2500, the flow is steady, with central symmetry, while time fluctuations appear at larger Reynolds numbers, and the central symmetry is recovered only for the mean flow.

A first conclusion from this preliminary study is the generation of an axial speed close to 1/3 of the container speed, and approximate equipartition of transverse and axial velocities which is obtained for $pr = 0.1$, at least when the length of the container is equal to its diameter. The azimuthal energy decreases when the Reynolds number increases: this is a consequence of driving through viscous coupling with the cylindrical curved wall. If needed for dynamo action, more azimuthal energy could be generated inertially, using for example baffles fixed on both ends.

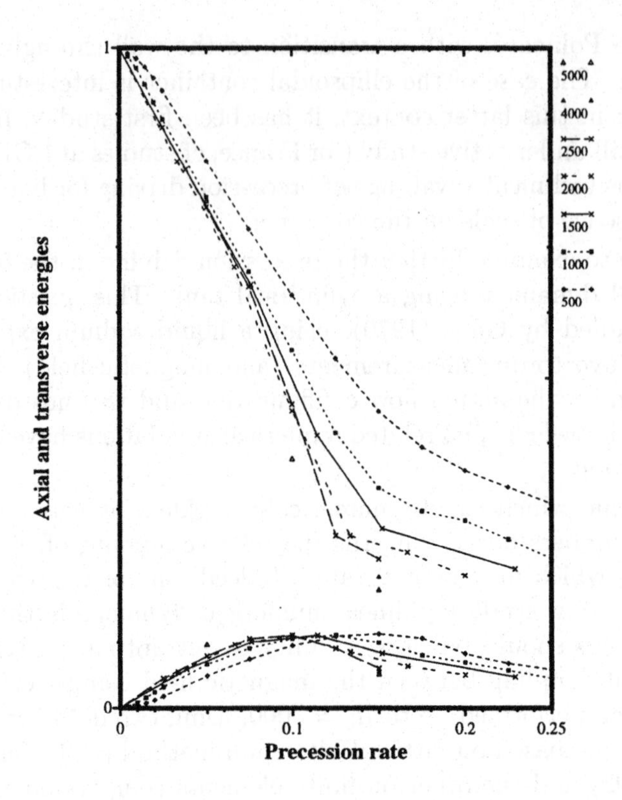


*Figure 1.* Variation with precession rate (pr between 0 and 0.25) of axial (lower curves) and transverse (upper curves) mean kinetic energies for different Reynolds numbers (between 500 and 5000). The energies are measured relatively to the solid body rotation energy

The numerical simulations cannot reach sufficiently large Reynolds numbers to produce a fully turbulent flow, from which the dissipated power and the power coefficient $k$ (see section 2.2 above) could be derived. These quantities may better be estimated using a water experiment which easily achieve Reynolds number above $10^5$, where asymptotic laws could be verified.

## 4. Water experiment

A water experiment has been designed and is currently under realization, with two main goals:

(i) estimation of flow characteristics which could play a role in the design of a large scale experiment for the study of large $R_m$ flows using liquid sodium. This corresponds to practical questions such as the determination of the power coefficient k and the optimization of the precession rate and aspect ratio for dynamo action, with help of a numerical code (see below).

(ii) study of the influence of rotation on the larger scales of the turbulent flow. This more fundamental set of questions takes advantage of precession acting as a volume force on the flow, while the moving grids used generally in this context give rise to an absence of homogeneity of the turbulent features. This is why the device has been named ATER (agitateur pour la turbulence en rotation).

The tank is a fully transparent (PMMA) cylinder, with a diameter of 30 cm and a ratio of length/diameter which may be adjusted between 4/3 and 1. Is is rotating around its axis (horizontal) up to 10 turns/s, and precession about a vertical axis may be fixed up to 1 turn/s. Velocities in the cylinder's sections will be obtained by particle image velocimetry (PIV), using a CCD camera corotating with the cylindrical container (optical axis coincide with cylinder axis) and a sheet of light orthogonal to the cylinder's axis and produced in a box fixed in the precessing frame.

## 5. Numerical approaches

Integration in time of the linear evolution equation of the magnetic field is used for the determination of the dynamo properties, as for example critical magnetic Reynolds numbers. Contrary to the eigenvalue method, direct simulations allows to study the consequences of time varying flows and time dependent external magnetic fields which may be used experimentally to estimate $R_m^\star$ from flow regime at lower $R_m$.

This approach has been used for example to optimize the Cadarache experiment (see Marié et al, these proceedings), with a code in cylindrical geometry, assuming axial periodicity of the flow and of the magnetic field. It has been verified that the conclusions were consistent with a more realistic, non periodic solution (F. Stefani, priv. communication), at least for the more important cases which could be compared.

The flow generated by precession is not compatible with axial periodicity: it leads to a strong poloidal circulation in a plane containing the rotation axis and orthogonal to the precession axis. Reproducing this flow by axial translation gives rise to a strong unrealistic shear on both ends of the container's ends. The use of a numerical code for MHD flows at large $R_m$ in a finite cylinder is thus required.

In order to be able to study the non linear MHD regime and to interpret experimental results, we have chosen to extend the existing hydrodynamical code based on a finite elements method, already used to study the precessing hydrodynamical flow. Another advantage of this code in relation with experimental devices is the easy implementation of spatial variations of conductivity of magnetic permeability of materials. The main question to consider first is the choice of the magnetic variables , electric and magnetic fields versus vector and scalar potentials, which are the most convenient to implement the boundary conditions relevant for an insolating external medium. To avoid the adjunction of a supplementary insolating envelope, the Laplace equation for the external scalar potential of the magnetic field may be solved with a surface element method (Guermond and Fontaine,1991). The relative merits of these approaches are currently examined using the Ponomarenko dynamo as a testing bench.

## 6. Conclusion

Dynamo studies have concentrated since a long time on the production of dynamos or antidynamo flows, leaving aside the complexities of the actual configurations of natural dynamos. The dynamo challenge has quite recently acquired new dimensions. After the first experimental evidences of dynamo action in laboratory, there is a need to design fluid dynamos at larger $R_m$ and without internal walls, and build *MHD wind tunnels* in order to study the nonlinear saturation regime under operating conditions closer to the ones of natural dynamos. The main obstacle to such experiments is a power constraint (kinetic energy injection and corresponding heat removal). MHD numerical codes adapted to the container geometry are also still needed to be able to predict the flow efficiency as a dynamo.

To make some progress in these directions, we propose to reexamine the case of precession forcing. With the 150 kW of electrical power which are needed to drive the Cadarache experiment ($R_m = 70$), if a driving design may be found such that its power coefficient is reduced by a factor 4, say, using a container radius L = 1m (instead of L =0.2m) would lead to a flow at $R_m = 190$. In the case of precession driving, assuming that the maximal speed of the flow is 1/3 of the container rotational speed, this would correspond to a centrifugal pressure of about 33 atmospheres, which is quite technically feasible. The dynamo properties of this class of precessing flows remains an open question, to be tackled numerically.

## References

1. d'Humières, D. (1992) Generalized lattice-Boltzmann equations in "Rarefied Gas Dynamics : Theory and Simulations", *Progress in Astronautics and Aeronautics*,**159**, B.D.Shizgal and D.P. Weaver edts, (AIAA, Washington, DC)
2. Fauve S., Laroche,C., and Castaing, B.,(1993) *J.Phys.II*,**3**,p 271

3. Gans, R.F. (1970) On the hydromagnetic precession in a cylinder, *J.Fluid. Mech.*,**45**, pp 111-130
4. Guermond, J.L. and Quartapelle,L. (1997) Calculation of incompressible viscous flows by an unconditionally stable projection FEM, *J. Comput. Phys.* **132**,pp 12-33
5. Guermond, J.L. and Fontaine,S. (1991) A discontinuous h-p Galerkin approximation of potential flows *Rech. Aérosp.* **4**,pp 38-49
6. Léorat, J., Pouquet,A. and Frisch, U. (1981) Fully developped MHD turbulence near critical magnetic Reynolds number, *J.Fluid. Mech.*,**104**, pp 419-443
7. Malkus, W.V.R. (1968) Precession of the Earth as the cause of geomagnetism, *Science*,**160** , pp 259-264
8. Mordant,N.,Pinton,J.F. and Chilla, (1997) *J.Phys.II*,**7**,p 1
9. Pinton,J.F., and Labbé, R.,(1994) *J.Phys.II*,**4**,pp 1461-1468

# MHD IN VON KÁRMÁN SWIRLING FLOWS

*development and first run of the VKS experiment*

L. MARIÉ, J. BURGUETE, A. CHIFFAUDEL*, F. DAVIAUD, D. ERICHER, C. GASQUET
*DSM/DRECAM/SPEC, CEA/Saclay, (*) C.N.R.S.*
*91191 Gif sur Yvette cedex, France*

F. PETRELIS, S. FAUVE
*École Normale Supérieure*
*25, rue Lhomond, F-75005, France*

M. BOURGOIN, M. MOULIN, P. ODIER, J.-F. PINTON
*École Normale Supérieure de Lyon, CNRS*
*46, allée d'Italie, F-69364 Lyon, France*

A. GUIGON, J.-B. LUCIANI, F. NAMER
*DRN/DER/STPI, CEA/Cadarache*
*13108 Saint Paul les Durance, France*

AND

J. LÉORAT
*Observatoire de Paris-Meudon*
*92195 Meudon, France*

## 1. Introduction

### 1.1. MOTIVATION

The magnetism of many astrophysical objects, such as various stars or planets, galaxies, the intergalactic medium, etc, is attributed to the motion of conducting fluid in their interiors. It has been first proposed by Larmor [17] that a flow of conducting fluid generates the magnetic field of the sun by maintaining the corresponding electric current against ohmic dissipation. Such a generation of electromagnetic energy from mechanical work using a self-excited dynamo has been known since Siemens [34] and is the most basic mechanism of electrical engineering. However, in industrial dynamos the path of the electric currents are constrained by a complex wiring, which even in the most elementary device, the homopolar dynamo [3], breaks mir-



*P. Chossat et al. (eds.), Dynamo and Dynamics, a Mathematical Challenge,* 35–50.

ror symmetry. In addition, magnetic field lines are usually canalized using a high magnetic permeability material. No such well controlled external constraints on the field or on the current lines exist in "natural" dynamos, and for a long time, it has been far from obvious that the dynamo effect was the correct explanation for solar or earth magnetism. It has been even shown that a lot of flow and / or field configurations with enough symmetries cannot behave as fluid dynamos [for a review on anti-dynamo theorems, see Kaiser et al., these proceedings].

The first simple example of an homogeneous dynamo (i. e. in a medium of constant electrical conductivity and thus no path of least resistance) was given by Herzenberg [14]. It consists of 2 or 3 rotating solid spheres embedded in a static medium of the same conductivity with which they are in perfect electrical contact. A slightly different version of the Herzenberg dynamo was operated experimentally by Lowes and Wilkinson and may be considered as the first experimental demonstration of a homogeneous dynamo [20]. In a simply connected domain, the simplest "flow" leading to dynamo action was found by Ponomarenko [33]. It consists of a cylinder in solid body rotation and translation along its axis, embedded in an infinite static medium of the same conductivity with which it is in perfect electrical contact. Each point of the cylinder thus follows an helical path. Another simple flow has been found by G. O. Roberts [35]. It is a spatially periodic flow that consists of a two-dimensional array of helical eddies. It should be noted that the three velocity fields, $\vec{v}$, quoted above, have a non zero helicity $\langle \vec{v} \cdot \vec{\nabla} \times \vec{v} \rangle$, where $\langle \cdot \rangle$ stands for the spatial average. Although this is not a necessary condition for dynamo action, it has been shown by Parker that the non zero helicity of cyclonic eddies leads to an efficient dynamo mechanism [28].

Very recent experiments using a Ponomarenko type [the Riga experiment [11, 12], these proceedings] or a Roberts type [the Karlsruhe experiment [37], these proceedings] flow of liquid sodium, have provided the first laboratory models of fluid dynamos. However, an experimental demonstration of the dynamo effect in an unconstrained turbulent flow of liquid metal is still lacking. This is the main objective of our experiments using von Kármán swirling flows of liquid gallium and sodium. The possibility of generating a dynamo action in a flow whose large scale velocity field comes close to the geometries considered by Dudley and James [8] has been suggested by Roberts and Jensen [34]. This has led to experiments where the fluid is confined in a closed tank and the flow is produced by the motion of 'stirrers'. One such experiment is being run by the Maryland group and has yielded results on MHD turbulence [29][Maryland group, these proceedings]. Another is planned by the group in Madison [the Madison experiment, these proceedings]. Our motivations for studying the dynamo in the von Kármán geometry are the following ones:

*1) Effect of turbulence on the dynamo onset*
In the absence of strong geometrical constraints, any flow of liquid metal is fully turbulent before possibly displaying dynamo action. Indeed, the magnetic Prandtl number, $Pm = \mu_0 \sigma \nu$, where $\mu_0$ is the magnetic permeability of vacuum, $\sigma$ is the electric conductivity of the fluid and $\nu$ is its kinematic viscosity, is smaller than $10^{-5}$ for all electrically conducting liquids. Since the dynamo action requires a large enough magnetic Reynolds number, $Rm = \mu_0 \sigma L V$, where $V$ is the fluid characteristic velocity and $L$ is the flow characteristic large scale, one expects to observe the dynamo effect when the flow kinetic Reynolds number, $Re = VL/\nu$, is larger than $10^6$. The role of turbulent fluctuations at such large Reynolds numbers may be twofold: on one hand, they decrease the effective electrical conductivity and thus inhibits dynamo action by increasing Joule dissipation. On the other hand, they may generate a large scale magnetic field through the "alpha effect" or higher order similar effects [15, 24]. Experiments are the only way to study the role of turbulent fluctuations because direct numerical simulations cannot be performed at such high Reynolds number.

*2) Saturation of the magnetic field*
The saturation mechanisms of the growth of the magnetic field above the dynamo onset should strongly depend on the geometrical constraints applied to the flow. Indeed, an unconstrained flow is more easily perturbed under the action of the Lorentz force and this is likely to affect the post-bifurcation regime. A problem of fundamental interest is to determine the mean magnetic energy and the mean Joule dissipation related to their kinetic counterparts in the supercritical saturated regime. Again, this can be studied only experimentally because direct simulations of the dynamical dynamo problem with realistic values of the magnetic Prandtl number cannot be performed.

*3) Dynamical and statistical properties of the generated magnetic field*
Magnetic fields of astrophysical objects may be found to be almost time-periodic, like in the sun, or nearly stationary i. e. very slowly varying, like for the earth magnetic field between two successive reversals. It is tempting to connect this temporal behavior with the nature of the dynamo bifurcation which can be either a stationary bifurcation or a Hopf bifurcation in the simplest generic cases. It should be noted that both a stationary bifurcation (Karlsruhe experiment), and a Hopf bifurcation (Riga experiment) have been observed so far. With these simple geometry flows, it is possible to guess the nature of the bifurcation using symmetry considerations. This is less obvious in the case of a fully developed turbulent flow and it would be interesting to try to understand how is determined the dynamical regime above the dynamo onset. Another interesting question with flows involving

relative turbulent fluctuations as large as 50%, is to determine whether the dynamo is generated mostly from the mean flow or predominantly from the turbulent fluctuations. Finally, the study of the statistical properties of magnetohydrodynamic turbulence without an externally applied field, and the observation of the dynamics of the large scale magnetic field (reversals) are of fundamental interest and may also improve our understanding of solar or terrestrial magnetism.

## 1.2. WHY THE VON KÁRMÁN FLOW GEOMETRY ?

The choice of an optimum configuration to perform a turbulent dynamo experiment is not an easy task. As said above, if the flow is not confined by internal boundaries, it is strongly turbulent when the dynamo action may occur. In the case of confined flows, turbulent fluctuations are of course also generated at large Reynolds numbers, but they are restricted to small scales. The large eddies that generate the magnetic field in the Karlsruhe [37] or Riga [11, 12] experiments are quenched by the internal boundaries and thus forced to act coherently in time. This is not the case for unconfined flows in which the large scale flow may have undergone several bifurcations before the dynamo onset. Thus, both the geometry of the large eddies as well as their time dependence are difficult to predict. Since direct numerical simulations are not possible at kinetic Reynolds numbers of order $10^6$, the efficiency of a given set-up with respect to dynamo action cannot be even roughly evaluated before having characterised the flow using water experiments. However, there are general arguments that have motivated our choice of von Kármán type flows. We will first shortly describe these flows and then discuss the motivations which have led us to use them in a dynamo experiment.

Flows that are generated between two co-axial rotating discs (see figure 1 and 4) have been called "von Kármán swirling flows" [39]. Mean velocity profiles in cases where the flow is confined inside cylindrical walls have been measured since the late fifties [30, 31, 38, 36, 2]. In the case of co-rotating discs, one observes that the flow can be split into two parts: (i) a central core, nearly in solid body rotation at an average angular speed, the radius of which is fixed by the distance between the discs; (ii) an external flow driven outward near the discs by the centrifugal forces, and consequently, inward in the mid-plane. At high Reynolds numbers the whole flow is turbulent.

Kármán swirling flows generated with counter-rotating discs strongly differ from the co-rotating case: a time average of the velocity field (see figure 1) shows the existence of differential rotation and meridian recirculation loops. No coherent stationary average flow can be observed, although slowly drifting coherent structures have been detected in the median region [5]. In the central region, a well defined average motion is not observed and large and random turbulent velocity fluctuations occur [31, 32, 25]. Visualisation using water seeded with air bubbles, shows intermittent formation of filaments of bubbles that are ascribed to vorticity concentrated on tube-like structures [7, 9, 4, 6].

The first motivation to try a dynamo experiment using von Kármán flows relied on their efficiency to amplify and concentrate vorticity. Pressure measurements showed that localized vortex involves velocity increments on their core size of the order of the integral velocity [9] and it has been further shown that the typical core size is of the order of the flow Taylor microscale [6]. Although we know that the Elsasser analogy between vorticity and magnetic field cannot be used without caution, a flow which is an efficient amplifier of vorticity is likely to be a good candidate for a dynamo experiment.

A second argument is the existence of local helicity in the vicinity of the rotating discs that generate strong swirling flows of identical helicity (respectively opposite) in the contra-rotating case (respectively co-rotating). Although not necessary, the existence of helicity or the weaker condition of the absence of mirror symmetry, are known to be in favor of dynamo action.

Finally the amount of turbulent fluctuations with respect to the mean flow can be easily modified in the von Kármán geometry. Turbulent fluctuations are much larger in the counter-rotating case and it is also easy to get a time dependent large scale mean flow by rotating the two discs at different angular velocities [36]. These aspects are obviously interesting in order to study the effect of turbulent fluctuations on the dynamo onset. Indeed, the mean flow has been measured and the threshold of dynamo action resulting from the mean flow alone has been computed using a kinematic code [2] [section 2, below].

There are of course other well-known turbulent flows in closed geometries that may be considered. Thermal convection is certainly of interest in astrophysical or geophysical contexts but it is probably a very inefficient way to get a dynamo on the laboratory scale. Flows generated by inertial forces may fall in the same category [21]. If one is interested by scale separation between the velocity field and the magnetic field, Couette-Taylor flow may be tried as a possible way to achieve such a configuration without internal boundaries [18].

## 2. Optimization

Self-excitation is expected if the magnetic Reynolds number $Rm$ exceeds a critical value $Rm^c$. In order for this to be achieved experimentally, one must optimize the flow configuration to have the lowest possible threshold and maximize the range of accessible $Rm$ in the set-up. The latter is fixed by the amount of mechanical power $P$ input into the flow: due to the low magnetic Prandtl number $Pm$ of all liquid metals, even moderate $Rm$ flows are strongly turbulent. As a result, given the characteristic length scale $L$ of the flow, $P$ scales as $P \sim \rho L^2 U^3$, so that $Rm^{\max} \sim \mu\sigma(PL/\rho)^{1/3}$, and is thus weakly influenced by size or power. Much stronger variations result from the flow geometry. Given the choice of the von Kármán flow class, one optimizes the entrainment device (poloidal vs. toroidal velocities) and boundary conditions (insulating or conducting outer shell). The main characteristics of the velocity field are measured for various discs drives and configurations in a water experiment and then introduced in a kinematic dynamo computer simulation. The water setup is a half-scale model, in a cylindrical ($R$ = 10 cm) vessel – at 50°C, the viscosity of water is closed to that of sodium at 120°C. Both global (power consumption) and local velocity measurements are made. Power is measured via torque and global temperature increase measurements. For large kinetic Reynolds numbers, the results exhibit a variation of the power as $P \sim \Omega^3$, with $\Omega$ the rotation frequency, in agreement with scaling predictions. The mean velocity field is obtained via Laser Doppler Anemometry and pulsed Doppler Ultrasonic Velocimetry – typical results are shown in figure 1. The local *rms* velocity fluctuations (relative to the mean) give a typical turbulence rate of 50 %.

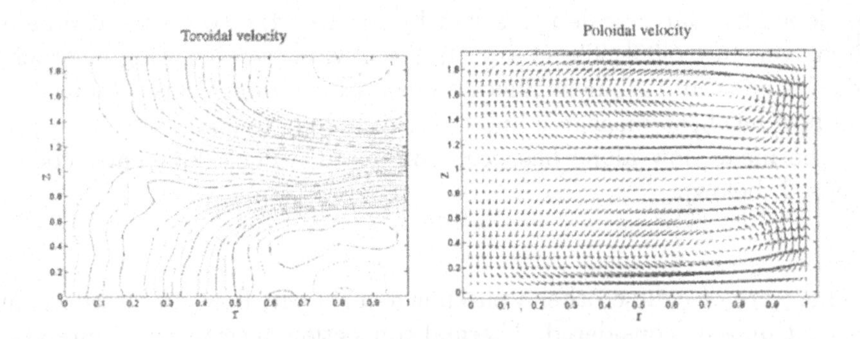


*Figure 1.* Mean velocity field in water experiment: (a) toroidal and (b) poloidal component of the velocity in the meridional plane. The abscissa corresponds to the normalized radius $r/R$, and the ordinate to the axial direction ($[0, 2] \equiv [\text{bottom}, \text{top}]$).

A kinematic dynamo code is developed in a periodic cylindrical domain, with a pseudospectral scheme in the azimuthal and axial directions and finite differences in the radial direction [19]. The conductivity of fluid $\sigma$ inside the cylinder is supposed to be uniform and the external medium is insulating. The influence of a layer of conducting fluid at rest surrounding the experiment was examined. The magnetic field equation,

$$\partial_t \vec{B} = \nabla \times (\vec{U} \times \vec{B}) + \frac{1}{\mu\sigma} \Delta \vec{B} \tag{1}$$

where $\vec{U}$ is the time averaged velocity field measured in the water prototype. It is integrated in terms of the magnetic field components:

$$\vec{B}(r, \phi, z, t) = \sum_{n,m} \vec{B}^{n,m}(r, t) \exp[i(m\phi + n\pi z)] \tag{2}$$

The energy evolution of mode $(m, n)$,

$$E^{m,n}(t) = \frac{1}{\mu} \int_V dV |B^{m,n}(r, t)|^2 \propto e^{pt} \tag{3}$$

is recorded and self-excitation is achieved if $\mathcal{R}(p) > 0$. The results reveal an extreme sensitivity to the position of the zero of the poloidal velocity, the maximum of the toroidal velocity and the poloidal to toroidal ratio $(P/T)$ [2]. In particular, the growth rate is maximum for $P/T \simeq 0.7$ (cf. figure 2(a)).

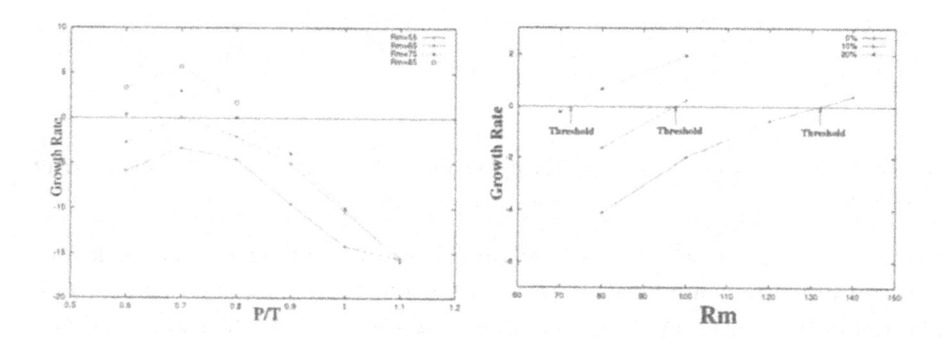


*Figure 2.* (a) Energy growth rate as a function of the poloidal to toroidal ratio P/T for various magnetic Reynolds numbers: $(+) : Rm = 55$; $(\times) : Rm = 65$; $(*) : Rm = 75$; $(\square) : Rm = 85$. (b) Energy growth rate as a function of the magnetic Reynolds number for various boundary condition. % indicates the proportion of sodium at rest surrounding the flow: $(+) : 0\%$, $(\times) : 10\%$, $(*) : 20\%$.

$Rm^c$ is also quite sensitive to the boundary condition: as shown in figure 2(b), a layer of sodium at rest of width 20% $R$ surrounding the experiment is highly favorable. Below threshold, the study of the system response to an externally applied field exhibits a divergence of the relaxation time and of the saturation energy.

The first configuration was chosen according to the results of the optimization process: impellers that yield a poloidal to toroidal ratio $\sim 0.7$, an efficiency ($U_{\text{flow}}/U_{\text{disc}} \sim 0.9$) and copper walls ($\sigma_{\text{Cu}} \sim 4\sigma_{\text{Na}}$). Under these conditions the expected threshold is $Rm^c \sim 70$, for a maximum of $Rm^{\max} \sim 55$ with 150kW of mechanical power input. It should be emphasized that this corresponds to the dynamo action of the mean flow as if it were acting alone. No quantitative estimate of the role of the turbulent fluctuations is performed with this analysis.

## 3. Experimental set-up and hydrodynamic measurements

*Sodium loop :* it is shown in figure 3. It is meant to be a versatile facility that can handle various flow configurations with a maximum sodium flow volume of about 300 liters. It is equiped with 2 electrical AC motors of 75kW each and a corresponding cooler is soon to be installed.

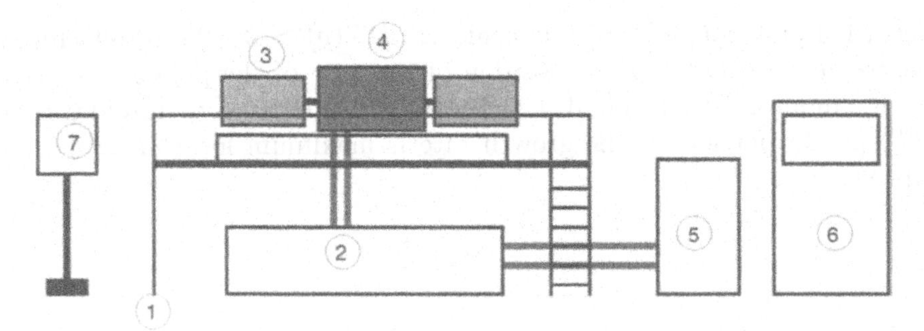


*Figure 3.* Sodium loop. (1) experimental platform, (2) sodium tank (270 liters), (3,4) motors and flow vessel, (5) sodium purifying unit, (6) control unit, (7) argon circuit command.

*Flow :* The gallium set-up has been described in details in earlier work [27]. The sodium flow configuration is of the same type – recalled in figure 4. The gallium experiment has a volume of about 6 litres with $2R \sim 20$ cm, insulating (steel) boundary conditions, and uses $2\times 11$kW motors to drive the flow. The sodium setup holds 70 liters with $2R \sim 40$ cm, has conducting (copper shell) boundary conditions and uses $2\times 75$kW motors. The rotation rates of the driving discs are equal and opposite, adjustable in the range $\Omega \in [0 - 50]$ Hz.

Both set-up are equipped with a piezoelectric pressure transducer mounted flush with the cylindrical wall. Figure 5(a) shows an example of the fluctuations in time of the pressure measured at the flow wall. The sudden drops are ascribed to concentrated vortex filaments [9, 1, 4] that have visualized

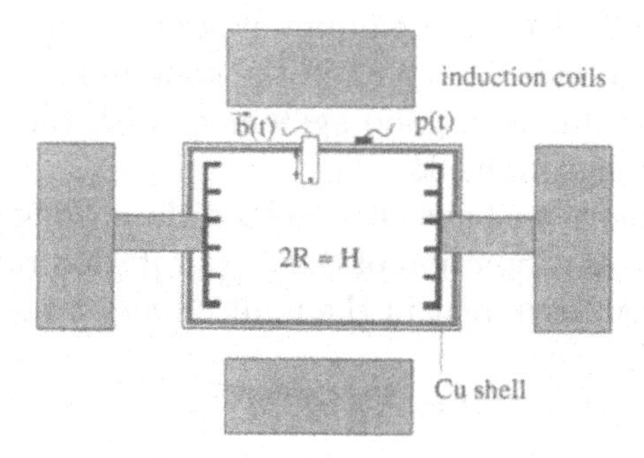


*Figure 4.* Experimental set-up. The magnetic induction coils can produce an applied field of about 20 Gauss inside the flow, either parallel to the rotation axis or perpendicular to it. The magnetic field is measured locally *in situ* using a Hall probe located in the median plane, at a distance $D \sim R/2$ from the rotation axis.

using water seeded with air bubbles [7]; their core size has been measured acoustically [6] and is found to be of the order of the Taylor microscale. The $rms$ intensity of the pressure fluctuations varies as the square of the rotation rates of the discs, as shown in figure 5(b). This yields a measurement of the intensity of the $rms$ velocity fluctuations [9, 25]:

$$p_{rms} \sim \frac{1}{2}\rho u_{rms}^2. \tag{4}$$

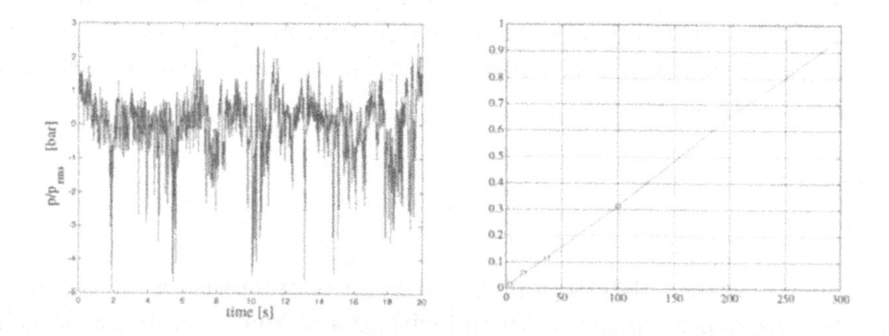


*Figure 5.* (a) time variation of the pressure measured at the flow wall ($\Omega = 17$ Hz); (b) evolution of the $rms$ value of the pressure fluctuations with the discs rotation rate.

This, in turn, gives an estimate of the efficiency of the discs driving the flow, evaluated as the ratio of the $rms$ velocity fluctuation to the disc rim speed:

$$K_u = \frac{u_{rms}}{U_{\text{rim}}} = \frac{\sqrt{2p_{rms}/\rho}}{2\pi R\Omega} \tag{5}$$

We obtain $K_u = 0.12$ for the gallium experiment using flat discs bearing an etched pattern and $K_u = 0.83$ in the sodium setup using discs with curved blades. This value is in good agreement with the optimization process and the water measurements.

For each experiment, we have also checked the scaling of the power consumption of the flow as a function of the discs rotation rate. As stated previously, a dimensional argument in the limit of very large kinetic Reynolds numbers yields:

$$P = K_P \rho R^5 \Omega^3 \quad , \tag{6}$$

where $K_P$ is a dimensionless factor that depends on the geometry of the cell and of the shape of the driving discs. To obtain $P$, we monitor the current and voltage in the driving motors or we record the temperature drift inside the flow when the external cooling is turned off. Both methods give results in very good agreement. We obtain $K_P = 2.57$ in the gallium set-up (discs with a 5 mm deep etched pattern), and $K_P = 33.9$ in the sodium experiments with the discs as defined by the optimization procedure.

*Magnetic measurements :* Induction coils are placed with their axis either aligned with the motors rotation axis or perpendicular to it – cf. figure 4. As a result, one can apply to the flow a steady magnetic field $\vec{B}_0$ with strength in the range 1-20 Gauss. This field does not modify the flow, since the interaction parameter $N = B_0^2/\rho\mu_0 U^2$ is of the order of $N \sim 10^{-5}$. However, it is distorted by the flow motion that generates an induced field $\vec{b}$ results. We measure the three components of the local magnetic field inside the flow using a calibrated and temperature compensated 3D Hall probe (F.W. BELL). Its dynamical range is 65 dB and time resolution 20 $\mu$s. The signal is digitized using a 16-bit data acquisition card (DataTranslation) and stored on PC.

## 4. Results

### 4.1. INDUCTION

Let us begin with a description of the main characteristics of the magnetic measurements, made *in situ* and with an externally applied field. The equation governing the magnetic field $\vec{b}$ generated by the flow, reads:

$$\partial_t \vec{b} - \frac{1}{\mu\sigma} \Delta \vec{b} = (\vec{B}_0.\vec{\nabla})\vec{u} - (\vec{u}.\vec{\nabla})\vec{b} + (\vec{b}.\vec{\nabla})\vec{u} \quad . \tag{7}$$

At low $Rm$, the first term on the right hand side of the equation for $\vec{b}$ is dominant and the induced field is thus also proportional to $Rm$. At larger $Rm$, the two last terms of the equation for $\vec{b}$ are not negligible and the induced field is no longer proportional to $Rm$.

When the magnetic Reynolds number is low enough, the induction equation is dominated by the stretching term (the 'quasistatic' approximation [13, 23]) and reads:

$$(\vec{B}_0.\vec{\nabla})\vec{u} + \frac{1}{\mu\sigma}\Delta\vec{b} \approx 0 \, . \tag{8}$$

For an external field $\vec{B}_0$ applied along direction $\hat{\jmath}$, the $i$-th component of the induced field results from the stretching by the velocity gradient $\partial_j u_i$. At low $Rm$, induction is directly linked to the velocity gradient tensor $[\partial_j u_i]$; in addition, the amplitude of the induced field is directly proportional to the magnitude of the applied field (the ratio of the two provide an *intrinsic* definition of the magnetic Reynolds number – cf. [22]).

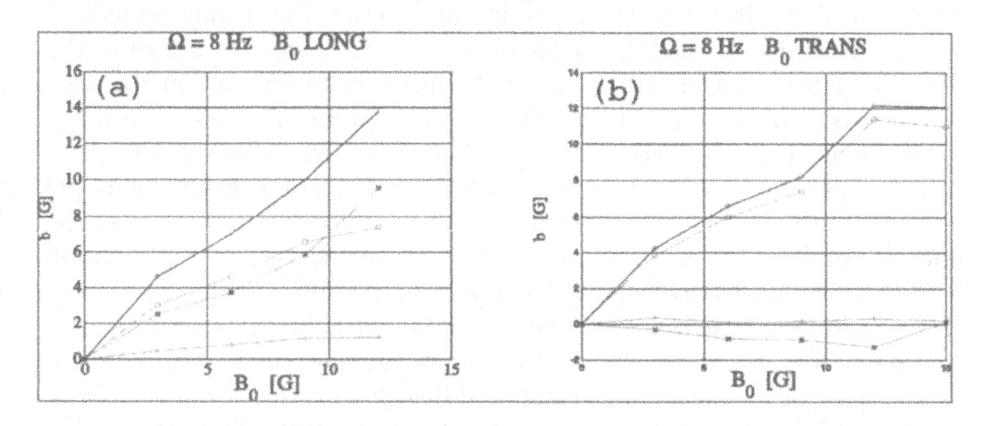


*Figure 6.* Induced field as a function of the applied field, for counter rotating discs at $\Omega = 8$ Hz. (a) $B_0$ is applied along the axis of rotation; (b) $B_0$ perpendicular to the axis of rotation. Symbols are: (○) axial component $b_z$ of the induced field, (∗) azimuthal component, (+) radial component and (◇) induction magnitude $||\vec{b}||$. The measurement probe is located near the mid-plane, 10 cm from the axis of rotation.

An example of the variation of the average components of the induced magnetic field with the applied field is given in figure 6. In each case, the magnitudes of the axial and azimuthal induced fields are of the same order and both larger than that of the radial field. This is in agreement with the averaged velocity profile measurements (see figure 1. The total induced field is of the same order as the applied one:

$$\frac{\partial||\vec{b}||}{\partial B_{0,\mathrm{LONG}}} \sim 0.9 \qquad \frac{\partial||\vec{b}||}{\partial B_{0,\mathrm{TRANS}}} \sim 1.2 \ , \tag{9}$$

where $B_{0,\mathrm{LONG}}$ and $B_{0,\mathrm{TRANS}}$ are the externally applied fields along the axis of rotation or in the direction perpendicular to it (cf. figure 2).

For comparison, measurements in the gallium experiment at a comparable rotation rate gives a maximum value $\partial||\vec{b}||/\partial B_{0,\mathrm{LONG}} \sim 0.1$, while $\partial||\vec{b}||/\partial B_{0,\mathrm{TRANS}}$ is of the order of a few percent [22]. Recalling that this ratio is a definition of an intrinsic magnetic Reynolds number, the observed increase in induction effects when going from the gallium to the sodium experiment is attributed to the changes in the fluid electrical conductivity ($\sigma_{\mathrm{Na}} \sim 2.2\sigma_{\mathrm{Ga}}$), size of the experiment ($L_{\mathrm{Na}} \sim 2L_{\mathrm{Ga}}$) and power in the driving motors ($P_{\mathrm{Na}} \sim 7P_{\mathrm{Ga}}$). In addition, the impellers profile has been optimized for a ratio of poloidal to toroidal velocity close to 1 so that the induction have the same magnitude in the axial and transverse directions.

## 4.2. TIME EVOLUTION AND SPECTRUM

The data shown in the previous section are averaged over long periods of time. In fact, due to the large value of the kinetic Reynolds number of the flow, the flow is strongly turbulent, with *rms* velocity fluctuations (related to the mean) in the range 30 to 50% (see the experimental set-up section).

We show in figure 7 the time evolution of the axially induced field for a transverse $B_0$, at a rotation rate of $\Omega = 8$ Hz. Large fluctuations are observed with a level comparable to that of the velocity fluctuations of the flow. In addition, they occur over slow time scales: the *rms* fluctuation level is nearly unchanged if the signal is low-pass filtered below $\Omega$ or even $\Omega/10$. Such long time scales can be associated with 'global' fluctuations of the mean flow which are known to exist in this geometry [32]; it is also of the order of magnitude of magnetic diffusion over the flow size ($\tau_{\mathrm{diff}} = \mu_0 \sigma R^2 \sim 1$ s). Advective time scales are much faster: for instance the advection of fluid past the Hall probe is of the order of $\tau_{\mathrm{adv}} \sim d_{\mathrm{probe}}/u_{rms} \sim (d_{\mathrm{probe}}/2\pi R)T \sim 0.024T$ ($T \equiv \Omega^{-1}$ is the integral time scale of the flow).

These time scales have a corresponding meaning in the correlation functions and power spectra of the induction field. Correlation is shown in figure 8. One observes that the autocorrelation decreases with a characteristic time of order $T$ and is null for time lags larger than about $10T$. It shows in particular that the drops in the induced field observed in figure 7 occur as decorrelated events. The components of the induced field are also cross-correlated (dashed lines in the main figure 8: at $\Omega = 8$ Hz, the axial and azimuthal components are correlated, whereas at $\Omega = 17$ Hz the axial and radial components are correlated. The induced field is also correlated to the pressure fluctuations (inset of figure 8):

$$\langle b(t)p(0)\rangle_{t=0} \sim \pm 0.2 \ , \tag{10}$$

a quite significant value considering that the measurements are made at points located some 15 cm away in a flow where the Reynolds number is

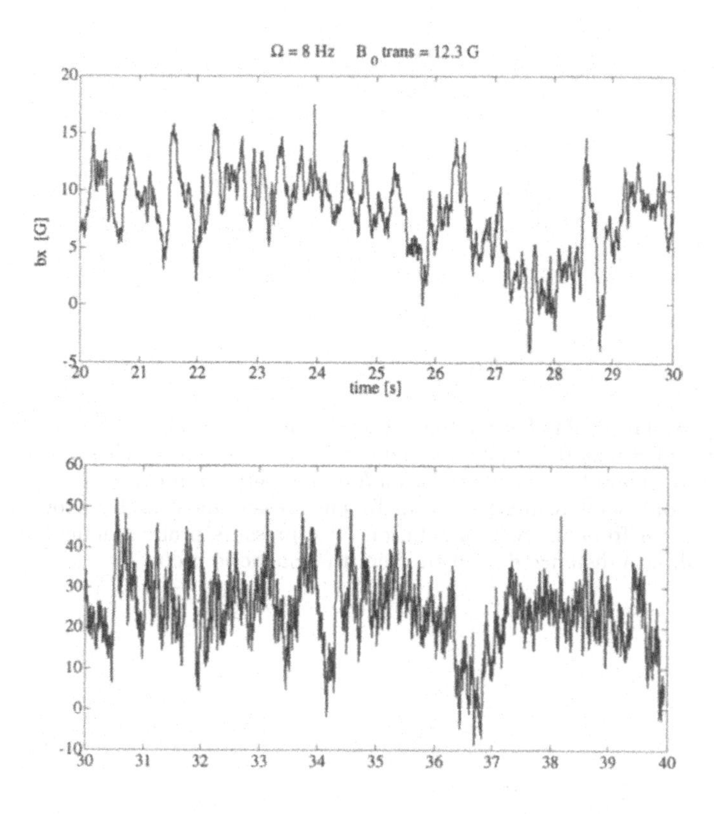


*Figure 7.* Time variation of the axial component of the induced field inside the flow, for a transverse applied field of magnitude $B_0 = 12.3$ G, at a rotation rate $\Omega = 8$ Hz (upper) and $\Omega = 17$ Hz (lower). The measurement probe is located near the mid-plane, 10 cm from the axis of rotation.

larger than $10^6$. We thus attribute the slow variations in the magnitude of the induction to changes in the large scales of the hydrodynamic flow.

Spectra (Fourier transform of the auto correlation functions) are shown in figure 9. The curves are very similar for the three components of the induced field. Three regimes are observed: (i) for frequencies lower than about $\Omega/10$, the spectral content is flat; (ii) for frequencies between $\Omega/10$ and $\Omega$ there is a power law behavior with an exponent close to $-1$; (iii) for frequencies higher than $\Omega$ the spectra decay algebraically with a slope close to $-11/3$. This last regime corresponds to the turbulent fluctuations, in agreement with Kolmogorov K41 phenomenology [23], provided that a Taylor hypothesis is valid for the local field measurement. This observation is in agreement with measurements made in the Gallium experiment [27]. The power law regime $\tilde{b}^2(f) \propto f^{-1}$ was not readily observed in the Gallium set-up; it is consistent with measurements of the Karlsruhe experiment [37], and with numerical results reported in [10].

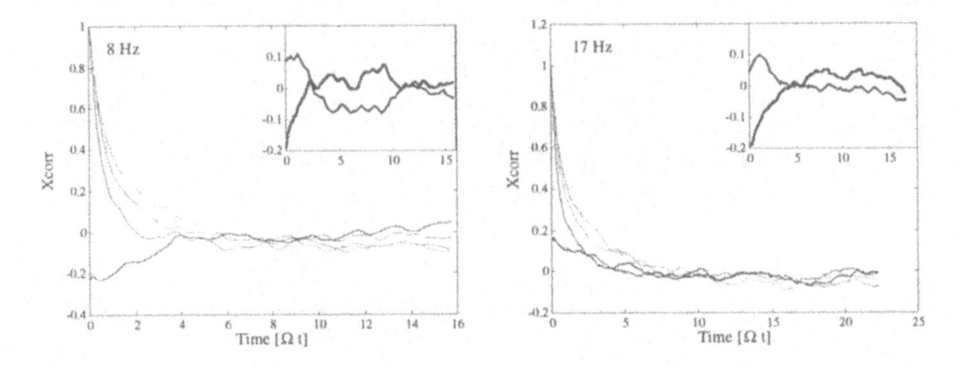


*Figure 8.* Correlation of induction fluctuations. Autocorrelation functions for each of the 3 components of B-field (solid line) and largest $\langle B_i B_j \rangle$ cross correlation function (dashed line). The inset shows the cross correlation function between pressure and magnetic field component (axial: solid; azimuthal: dashed). The measurement probe is located near the mid-plane, 10 cm from the axis of rotation. The pressure probe is mounted flush with the inner wall, at a distance $d = 10$ cm from the magnetic probe.

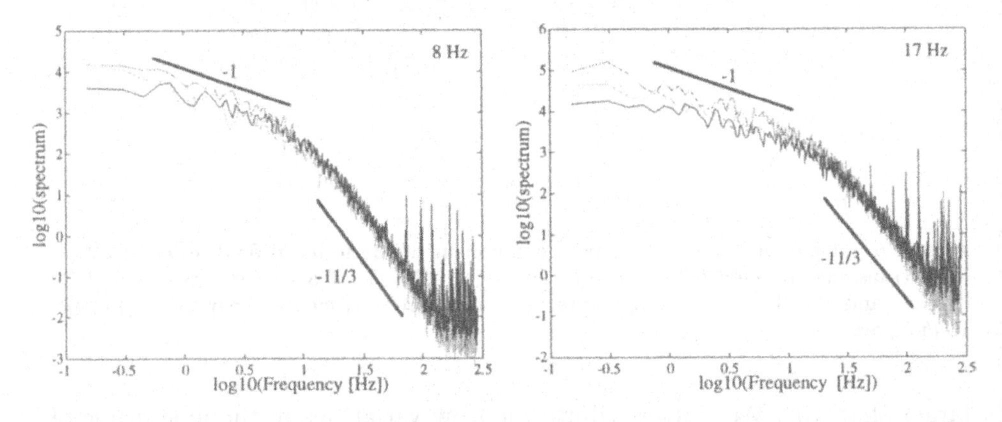


*Figure 9.* Power spectrum of the components of the magnetic induction. The measurement probe is located near the mid-plane, 10 cm from the axis of rotation.

## 4.3. EVOLUTION WITH THE MAGNETIC REYNOLDS NUMBER

Experiments with a steady externally applied field have been made for rotation frequencies of the impellers between 0 and 24 Hz. In the case where $\vec{B}_0$ is perpendicular to the axis of rotation, the corresponding evolution of the average magnitude of the induced field $|\vec{b}|$ and of its *rms* fluctuations with the magnetic Reynolds number are shown in figure 10.

At low discs rotation speed, the behavior is linear: the induced field is proportional to the magnetic Reynolds number. As argued previously, such a linear behavior is expected at low $Rm$ and the values of the induction strength obtained are in agreement with our optimization procedure.

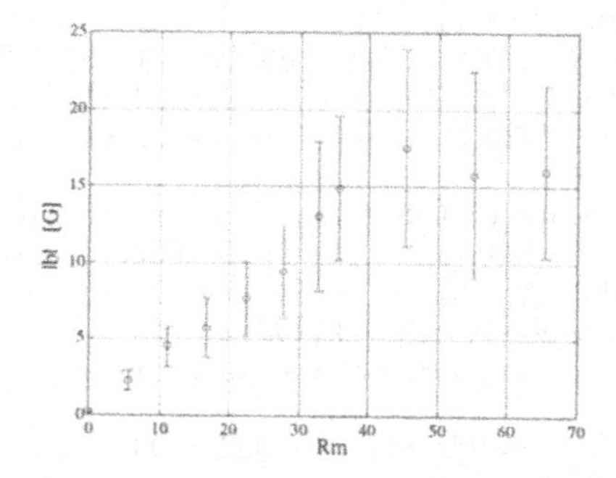


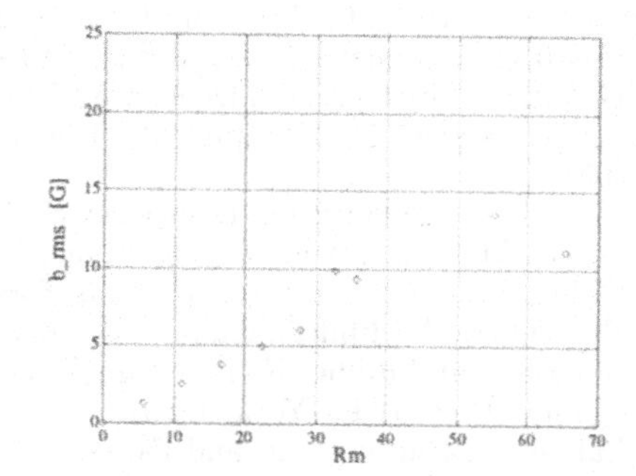


*Figure 10.* Induction for an applied transverse field, $B_0 = 12.3$ G. (a) Variation of the average induced magnetic field with the magnetic Reynolds number of the flow; (b) *rms* variation of the magnitude of $\vec{b}$ (it accounts for the error bars in (a)). In the calculation of $Rm$, the variations of the electrical conductivity with the sodium temperature have been accounted for. The measurement probe is located near the mid-plane, 10 cm away from the axis of rotation.

At higher rotation rates, one observes a change in behavior: despite the increase in the discs rotation rate, the amplitude of the induced field saturates. As the magnitude of the induced field reaches the magnitude of the applied one, it is not surprising to observe non linear effects. For instance, when a transverse field is applied as in figure 10, an axial field component is induced (cf. figure 6(b)); this may in turn generate a transverse induced field as in figure 6(a). If this is the case in the above figure, the non-linearity tends to decrease the overall magnitude of the induced field. Further measurements are needed establish clearly these effects and their link to a possible dynamo action in this geometry.

## 5. Concluding remarks

A sodium experimental platform has been established in a CNRS/ENS/CEA collaboration, that has allowed new MHD measurements on von Kármán swirling flows. The results reported here are based on the experiment first run. Further progress are expected as the experiment proceeds.

## References


1. Abry, P. , Fauve, S., Flandrin, P. and Laroche C., *J. Phys. II France* **4**, 725-733 (1994).
2. Burguete J., Daviaud F., Léorat J., preprint, (2000).
3. Bullard E.C., *Proc. Camb. Phil. Soc*, **51**, 744-76, (1955).
4. Cadot, O., Douady, S. and Couder, Y., *Phys. Fluids* A**7**, 630-646 (1995).
5. Daviaud F., Chiffaudel A., Mari L., private communication.
6. Dernoncourt, B., Pinton, J. F. and Fauve, S., Physica D, **117**, 181-190, (1998).
7. Douady, S., Couder, Y. and Brachet, M.-E., *Phys. Rev. Lett.* **67**, 983-986 (1991).
8. Dudley N.L., James R.W., *Proc. Roy. Soc. Lond.*, **A425**, 407,(1989).

9. Fauve S., Laroche C., Castaing B., *J. Phys. II*, **3**, 271, (1993).
10. Frisch U., Pouquet A., Léorat J., Mazure A., *J. Fluid Mech.*, **68**, 769, (1975).
11. Galaitis A., Lielausis O., Dement'ev S., Placatis E., Cifersons A., Gerbeth G., Gundrum T., Stefani F., Chrsiten M., Hänel H., Will G., Phys. Rev. Lett., **84**, 4365, (2000).
12. Galaitis A., Lielausis O., Dement'ev S., Placatis E., Cifersons A., Gerbeth G., Gundrum T., Stefani F., Chrsiten M., Hänel H., Will G., arXiv:physics/0010047, (2000).
13. Golitsyn, G. S., *Sov. Phys. Dokl.*, **5**, 536-539, (1960).
14. Herzenberg A., Philos. Trans. Roy. Soc. London A**250**, 543 (1958).
15. Krause F. and Rädler K.-H., *Mean field magnetohydrodynamics and dynamo theory*, Pergamon Press (New-York, 1980).
16. Labbé, R., Pinton, J.-F. and Fauve, S., *Phys. Fluids*, **8**(4), 914-922, (1996).
17. Larmor J., *Rep. Brit. Assoc. Adv. Sci*, 159-160, (1919).
18. 'Generation of magnetic field in the Couette-Taylor system', these proceedings.
19. Léorat J., Proc. Pamir Conference, Aussois(France), (1998).
20. Lowes F. J. and Wilkinson I., Nature **198**, 1158 (1963); **219**, 717 (1968).
21. Malkus W.V.R., *Science*, **60**, 259-264, (1968).
22. Martin A. , Odier P., Pinton J.-F., Fauve S., *Euro. J. Phys.*, to appear, (2000).
23. Moffatt H. K., *J. Fluid Mech.*, **11**, 625, (1961).
24. Moffatt H. K. , *Magnetic field generation in electrically conducting fluids*, Cambridge University Press (Cambridge 1978).
25. Mordant N., Pinton J.-F., Chillà F., *J. Phys. II*, **7**, 1, (1997).
26. Nore C., Brachet M.-E., Politano H., Pouquet A., *Phys. Plasmas*, **4**, 1, (1997).
27. Odier P. , Pinton J.-F., Fauve S., *Phys. Rev. E*, **58**, 7397, (1999).
28. Parker E. N., Astrophys. J. **122**, 293 (1955).
29. Peffley N.L., Cawthrone A.B., Lathrop D.P., Phys. Rev. E, **61**(5), 5287 - 5294, (2000).
30. Picha, K. G., *PhD Thesis, U. of Minnesota* (1957).
31. Picha, K. G. and Eckert, *Proc. 3rd US Natl. Cong. Appl. Mech.*, 791-798 (1958).
32. Pinton J.-F., Labbé R., *J. Phys.II (France)*, **4**, 1461-1468, (1994).
33. Ponomarenko Yu. B., J. Appl. Mech. Tech. Phys. **14**, 755 (1972).
34. Roberts P.H. and Jensen T. H. , Phys. Fluids B **5**, 1408 (1993) and references therein.
35. Roberts G. O., Phil. Trans. Roy. Soc. London A **271**, 411 (1972).
36. Simand C., Chillá F., Pinton J.-F., in "Vortex structure and dynamics",, 285-292, LNP555, Spinger, (2000).
37. Stieglitz R., Müller U., Naturwissenschaften, **87**(9), 381-390, (2000).
38. Welsh, W. E. and Harnett, J. P., *Proc. 3rd US Nat. Cong. Appl. Mech.* (1958).
39. Zandbergen P. J., Dijkstra D., *Ann. Rev. Fluid Mech.*, **19**, 465, (1987).

# DYNAMO ACTION IN A FORCED TAYLOR-GREEN VORTEX

C. NORE
*Université Paris-Sud, L.I.M.S.I-C.N.R.S, BP 133 91403 Orsay*
M.E. BRACHET
*LPS-ENS 24 Rue Lhomond, 75231 Paris, France*
AND
H. POLITANO AND A. POUQUET
*Observatoire de la Côte d'Azur BP 229, 06304 Nice, France*

**Abstract.** Dynamo action is demonstrated numerically in the forced Taylor-Green (TG) vortex [1] made up of a confined swirling flow composed of a shear layer between two counter–rotating eddies and corresponding to a standard experimental set–up in the study of turbulence called the von Kármán (VK) swirling flow. The critical magnetic Reynolds number above which the dynamo sets in depends crucially on the fundamental symmetries of the TG vortex which can be broken by introducing a scale separation in the flow, or by letting develop a small non–symmetric perturbation which can be either kinetic and magnetic, or only magnetic [2]. We present cases where a long term magnetic field produced by dynamo action saturates. Implications of our results to VK sodium experiments are discussed.

## 1. Introduction

The generation of magnetic fields can be linked to a stirring mechanism that is often associated with convective motions, as in the Earth's mantle or the solar granulation, or to a combination of a large-scale shear and rotation, as in our Galaxy. Such swirling motions constitute one of several standard turbulent flows studied recently in a systematic way both in the laboratory where it is called the von Kármán flow (hereafter referred to as VK) [3, 4, 5] and in numerical computations using the so-called Taylor-Green (TG) vortex [1]. The relation between the VK flow and the TG vortex is a similarity in overall geometry: a shear layer between two counter–rotating



*P. Chossat et al. (eds.), Dynamo and Dynamics, a Mathematical Challenge,* 51–58.

eddies. The TG vortex, however, is periodic with free-slip boundaries while the VK flow is contained inside a tank between two counter–rotating disks. In search of yet higher Reynolds numbers for experiments, a possible venue is to resort to liquid metals; one such experiment of VK flow using liquid gallium [6] has demonstrated the generation through differential rotation of a toroidal magnetic field from a poloidal field but did not show dynamo effect, probably because magnetic Reynolds numbers were too small. A new experiment using liquid sodium is being carried out [7] in which the magnetic Reynolds number may be close to the critical value above which a dynamo sets in.

The primary objective of this article is to establish under what conditions the dynamo action is produced in the TG vortex. The governing equations, the TG flow symmetries, the forcing term and the possible symmetry breakings are described in §2. Parametric studies of thresholds and preliminary results on saturation are presented in §3. The experimental implications of our results are then discussed.

## 2. Definition of the forced Taylor-Green vortex

The magnetohydrodynamic (MHD) equations for incompressible fluids with $\nabla \cdot \mathbf{v} = 0$ and $\nabla \cdot \mathbf{b} = 0$, and with $\rho_0 = 1$ the constant density, read:

$$\begin{aligned} \frac{\partial \mathbf{v}}{\partial t} + \mathbf{v} \cdot \nabla \mathbf{v} &= -\nabla P + \nu \nabla^2 \mathbf{v} + \mathbf{j} \times \mathbf{b} + f(t) \mathbf{v}^{TG}, \\ \frac{\partial \mathbf{b}}{\partial t} &= \text{curl } (\mathbf{v} \times \mathbf{b}) + \eta \nabla^2 \mathbf{b} \ . \end{aligned} \tag{1}$$

Here, $\mathbf{v}$ is the velocity, $\mathbf{b}$ the Alfvén velocity (with $\mathbf{B} = \sqrt{\mu_0 \rho_0}\ \mathbf{b}$ the magnetic induction, where $\mu_0$ is the vacuum permittivity), $P$ the pressure, $\nu$ the kinematic viscosity, $\eta$ the magnetic diffusivity, and $\mathbf{j} = \nabla \times \mathbf{b}$ the current density. The driving term $f(t)\mathbf{v}^{TG}$ with $\mathbf{v}^{TG} = (\sin(k_0 x)\cos(k_0 y)\cos(k_0 z),$ $-\cos(k_0 x)\sin(k_0 y)\cos(k_0 z),\ 0.)$ and $k_0$ an odd wavenumber is introduced in order to balance the energy dissipation. The choice of the volume forcing term $f(t)$ is discussed in section §2.3. The two nondimensional parameters are the kinetic Reynolds number $R^v = V_0 L_{int}/\nu$ and the magnetic Reynolds number $R^m = V_0 L_{int}/\eta$ where $V_0$ is the *r.m.s.* velocity in the stationary non–magnetic phase and $L_{int}$ the integral scale of the turbulent flow. The ratio of $R^m$ and $R^v$ is the magnetic Prandtl number $P^m = \nu/\eta$.

### 2.1. FLOW SYMMETRIES

The $\mathbf{v}^{TG}$ velocity presents a number of symmetries dynamically compatible with the equations of motion [1], *i.e.* an initial data with the same symmetries leads to a symmetric solution $\mathbf{v}_s$. The symmetries of $\mathbf{v}_s$ amount, with

$k_0 = 1$, to the expansion $\mathbf{v}_s = \sum_{m,n,p} \Big(\hat{u}_{sx}(m,n,p,t) \sin mx \cos ny \cos pz,$ $\hat{u}_{sy}(m,n,p,t) \cos mx \sin ny \cos pz, \hat{u}_{sz}(m,n,p,t) \cos mx \cos ny \sin pz\Big)$ where $\hat{\mathbf{u}}_s(m,n,p,t)$ vanishes unless $m,n,p$ are either all even or all odd integers. The expansion coefficients obey the additional relations : $\hat{u}_{sx}^{(r)}(m,n,p) = (-1)^{r+1}\hat{u}_{sy}^{(r)}(n,m,p)$ and $\hat{u}_{sz}^{(r)}(m,n,p) = (-1)^{r+1}\hat{u}_{sz}^{(r)}(n,m,p)$, where $r = 1$ when $m,n,p$ are all even and $r = 2$ when $m,n,p$ are all odd. The corresponding symmetries of $\mathbf{v}_s$ in physical space are rotational symmetries: of angle $\pi$ around the axis $(x = z = \pi/2)$ and $(y = z = \pi/2)$ ; and of angle $\pi/2$ around the axis $(x = y = \pi/2)$. There are also planes of mirror symmetry: $x = 0, \pi$, $y = 0, \pi$, $z = 0, \pi$. The velocity and the vorticity $\omega_s = \nabla \times \mathbf{v}_s$ are respectively parallel and perpendicular to these planes that form the sides of the so-called *impermeable box* which confines the flow. The kinetic helicity $h_s(\mathbf{x}) = \mathbf{v}_s \cdot \omega_s$ is *anti-symmetric* with respect to the planes of mirror symmetries. Thus, the total helicity of the TG flow $< h_s(\mathbf{x}) >\equiv 0$ when integrated over the whole periodic box $[0, 2\pi]^3$. However, the helicity in the impermeable box is non zero: the eddy at the top of the impermeable box entrains an aspirating motion upward with velocity and vorticity anti–parallel, and similarly for the counter–rotating eddy at the bottom.

## 2.2. DETERMINATION OF THE VOLUME FORCING TERM

The purely kinetic ($\mathbf{b} = 0$) forced Navier-Stokes problem is characterized by the energy budget: $d < \mathbf{v}^2/2 >/dt =< -\nabla P \cdot \mathbf{v} > + < \nu\nabla^2\mathbf{v} \cdot \mathbf{v} > + < f(t)\mathbf{v}^{TG} \cdot \mathbf{v} >= 0 - \epsilon_d(t) + \epsilon_i(t)$, where $< >$ stands for the spatial average and $\epsilon_d$ and $\epsilon_i$ for the dissipation and injection powers. Averaging over temporal fluctuations in stationary state, we can check that $0 = -\overline{\epsilon_d(t)} + \overline{\epsilon_i(t)}$. In a preliminary stage, we have tested three types of forcing: (i) constant velocity $\mathbf{v}^{TG}$ with $f(t)$ set by the condition that the $(k_0, k_0, k_0)$ Fourier components of the velocity remain equal to $\mathbf{v}^{TG}$; (ii) constant force $f(t) = 0.12$; (iii) constant injection power $\epsilon_i(t) = 0.03$. The numerical results of Figs. 1 and 2 are obtained with a symmetric forced code: all the TG symmetries (for $k_0 = 1$) are implemented within this code [1] which is used to save computer resources. Note that the kinetic helicity in the impermeable box is negative most of the time (Fig. 2). We have chosen to use the constant velocity forcing in the dynamo problem because of its simplicity.

## 2.3. SYMMETRY BREAKING

The TG symmetries can be *spontaneously broken* in two ways: either the symmetry of the velocity field itself is broken, in the sense that a small non–symmetric component of the initial velocity field will grow and eventually

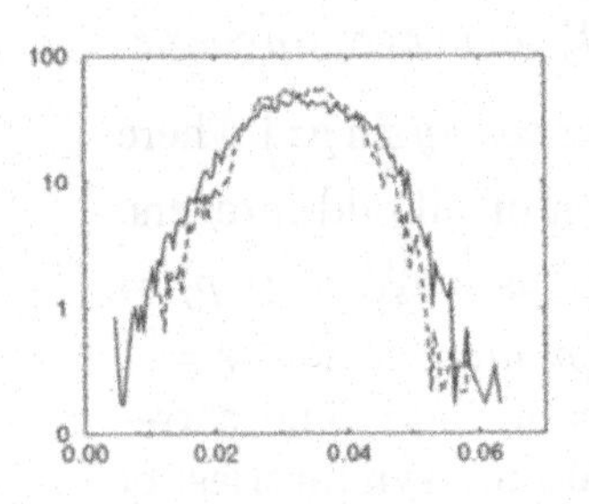


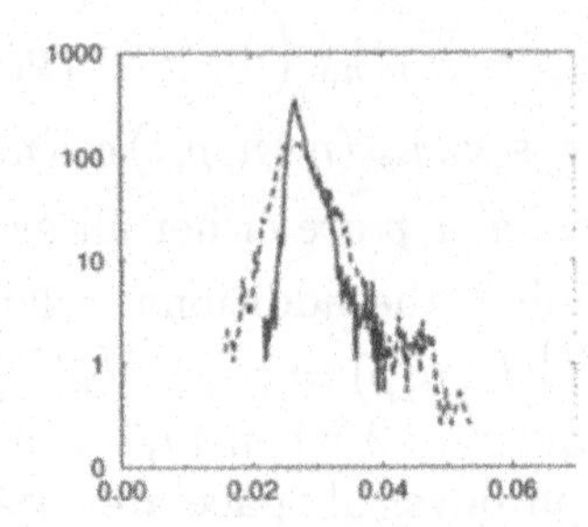


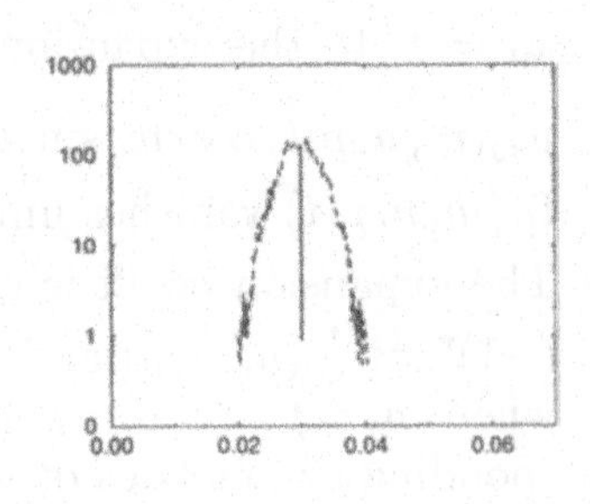


*Figure 1.* Histograms of dissipation (solid line) and injection (dashed line) powers for different forcings: (a) constant velocity $\mathbf{v}^{TG}$; (b) constant force; (c) constant injection power.

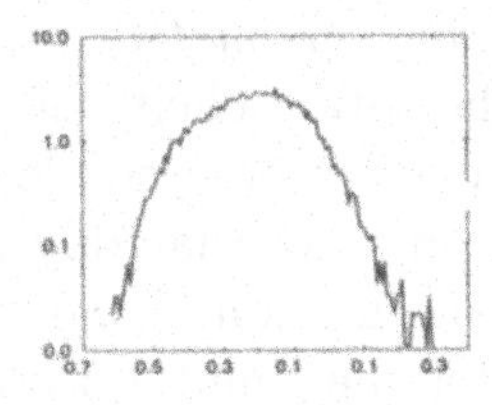

*Figure 2.* Histogram of kinetic helicity in the impermeable box (constant velocity forcing).

break the symmetry of the solution, or, even with a TG symmetric velocity field, non symmetric magnetic modes can be unstable.

In order to take these possibilities into account, two $2\pi$–periodic pseudo–spectral codes are used. Both are de–aliased by the 2/3 rule. The wavenumbers are thus integers in the range $[k_{min} = 1, k_{max} = N/3]$, where $N^3$ is the number of grid points. The first code is a standard periodic one. The second code uses the symmetries of the TG vortex for both the velocity and the magnetic field, *i.e.* the fields are parallel to the sides of the impermeable box. Thus, $\mathbf{j}$ is perpendicular to these sides which physically correspond to perfectly conducting walls.

Another way of breaking symmetry is to force (1) at $k_0 > 1$. Within the symmetry–conserving algorithm of a symmetric code, only modes with wavevector $(k_x = p, k_y = q, k_z = r)$, with $(pqr)$ jointly even or odd can be excited [1]. When $k_0 = 1$, only larger modes are generated from the initial data, whereas when $k_0 = 3, 5, 7$, *smaller* and larger modes can be created. Thus, in the $k_0 > 1$ case, there is a possibility *within a symmetric TG code* for symmetry breaking of the initial flow by smaller or larger and odd or even modes.

## 3. Dynamo action in a forced TG flow

### 3.1. INITIALIZATION AND DATA ANALYSIS

In a preliminary phase to the dynamo runs described henceforth, the kinetic energy is let to settle to its stationary values under the action of the TG forcing in both codes; a magnetic field seed is then introduced with initially $\beta_0 = < \mathbf{b}^2 > / < \mathbf{v}^2 > \ll 1$. The results of a series of computations at resolutions of $64^3$, $128^3$, $200^3$, $256^3$, $400^3$, $512^3$ points, corresponding to a wide range of magnetic Reynolds numbers $R^m$ are summarized in Fig. 3 giving the growth rate $\sigma$ of the dynamo field as a function of $R^m$, with $< \mathbf{j}^2/2 > (t) \sim e^{\sigma t}$. The precision of the computation measured by the logarithmic decrement $\delta(t)$ defined from a fit of energy spectra in the near dissipative range as $E(k,t) \sim exp(-2\delta(t)k)$ is such that for all times $\delta(t)k_{max} \sim 2$, a standard condition for computations of turbulent flows. A symmetric run of $\sim 25\ \tau_{NL}$, where $\tau_{NL} = L_{int}/V_0 \sim 0.6$ is the turn–over time, takes 15,400 seconds of NEC-SX5 at $N^3 = 512^3$.

### 3.2. PARAMETRIC STUDY OF THRESHOLDS AND SATURATION

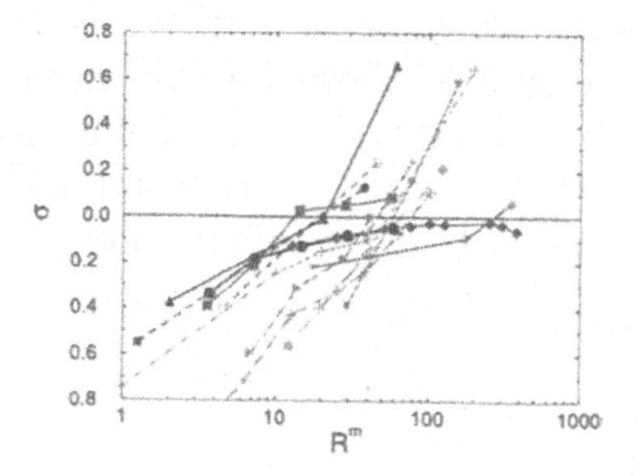


*Figure 3.* Growth–rates $\sigma$ of the square magnetic current as a function of $\log R^m$. Symbols for each computation differ according to the type of run: symmetric TG runs with $k_0 = 1$ for $R^v = 14.3$ are represented by circles and solid line, $R^v = 51$ by diamonds and solid line, $R^v = 175$ by stars and solid line; symmetric TG runs with $k_0 = 3$ for $R^v = 12.75$ are represented by crosses and dash line, $R^v = 27.5$ by right triangles and dash line, $R^v = 88$ by stars and dash line, $R^v = 122$ by dots and solid line, $R^v = 153$ by down triangles and solid line, $R^v = 196$ by crosses and alternate dash line; symmetric TG runs with $k_0 = 5$ are for $R^v = 12.6$ (dots and dash line), $R^v = 45.6$ (open dots and dash line); symmetric TG runs with $k_0 = 7$ and $R^v = 20.5$ are represented by up triangles and solid line; and squares and solid line are for the non–symmetric runs with $k_0 = 1$ and $R^v = 14.3$. Growth rates for $< \mathbf{b}^2 >$ are identical to those displayed here.

#### 3.2.1. *Breaking symmetry for $k_0 = 1$ in the general periodic code*

A simple test of the dynamical constraints imposed by the TG symmetries can be done by performing identical runs at $k_0 = 1$ with (a) the general periodic code and (b) the TG symmetric code. We compare the results in Fig. 3 . Initial conditions are nearly identical: in the general periodic code a non–symmetric perturbation (1 % of the TG kinetic energy) is added to the velocity in the shell of the TG wavevector. The two runs are performed with $R^v = 14.3$ on a grid of $64^3$ points. In the non–symmetric run (*squares and solid line*), the resulting velocity settles at a larger kinetic energy than that at which the basic TG flow settles, namely $E^v \sim 0.28$ instead of 0.17 (Fig. 4 a). Nonlinear saturation sets in for $< \mathbf{b}^2 > / < \mathbf{v}^2 > \sim 100\%$ and $< \mathbf{j}^2 > / < \omega^2 > \sim 300\%$ corresponding to a ratio for the magnetic and viscous dissipations of $\sim 75\%$. As seen in Fig. 3, the growth rate remains negative in the symmetric case (*circles and solid line*), whereas $R_c^m \sim 10$ in the general periodic case.

Whereas in the above non–symmetric runs, symmetries on both the velocity and the magnetic field have been initially broken, we now perform the more stringent test of using a (very small) non–symmetric magnetic seed, but initially imposing all symmetries to the TG velocity, to within round–off errors. In that case, with $R^m = 41$ and $R^v = 10.3$ (Fig. 4 b), there is at first a weak growth of the magnetic field. At $t \sim 220$, there is a spontaneous breaking of the symmetry of the TG velocity field, with an increase in kinetic energy which settles, after a transient, at $E^v = 0.194$. This change in the velocity field allows for a substantial increase (roughly, a quadrupling) of the growth rate of the magnetic energy, before nonlinear saturation sets in for $< \mathbf{b}^2 > / < \mathbf{v}^2 > \sim 21\%$ and $< \mathbf{j}^2 > / < \omega^2 > \sim 100\%$. The ratio of the magnetic and viscous dissipations is then $\sim 25\%$.

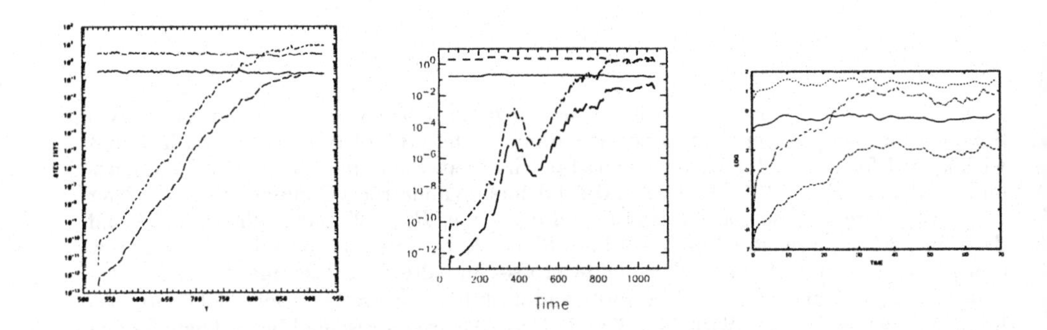


*Figure 4.* Temporal evolution of the kinetic (solid line) and magnetic (long dash line) energies as well as that of the square vorticity $< \omega^2 >$ (dash) and the square current $< \mathbf{j}^2 >$ (alternate dash) for runs with $k_0 = 1$, (a) with $k_0 = 1$, $R^v = 14.3$ and $R^m = 57.2$ with, at $t = 0$, a perturbed TG velocity field in the periodic code; (b) with $k_0 = 1$, $R^v = 10.3$ and $R^m = 41$ with, at $t = 0$, a pure TG velocity field in the periodic code; (c) with $k_0 = 3$, $R^v = 27.5$ and $R^m = 82.5$ with, at $t = 0$, a perturbed TG velocity field in the symmetric code.

The growing non–symmetric magnetic field is found to have Fourier components of the type $(0, 0, 1)$ (data not shown). Because of the relations $\nabla \cdot \mathbf{b} = 0$ and $\mathbf{j} = \nabla \times \mathbf{b}$, the corresponding physical space fields have the form $\mathbf{b} = \Big((a_x, a_y, 0)\ \exp(iz) + c.c.\Big)$ and $\mathbf{j} = \Big((-i\, a_y, i\, a_x, 0)\ \exp(iz) + c.c.\Big)$, where $c.c.$ denotes the complex conjugate. Thus, in a given horizontal cut $z = const.$, both the magnetic field and the current are constant horizontal vector fields, perpendicular to each other. Note that such a *slab geometry* is forbidden in a symmetric TG code because of overall rotational invariance.

3.2.2. *Results for the TG symmetric code*
Figure 3 shows that, with $k_0 = 1$, it is only for the largest performed kinetic Reynolds number $R^v \sim 175$ that dynamo is obtained with the TG symmetric code. That defines a critical magnetic Reynolds number $R_c^m(k_0 = 1) \sim 250$. A forcing with $k_0 > 1$ is seen to reduce the critical magnetic Reynolds number, namely $R_c^m(k_0 = 3) \sim 45 - 70$, $R_c^m(k_0 = 5) \sim 20$ and $R_c^m(k_0 = 7) \sim 20$. The drastic change of behaviour when choosing $k_0 > 1$ may be due to a combination of two factors. First, spontaneous symmetry breaking can take place as mentioned before. Secondly, scale-separation may be favorable (for a similar effect in ABC dynamo, see [8]).

We have studied saturation for $k_0 = 3$. A typical temporal evolution of energies is shown on Fig. 4 (c) where $R^v = 27.5$, $R^m = 82.5 > R_c^m \sim 45$. The saturation sets in for $< \mathbf{b}^2 > / < \mathbf{v}^2 > \sim 2\%$ and $< \mathbf{j}^2 > / < \omega^2 > \sim 32\%$. The ratio of the magnetic and viscous dissipations is then around 11%.

## 4. Discussion

We have shown in this article that dynamo action occurs in the Taylor-Green vortex. Its structure, which can be imposed numerically at all times, allows for substantial savings in computing both CPU–wise and memory–wise. By breaking its basic symmetries, the critical magnetic Reynolds number above which dynamo action develops is drastically reduced. This may happen in several ways, for example when implementing a scale separation within a TG code.

When comparing with an experimental set–up, one has to take into account the fact that the magnetic Prandtl number (say, of liquid sodium) is much smaller than unity, a regime unattainable with direct numerical simulations. However, a simple examination of the MHD equations setting $P^m \ll 1$ but keeping $R^m > 1$ indicates that in the presence of an external large–scale magnetic field $\mathbf{B_0}$, an equilibrium in the induction equation is rapidly established, namely $\eta \Delta \mathbf{b} \sim \mathbf{B_0} \cdot \nabla \mathbf{v}$; hence, in amplitude, $b \sim B_0 R^m$, similar to the case of low thermal Prandtl number convection [9]; this suggests that a dynamo mechanism may work as well in the low

$P^m$ regime, granted $R^m$ be sufficiently high. Keeping this in mind, there are several implications of our results to experimental set–ups. First, note that the kinetic helicity inside the impermeable box is non zero (Fig. 2) but its integral over the whole flow vanishes because, numerically, of the presence of mirror images of the impermeable box. Experimentally, only the impermeable box is of relevance, and thus the experimental TG flow is helical and likely a good candidate for dynamo action.

Finally, the present computations indicate that the regime of magnetic Reynolds numbers reachable experimentally for VK flows may be close to criticality. The results obtained with the periodic code show that, when allowing for a dipolar magnetic field to loop outside the vessel, together with a large–scale current looping in the orthogonal direction (*slab mode*), the critical magnetic Reynolds number for dynamo action is considerably lowered ($R_c^m \sim 10$). This type of circulation can be achieved experimentally by using materials with different conductibility at the wall, closing $\mathbf{j}$ with a conductor and $\mathbf{b}$ with a ferromagnetic material. Also scale-separation seems to be favourable for dynamo but the experimental set-up needs to be defined.

*All computations reported in this paper have been performed on the C98 and the NEC-SX5 of IDRIS (Orsay).*

## References

1. M. E. Brachet, D. I. Meiron, S. A. Orszag, B. G. Nickel, R. H. Morf & U. Frisch, *J. Fluid Mech.* **130**, 411 (1983).
2. C. Nore, M. Brachet, H. Politano & A. Pouquet, *Physics of Plasmas* **4**, p. 1 (1997).
3. S. Douady, Y. Couder & M. E. Brachet, *Phys. Rev. Lett.* **67**, 983 (1991).
4. J. Maurer, P. Tabeling & G. Zocchi, *Europhys. lett.* **26**, 31 (1994).
5. S. Fauve, C. Laroche & B. Castaing, *J. Phys. II, France* **3**, 271 (1993).
6. P. Odier, J.-F. Pinton & S. Fauve, *Phys. Rev. E* 58, 6, 7397 (1998).
7. F. Daviaud, S. Fauve, J.-F. Pinton (2000).
8. B. Galanti, P.L. Sulem & A. Pouquet, *Geop. Astrop. Fluid Dyn.* **66**, 183 (1992).
9. K. Kumar, S. Fauve & O. Thual, *Journal de Physique* **II 6**, 945 (1996).

# ON THE POSSIBILITY OF AN HOMOGENEOUS MHD DYNAMO IN THE LABORATORY

R. O'CONNELL, ROCH KENDRICK, MARK NORNBERG, ERIK SPENCE, ADAM BAYLISS AND C.B. FOREST
*Department of Physics, University of Wisconsin, Madison, WI 53706*

## 1. Introduction

The cause of spontaneous generation of magnetic fields in conducting bodies (such as plasmas) is a longstanding, major problem in plasma astrophysics, geophysics, and laboratory plasmas. It is observed that magnetic fields exist in the Earth, Sun and other stars (and perhaps in galaxies), that cannot be explained as surviving primordial fields, and generally believed that such magnetic fields are generated by plasma flow (or flow of liquid metal for the Earth). The question of how magnetic fields are generated by unconstrained flows of conducting fluids and plasma is referred to as the "dynamo" problem; theoretical research into dynamo mechanisms has been actively pursued for several decades. However, until quite recently our probing of the dynamo problem has been limited to analytic calculations, numerical modelling and observational studies; experimental validation (the critical test for any theory) of aspects of the theory and experimental studies of laboratory dynamos have been scarce.

In the kinematic dynamo problem one looks for growing solutions to the magnetic induction equation:

$$\frac{\partial \mathbf{B}}{\partial t} = \nabla \times (\mathbf{V} \times \mathbf{B}) + \frac{1}{\sigma \mu_0} \nabla^2 \mathbf{B}, \tag{1}$$

where $\mathbf{B}$ is the magnetic field and $\mathbf{V}$ is the velocity field. This partial differential equation is linear in magnetic field, is derived from Faraday's law and the Lorentz force generation of current by motion of a conductor across a magnetic field. Thus, the physical principles of the induction equation have been thoroughly tested by experimental validation and are not in question. The untested part of the kinematic dynamo problem is



*P. Chossat et al. (eds.), Dynamo and Dynamics, a Mathematical Challenge,* 59–66.

a geometric and topological question where one attempts to differentiate between those geometries of insulators, conductors, liquids, solids, baffles, and flow topologies (usually turbulent) which lead to dynamos and those which don't.

There has been some experimental progress in identifying real flow geometries and topologies which lead to dynamo action. Notably, Lowes and Wilkinson constructed a kinematic dynamo (based on a theorem by Herzenberg) which was homogeneous electrically by embedding two rotating, conducting spheres in a larger conducting block, and observed a growing magnetic field[1, 2, 3]. By using liquid metal, rather than solid conductors, new flow topologies are possible as demonstrated by by recent experiments in Karlsruhe and Riga. Both of these experiments have demonstrated kinematic dynamo action using helical flows of liquid sodium–flows which cannot be realized by rotating solid conductors[4, 5]. The flows were, however, constrained by helical baffles and pipes which separated the sodium into a number of different regions. So, although these experiments could be considered as electrically homogeneous, they were not mechanically homogeneous and cannot address the the second question relating to the saturation and large scale backreaction of the self-generated magnetic field on the flow.

No experiments have yet demonstrated dynamo action in a geometry in which the velocity field is unconstrained by pipes and baffles on the scale for which the magnetic field is generated. The goal of this study is to investigate the possibility of constructing an *electrically* and *mechanically homogeneous dynamo* in which the geometry is simply connected and surrounded by a vacuum region, and for which the flow is driven but unconstrained and therefore free to respond in all directions to the lorentz forces induced by the generated magnetic fields[6].

One such experiment using a spherical geometry is being constructed at the University of Wisconsin. Similar experiments are under construction at the CNRS, Caderache[7] and in operation at the University of Maryland[8]. The Madison dynamo experiment consists of a 1 m diameter, spherical vessel filled with liquid sodium with 150 kW of mechanical energy input from propellers. The flow geometries have been motivated in part by theoretical kinematic dynamos studies in a sphere which have shown that simple, axisymmetric flows lead to self-excited magnetic eigenmodes[9, 10, 11, 12]. These studies indicate that magnetic eigenmodes grow when the magnetic Reynolds number

$$Rm = \mu_0 a \sigma V_0 \tag{2}$$

exceeds a critical value $Rm_{crit}$ ; $a$ is a scale length taken to be the radius for spherical geometry, $\sigma$ is the fluid conductivity ($10^7$ mho/m for liquid sodium), and $V_0$ is taken to be the peak speed of the mean flow. The main difficulties in executing such an experiment is to safely accumulate a sufficient volume of heated liquid metal and to drive it at a sufficiently high speed so that $Rm > Rm_{crit}$. The required value of $Rm_{crit}$ depends upon the the details of the velocity field; thus, target velocity fields with low values of $Rm_{crit}$ are more easily acheived in an experiment since the required inventory of liquid sodium and required velocity can be minimized.

## 2. Spherical kinematic dynamos with low $Rm_{crit}$

Eigenmode analyses and numerical solutions of kinematic dynamos in spherical geometries have been studied by a number of authors[13, 9, 10, 12]. Each study is similar in that they all use spherical harmonic expansions of vector potentials to represent magnetic fields, and then search for growing eigenfunctions for the magenetic field. These studies have provided strong motivation for the experiment proposed here, in that they give examples of axisymmetric velocity fields which lead to kinematic dynamos (positive growth rates). The main flow topologies considered here identical with the axisymmetric roll flows proposed by Dudley and James. The simplicity of the flows, and the low $Rm_{crit}$ values required for self-excitation make these flows well suited to experiments. We have revisited this problem with the goal of finding kinematic dynamo solutions which self-excite at lower $Rm$.

The kinematic dynamo problem searches for velocity fields which are linearly unstable to magnetic perturbations and thereby generate exponentially growing magnetic fields, i.e. $\mathbf{B} \propto e^{\lambda t}$. If a laminar velocity field is given (ignoring Euler's equation), Eq. 1 becomes a linear eigenvalue equation for $\lambda$,

$$\lambda \mathbf{B} = Rm \nabla \times (\mathbf{V} \times \mathbf{B}) + \nabla^2 \mathbf{B}, \tag{3}$$

where the equation has been recast in terms of normalized variables and the characteristic resistive diffusion time $\tau_\sigma = \mu_0 \sigma a^2$ has been used. The solution to Eq. 3 consists of a series of discrete eigenmodes

$$\mathbf{B} = \sum_i \mathbf{B}_i(x) \exp(\lambda_i t), \tag{4}$$

each characterized by different complex eigenfrequencies $\lambda_i$. The kinematic dynamo problem is primarily concerned with finding the least damped (fastest growing) eigenmode. If $\mathrm{Re}(\lambda_i) > 0$ the system is a dynamo, a velocity field with a linearly unstable magnetic eigenmode.

The method used here is similar to the original work by Bullard and Gellman[13]; a vector potential is defined for each the velocity and the magnetic field in terms of spherical harmonic functions. A finite difference is used to create a banded linear system; the complex eigenvalues and eigenmodes are determined using standard linear algebra techniques.

The axisymmetric dynamos studied by Dudley and James consisted of three examples with $m = 0$. Specifically, they considered flows described by two profile functions: `t1s1` $[s_0^1(r), t_0^1(r)]$, `t1:s2` $[s_0^2(r), t_0^1(r)]$, and `t2:s2` $[s_0^2(r), t_0^1(r)]$ combinations. The profile functions were specified to be simple trigonometric functions in $r$. We have generalized these radial profile functions forms suitible for parametric searchs. The flow field is parametrized for each mode by the position of the poloidal null ($r_s$) or the position of peak toroidal velocity ($r_t$), the shear ($w$), the ratio of the toroidal to poloidal flow ($\epsilon$), the mode amplitudes ($a$) and a boundary layer widths ($l$) to implement the non-slip boundary condition, as shown here:

$$s, t(t) = a_{s,t} \times r^2 \exp\left[-\frac{l_{s,t}}{1-R} - \left(\frac{r - r0_{s,t}}{w_{s,t}}\right)^2\right] \qquad (5)$$

The optimizations are performed by specifying the topology to be optimized (choice of $n$ and $m$ functions for velocity field) and then solving eigenvalue problem, searching for combinations of parameters which maximize the growth rate at a given $R_m$. A simplex search algorithm[14] was used to find the optimized solutions. Configurations with up to 50% better performance compared with the Dudley-James flows have been found. A lower $R_{m_{crit}}$ was found for all three configurations, as shown in table 1. Laboratory experiments show that T4 velocity components are significant, so optimization studies were extended to analyse T2T4S2 flows. The optimizations for the T2T4S2 flows were resticted to approximately the same ratio of T2:T4 found in the experiment. The optimization shows that for this regime it would be better to remove the T4 component entirely.

| Type | Dudley &James | **Opt.** | Amp. | Shift | Width | Edge |
|---|---|---|---|---|---|---|
| `t1:s1` | 150 | **72** | 1:0.12 | 0.207:0.2 | 0.256:0.311 | 0.03 |
| `t1:s2` | 95 | **67** | 1:0.07 | 0.36:0.31 | 0.31:0.385 | 0.054 |
| `t2:s2` | 55 | **47** | 1:0.14 | 0.5:0.46 | 0.32:0.32 | 0.05 |
| `t2t4:s2` | n/a | **66** | 1,0.37:0.17 | 0.64,0.38:0.47 | 0.4,0.3:0.5 | 0.05 |

TABLE 1. *Critical magnetic Reynolds numbers for the Dudley and James flows and also for the optimised flows.*

## 3. Water Experiments

The optimized axisymmetric flows proposed as kinematic dynamos from above, are acceptable flows for experiments, ie. they have growing eigenmodes magnetic Reynolds numbers low enough to be achievavble in the laboratory[1]. However, the only physical constraint imposed by the theory from above is that the flows is that they are divergence free–they do not necessarily correspond to physically realistic solutions of the Euler equations in an unconstrained sphere. This is the realm of experimentation: experiments have been used to determine whether kinematic dynamo velocity fields can be produced in the laboratory. This has been accomplished by driving flows in a sphere using a impellors and measuring the velocity fields directly. We use water rather than sodium. From the measured velocity fields the corresponding growth/damping rate are estimated using the same solution technique as above, assuming that the water is replaced by sodium. The experimental device is a 1 m diameter, spherical, stainless

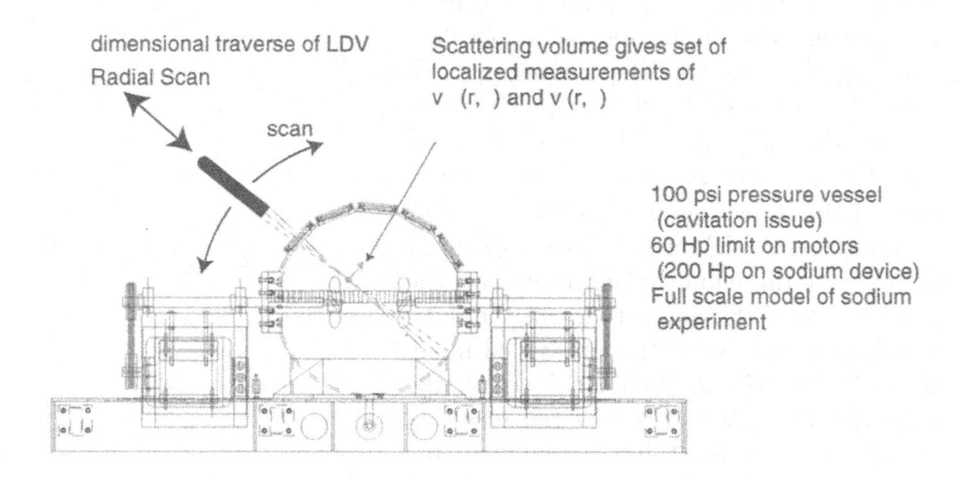


*Figure 1.* The range of radial profiles scanned by the experiment over time due to large scale turbulent flucuations. The upper and lower bounds show one standard deviation from the mean profile.

steel vessel. A 40 kW variable-speed AC inverter drives two independent 45kW motors which in turn drive 2" diameter shaft/impeller assemblies. For counter rotating T2S2 toroidal flows two left-handed square pitch 12" diameter impellers are used, left and right hand combinations are used for co-rotating flows. The impellers are fitted with Kort rings to prevent excessive radial flow at the impeller tips. Due to the large input power to the device we use water cooling via spherically shaped copper plates on the surface. The cooling plates maintain the water at $\approx 50C$ (at which temperature the viscosity and density are the same as sodium at $\approx 110C$) thus the Reynolds number of a given water flow is the same as the equivalent

[1]The Madison dynamo experiment has scale size 0.5m, $Rm = 6 \times V_{peak}$[m/s]

sodium experiment. Cavitation which occurs when the vapour pressure of the fluid is greater than the local pressure (usually at the large pressure drop at the tip of the impellers) has the potential to damage, and severly restrict our peak velocity. The experiment has been successfully pressurized to 40psi to deter the onset of cavitation, and can be pressurized further to 150psi.

Flows corresponding to $R_m \approx 60$ have been measured which, when scaled to $R_m = 120$ are predicted to generate dynamo fields. All of the flows are performed at high hydromagnetic Reynolds numbers of order Re$\approx 10^7$ and are therefore very turbulent. The velocity varies as a function of time - even over resistive timescales ($\approx 3s$). This has a profound effect on the character of the magnetic field produced by the system by making the theoretical concept of static mean flow profile somewhat obsolete - on longer timescales the average growth rate will corresond to the mean flow profile but the magnetic field evolves on resistive timescales, therefore the velocity variation on this timescale defines the character of the magnetic field evolution. To quantify the effect of these flucuations in terms of the growth rate of the magnetic field, the measured mean flow velocities are varied many times in the Monte-Carlo analysis by the measured fluctuation levels. Each resulting set of velocities is fit by the same fitting procedure, yielding the range of radial profiles shown in figure 1. The calculated flow profiles are input to the eigenvalue code to evaluate the least damped or highest growing mode. This results in a range of growth damping rates for a given input power, as shown in the figure. Supporting evidence for this behaviour has been seen at other laboratories - the dynamo experiment at Maryland has seen large changes in the damping rate of its flows, also the alpha experiment in Riga achieved dynamo at a much lower critical Rm than expected from theory. It is still desirable to have a flow which over a long period of time so that the probablity of having a positive growth rate is high. At higher Rm values, or better mean flows, this probability increases. Small differences in the mechanical shape of the impellers can cause small asymmetries in the flows, which manifests as a non-zero T1/S1 modes. These modes in turn can have a very detrimental effect on the dy-

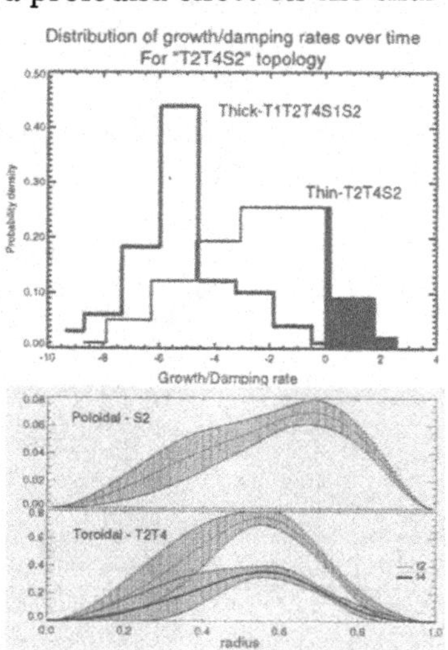


*Figure 2.* : Top: histogram of growth rates for the range of flow configurations measured for a given propeller geometry. The effect of the T1/S1 component is shown. The radial profiles for the purely T2T4S2 are shown in the lower graph.

namo efficiency of the flow. Typically, it is found that flow with T1/S1 modes whose magnitudes are $\approx$ 10% of the total flow can halve the efficiency of the flow doubling the value of the $R_{mcrit}$. In the figure, the average growth/damping rate is shown with the T1/S1 components present and removed. For the ideal simulation, the velocity would be measured simultaneously at all points in the sphere and the real growth/damping rate could be plotted as a function of time, in this analysis it is inherently assumed that there is no correlation between the velocities at different points in the sphere and that the axisymmetry holds over the resistive timescale. The measured self-similar correlation time is used in the evolution of the velocity so that the character of the velocity evolution matches the measured data.

The power needed to drive higher magnetic Reynolds numbers is expected to scale like ${R_m}^3$ - this has been confirmed by experiment and by extrapolation points to an upper $R_m$ limit of $\approx$ 120 beyond which the input power requirement is $\approx 200kW$ and becomes increasingly overwhelming due to the cubic dependence. These figures are worst case due to the improved

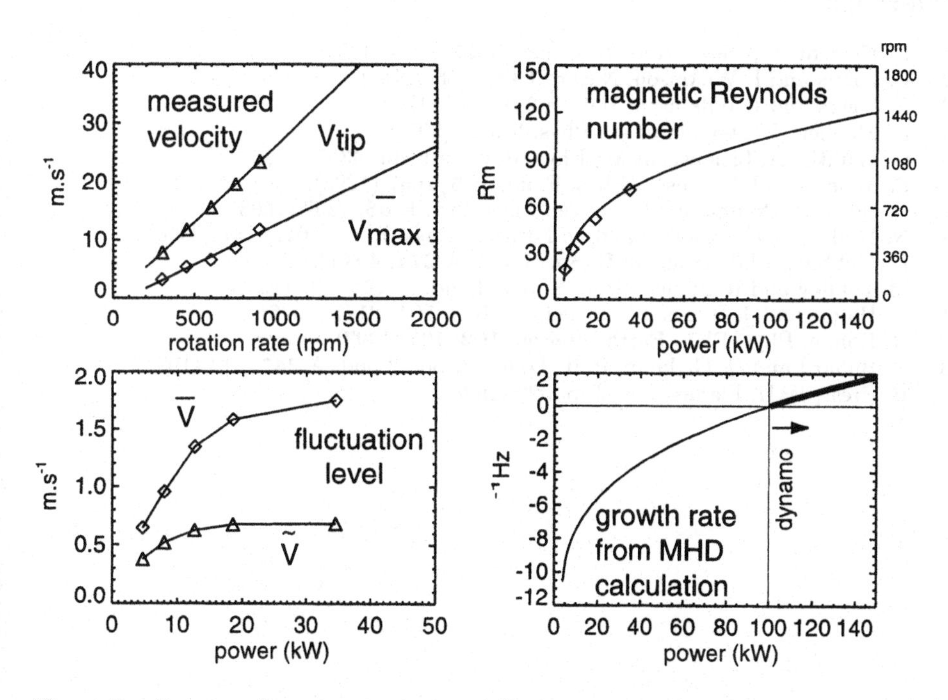


*Figure 3.* Power scaling measurements: a) Peak measured velocity, with propeller tim speed as benchmark, as a function of rotation rate. b) Magnetic Reynolds number assuming the water had the same conductivity as sodium versus input mechanical power. c) Volume averaged mean velocity and fluctuation level as a function of input power. d) Growth/damping rate of the least damped mode as a function of input power.

efficiencies of the inverter, motors and drivetrain of the sodium device over the water device. The actual input power to the water as measured by the heat rise over time is one third of the mechanical input power.

## 4. Conclusions

Extensive numerical studies have identified simple spherical flows with critical magnetic Reynolds numbers low enough to be achieved in the laboratory. Experiments have been performed investigated the feasibility of actually reproducing the flows in water - which in the low B limit has the same fluid properties as sodium. In this experiment the resistive timescale is of the order of the self-correlation time for the flows, hence it is expected that the growth rate will change on a resistive timescale. Using measured data for the velocity flow field, its flucuation level and its self-correlation time, we have numerically modelled the expected magnetic field from the experiment. The velocity fields are predicted to generate intermittant magnetic field.

## References

1. A. Herzenberg, Phil. Trans. Roy. Soc. A **250**, 534 (1958).
2. F. Lowes and I. Wilkinson, Nature **198**, 1158 (1963).
3. F. Lowes and I. Wilkinson, Nature **219**, 717 (1968).
4. R. Mueller U., Steiglitz et al., Phys. Fluids (2000).
5. O. Gailitis, A. Lielausis at al., Phys. Rev. Lett. **84**, 4365 (2000).
6. P. Roberts and T. Jensen, Phys. Fluids B **5**, 2657 (1993).
7. P. Odier, J. Pinton, and S. Fauve., Phys. Rev. E **58**, 7397 (1998).
8. N. Peffley, A. Cawthorne, and D. Lathrop, Phys. Rev. E **61**, 5287 (2000).
9. D. Gubbins, Phil. Trans. of R. Soc. Lond. A **274**, 493 (1973).
10. M. Dudley and R. James, Proc. R. Soc. Lond. A **425**, 407 (1989).
11. R. Holme and J. Bloxham, J. Geophys. Res. **101**, 2177 (1996).
12. R. Holme, Phys. Plan. Earth Interiors **102**, 105 (1997).
13. E. Bullard and H. Gellman, Phil. Trans. R. Soc. Lond. A **247**, 213 (1954).
14. B. Press, W.H. Flannery et al. p. 329 (1988).

# SATURATION OF A PONOMARENKO TYPE FLUID DYNAMO

A. NUNEZ, F. PETRELIS, S. FAUVE
*Laboratoire de Physique Statistique, Ecole Normale Supérieure*
*24 rue Lhomond, 75005 Paris France*

## 1. Introduction

The kinematic dynamo problem is rather well understood in the case of laminar flows [1]. Several simple but clever examples have been found in the past [2, 3, 4, 5] and more realistic geometries can be easily studied numerically [6]. However, most flows of liquid metal are fully turbulent before reaching the dynamo threshold: indeed, the magnetic Prandtl number, $Pm = \mu_0 \sigma \nu$, where $\mu_0$ is the magnetic permeability of vacuum, $\sigma$ is the electric conductivity and $\nu$ is the kinematic viscosity, is smaller than $10^{-5}$ for all liquid metals. Since the dynamo action requires a large enough magnetic Reynolds number, $Rm = \mu_0 \sigma L U$, where $U$ is the fluid characteristic velocity and $L$ is the characteristic scale, one expects to observe the dynamo effect when the kinetic Reynolds number, $Re = UL/\nu$, is larger that $10^6$. The kinematic dynamo problem with a turbulent flow is much more difficult to solve. A theoretical approach exists only when the magnetic neutral modes grow at large scale. It has been shown that the role of turbulent fluctuations may be twofold: on one hand, they decrease the effective electrical conductivity and thus inhibits dynamo action by increasing Joule dissipation. On the other hand, they may generate a large scale magnetic field through the "alpha effect" or higher order similar effects [7]. Consequently, it is not known whether turbulent fluctuations inhibits or help dynamo action. More precisely, for a given configuration of the moving solid boundaries generating the flow, the behavior of the critical magnetic Reynolds number $Rm_c$ for the dynamo threshold, as a function of the flow Reynolds number $Re$ (respectively $Pm$) in the limit of large $Re$ (respectively small $Pm$), is not known.

Another important open question concerns the prediction of the saturation level of the amplitude of the magnetic field. This problem has been considered several times in the past, but with very unrealistic values of the



*P. Chossat et al. (eds.), Dynamo and Dynamics, a Mathematical Challenge,* 67–74.

fluid parameters that cannot be achieved in laboratory experiments. Phenomenological descriptions or perturbative calculations of the saturation of the "alpha effect" have been performed[9, 10, 11, 12, 13] for a large scale growing magnetic field with the assumption $Pm$ of order one or very large. The same assumptions (scale separation and large $Pm$) have been used in models of dynamically consistent convective dynamos [14, 15, 18, 16]. The only case considered so far without the assumption of scale separation concerns the saturation of a Ponomarenko type fluid dynamo[19]. However, the study has been performed in the limit of large $Rm$ for which a lot of magnetic modes are strongly unstable. Or goal here is to study the saturation of the the first unstable magnetic mode in the vicinity of the bifurcation threshold $Rm_c$.

## 2. A fluid in solid body rotation and translation up to the dynamo threshold

We use the following simple idea in order to be able to study the saturation of the magnetic field analytically: we consider the simplest possible flow, i. e. a fluid in solid body rotation and translation. This is the only way to avoid turbulence at dynamo onset. This may look unrealistic but an experimental configuration is possible. Consider for example the Herzenberg dynamo [2]: it consists of 2 or 3 rotating solid spheres embeded in a static medium of the same conductivity with which they are in perfect electrical contact. A slightly different version of the Herzenberg dynamo was operated experimentally by Lowes and Wilkinson using two cylinders instead of spheres [8]. Now, assume that one of the cylinders is hollow and filled with liquid metal. The flow will remain in solid body rotation up to the dynamo threshold. Above threshold the Lorentz force will slow down the fluid and modify the flow, thus leading to saturation of the magnetic field.

We study here an even simpler configuration found by Ponomarenko[4]: It consists of a cylinder of radius $R$, in solid body rotation at angular velocity $\omega$, and translation along its axis at speed $V$, embeded in an infinite static medium of the same conductivity with which it is in perfect electrical contact. In the same way as above, we consider that the cylinder is hollow and filled with a liquid metal with the same electrical conductivity. The kinematic dynamo problem is thus the same as the one studied by Ponomarenko. However, above the dynamo threshold the flow is modified by the Lorentz force and we show that this saturates the growth of the magnetic field.

Using respectively, $R$, $\mu_0\sigma R^2$, $(\mu_0\sigma R)^{-1}$, $\rho(\mu_0\sigma R)^{-2}$ and $\sqrt{\mu_0\rho}/\mu_0\sigma R$, as units for length, time, velocity, pressure and magnetic field, the governing equation for the velocity field, $\vec{v}(\vec{r},t)$, and the magnetic field, $\vec{B}(\vec{r},t)$, are

$$\nabla \cdot \vec{B} = 0 \tag{1}$$

$$\frac{\partial \vec{B}}{\partial t} = \nabla \times (\vec{v} \times \vec{B}) + \Delta \vec{B} \tag{2}$$

$$\nabla \cdot \vec{v} = 0 \tag{3}$$

$$\frac{\partial \vec{v}}{\partial t} + (\vec{v}.\nabla)\vec{v} = -\nabla \tilde{P} + Pm\, \Delta \vec{v} + (\vec{B} \cdot \nabla)\vec{B} \tag{4}$$

where $\tilde{P}$ is the sum of the hydrodynamic and magnetic pressures and $\rho$ is the fluid density. $Pm$ is the magnetic Prandtl number, $Pm = \mu_0 \sigma \nu$, and the boundary conditions for $\vec{v}$ involves the magnetic Reynolds number, $Rm = \mu_0 \sigma R \sqrt{(V^2 + (R\omega)^2}$ and the Rossby number, $Rb = V/R\omega$.

Below the dynamo threshold, $\vec{B} = 0$ and solid body rotation is solution of equations (3, 4). The corresponding kinematic dynamo problem has been solved by Ponomarenko[4]. Using cylindrical coordinates, he considered unstable modes of the form

$$\vec{B}(\vec{r}, t) = \vec{b}(r) \exp i(m\theta + kz + \omega_0 t) + c.c., \tag{5}$$

where c. c. stands for complex conjugate. We get from (2) that $\vec{b_p}(r)$ is an eigenmode of the operator $L$, defined by

$$L\vec{b_p} = i\left(\omega_0 + \mu\Gamma(r)\right)\vec{b_p} - \Delta\vec{b_p} + (boundary\ terms)\vec{b_p} = 0, \tag{6}$$

with $\mu = m\omega + kV$, $\Gamma(r) = 1$ if $r < 1$ and zero if $r > 1$, $\Delta$ results from the Laplacian operator applied to (5).

This formulation is of course equivalent to Ponomarenko's one and the boundary terms are the mathematical translation of the discontinuity in the derivative of $\vec{b_p}$ induced by the discontinuity of velocity at the boundary. The interest of this formulation will become clear in the nonlinear analysis.

The critical magnetic Reynolds number $Rm_c(Rb, m, k, \omega_0)$ reaches a minimum $Rm_c = 17.722$ for $Rb = 1.314$, $k = -.388$, $m = 1$, $\omega_0 = 0.410$. The growth rate above $Rm_c$ is $\alpha = (0.0268 + 0.00174\, i)(Rm - Rm_c)$.

## 3. Nonlinear saturation of the growing magnetic field

If the magnetic field saturates at a small amplitude just above the dynamo threshold, we see from (4) that the velocity perturbation that results from the Lorentz force is proportional to the square of the field amplitude. We thus choose the scalings: $Rm = Rm_c\,(1 + \Lambda\epsilon)$, $\tilde{P} = (\tilde{P}_f + \epsilon\tilde{P}_1 + \epsilon^2\tilde{P}_2 + \cdots)$, $\vec{v} = \vec{v}_f + \epsilon\vec{v}_1 + \epsilon^2\vec{v}_2 + \cdots$, $\vec{v}_f = \vec{v}_0\ (1 + \Lambda\epsilon)$, $\tilde{P}_f = \tilde{P}_0\ (1 + \Lambda\epsilon)$, $\vec{B} = \sqrt{\epsilon}\,(\vec{B}_0 + \epsilon\vec{B}_1 + \epsilon^2\vec{B}_2 + ...)$, $T = \epsilon t$, where $\epsilon$ is a small parameter representing the distance from criticality and $\Lambda$ is of order one. $v_f$ (respectively $\tilde{P}_f$) is the velocity (respectively the pressure) in the absence of magnetic field and $v_0$ (respectively $\tilde{P}_0$), is the value of the corresponding field at onset. $T$ is the slow time that describes the growth of the magnetic field.

At first order we get

$$L\vec{B}_0 = 0, \tag{7}$$

$$\frac{\partial \vec{v}_f}{\partial t} + (\vec{v}_f.\nabla)\vec{v}_f = -\nabla \tilde{P}_f + Pm\, \Delta \vec{v}_f, \tag{8}$$

L being the operator defined by equation (5). We thus have

$$\vec{B}_0(t,T) = A(T)\vec{B}_p + c.c. = A(T)\vec{b}_p \exp i\,(m\theta + kz + \omega_0 t) + c.c., \tag{9}$$

where $\vec{b}_p$ is Ponomarenko's eigenmode. $\vec{v}_f$ represents solid-body rotation and translation and is thus solution of Navier-Stokes equation without magnetic field.

At second order we get

$$L\vec{B}_1 = -\frac{\partial \vec{B}_0}{\partial T} + \Lambda \nabla \times (\vec{v}_0 \times \vec{B}_0) + \nabla \times (\vec{v}_1 \times \vec{B}_0) \tag{10}$$

$$\frac{\partial \vec{v}_1}{\partial t} + (\vec{v}_0.\nabla)\vec{v}_1 + (\vec{v}_1.\nabla)\vec{v}_0 = -\nabla \tilde{P}_1 + Pm\, \Delta \vec{v}_1 + (\vec{B}_0.\nabla)\vec{B}_0 \tag{11}$$

From equation (11), we can calculate the perturbation in velocity induced by the magnetic field (see appendix B).

Using the solvability condition for equation (10), we get the amplitude equation for $A(T)$. Let $\vec{C}$ be in the kernel of $L^\dagger$ the adjoint operator of $L$ (see appendix A). We have

$$\left\langle \vec{C}|\vec{B}_p \right\rangle \frac{dA}{dT} = \Lambda \left\langle \vec{C}|\nabla \times (\vec{v}_0 \times \vec{B}_0)\right\rangle + \left\langle \vec{C}|\nabla \times (\vec{v}_1 \times \vec{B}_0)\right\rangle, \tag{12}$$

which is of the form

$$\frac{dA}{dt} = \alpha A + \beta |A|^2 A. \tag{13}$$

We thus find the normal form of a Hopf bifurcation. Although this is obvious from symmetry considerations, we note that the calculation of the coefficients requires the solvability condition which cannot be easily guessed as in most examples of nonlinear oscillators or pattern forming instabilities.

The first term on the right hand side of equation (12) gives the linear growth rate

$$\alpha = (0.0268 + 0.00175\,i)(Rm - Rm_c), \tag{14}$$

in very good agreement with Ponomarenko's stability analysis.

The second term on the right hand side of equation (12) traces back to the magnetic retroaction on the velocity field. $Pm$ being very small for all liquid metals, we approximately have (see appendix B)

$$\beta \approx \beta_0 = \frac{1}{Pm}(-0.0034 - 0.0015\,i). \tag{15}$$

Thus, the bifurcation is supercritical ($Re(\beta_0) < 0$) and the amplitude saturates. This gives for the magnetic field in the M.K.S.A. unit system

$$\vec{B}_{sat} \approx \frac{2.81}{R}\sqrt{\frac{\rho\nu}{\sigma}}\sqrt{Rm - Rm_c}\,Re(\vec{B}_p). \tag{16}$$

We have for the velocity perturbation,

$$\vec{v}_{sat} \approx \frac{7.88}{\mu_0 \sigma R}(R_m - Rm_c)\,\vec{v}_1 \tag{17}$$

where $\vec{v}_1$ is the zero frequency component of the solution of equation (11) (see figure 1).

It may look surprising that the field saturates at a larger value when the viscosity is large whereas the velocity perturbation does not depend on the viscosity. This is due to the fact that the Lorentz force is balanced by the viscous term in the equation for the velocity perturbation (11).

For a turbulent flow, we expect a different balance between $\vec{v}_1$ and $\vec{B}_0$. Indeed, the saturated field amplitude should not depend any more on the kinematic viscosity in the large $Re$ limit. Dimensional analysis then gives

$$B_{sat} \propto \sqrt{\frac{\rho}{\mu_0 \sigma^2 R^2}}\sqrt{\frac{Rm - Rm_c}{Rm_c}}, \tag{18}$$

which is larger than the above laminar scaling by a factor $Pm^{-1/2}$. The later scaling is likely to be appropriate for the "Karlsrhue" and "Riga" experiments (see these proceedings) and gives a magnetic field of the order of 100 gauss 10% above threshold.

There is of course never energy equipartition close to the dynamo threshold since the kinetic energy is finite whereas the magnetic energy tends to zero. In our example, we have for Joule dissipation, $P_j \propto B_{sat}^2 \propto \frac{\rho\nu}{\sigma}(R_m - Rm_c)$ whereas for viscous dissipation, $P_\nu \propto v_{sat}^2 \propto (Rm - Rm_c)^2$. Thus, close to threshold, most of the mechanical power is used to create the magnetic field. This is due to the nature of the basic flow. In turbulent flows, it would be interesting to check whether Joule dissipation scales like the additional viscous dissipation that results from the perturbed velocity field above dynamo threshold.

## Appendix A: the adjoint problem

For $\vec{B}_a$ of the form given in (5), we define the scalar product, $\left\langle \vec{B}_a | \vec{B}_b \right\rangle = \int_0^\infty \vec{b}_a^*(r) \cdot \vec{b}_b(r)\ r dr$, where $\vec{b}_a^*$ is the complex conjugate of $\vec{b}_a$. With this definition and for $\vec{C} = \vec{c} \exp i(m\theta + kz + \omega_0 t)$, we have

$$L^\dagger \vec{c} = -i(\omega_0 + \mu\Gamma(r))\vec{c} - \Delta\vec{c} + (boundary\ terms)^\dagger \vec{c} \tag{19}$$

Except for the boundary terms, $L^\dagger$ is obtained from $L$ with the transformations: $\omega_0 \to -\omega_0$, $\mu \to -\mu$, i. e. by changing the signs of all velocities. However, the boundary terms dramatically change the form of the eigenvectors of $L^\dagger$: indeed, the eigenvector $\vec{c}$ has no component in the $z$ direction and is not even divergenceless. Thus, the adjoint problem of a kinematic dynamo problem may be not a dynamo problem, as already observed by Roberts [20].

## Appendix B: calculation of the velocity perturbation

We calculate the velocity perturbation $\vec{v}_1$ by solving equation (11). This is a linear second order equation for $\vec{v}_1$, with the forcing term $(\vec{B}_0.\nabla)\vec{B}_0$. The response $\vec{v}_1$ involves a zero frequency component and two oscillatory components at frequencies: $\pm 2(m\theta + kz + \omega_0 t)$. The non-zero contributions to the scalar product $\left\langle \vec{C} | \nabla \times (\vec{v}_1 \times \vec{B}_0) \right\rangle$ come from the zero frequency and the second harmonic components of $\vec{v}_1$. We call $\beta_0$ (respectively $\beta_d$) their contribution to the value of the coefficient $\beta$ in (13).

We first calculate the response at zero frequency. Since we have to consider velocity and pressure fields that are only functions of $r$, the resolution is easy. Equation (3) implies $v_{1r} = 0$ and the equations for the other components of $\vec{v}_1$ are decoupled. The boundary conditions are, the non-slip condition, $\vec{v}_1(1) = 0$, and, in order to keep the velocity field smooth, $v_{1r}(0) = 0$, $v_{1\theta}(0) = 0$, $v'_{1z}(0) = 0$. We solve the equations for $v_{1\theta}(r)$ and $v_{1z}(r)$ with Mathematica (see figure 1). Note that $\vec{v}_1$ and thus $\beta_0$ are inversely proportional to $Pm$ because the right hand side of (11) does not contribute to the zero frequency response.

The calculation of the harmonic two response is more complicated because two components of the velocity field $v_{1r}$, $v_{1\theta}$ and the pressure are

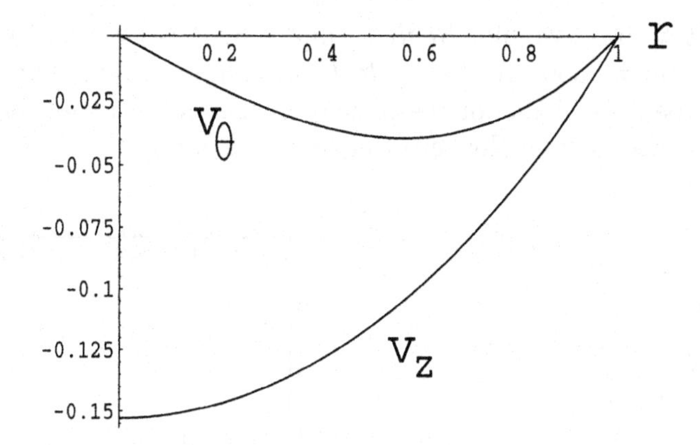


*Figure 1.* Zero frequency velocity perturbation for $Pm = 1$.

coupled. Taking the divergence of (11), we get

$$\Delta\tilde{P}_1 - 2\omega\left(\frac{v_{1\theta}}{r} + v'_{1\theta}\right) + 4im\omega\frac{v_{1r}}{r} - \nabla\cdot(\vec{B}_0.\nabla)\vec{B}_0 = 0. \qquad (20)$$

The boundary conditions for the pressure are $\tilde{P}(0) = 0$ and $\tilde{P}(\infty) = 0$. We solve (20) outside the cylinder where the velocity field is zero, and then solve (11) inside the cylinder using the continuity of the pressure at the boundary. The velocity and pressure fields are displayed in figure 2 for $Pm = .2$.

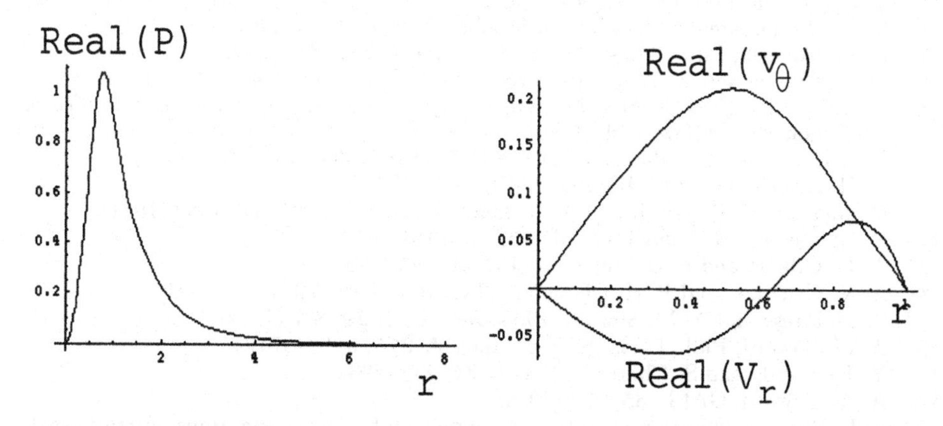


*Figure 2.* Real part of the complex amplitude of the harmonic-two response of the pressure and velocity for $Pm = 0.2$.

We observe that the harmonic-two response of the fluid velocity is locally enhanced on one part of the oscillation cycle, which may explain that $Re(\beta_d)$ is positive. Contrary to $\beta_0$, $\beta_d$ is not inversely porportional to $Pm$. In order to compare it with $\beta_0$, we plot $Real(\beta_d)$ and $-Real(\beta_0)$ versus $Pm$ in figure 3. For small $Pm$ we observe that $-Real(\beta_0)$ is much larger than $Real(\beta_d)$. Thus the retroaction of the magnetic field on the amplitude of the unstable mode mostly results from the zero frequency response.

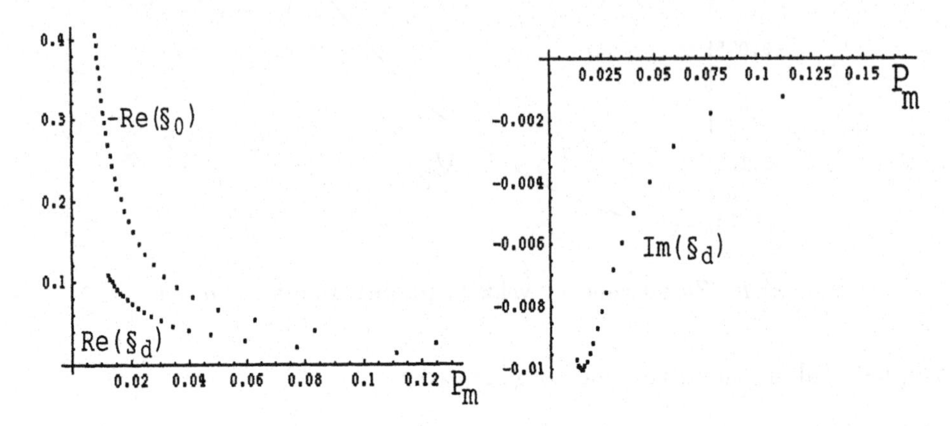


*Figure 3.* $-Re(\beta_0)$ and $Re(\beta_d)$ versus $Pm$ and $Im(\beta_d)$ versus $Pm$.

## References


1. H. K. Moffatt, *Magnetic field generation in electrically conducting fluids*, Cambridge University Press (Cambridge 1978).
2. A. Herzenberg, Philos. Trans. Roy. Soc. London **A250**, 543 (1958),
3. D. Lortz, Plasma Phys. **10**, 967 (1968).
4. Yu. B. Ponomarenko, J. Appl. Mech. Tech. Phys. **14**, 775 (1973).
5. G. O. Roberts, Phil. Trans. Roy. Soc. London A **271**, 411 (1972).
6. M. L. Dudley and R. W. James, Proc. Roy. Soc. London A **425**, 407 (1989).
7. F. Krause and K.-H. Radler, *Mean field magnetohydrodynamics and dynamo theory*, Pergamon Press (New-York, 1980).
8. F. J. Lowes and I. Wilkinson, Nature **198**, 1158 (1963); **219**, 717 (1968).
9. R. H. Kraichnan, Phys. Rev. Lett. **42**, 1677 (1979).
10. M. Meneguzzi, U. Frisch and A. Pouquet, Phys. Rev. Lett. **47**, 1060 (1981).
11. F. Krause and R. Meinel, GAFD **43**, 95 (1988).
12. A. D. Gilbert and P. L. Sulem, GAFD **51**, 243 (1990).
13. A. V. Gruzinov and P. H. Diamond, Phys. Rev. Lett. **72**, 1651 (1994).
14. S. Childress and A. M. Soward, Phys. Rev. Lett. **29**, 837 (1972).
15. A. M. Soward, Phil. Trans. R. Soc. Lond. **A 275**, 611 (1974).
16. Y. Fauterelle and S. Childress, GAFD **22**, 235 (1982).
17. A. M. Soward, GAFD **35**, 329 (1986).
18. F. H. Busse, Generation of planetary magnetism by convection, Phys. Earth Planet. Inter. **12**, 350-358 (1976).
19. A. P. Bassom and A. D. Gilbert, J. Fluid Mech. **343**, 375 (1997).
20. P. H. Roberts, in *Lectures on solar and planetary dynamos* , chap. 1, M. R. E. Proctor and A. D. Gilbert eds., Cambridge University Press (Cambridge, 1994).

# DYNAMO ACTION DUE TO EKMAN LAYER INSTABILITY

YANNICK PONTY
*Observatoire de la Côte d'Azur, Laboratory Cassini -*
*CNRS UMR 6529, B. P. 4229, 06304 Nice Cedex 4, France*

AND

ANDREW GILBERT AND ANDREW SOWARD
*School of Mathematical Sciences, University of Exeter,*
*North Park Road, Exeter, Devon EX4 4QE, UK*

## 1. Introduction

The Ekman layer becomes hydrodynamically unstable at sufficiently large Reynolds number *Re*. For the case of purely vertical rotation, the Ekman layer instability has been studied experimentally by Faller [4] and Caldwell & Van Atta [1], and numerically by Faller & Kaylor [5] , Lilly [10], Melander [11] and Ponty *et al.* [13]. The linear and nonlinear behaviour of Ekman–Couette instabilities in a plane layer has been discussed by Hoffmann *et al.* [8]. The transition between the Taylor–Couette instability and the Ekman layer instability is explored in Hoffmann & Busse [9]. Two different Ekman layer instabilities are distinguished in these studies, which for historical reasons are now referred to as types I and II. Type II occurs when the Reynolds number $Re^*$ defined using the Ekman layer thickness, exceeds the experimentally measured value of $Re_c^* \simeq 56.7$ (or 124.5 for type I). We will focus on the type II travelling wave, which has the smaller critical Reynolds number and so is easier to study numerically.

We simulated the finite amplitude development of the Ekman instability with our nonlinear numerical code subject to two-dimensional restrictions. Within that framework, we reach Reynolds numbers (based on the depth of the layer) of up to 800 and find that the saturated flow remains steady in a moving frame. Since such flow has no chaotic particle paths, any resulting dynamo cannot be fast. Nevertheless, we have found robust slow dynamo action which we now discuss.



*P. Chossat et al. (eds.), Dynamo and Dynamics, a Mathematical Challenge,* 75–82.

## 2. Governing equations

### 2.1. DIMENSIONLESS EQUATIONS

We consider a Cartesian fluid layer of depth $h$, rotating with angular velocity $\mathbf{\Omega} = \Omega\hat{\mathbf{\Omega}} = \Omega(\cos\vartheta\,\hat{\mathbf{z}} + \sin\vartheta\,\hat{\mathbf{y}})(\Omega \geq 0)$, which models a thin shell locally at co-latitude $\vartheta$ ( $\hat{\mathbf{x}}$ East; $\hat{\mathbf{y}}$ North; $\hat{\mathbf{z}}$ vertical). The top and bottom boundaries are rigid; the top boundary is fixed, while the bottom moves with velocity $\mathbf{U}_0 = U_0\hat{\mathbf{x}}$. The fluid has viscosity $\nu$ and magnetic diffusivity $\eta$. Length, time and velocity are non-dimensionalised using the depth $h$ and the viscous time-scale $h^2/\nu$. After non-dimensionalisation, the velocity at the bottom boundary becomes $\mathbf{U} = (U_0 h/\nu)\,\hat{\mathbf{x}} = Re\,\hat{\mathbf{x}}$; the basic steady, non-magnetic, equilibrium state depends only on the vertical coordinate $z$. The absence of an imposed horizontal pressure gradient ensures that the bulk of the basic horizontal shear $Re\,\mathbf{\Lambda}(z)$ driven by the differential motion of the plane parallel boundaries is concentrated in an Ekman layer attached to the bottom boundary. We calculate the Ekman profile $\mathbf{\Lambda}(z)$ analytically and note that the Ekman boundary layer thickness is $1/\sqrt{\tau\cos\vartheta}$.

Once the steady shear is disturbed, we write $\mathbf{U} = Re\,\mathbf{\Lambda}(z) + \mathbf{u}$. We assume spatial periodicity in the unbounded horizontal direction. The magnetic field $\mathbf{B}$ is taken to obey insulating boundary conditions. Accordingly the no-slip boundary conditions imply that $\mathbf{u} = 0$ on $z = 0$ and 1. The governing equation become

$$\partial_t\mathbf{u} + \mathbf{u}\cdot\nabla\mathbf{u} + Re(\mathbf{\Lambda}\cdot\nabla\mathbf{u} + u_z\partial_z\mathbf{\Lambda}) + \tau\hat{\mathbf{\Omega}}\times\mathbf{u} = -\nabla\Pi + \nabla^2\mathbf{u} + (\nabla\times\mathbf{B})\times\mathbf{B},$$

$$\partial_t\mathbf{B} + \mathbf{u}\cdot\nabla\mathbf{B} - \mathbf{B}\cdot\nabla\mathbf{u} + Re(\mathbf{\Lambda}\cdot\nabla\mathbf{B} - B_z\partial_z\mathbf{\Lambda}) = P_m^{-1}\nabla^2\mathbf{B}.$$

The dimensionless parameters employed are

$$\tau = 2\Omega h^2/\nu\,, \quad P_m = \eta/\nu\,, \quad Re = U_0 h/\nu,$$

namely the square root of Taylor number, the magnetic Prandtl number and the Reynolds number respectively. With a suitable choice of units for the magnetic field, we avoid the introduction of any additional coefficient in a front of the Lorenz force term.

### 2.2. TWO-DIMENSIONAL FORMULATION

At the onset of fluid instability, the fluid motion takes the form of rolls with a specific orientation. This flow is two-dimensional independent of the coordinate $\bar{y}$ along the roll axes. Our main assumption is that the two-dimensionality at onset is maintained in the fully developed nonlinear state, so that the velocity has the restricted functional form $\mathbf{u}(\bar{x}, z, t)$, where $\bar{x}$ is the horizontal coordinate normal to the roll axis.

To investigate the possibility of dynamo action in our $\bar{y}$-independent fluid flow, we consider magnetic field described by $\bar{y}$-dependent normal modes of the form $\mathbf{B} = \mathbf{b}(\bar{x}, z, t)\exp(il\bar{y})$ $(l \neq 0)$. The Lorenz force is projected on the zero-$l$ mode of the momentum equation by taking into account only the beating of conjugate magnetic modes $l = \pm 1$. Our simplifying two-dimensional assumption is important because it allows us to study dynamo action at large magnetic Reynolds number with high numerical resolution.

Since $\mathbf{u}$ is independent of $\bar{y}$, we may write $\mathbf{u} = -\partial_z\psi\,\hat{\bar{\mathbf{x}}} + v\,\hat{\bar{\mathbf{y}}} + \partial_{\bar{x}}\psi\,\hat{\mathbf{z}}$ $= \nabla\times(\psi\hat{\bar{\mathbf{y}}}) + v\hat{\bar{\mathbf{y}}}$. Furthermore, since the flow is steady in a frame co-moving with the rolls, it is helpful to define the total stream function

$$\Psi(\bar{x}, z) = \psi(\bar{x}, z) - Re\,P \int_0^z \Lambda(\xi).\hat{\bar{x}}\,d\xi + U_{\text{roll}}\,z.$$

In that moving frame (relative velocity $U_{\text{roll}}$), partricles follow the stream lines $\Psi =$constant and the components of the total velocity are

$$U - U_{\text{roll}} = -\partial_z\Psi, \qquad V(\bar{x}, z) = v(\bar{x}, z) + Re\,P\,\mathbf{\Lambda}(z).\hat{\bar{y}}, \qquad W = \partial_{\bar{x}}\Psi.$$

## 3. Numerical method and diagnostics

The solution of the magnetohydrodynamic system is achieved numerically using a time-stepping pseudo-spectral code with the collocation-tau method (see Ponty *et al.* [14] for further details).

The results of our simulations, especially from the kinematic dynamo viewpoint, depend on the magnetic Reynolds number $Rm$. Thus we introduce the root-mean-square value $\mathcal{U}$ of the non-dimensional total velocity $\mathbf{U}$, averaged over both space and time, and define $Rm = P_m\mathcal{U}$. We also introduce turn-over time-scale $\lambda = \sigma/\mathcal{U}$ measured by the real magnetic field growth rate $\sigma$ based on the viscous time scale.

## 4. Fluid topology

We present one example of dynamo in flows resulting from the Ekman instability. We take $\vartheta = 67.5°$, $\tau = 100$ and $k_c = 4.30$; our Reynolds number is $Re = 250$, which is to be compared to the critical value $Re_c \simeq 138$ for the instability. The results are illustrated in figures 1(a,b), which give the $\bar{y}$-velocity $V$ and total stream function $\Psi$ in the co-moving frame. The stream lines of the ensuing finite amplitude flow have the cats' eye configuration, which is well known to occur at critical levels in other shear flows, e.g., the Kelvin–Helmholtz instability. In the $(\bar{x}, z)$-plane, this stream line topology contains elliptic and hyperbolic stagnation points. The exponential stretching at the stagnation points, and the differential rotation around the elliptic points can stretch out field, contributing to a dynamo process.

## 5. Kinematic dynamo

Different kinematic dynamo mechanisms are found, which largely depend on the size of the wave number $l$. Figure 2 shows the growth rates $\sigma$ and $\lambda$ plotted against the wave number $l$ for $P_m = 40$, when $Rm \simeq 3600$. The curve is complicated, having many peaks and windows of dynamo and non-dynamo action as $l$ is varied.

Below the wave number $l \simeq 2.0$, the magnetic field shown in figure 1(c) for $l = 1.2$ is concentrated around the principal vortices of the flow displayed in figure 1(b). The dynamo process appears to correspond to the Ponomarenko dynamo [12] (see Gilbert [6], Ruzmaikin, Sokoloff & Shukurov [16], Gilbert & Ponty [7]). Field directed in or out of the eddy is stretched by differential rotation on helical streamlines to generate field directed along the streamlines. Diffusion of this field in curved geometry generates field across streamlines so closing the dynamo loop and leading to magnetic field amplification. To confirm this picture, the magnetic field, which takes the form of two spiralling tubes, is visualised in three dimensions in figure 3.

Above the wave number $l \simeq 2.0$, the magnetic field is localised along the separatrices and the stagnation points play a crucial role. The magnitude of the magnetic field is shown in figure 1(d) for $l = 9.0$, where the magnetic growth rate is a maximum in the figure 2. We see that the dominant magnetic mode has field localised in sheets along the separatrices of the flow in the $(\bar{x}, z)$-plane (using the stream function $\Psi$ for the appropriate moving frame). The sheets intersect at the hyperbolic stagnation points. Our simulation represents the first example of a dynamo effect obtained in a cats' eyes configuration resulting from hydrodynamic instability.

The dynamo mechanism, just described, has some similarities with that of the periodic G.O Roberts [15] cellular flow investigated analytically in the large-$Rm$ limit by Childress [2] and Soward [17]. In this case the field is also associated with the stagnation points and separatrices, and is amplified by a steady stretch–fold–shear mechanism. Dynamo action is also considered for doubly periodic rows of cats' eyes by Childress & Soward [3], who compute an $\alpha-$effect associated with boundary layers on the separatrices. Their calculations involve averaging over the infinite plane, and it remains to be seen how their results relate to dynamo action in our row of cats' eyes in a plane layer with insulating boundaries.

In our example of a saturated Ekman instability, we have identified two different kinematic dynamo mechanisms which compete with each other. The stretch–fold–shear mechanism in the steady flow is the more efficient one, preferring modes with short scale in the $\bar{y}$ direction, but the Ponomarenko mechanism is also there, subdominant, preferring large-scale fields. Note that regardless of the wavenumber $l$ in the $\bar{y}$-direction the two mechanisms tend to amplify fields of different length-scales in the $(\bar{x}, z)$-plane for large $Rm$, as seen in figure 1(c,d). For $l = O(1)$ the Ponomarenko dynamo amplifies field on $O(Rm^{-1/4})$ length-scales (Gilbert & Ponty [7]), while the fields associated with the hyperbolic stagnation points localise on $O(Rm^{-1/2})$ length-scales (Childress [2]) – harder to resolve numerically.

## 6. Nonlinear regime

Solutions in the fully nonlinear regime have been computed. In figure 4 the magnetic energy and the kinetic energy are plotted versus time for a particular set of parameter values. In figure 4a, the initial linear kinematic regime is clearly distinguished from the subsequent nonlinear saturation. The magnetic energy and the kinetic energy appear to settle down to steady values but with slowly decaying transient oscillations. Throughout the range of wave number $l$ displayed in figure 2, the system converges to the same kind of magnetohydrodynamic solution; here the saturated magnetic field is located around the separatrices, just as in large wave number kinematic case (see figure 1d). It means that a short magnetic length scale along the roll $\bar{y}$ axis is then preferred.

## 7. Discussion

The equilibrated Ekman layer instability flows have the cats' eyes configuration and are steady in a co-moving frame. Here kinematic dynamo action may occur by the Ponomarenko [12] dynamo mechanism or with fields associated with hyperbolic stagnation points and their connecting separatrices. The Ponomarenko mechanism is now well-understood: asymptotic high-$Rm$ growth rates may be obtained in cases such as those seen in the simulations with an arbitrary flow profile (Gilbert & Ponty [7]).

Kinematic dynamo action associated with hyperbolic stagnation points and separatrices is rather more complicated. Though dynamo action in flows with cats' eyes in the doubly-periodic plane has been studied by Childress & Soward [3], there are non-trivial geometrical complications that arise in the flows of our plane layer model. In particular the layer is periodic in only one direction, along the cats' eyes, and the velocity perpendicular to the plane of the cats' eyes is not constant on stream lines. It remains an interesting asymptotic problem to obtain high-$Rm$ growth rates in this more general situation.

In the nonlinear regime, a stable nonlinear saturation is preferred, with the magnetic field located along the cats' eyes separatrices. Interestingly, the feed back of the Lorentz force does not appear to destroy the separatrix topology, which leads to an efficient kinematic dynamo process. Perhaps these features render the steady equilibrium robust.

In conclusion, dynamo action due to the Ekman layer instability provides a nice magnetohydrodynamic system for analytic study. Indeed we intend to undertake further detailed investigations of kinematic growth rates and the nonlinear equilibrium in ongoing studies.

## Acknowledgments

Y.P. gratefully acknowledges the support of a Leverhulme Fellowship, grant no. F/144/AH, at the University of Exeter between August, 1997 and August, 2000. The numerical calculations were performed using the computing facilities of the laboratory Cassini, Observatoire de la Côte d'Azur (France), provided by the program "(SIVAM)" and the computing facilities of the parallel computer Ceres at the University of Exeter.

## References

1. Caldwell, D.R. & Van Atta, C.W. 1970 Characteristics of Ekman boundary layer instabilities. *J. Fluid. Mech.* **44**, 79–95.
2. Childress, S. 1979 Alpha-effect in flux ropes and sheets. *Phys. Earth Planet. Int.* **20**, 172–180.
3. Childress, S. & Soward, A.M. 1989 Scalar transport and alpha-effect for a family of cat's-eyes flows. *J. Fluid Mech.* **205**, 99–133.
4. Faller, A.J. 1963 An experimental study of the instability of the laminar Ekman boundary layer. *J. Fluid Mech.* **15**, 560–576.
5. Faller, A.J. & Kaylor, R.E. 1966 A numerical study of the instability of laminar Ekman boundary layer flow. *J. Atmos. Sci.* **23**, 466–480.
6. Gilbert, A.D. 1988 Fast dynamo action in the Ponomarenko dynamo *Geophys. Astrophys. Fluid Dynam.* **44**, 241–258.
7. Gilbert, A.D. & Ponty, Y. 2000 Slow Ponomarenko dynamos on stream surfaces. *Geophys. Astrophys. Fluid Dynam.*, in press.
8. Hoffmann, N.P., Busse, F.H. & Chen, W.-L. (1998) Transition to complex flows in the Ekman–Couette layer. *J. Fluid Mech.* **366**, 311–331.
9. Hoffmann, N.P. & Busse, F.H. (1999) Instabilities of shear flows between two coaxial differentially rotating cones. *Phys. Fluids* **11**, 1676–1678.
10. Lilly, D.K. (1966) On the instability of Ekman boundary flow. *J. Atmos. Sci.* **23**, 481–494.
11. Melander, M.V. (1983) An algorithmic approach to the linear stability of the Ekman layer. *J. Fluid Mech.* **132**, 283–293.
12. Ponomarenko, Y.B. (1973) On the theory of hydromagnetic dynamos *Zh. Prikl. Mekh & Tekh. Fiz.* **6**, 47–51.
13. Ponty, Y., Gilbert, A.D. & Soward, A.M. (2000) Rotating convective instability in an Ekman layer driven by a shear flow. in preparation.
14. Ponty, Y., Gilbert, A.D. & Soward, A.M. (2000) Kinematic dynamo in large magnetic Reynolds number flows driven by shear and convection *J. Fluid. Mech.*, submittted.
15. Roberts, G.O. (1972) Dynamo action of fluid motions with two-dimensional periodicity. *Phil. Trans. R. Soc. Lond.* A **271**, 411–454.
16. Ruzmaikin, A.A., Sokoloff, D.D. & Shukurov, A.M. (1988) A hydromagnetic screw dynamo. *J. Fluid Mech.* **197**, 39–56.
17. Soward, A.M. (1987) Fast dynamo action in a steady flow. *J. Fluid Mech.* **180**, 267–295.

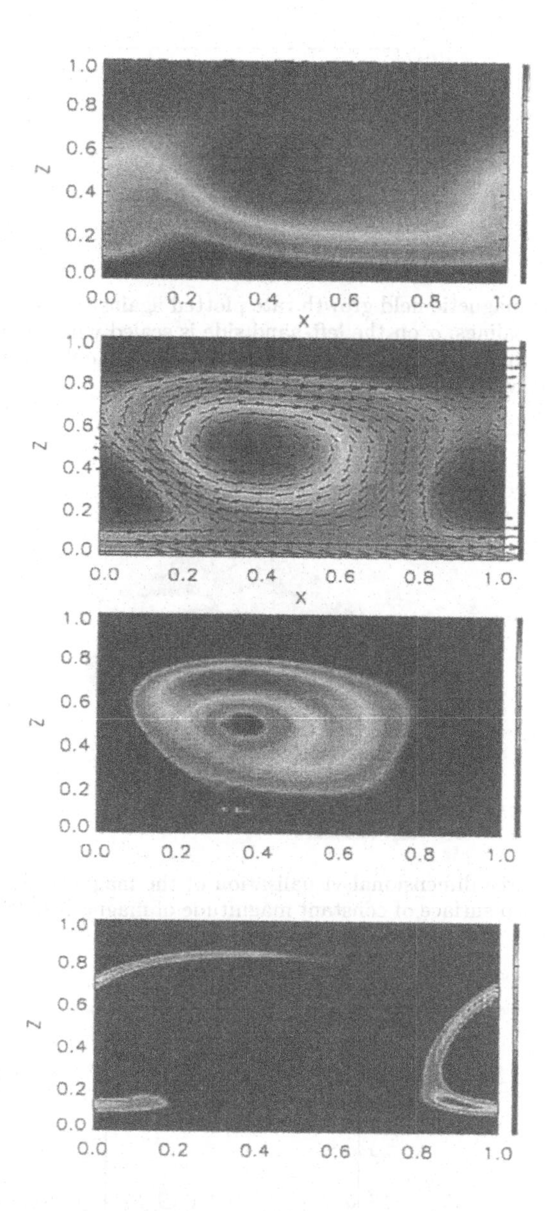


*Figure 1.* A flow resulting from a saturated Ekman layer instability drives a dynamo for $\vartheta = 67.5°$, $\tau = 100$, $\epsilon = 79.28°$, $k_c = 4.64$, $Re = 250$ and $P_m = 40$, corresponding to $Rm \simeq 3600$. It is shown a as time-snapshot in the $(\bar{x}, z)$ plane of (a) the total velocity $V$ along the $\bar{y}$-axis, (b) the total stream function $\Psi$, (c) the magnitude $B$ of the magnetic field in the case $l = 1.2$, and (d) the magnitude $B$ of the magnetic field in the case $l = 9.0$.

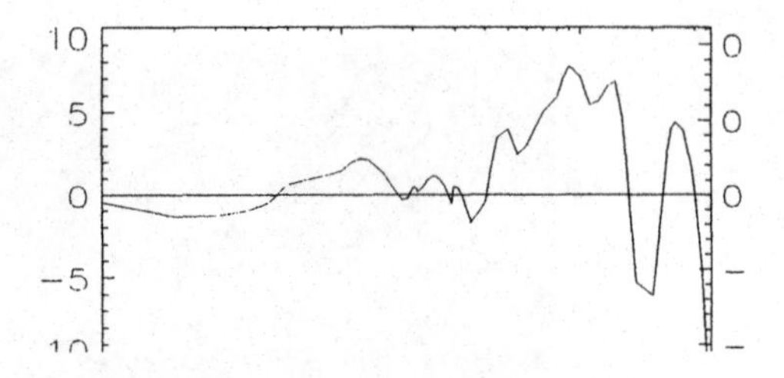


*Figure 2.* The magnetic field growth rate plotted against the wave number $l$ along the $\bar{y}$ axis, with two scalings: $\sigma$ on the left-hand side is scaled with the thermal diffusion time, and $\lambda$ on the right-hand side with the turn-over time. The flow is an equilibrated Ekman layer instability with parameters as in figure 1.

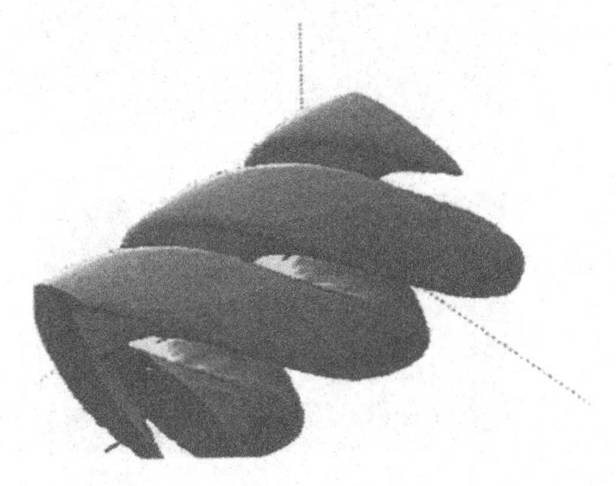

*Figure 3.* A three dimensional visualisation of the magnetic field in the case shown in figure 4(c). An iso-surface of constant magnitude of magnetic field is plotted with respect to $(\bar{x}, \bar{y}, z)$ axes.

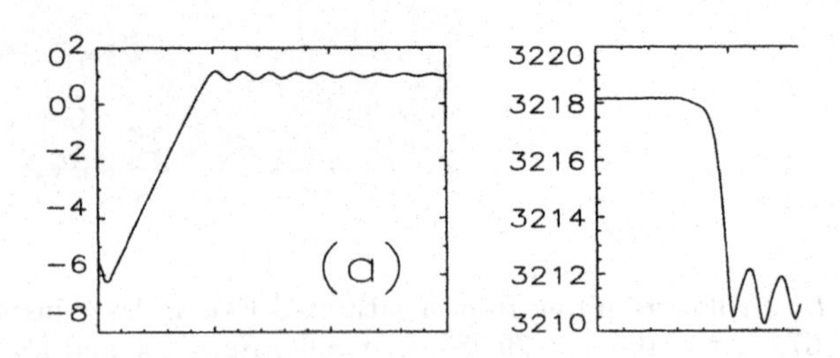


*Figure 4.* (a) The magnetic energy is plotted in log scale versus the running time for the same parameter values as in figure 1. (b) The kinetic fluid velocity associated is also presented.

# HUNTING FOR DYNAMOS: EIGHT DIFFERENT LIQUID SODIUM FLOWS

WOODROW L. SHEW, DANIEL R. SISAN AND DANIEL P. LATHROP
*University of Maryland*
*Department of Physics*
*Institute for Plasma Research Bldg 223*
*College Park, MD, 20742, USA*

**Abstract.** In attempting to create a laboratory scale dynamo, an experimentalist is faced with a daunting question: What sort of flow can I produce that will yield a dynamo? We present eight variations of a flow motivated by the $s2t2$ flow numerically studied by Dudley and James [1]. Pulse decay measurements of an externally applied magnetic field are used to quantify the approach to transition to dynamo action.

## 1. Introduction

"Dynamo and Dynamics, a Mathematical *and Experimental* Challenge": this is perhaps a more appropriate title for a conference addressing the state of the dynamo problem today. In the past few decades, advances in technology for handling liquid sodium have lead to great advancements in efforts to address the dynamo problem experimentally. Around the world there are groups attacking the problem with liquid sodium experiments: Riga, Latvia [2]; Karlsruhe, Germany [4]; Cadarache, France [5]; Madison, WI, U.S. [6]; U of Maryland, MD, U.S. [7]; and Soccoro, NM, U.S. [3]. The Karlsruhe and Riga groups have succeeded in self-generating magnetic fields in guided flows.

In designing an experimental dynamo, one is concerned with two essential ingredients: high magnetic Reynolds number $R_m = Ul/\eta$ and a good flow geometry. Achieving a high $R_m$ is a relatively unambiguous goal. Sodium has the highest conductivity of any liquid conductor and hence the lowest magnetic diffusivity $\eta$ ($\eta = 830cm^2/s$ for Na at 120° C). The characteristic velocity $U$ is limited by power consumption and cooling ca-



*P. Chossat et al. (eds.), Dynamo and Dynamics, a Mathematical Challenge,* 83–92.

pabilities. The size $l$ of the experiment is limited only by the courage and funding of the experimentalist.

Choosing an advantageous flow geometry, on the other hand, is not as clear a problem for unconstrained flows. There are suggestions to be taken from numerical and analytic studies and qualitative reasoning: helicity in the flow is good, a near unity ratio of toroidal to poloidal components of the flow is good, hyperbolic points are good, purely toroidal flow is bad [8]. The problem is that a laboratory scale dynamo with the sought high $R_m$ is necessarily highly turbulent. This is evident in the relationship between the hydrodynamic Reynolds number $R = Ul/\nu$ and the magnetic Reynolds number: $R = R_m/Pr_m$. The magnetic Prandtl number $Pr_m$ is very small for all liquid metals ($Pr_m = 8.3 \times 10^{-6}$ for sodium at 120° C). Therefore a flow with $R_m >> 1$ has $R >> 10^5$; it is turbulent. Many of the theoretical suggestions are based on analytical, time independent velocity fields which are not stable solutions to the Navier-Stokes equation for the high Reynolds number flows of our experiments. Without highly constraining the flow physically, the theoretical flow geometries are unattainable. At best, we could ignore the turbulence and try to match the mean flow field to theory. The design of our apparatus was motivated by just such an attempted match. Modifications and adjustments to this apparatus were made to explore eight different flow geometries.

## 2. Apparatus

The liquid sodium is contained in a hollow 31.2 cm diameter sphere made of 0.95 cm thick 304 non-magnetic stainless steel. The flow is driven by propellers mounted on coaxial shafts that enter through opposite ends of the sphere. The original propellers (Figure 7a) were three lobed 6.35 cm radius marine propellers. The axial positions of the propellers are adjustable. An additional influence on the flow field is provided by four pole to pole baffles that extend 1.5 cm into the sodium. The propellers are belt driven with two 7.5 kW electric motors at rotation rates from 10 to 120 Hz. The rotation rates are monitored with optical pickups. Each motor's speed and direction is independently controlled digitally. We also record from the drives the power delivered to the motors.

At the highest rotation rates, the propellers impart 15 kW of mechanical power to the sodium which is dissipated as heat. Chilled hexane is pumped through a system of copper tubing wrapped around the outside of the sphere to maximally cool 8 kW. At the highest rotation rates, our cooling system is overwhelmed and we are limited to run for short periods. For low rotation rates the sodium is kept above the melting point with tubular cartridge heaters also wrapped around the vessel. The heaters supply up to

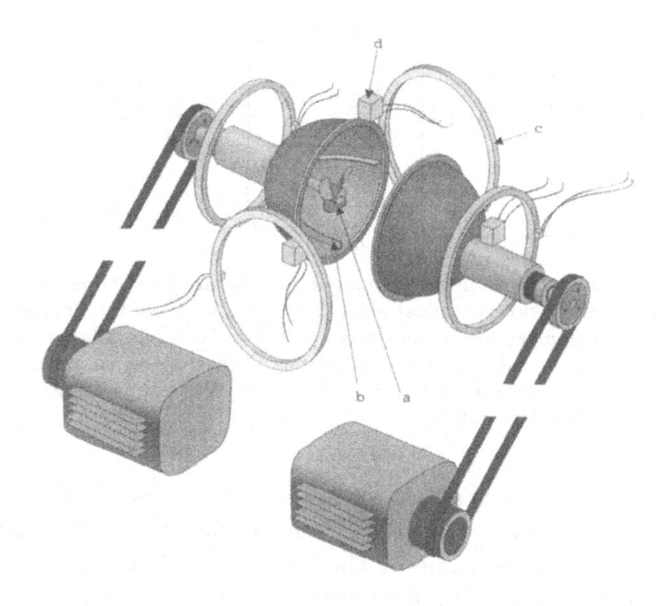


*Figure 1.* An exploded view of the experiment shows the propellers (a) and poloidal baffles (b) inside the sphere. The Helmholtz coils (c) and hall probes (d) are located outside the sphere.

2.8 kW of heat. Good thermal contact is insured with thermally conductive putty. The heating and cooling systems are computer controlled to keep the temperature within 5° C of 120° C. The temperature is monitored with a thermocouple that protrudes 1 cm into the sodium. The sphere is plumbed to the sodium storage tank. Pressurized nitrogen is used to transfer the sodium to and from the sphere. The transfer pipe is open during a run to allow for small variations in the volume due to temperature variations.

We define the magnetic Reynolds number for our apparatus, $R_m = Ul/\eta = \Omega ba/\eta$ where $U = \Omega b$ is the tip speed of the propeller of radius $b$, and $a$ is the radius of the sphere. With 6.35 cm radius propellers, we can achieve magnetic Reynolds numbers from 7 to 80, though we can only run for short periods at the values higher than 70 due to limited cooling capability.

## 3. Pulse Decay Measurements

For an appropriate flow field and high enough $R_m$ we expect a transition to a state where the magnetic field is unstable to growth of small perturbations: a self-generating dynamo. We quantify the distance to this transition using pulse decay measurements. An external magnetic field is applied for a short time (1 s) with two pairs of Helmholtz coils. One pair of coils is coaxial with

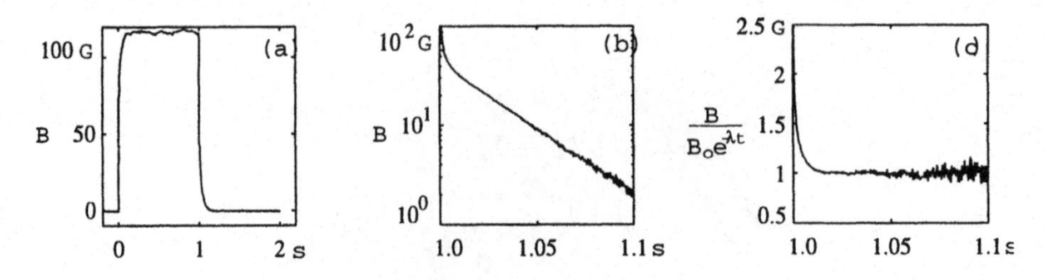


*Figure 2.* Externally applied pulses of magnetic field are used to estimate the decay rate $\lambda$ for the magnetic field. A typical pulse (a) at $R_m = 42$ shows fluctuations during the pulse and the decay after the current is cut to the coils. A magnified view (b) plotted semi-logarithmically after the external current has been stopped shows the asymptotic exponential decay and slope used for $\lambda$. Part (c) shows the same data as in (b) divided by the estimated exponential function.

the propellers. The other is at a right angle to this axis. The current to the coils is cut rapidly ($< 1$ ms) with IGBT semiconductor switches. Hall probes located outside the sphere, one at the "north pole" and the other at the equator, monitor the magnetic field of the pulse and the magnetic response of the flowing bulk of sodium. The left plot in Figure 2 shows a typical pulse as measured by one of the hall probes. A decay rate $\lambda$ is extracted from the data during the decay of the pulse. The slope of the middle plot in Figure 2 is the decay rate. The applied magnetic field pulses are small enough that they do not significantly alter the flow field. In this situation the Navier-Stokes equation decouples from the applied B field,

$$\frac{\partial \vec{v}}{\partial t} + (\vec{v} \cdot \vec{\nabla})\vec{v} = -\frac{1}{\rho}\vec{\nabla}P + \nu\nabla^2\vec{v} \quad , \tag{1}$$

and the dynamics of the magnetic field is that of a passive vector field and is determined only by the induction equation,

$$\frac{\partial \vec{B}}{\partial t} = \vec{\nabla} \times \vec{v} \times \vec{B} + \eta\nabla^2\vec{B} \ . \tag{2}$$

We can recast this equation,

$$\frac{\partial \vec{B}}{\partial t} = L(t)\vec{B} \tag{3}$$

where $L(t)$ is the linear operator that is responsible for changes in $\vec{B}$. Its time dependence comes from the time dependence of the velocity field $\vec{v}$. Expanding $\vec{B}$ in some convenient basis (say vector spherical harmonics),

$$\vec{B}(\vec{x}, t) = \sum_{i=1}^{\infty} b_i(t)\vec{B}_i(\vec{x}) \tag{4}$$

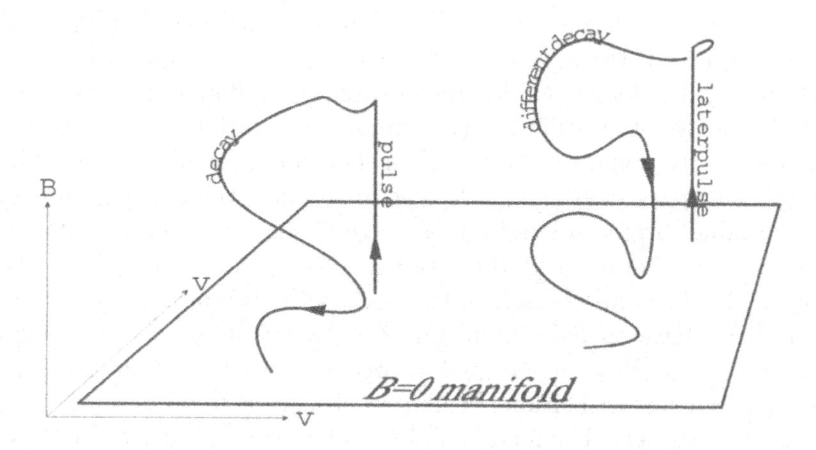


*Figure 3.* At the same rotation rate we observe different decay rates. The differences are due to changes in the flow field which introduce trajectories in phase space that have slower rates of return to the $B = 0$ manifold. Ultimately, we search for a field that will support trajectories that do not return to $B = 0$, i.e. a self-generating dynamo.

the magnetic field then evolves according to,

$$\vec{B}(\vec{x},t) = \sum_{i=1}^{\infty} \exp\left[\int L(t)\,\mathrm{dt}\right] b_i(0)\vec{B}(\vec{x}). \tag{5}$$

We can see (Figure 2) that for a single measurement of the decay rate, the fluctuations from the time dependence of $L$ are small compared to the exponential decay of the pulse. The field then decays according to

$$\| b_i(t) \| \sim \| b_i(0) \| \exp[\lambda_N t], \tag{6}$$

where the Lyapunov exponent $\lambda_N$ may depend upon the initial condition $b_i(0)$. In general, there are an infinite number of Lyapunov exponents for a spatially extended system, but dissipation causes most to be very quickly decaying modes. For most initial conditions we expect to obtain the $\lambda_{max}$, the largest (slowest decaying) mode for the system.

$$\lambda_N = \lim_{t\to\infty} \frac{1}{t} \ln \frac{\| b_i(t) \|}{\| b_i(0) \|} \tag{7}$$

However, our measurements are not infinite time estimates, rather are estimated over a tenth of a second or so. When the same measurement is repeated later at the same rotation rate, different estimates for $\lambda_N$ are found. This indicates the dependence of $L$ on the erratically evolving flow field. The full system has a high dimensional phase space composed of two subspaces, one for the velocity field and one for the magnetic field. Below the transition to self-generation, $B = 0$ is a stable invariant manifold in this phase space. If the system starts with a nonzero magnetic field, it will quickly return to $B = 0$. At different times, i.e., different regions of $B = 0$ plane, the decay rate is different (Figure 3). Intermittency at the transition would be due to some of these regions becoming locally unstable before the whole subspace does [7, 9, 12]. The long time average Lyapunov exponent is obtained from an ensemble average of short time exponents. In our experiment, we average say 10 decay rates for each rotation rate. The geometry of the flow and the value of $R_m$ affect the vector field governing the dynamics in phase space. Our ultimate objective is to create a flow field which causes the $B = 0$ manifold to become unstable. Our experiments do not achieve this transition, rather we observe a trend towards it as we increase the magnetic Reynolds number. The faster we drive the flow, the slower the slowest decaying mode.

## 4. Practical Issues of Decay Measurements

During discussions at the conference, Lèorat raised issues regarding decay measurements and the possibility of transient behavior interfering with the measurement of decay rates. After applying a pulse of magnetic field, one must wait for faster decaying modes to die prior to the estimation of the slowest decaying mode. Lèorat used advective time scales for the estimate and concluded that transients might yet be a problem. These remarks address this concern. The time scale for the transient is set by the subdominant Lyapunov exponents. Without knowledge of the exponent separation, we take as a likely overestimate of that time $\tau_{max} = 1/\lambda_{max} = 30$ ms. We typically use a time span to obtain decay rate estimates for $\lambda_{max}$ of $2\tau_{max} < t < 4\tau_{max}$. Empirically, $2\tau_{max}$ appears adequate for the transients to decay and reveal the slowest decaying mode.

It is also necessary to have sufficient signal to noise ratio in order to obtain decay measurements. One needs to observe the field falling two decades and still have limited noise in the tail. For this, a ratio of at least $10^3$ for the pulse height to noise level is necessary.

## 5. Eight Different Flows

The setup described in Section 1 was motivated by the $s1t1$ and $s2t2$ flows studied numerically by Dudley and James [1]. Although motivated by $s2t2$ flows, the flows in our experiment do not match the Dudley and James flows even as mean fields. A better picture of the velocity fields was obtained with ultrasound velocity measurements [11] of water flows in the same apparatus (Figure 4). Since water has nearly the same viscosity as liquid sodium the

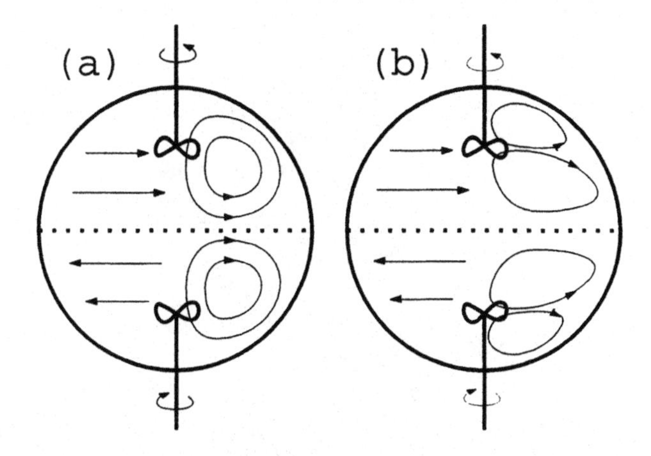


*Figure 4.* The design of the experiment was motivated by the $s2t2$ Dudley and James flow (a). The flow we actually have is different (b). The propellers tend to fling the sodium radially away from the center rather than pump in circles like the $s2t2$ flow.

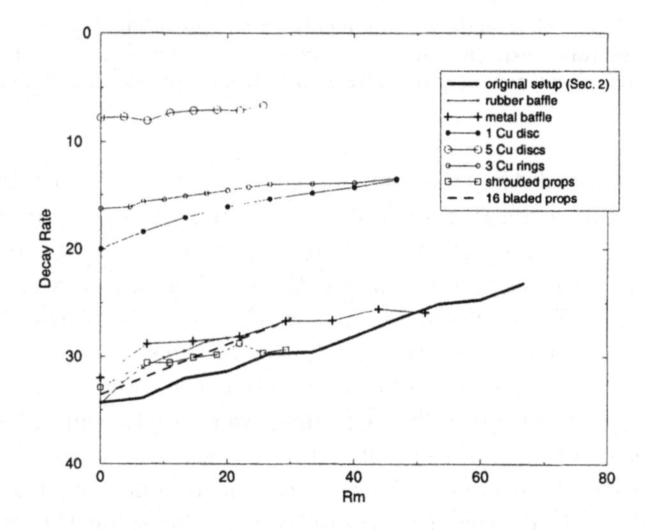


*Figure 5.* This plot summarizes the observed trends in decay rates $\lambda$ for all the different geometries tried.

water flows are a good model of the sodium flows when Lorenz forces are very small.

Our typical sodium experiment consists of taking pulse decay measurements for different rotation rates and positions of the propellers. The bold line in Figure 5 presents the average decay constant $\lambda$ for each of ten rotation rates with the propellers positioned half way between the center of the sphere and the poles.

With the experiment as described in section 1, we observe a shift of

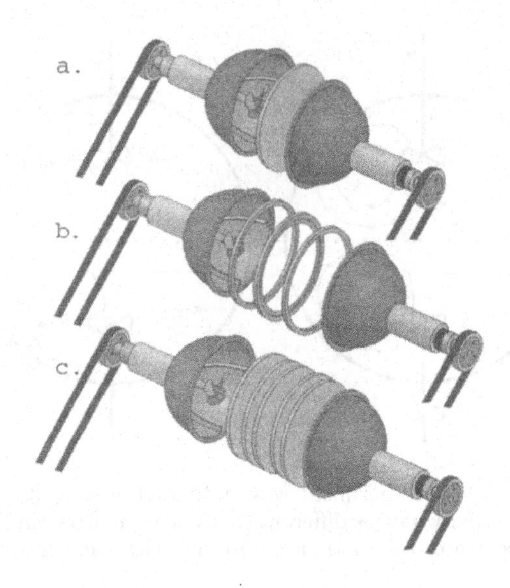


*Figure 6.* Exploded view showing modifications to the original setup: one Cu disc a), three Cu rings (one with an equatorial baffle) b), and five Cu discs c). The copper increases the static decay rate for the system as well as changes the flow dynamics.

thirty percent towards the transition at the fastest propeller rotation speeds. Static sodium has a decay rate $\lambda = -35s^{-1}$ and at $R_m = 70$ this shifted to about $-23s^{-1}$. Three new geometries were motivated by attempting to boost the $\lambda$ for static sodium up so that the previously observed shift might be enough to reach transition, i.e. $\lambda > 0$. The boost is provided by sandwiching solid copper between the two halves of the sphere. Three different configurations were tried: five solid discs (Figure 6c); one solid disc (Figure 6a), and three rings (Figure 6b). The discs were all 1.9 cm thick and 38.1 cm in diameter. The rings were the same dimensions as the discs with a 31.2 cm diameter hole cut in the center. The middle ring had a 1.5 cm baffle around its inside surface. In addition to boosting the static decay rate of the system, the Cu modifications dramatically alter the flow dynamics. The solid discs separate the two halves of the volume. The large gap between hemispheres for the 5 disc case greatly decreases the magnetic influence that one side has on the other. In the three ring case, as well as increasing the volume of the sphere (actually closer to a ellipsoid now), the middle ring had an equatorial baffle that extended 1.3 cm into the flow at the equator. Pulse decay measurements were again taken at different $R_m$ values (Figure 5). The changes in geometry do not gave a smaller shift in decay rates than observed for the original geometry (Section 2).

The equatorial baffle on the middle copper ring was motivated by the results of an earlier experiment. In the earlier experiment, a rubber gasket

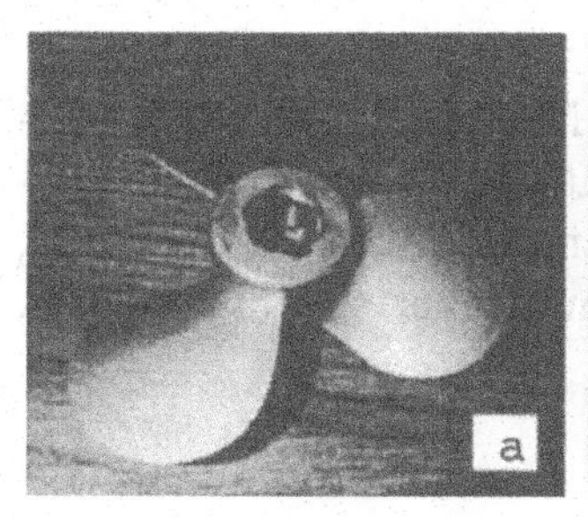

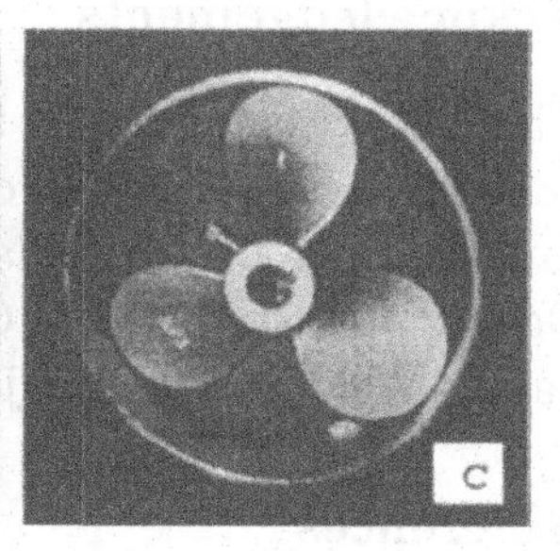


*Figure 7.* Photographs of three different propellers used: a 6.35 cm radius marine propeller a), a 6.35 cm radius 16 bladed jet turbine impeller b), and a shrouded 7.62 cm radius marine propeller c).

was used to seal the union of the two halves of the sphere. When the gasket was compressed it deformed and extended into the flow (similar to the purposeful equatorial baffle on the Cu ring) forming a flexible, wavy equatorial baffle. We imagined this encouraged a well defined in-flow at the equator. The result was a slightly more pronounced trend toward transition. Figure 5 shows the rubber baffle data. The encouraging trend towards zero with the rubber baffle motivated us to try another experiment with an equatorial baffle. This time a .05 cm thick stainless steel baffle extending 1.5 cm into the flow was used. The result (Figure 5) was less dramatic than with the rubber baffle.

Two different propellers were also tried: a shrouded 7.62 cm radius marine propeller (Figure 7c) and a 6.35 cm radius 16 bladed titanium impeller (Figure 7b). The shrouded marine propeller was motivated by the $s2t2$ Dudley and James flow and unpublished remarks of Cary Forest [6]. The shroud discourages the radial flinging as seen with the 6.35 cm radius unshrouded propellers of the original setup. The turbine impeller moves the fluid more efficiently. The hope was that we might increase the effective magnetic Reynolds number based on mean velocity as well as test a different flow geometry. The pulse decay data for the shrouded propeller and 16 bladed impeller are also shown in Figure 5.

## 6. Conclusion

Experimental observations have quantified the approach to dynamo action in eight different liquid sodium flows. Each flow showed a trend towards the transition to dynamo action with increasing $R_m$. The most dramatic progress towards transition was observed when 6.35 cm radius marine propellers were used to drive the sodium with poloidal baffles only.

## Acknowledgements

This work was funded by the NSF EAR-9903958, NSF EAR-9903162, DMR-9896037 and the Research Corporation. We gratefully acknowledge assistance from Kurram Gillani, Edward Bolton, James Drake, Jay Fineberg, Donald Martin, Edward Ott, Yasushi Takeda, Paul Roberts, John Rodgers, James Weldon, and Benjamin Zeff.

## References

1. M.L. Dudley and R.W. James, Proc. R. Soc. Lond. A **425**, 407 (1989).
2. A.K. Gailitis, B.G. Karasev, I.R. Kirillov, O.A. Lielausis, S.M. Luzhanskii and A.P. Ogorodnikov, Magnetohydrodynamics **23**, 349 (1987).
3. H. Beckley, S. Colgate, R. Ferrel, V. Romero, and D. Westpfahl, Bull. Am. Phys. Soc. **2057**, 43 (1998)
4. K-H Radler, E. Apstein, M. Rheinhardt, and M. Schuler, Stud. Geophys. Geod. **224**, 42 (1998).
5. P. Odier, J.-F. Pinton, and S. Fauve, Phys. Rev. E **58**, 7397 (1998).
6. C. Forest, http://aida.physics.wisc.edu
7. N.L. Peffley, A.B. Cawthorne, and D.P. Lathrop, Phys. Rev. E **5287**, 61 (2000).
8. WM. Elsasser, Phys. Rev. **106**, 69 (1946).
9. D. Sweet, E. Ott, J.M. Antonsen, Jr., D.P. Lathrop, Phys. Rev., submitted to Phys. Rev. E.
10. M. Ghil and S. Childress, *Atmospheric Dynamics, Dynamo Theory, and Climate Dynamics,* , Vol.60 of Topics in Geophysical Fluid Dynamics , edited by Fritz John, J. E. Mardsen, and Lawrence Sirovich (Springer-Verlag, 1987).
11. We would like to thank Yasushi Takeda for the loan of the ultrasound velocity measurement apparatus and helpful advice.
12. E Covas and Reza Tavakol, Phys. Rev. E **5434**, 60 (1999).

# THERMAL FLOW IN A ROTATING SPHERICAL GAP WITH A DIELECTROPHORETIC CENTRAL FORCE FIELD

B. SITTE, W. BRASCH, M. JUNK AND V. TRAVNIKOV
*Center of Applied Space Technology and Microgravity (ZARM)*
*University of Bremen, Am Fallturm, 28359 Bremen, Germany*

AND

C. EGBERS
*Department for Aerodynamics and Fluiddynamics*
*Brandenburg Technical University Cottbus*
*Postfach 10 13 44, 03013 Cottbus, Germany*

**Abstract.** The understanding of thermal convection in spherical gaps under a central force field is important for large scale geophysical motions. Neglecting the magnetic field, the dielectrophoretic force can be used to produce a central force field under microgravity conditions. In a space experiment, currently under construction, thermal convection in a rotating spherical gap with heated inner sphere and cooled outer sphere will be visualized by a Wollaston interferometer. High voltage is used to produce a dielectrophoretic central force field in the gap. The parameters are chosen in analogy to the convection in the earth's inner core. The experiment and its restrictions are presented as well as numerical predictions for the expected flows. The axial-symmetric flow is calculated on a staggered grid with a finite-volume method. The conjugate-gradient method with a preconditioner accelerates the approximation. In azimuthal direction a spectral analysis allows a three-dimensional simulation for spherical shells with a wide gap.

## 1. Introduction

The large scale motions of atmospheres of planets and in the convection zones of rotating stars are strongly influenced by Coriolis forces (due to rotation) and by buoyancy forces (due to gravity), which drive thermal cir-



*P. Chossat et al. (eds.), Dynamo and Dynamics, a Mathematical Challenge,* 93–100.

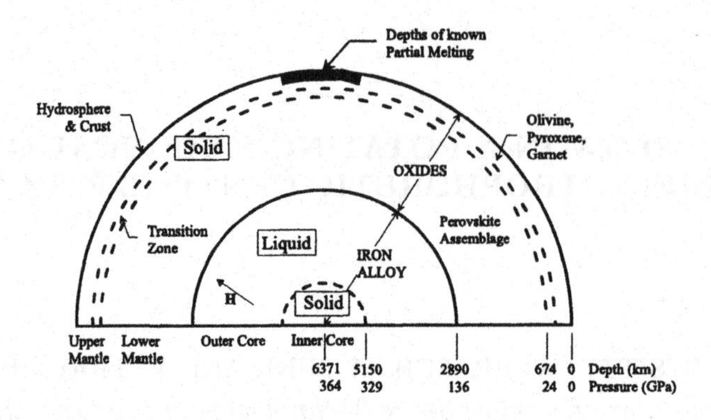


*Figure 1.* Schematic cross section of the earth. Thermal convective phenomena occur in the outer liquid core.

culation. The resulting flow structures show a rich variety of different types of instabilities, which depend strongly on different parameters as rotation rate, temperature gradient, gap width, material functions and others.

Figure 1 shows a schematic cross section of the earth. The convective motions of the molten iron alloy in the outer core, driven by the temperature difference between the inner core and the earth mantle, generate the main geomagnetic field. It is nearly impossible to investigate similar large scale geophysical motions in a laboratory experiment, simply because the earth's gravity inhibits every attempt to employ a radial central force field. Experiments can be performed for the case of fast rotation only, where the centrifugal force is dominant [1]. To overcome this restriction, the dielectrophoretic effect can be used to simulate a central force field similar to the gravity force of planets. To turn off the unidirectional gravitation in a spherical gap under terrestrial conditions, these experiments require an environment of microgravity. In preparation of long-term experiments in space, several experiments have been performed under laboratory conditions as well as under short-term microgravity conditions. Now, a long term experiment is set up to investigate the thermal convection in a fluid shell between two concentric spheres with and without rotation under a central dielectrophoretic force field. The inner sphere is heated, the outer is cooled. Besides its astro- and geophysical importance, this model is of basic interest for the understanding of the laminar-turbulent transition.

In preparation of the experiment, linear stability analysis has been performed to determine the critical Rayleigh and Taylor numbers for the onset of thermal convection under a simulated central force field. Nonlinear finite-amplitude convective motions have been studied numerically.

Further series of experiments with different gap widths and supercritical up to turbulent Taylor and Rayleigh numbers need a microgravity time of days or weeks, due to the long thermal diffusion times. Therefore, experiments with different diagnostics in the Experiment Container of the Fluid Science Lab of the ISS [5] are planned.

## 2. Theoretical Model

### 2.1. MATHEMATICAL FORMULATION

We consider the model of two concentric spherical shells with outer radius $R_2$, inner radius $R_1$ with a constant temperature difference between the two shells of $\Delta T$ (see figure 2). The geometrical non dimensional parameter is the radius ratio

$$\eta = \frac{R_1}{R_2} . \tag{1}$$

Using the thickness of the shell $d = R_2 - R_1$ as length scale, $d^2/\nu$ as time scale and $\Delta T \frac{\nu}{\kappa}$ as a scale of the temperature, the following equations of motion and heat equation in the non-dimensional form are obtained in the Boussinesq approximation:

$$\frac{\partial \mathbf{u}}{\partial t} + \mathbf{u} \cdot \nabla \mathbf{u} + \sqrt{Ta}\, \mathbf{k} \times \mathbf{u} = -\nabla \pi + Ra\, \gamma(r)\, \hat{\mathbf{r}}\, T + \nabla^2 \mathbf{u} \tag{2}$$

$$\nabla \mathbf{u} = 0 \tag{3}$$

$$Pr \frac{\partial T}{\partial t} + Pr\, \mathbf{u} \cdot \nabla T = \nabla^2 T \tag{4}$$

$\pi$ is the reduced pressure and $T$ is the temperature. The dimensionless parameters are the Rayleigh number $Ra = \frac{\alpha g_0 \Delta T\, d^3}{\kappa \nu}$, the Taylor number $Ta = \frac{4\Omega^2 d^4}{\nu^2}$ and the Prandtl number $Pr = \frac{\nu}{\kappa}$, where $\alpha$ is the thermal expansion coefficient, $\Delta T$ the temperature difference between inner and outer spherical shell, $g_0$ the constant of gravity at the outer shell, $\kappa$ the thermal diffusivity, $\nu$ the kinematic viscosity and $\Omega$ the angular velocity of rotation.

### 2.2. SIMULATION OF CENTRAL FORCE FIELD

In a laboratory experiment, the earth gravity always points downwards. There is no way to produce a central gravity force field in reasonable scales. However, the dielectrophoretic effect can be used to produce force field acting similar to a central gravity force field. In an electrical field, dipoles align to the field lines. In an inhomogeneous field, dipoles are dragged towards the stronger field region, independent of the polarization of the field. This dielectrophoretic effect can be utilized in spherical gap experiments. The inhomogeneous field is produced by using the inner and the outer sphere as a spherical capacitor. If the fluid has a dielectrical constant $\epsilon = \epsilon_0 \epsilon_r$ and if $\epsilon$ is significant larger than 1, then the dielectrophoretic force acts on every fluid molecule in the direction of the inner sphere. Since the dielectrical properties depend on the density of the fluid, a temperature dependence $\epsilon(T) = \bar{\epsilon}\,[1 - \chi\,(T - T_0)]$ can be defined similar to the definition of the thermal expansion coefficient $\alpha$. This results in a bouyancy force. In analogy to the gravity vector $\mathbf{g}$ an electrical acceleration vector

$$\mathbf{g}_{el}(\mathbf{r}) = g_0 \gamma(r) \hat{\mathbf{r}} = \frac{2\bar{\epsilon}}{\rho} \left( U \frac{R_2 R_1}{d} \right)^2 \frac{1}{r^5} \hat{\mathbf{r}} \tag{5}$$

can be defined and calculated as a function of the radial coordinate $\mathbf{r} = r\hat{\mathbf{r}}$. The central force is proportional to $r^{-5}$ and depends quadratic on the alternating potential $U$ across the fluid shell. An alternating potential is used to avoid electro-convection. This force has been used by Hart *et al.* to investigate thermal convection in a hemispherical small gap under microgravity conditions [6].

## 3. Preparation of microgravity experiments

On earth, the dielectrophoretic effect can not be used to simulate a central gravity force, because the effect is not strong enough to neglect the influence of the earth's gravity. In space, the central symmetry of $\mathbf{g}_{el}$ is not disturbed.

In preparation of future experiments on the International Space Station (ISS), an experimental investigation on stationary and time-dependent thermal convective phenomena between spherical shells under micro-gravity is being prepared. Because of long diffusion time scales these experiments require long investigation times in space, i.e. 50 up to 100 hours. Therefore an experiment in a NASA Get-Away-Special container with 1 week investigation time is planned. The autonomous container provides an experiment volume of about 141 liter including the energy source and the control units with a total permissible weight of 90.7 kg

The experimental cell is formed by an outer glass sphere, which can be cooled, and an inner sphere, which can be uniformly heated from within. Both spheres can be rotated as a rigid body (see figure 2). Thermal convective pattern occur due to the interaction of rotational and density effects. The central force field is generated by applying a high voltage of 12 kV rms between inner and outer sphere, in analogy to the earth's gravity field. Due to the action of the high voltage potential on the fluid in the gap, an optical observation technique for flow visualization and temperature measurements without tracer particles is used in this experiment. In the Get Away Special experiment, a Wollaston interferometry is used to visualize the thermal flows (see figure 3, left). Using the inner sphere as a mirror, a 90 degrees view area is mapped on a CCD camera.

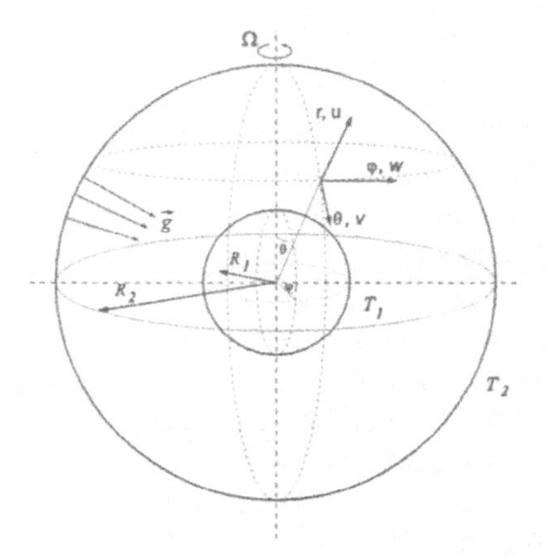

*Figure 2.* Sketch of the coordinate system (left) and picture of the experiment cell for the GAS mission (right)

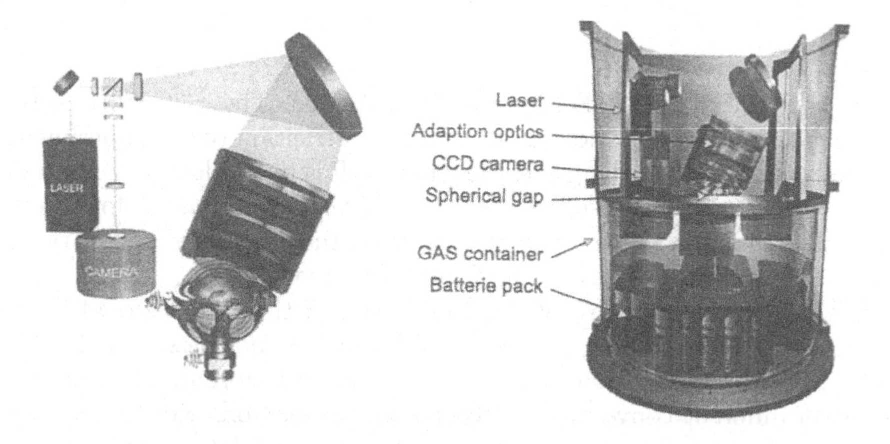


*Figure 3.* Illustration of the differential Wollaston interferometrie (left) and of the experiment structure for the GAS mission (right).

The experiment is currently under construction and is scheduled to fly on board of the space shuttle in early 2002. The parameters of the experiment are chosen in analogy to the earth's outer core (see table 1). This experiment is focussed on the basic thermal convection in a wide gap with radial force field, with and without rotation. For higher Rayleigh numbers, the temporal and spatial scales of the convection are of basic interest, as well as the transition to turbulent states. Besides this scientific interest, the GAS experiment proofs the technical feasibility of the dielectrophoretic force field in a wide gap in connection with a differential interferometry.

Not all parameters can be matched to the earth's outer core and the

TABLE 1. Parameters of the GAS experiment for Baysilone M3 silicone oil

| | |
|---|---|
| $R_1 = 14$ mm | Radius of inner sphere |
| $R_2 = 35$ mm | Radius of outer shell |
| $d = 21$ mm | gap width |
| $\eta = 0.4$ | Radius ratio |
| $0.5 < \Delta T = T_1 - T_2 < 10$ K | Temperature difference |
| $\nu = 3.0$ mm$^2$/s | Kinetic viscosity (25 Celsius) |
| $\bar{\rho} = 0.90$ kg/l | Specific density (25 Celsius) |
| $\alpha = 1.11 \times 10^{-3}$ K$^{-1}$ | Thermal expansion coefficient |
| $\kappa = 77 \times 10^{-9}$ | Thermal diffusion coefficient |
| $Pr = 38.8$ | Prandtl number (25 Celsius) |
| $0 < \Omega < 1.6$ rad/s | Angular velocity |
| $E = \frac{\nu}{\Omega d^2} > 4.25 \times 10^{-3}$ | Ekman number |
| $Ta = \frac{4\Omega^2 d^4}{\nu^2} < 2.2 \times 10^5$ | Taylor number |
| $10^4 < Ra = \frac{\alpha g_0 \Delta T d^3}{\kappa \nu} < 3 \times 10^7$ | Rayleigh number |

influence of the magnetic field is totally neglected. The $r^{-5}$ decay of the dielectrophoretic acceleration $\mathbf{g}_{el}$ is a major restricition of the experiment. For the case of $\eta = 0.4$, the $\mathbf{g}_{el}$ at the outer shell is two orders of magnitude smaller than at the inner sphere. However, Yavorskaya *et al.* [7] found evidence, that the qualitative differences between the flows for different radial force fields proportional to $r^{-5}$, $r^{-2}$ and $r$ are small.

Further experiments are planned on board of the International Space Station (ISS). Here, the aspect ratio will be varied and, based on the first results from the GAS experiment, much longer and more detailed investigations of different convection structures and transitions will be possible. Beside pure geophysical interests, the experimental validation of bifurcation theories in spherical geometry [3] is one of the main aims.

## 4. Numerical simulations

For space experiments, numerical simulations prior to the experiments are of high importance. Since the experiment and microgravity time is expensive, a solid theoretical and numerical basis is needed not only for the purpose of physical insight, but also to focus the experimental work on interesting parameter regions.

The three-dimensional nonlinear Boussinesq equations (equations 2-4) are solved numerically with finite–volumes and spectral analysis in azimuthal direction. For the axisymmetric problem we used a finite–volume method on a non-homogeneous staggered grid in the $(r, \vartheta)$–plane to solve the discretized equations in primitive variables. On a equidistant grid all terms are solved with second–order–accurate formula and the advective terms are used in the conservative form. For a better resolution of the boundary flow it is possible to stretch the radial component with the function $r' = R_1 + 0.5\,(1 - cos\pi(r - R_1))$. In one time step all variables $(u, v, w, p, T)$ are solved simultaneously with the conjugate gradient method BiCGSTAB [4]. A preconditioner is used to ensure the divergence free velocity field and to accelerate the iteration. The truncated matrix for the preconditioner only includes the diagonal terms of the original matrix of the discretisized equations, the pressure gradient and the divergence of the velocity. The Fourier coefficients of the $2\pi$-periodic function are calculated iterativ for every time step. The time is discretisized with a second-order Euler-backward method.

The critical Rayleigh number for the given boundary conditions $\eta = 0.4$ and $g \propto 1/r^5$ is $Ra_c = 182$ and the critical latitudinal wavenumber is $l_c = 3$. For twice the critical Rayleigh number we have simulated the convection flow. Figure 4b shows the temperature field in the middle of the fluid shell. It has a typical tetrahedral pattern as suggested by [2]. Figure 4a shows the transition from the conductive basic state to an intermediate state. The maximum Nusselt number and the maximum kinetic energy is reached after $t \approx 8$ but the pattern of convection is still changing until $t \approx 100$, then the stationary final state is reached shown in figure 4b.

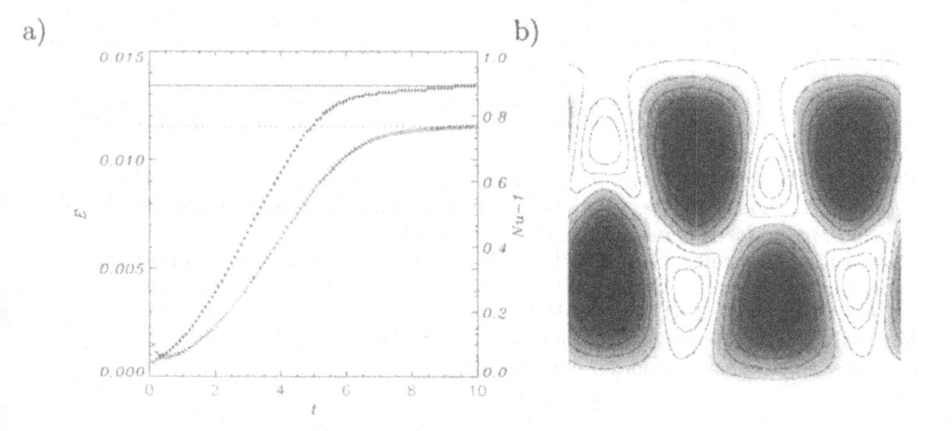


*Figure 4.* Kinetic energie and Nusselt number as a function of the nondimensional time (a) and the tetrahedral pattern (b). Pr = 35,0, $\eta$ = 0,4, $Ra$ = 400, resolution $N_T = 33 \times 33 \times 7$, (+)$\hat{=}$ $E$, (x) $\hat{=}$ $Nu$, $E_{max} = 0.0134$, $Nu_{max} = 1.769$

## 5. Conclusions

In combination with a dielectrophoretic central force field and microgravity conditions, a spherical gap apparatus offers the opportunity to study fundamental structures and transitions of thermal convection with geophysical relevance. Preparative experiments have been performed, a first wide gap spherical gap is set up for a long term investigation of thermal convection on board of the space shuttle. This experiment will be followed by a much more extensive study on board of the ISS.

For the theoretical analysis and for the prediction of the flow in the experimental investigation a new numerial code is developed. For an optimized resolution of the periodic flow in rotating states, a spectral method is used in the azimuthal direction. This programme is used to predict the flow patterns and to determine the interesting parameters for the experiments.

## 6. Acknowledgements

The financial support from the Bundesministerium für Bildung und Forschung (BMBF) and the Deutsches Zentrum für Luft- und Raumfahrt (DLR) is gratefully acknowledged.

## References

1. Busse, F.H. (1970) Thermal instabilities in rapidly rotating systems, *J. Fluid Mech.*, **Vol. 44**, pp. 441-460
2. Busse, F.H. and Riahi, N. (1982) Patterns of convection in spherical shells. Part 2, *J. Fluid Mech.*, **Vol. 123**, pp. 283-301
3. Chossat, P., Guyard, F. and Lauterbach, R. (1999) Generalized heteroclinic cycles in spherical invariant systems and their perturbations, *J. Nonlinear Sci.*, **Vol. 9**, pp. 479-524
4. Deng, G. B., Guilmineau, E., Piquet, J., Queutey, P., Visonneau, M. (1994) A new fully coupled solution of the Navier–Stokes equation, *Int. J. Num. Meth. in Fluids*, **Vol. 19**, pp. 605–639
5. Egbers, C., Brasch, W., Sitte, B., Immohr, J. and Schmidt, J.-R. (1999) Estimates on diagnostic methods for investigations of thermal convection between spherical shells in space, *Meas. Sci. Technol.*, **Vol. 10**, pp. 866-877
6. Hart, J.E., Glatzmaier, G.A. and Toomre, J. (1986) Space laboratory and numerical simulations of thermal convection in a rotating hemispherical shell with radial gravity, *J. Fluid Mech.*, **Vol. 173**, pp. 519-544
7. Yavorskaya, I.M., Fomina, N.I. and Belyaev, Yu.N. (1984) A simulation of central symmetry convection in microgravity conditions, *Acta Astronaut.*, **Vol. 11**, pp. 179-183

# PRELIMINARY MEASUREMENT OF THE TURBULENT MAGNETIC DIFFUSIVITY IN THREE DIMENSIONS

JEAN-CLAUDE THELEN
*Department of Astronomy and Astrophysics*
*University of Chicago*

AND

FAUSTO CATTANEO
*Department of Mathematics*
*University of Chicago*

**Abstract.** Turbulent transport plays an important role in the evolution of astrophysical magnetic fields. The underlying idea is that the small-scale, turbulent motions increase the magnetic diffusivity above the value due to the molecular diffusion . However, it has been shown that in two dimensions the turbulent diffusion of large-scale magnetic fields is reduced significantly once the large-scale field strength exceeds a critical value, which can be much less than equipartition. The question that naturally arises is whether this holds true in three dimensions or whether magnetic fields of equipartition energy are required in order to suppress the effective turbulent diffusion. In order to elucidate this question we performed numerical calculations based on the full, incompressible MHD equations, that allow us to measure the effective magnetic diffusivity for different values of the large-scale magnetic field strength.

## 1. Introduction

In most astrophysical situations the fluid flow is highly turbulent, i.e. there exists a random velocity superimposed on the mean velocity. This random velocity component plays a crucial role, not only in the overall dynamics of the flow, but also in the transport of any flow property.
It is a well known fact that, in the absence of any magnetic fields, turbulent motions can speed up the mixing of passive-scalar quantities, as for example



*P. Chossat et al. (eds.), Dynamo and Dynamics, a Mathematical Challenge,* 101–108.

temperature or the concentration of a contaminant. A similar mechanism is used to describe the diffusion of magnetic fields by turbulence (Moffatt 1983). Typically, it is assumed that the magnetic field does not exert any influence on the underlying flow as long as the large-scale field strength is smaller than equipartition. The large-scale magnetic field is only supposed to play a dynamical role when its energy is comparable to the kinetic energy. It has been shown by Cattaneo and Vainshtein (1991) that, at least in two dimensions, this is not the case. It was found that, for large magnetic Reynolds numbers $R_M$, the turbulent magnetic diffusivity is suppressed even for large-scale fields which are much weaker than equipartition. This implies that the magnetic field starts to play an important dynamical role before equipartition is reached. The next question is whether this result extends to three dimensions, or whether a large-scale field of equipartition energy is required in order to obtain suppression. This problem is the object of some debate. While Vainshtein and Cattaneo (1992) suggest, on heuristic grounds, that in three dimensions the suppression of the turbulent magnetic diffusivity should be even more severe than in two dimensions, Gruzinov and Diamond (1994) propose, using a closure argument, that it should not be suppressed at all.

In this paper, we present three dimensional simulations that allow us to measure both the turbulent magnetic diffusivity and the turbulent scalar diffusivity. In section 2 we briefly discuss the methodology used to measure the magnetic diffusivity and we describe the model. In section 3 we present the results and in section 4 we give a conclusion.

## 2. Mathematical Description

### 2.1. MEASURING THE TURBULENT DIFFUSIVITIES

In this section we briefly describe how the turbulent magnetic diffusivity $\eta_{eff}$ is measured as a function of the large-scale magnetic field strength.

Measuring $\eta_{eff}$ of a large-scale magnetic field in a two-dimensional domain is relatively easy since the induction equation reduces to a scalar equation. There exists an anti-dynamo theorem that states that dynamo action is not possible in two-dimensions and hence that the large-scale magnetic field must decay. Thus, it is possible to obtain $\eta_{eff}$ by determining the rate of decay of the large-scale magnetic field.

This method however, fails in three dimensions since dynamo action is now possible and the induction equation no longer reduces to a scalar equation. Instead, we developed a technique based on the much simpler problem of heat diffusion in a rod with time-dependent boundary conditions and thermal diffusivity $\kappa$. Consider a rod which is maintained at a constant temperature (say $T = 0$) at one end, while at the other end the temperature

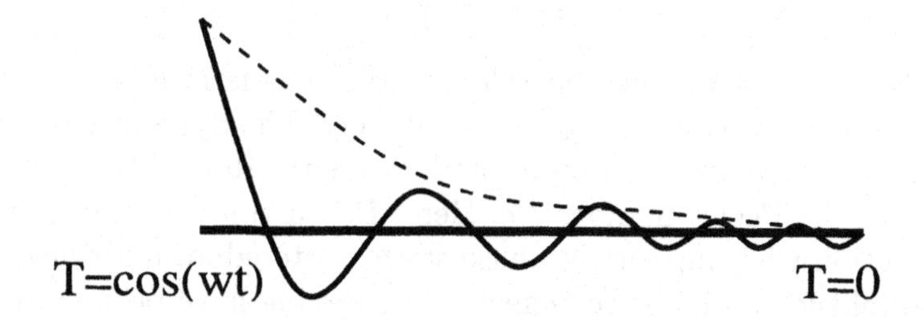


*Figure 1.* Schematic representation of the solution to the heat diffusion problem described in the text. Here the dashed line represents the envelope to the solution, the shape of which depends on the thermal diffusion $\kappa$.

oscillates periodically with a fixed frequency $\omega$. It is important to note that although the temperature distribution within the rod is time-dependent for all times the envelope of the solution eventually becomes stationary. The shape of this stationary envelope is determined by the value of the diffusion coefficient $\kappa$ (see figure 1). As $\kappa$ is increased the envelope becomes steeper, while decreasing $\kappa$ results in a more shallow slope. Notice that this characteristic behaviour of the envelope is not restricted to one-dimensional solids but that it also applies to scalar quantities in a three-dimensional domain filled with a fluid. Thus, it seems reasonable to assume that this behaviour even extends to the much more general case of turbulent diffusion of scalar and vector quantities.
Our aim is to exploit this behaviour of the envelope in order to measure the turbulent magnetic diffusivity $\eta_{eff}$ of the large-scale magnetic field. We introduce a large-scale, time-dependent, magnetic source function $\mathbf{S}$ into our computational domain. Physically we can think of this source function as electrical coils which have been inserted into the plasma and through which an oscillating current passes. This magnetic source results in a large-scale, time-dependent magnetic field whose steady envelope can be calculated. Since the strength of this magnetic field depends on the turbulent magnetic diffusivity $\eta_{eff}$ we can assume that the shape of the envelope is also a function of $\eta_{eff}$.

## 2.2. MODEL & EQUATIONS

We consider a three dimensional, periodic, layer containing an electrically conducting plasma with constant viscosity and (molecular) diffusivity. The evolution equations, in dimensionless form, can be written as:

$$(\partial_t - R_E^{-1}\nabla^2)\mathbf{U} + (\mathbf{U}\cdot\nabla)\mathbf{U} = -\nabla\mathbf{p} + \mathbf{J}\wedge\mathbf{B} + \mathbf{F} \tag{1}$$

$$(\partial_t - R_M^{-1}\nabla^2)\mathbf{B} = \nabla\wedge(\mathbf{U}\wedge\mathbf{B}) + \mathbf{S} \tag{2}$$

$$(\partial_t - Pe^{-1}\nabla^2)\theta + (\mathbf{U}\cdot\nabla)\theta = \mathbf{0} \tag{3}$$

$$\nabla \cdot \mathbf{U} = \nabla \cdot \mathbf{B} = 0 \tag{4}$$

Here **U**, **B**, **J** and $\theta$ denote the velocity, the magnetic field, the current density, and a scalar concentration respectively. The dynamics of this system are determined by the kinetic and magnetic Reynolds numbers $R_E$ and $R_M$ and the Peclet number $Pe$. Here, **F** is a time-dependent random forcing function with support in phase space centered around wavenumber 10. The amplitude and the frequency of the forcing have been chosen such that the ratio of the correlation time of the flow and the turn-over time is unity.
The extra term **S** in the induction equation (2) is a $z$-independent magnetic source function. We choose **S** to be of the form $\mathbf{S} = (0, 0, S_z)$ in order to obtain a large-scale, driven, magnetic field in the $\pm\hat{\mathbf{z}}$-direction. $S_z$ describes a periodic array of alternating magnetic sources and sinks. Since we are interested in the turbulent decay of the large-scale magnetic field we choose the length-scale of **S** to be large compared to the length-scale of the flow. Thus, in comparison to the flow the driven component of the magnetic field can be considered as having a large-scale.
$S_z$ oscillates in time with a fixed frequency $\omega_M$, which is long compared to the correlation time of the flow, and with a fixed amplitude $A_M$. By varying $A_M$ we increase or decrease the strength of the large-scale field and for each value we the compute the steady envelope of the time-dependent solution of the large-scale magnetic field. The shape of the envelope will allow us to determine qualitatively whether the turbulent magnetic diffusivity $\eta_{eff}$ increases or decreases as $A_M$ is varied. By comparing this envelope to that of the two dimensional solution of

$$\partial_t \bar{B}_z = \eta_{eff} \nabla^2 \bar{B}_z + S_z, \tag{5}$$

where $\bar{B}_z$ denotes the large-scale component of $B_z$, we obtain a quantitative measure of $\eta_{eff}$.

## 3. RESULTS

We computed the turbulent magnetic diffusivity $\eta_{eff}$ and the turbulent magnetic diffusivity of a passive-scalar $\kappa_{eff}$ for different values of the large-scale magnetic field strength, ranging from a kinematic value to the equipartition value. All the runs have been computed with a molecular magnetic diffusivity of $\eta_{mol} = 0.0025$, a molecular viscosity of $\nu_{mol} = 0.01$ and a molecular scalar diffusivity of $\kappa_{mol} = 0.0025$.
Figure 2 shows the time-averaged histories of the kinetic and the magnetic energies for different values of the driving amplitude of $A_M$ . The first thing to note is that even in the absence of any magnetic forcing ($A_M = 0$) we obtain a non-zero magnetic field. Thus our underlying turbulent flow acts

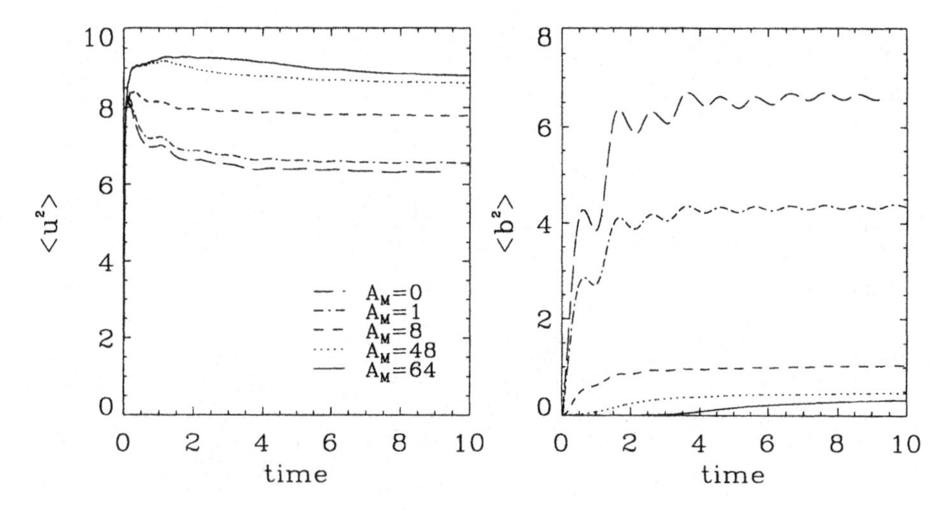


*Figure 2.* Time histories of $\langle u^2 \rangle$ and $\langle b^2 \rangle$ for different values of $A_M$, i.e for different values of the large-scale magnetic energy. $A_M = 64$ corresponds to a total magnetic energy of $\langle b^2 \rangle = 6.61$ and to a large-scale energy of $\langle B_0^2 \rangle = 2.20$.

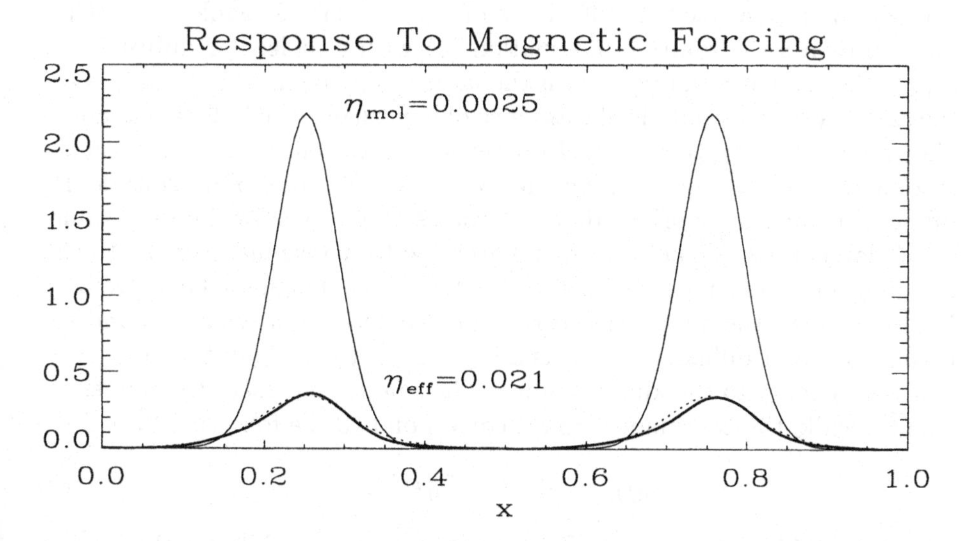


*Figure 3.* Response of our system (thick line) and that of equation (5) to the large-scale magnetic forcing for $\eta_{mol} = 0.0025$ (continuous line) and $\eta_{mol} = 0.021$ (dashed line).

as a dynamo. The intensity of the turbulence is only reduced by a factor of about 1.3 although the total magnetic energy increases by about one order of magnitude.

Figure 3 describes how the turbulent magnetic diffusivity was measured. It shows the response (thick line) of our system to the the large-scale magnetic driving for $A_M = 16$. The thin continuous line is the solution of equa-

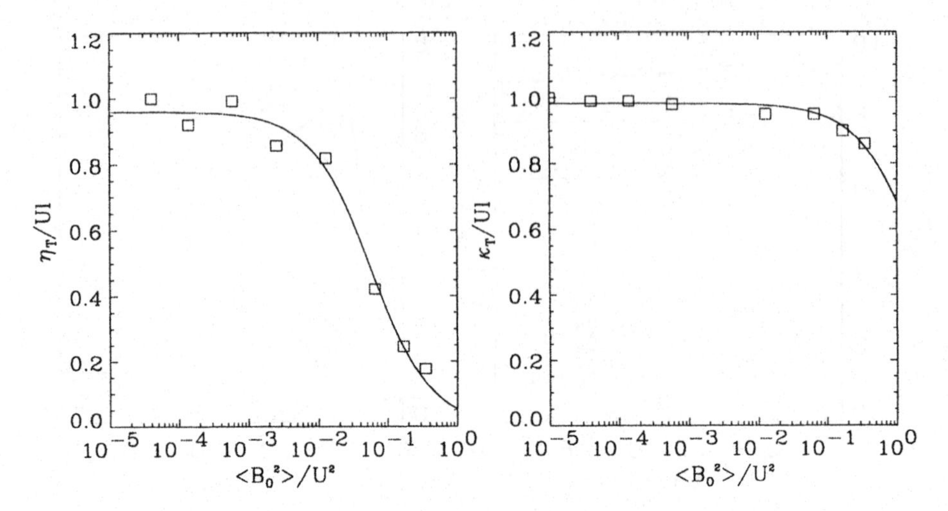


*Figure 4.* Normalized turbulent magnetic diffusivity $\eta_{eff}$ and turbulent scalar diffusivity $\kappa_{eff}$ as a function of the large-scale magnetic energy.

tion (5) with a molecular diffusivity of $\eta_{mol} = 0.0025$, while the dashed line represents the solution of equation (5) with a molecular diffusivity of $\eta_{mol} = 0.021$. Thus, in our system the large-scale magnetic field diffuses at the same rate as it would in the absence of any turbulence but with a molecular diffusivity of $\eta_{mol} = 0.021$. In other words, the effect of the turbulence is to increase the magnetic diffusivity by nearly one order of magnitude. By taking this measurement for different values of $A_M$, i.e. for different values of the large-scale magnetic field, we are able to determine how the turbulent magnetic diffusivity varies with the large-scale magnetic field strength. Figure 4 shows the normalized turbulent magnetic diffusivity $\eta_{eff}$ and the turbulent scalar diffusivity $\kappa_{eff}$ as a function of the ratio of the large-scale magnetic energy to the kinetic energy. The squares represent the numerical results, while the continuous curves correspond to the following fit

$$\eta_{eff} = \eta_0(1 + \eta_1 B_0^2/U^2)^{-1}, \tag{6}$$

with $\eta_0 = 0.96$ and $\eta_1 = 0.48$ for the magnetic case, while for the passive scalar case we obtain $\kappa_0 = 0.98$ and $\kappa_1 = 0.0458$. From this data it is clear that the transport properties of a three-dimensional flow are completely different for a passive scalar and the magnetic field . Let us first consider the diffusion of the passive scalar. In the absence of any magnetic field the turbulence increases the rate of diffusion above the molecular rate, as we mentioned earlier. In three dimensions the introduction of a weak field does not seem to have an effect on the turbulent rate of scalar diffusion, in sharp contrast to the two-dimensional case (Cattaneo and Vainshtein 1991, Cattaneo 1994) . It is only when the large-scale field reaches equipartition

energy that the diffusion rate starts to decrease. The slow decrease in $\kappa_{eff}$ which can be observed seems to be entirely due to a reduction in the intensity of the turbulence through the Lorentz force, since by going from $A_M = 0$ (no large-scale field) to $A_M = 64$ (strong large-scale field) both $< u^2 >$ and $\kappa_{eff}$ are reduced by about a factor of 1.3
The turbulent magnetic diffusivity, on the other hand, decreases by one order of magnitude for the same increase in the large-scale magnetic field strength and it approaches its molecular value as the large-scale magnetic field strength tends to its equipartition value. This decrease in $\eta_{eff}$ can no longer be explained solely by a reduction in the turbulence intensity. Thus we may conclude that $\eta_{eff}$ becomes suppressed once the large-scale field strength exceeds some critical value which is much less than equipartition energy.
This gives rise to the question as to why the turbulent scalar diffusivity is not suppressed. A probable explanation is that the anisotropy, introduced into the flow through the large-scale vertical magnetic field, plays an important role. A preliminary calculation shows that the passive scalar diffuses much quicker along the magnetic field lines than across the magnetic field lines. Here we released a blob of hot fluid in a region of strong vertical magnetic field. Thus the diffusion across the field lines is impeded to a larger extend than along the field lines. The magnetic field, on the other hand, cannot diffuse along the field lines and thus has to diffuse in the transverse direction. This implies that the magnetic field lines have to bend in order to bring together magnetic fields of opposite polarity, an action which is impeded by strong large-scale fields.
Another mechanism by which the mixing of a passive-scalar can be obtained is by interchanging magnetic field lines. This can be achieved by the $k_z = 0$ mode of the flow, which simply shifts the field lines around without bending them. However, in order to clarify this question further calculations need to be done.

## 4. Conclusion

It has been shown previously that a large-scale magnetic field has a considerable effect on the transport properties of a two-dimensional flow. In particular, the turbulent magnetic diffusivity and the turbulent scalar diffusivity are suppressed once the large-scale magnetic field strength exceeds a critical value which is much less than equipartition. In this paper we provide evidence that the turbulent magnetic diffusivity is also suppressed in three dimensions for large-scale magnetic fields which are considerably weaker than equipartition strength. Thus the magnetic field starts to play a dy-

namical role, both in two and three dimensions, long before the large-scale field reaches equipartition value. This is in agreement with the predictions of Vainshtein & Cattaneo (1991) which state that $\eta_{eff}$ is suppressed in both two and three dimensions.
The turbulent diffusivity of a passive scalar, on the other hand, does not become suppressed unless the large-scale magnetic field is of equipartition strength. The small reduction that can be seen we in $\kappa_{eff}$ is entirely due to a reduction in the intensity of the turbulence through the back-reaction of the field on the flow through the Lorentz force. This is because in three dimensions motions that interchange field lines can bring together oppositely directed field lines without bending them (Cattaneo 1994).

## References

1. Cattaneo F., 1994, ApJ,434,200
2. Cattaneo F. & Vainshtein S.I.,1991, ApJ, 376, L21
3. Gruzinov A.V. & Diamond P.H. ,1994, Phys. Rev. Let.,72,1651
4. Moffatt H.K., 1983, Rep. Prog. Phys,46,621
5. Vainshtein S.I. & Cattaneo F.,1992, ApJ, 393,165

# SATURATION MECHANISM IN A MODEL OF THE KARLSRUHE DYNAMO

A. TILGNER AND F.H. BUSSE
*Institute of Physics, University of Bayreuth*
*95440 Bayreuth, Germany*

The dynamo experiment in Karlsruhe (Busse et al., 1998) has been successful in generating a self sustained magnetic field (Stieglitz and Müller, 2000). The focus of theoretical investigations now shifts from the study of the conditions for the onset of dynamo action to the mechanisms responsible for saturation. One might suspect that due to the guiding mechanical structures present in this experiment, saturation simply happens in that the magnetic field grows until the Lorentz force reduces the volumetric flow rate to its critical value, without any significant change to the shape of the velocity field. Any pressure applied by the pumps in addition to the critical pressure would be balanced by the Lorentz force according to this scenario. Experimental observations show however that volumetric flow rates continue to increase as a function of the applied pressure even when a steady dynamo field is present, which indicates that the velocity profile inside individual spin generators changes in response to the magnetic field.

The study of the back reaction of the magnetic field on the velocity profile is complicated because of the intricate structure of the stainless steel pipes, bends and blades inside the sodium filled vessel. Drastic simplifications are necessary in order to adapt existing computer codes for this purpose. A tractable model assumes a sphere filled with sodium set into motion by a volume force chosen such that the flow resembles the experimental one. In this way, the experimental flow is reproduced without taking into account all the internal mechanical structure. The induction equation coupled to the full Navier-Stokes equation can then be integrated numerically. In order to prevent the flow from deviating too much from its columnar form, it must be hydrodynamically stable and therefore have a low enough Reynolds number. This assumption is crudely justified if one identifies the viscosity in the Navier-Stokes equation with a turbulent viscosity relevant for the time averaged velocity field. While turbulence in the Karlsruhe experiment greatly modifies the effective viscosity, it contributes



*P. Chossat et al. (eds.), Dynamo and Dynamics, a Mathematical Challenge,* 109–116.

little to the effective magnetic diffusivity because the molecular value of the latter exceeds the viscous diffusivity by a factor $10^5$. In this paper, we shall neglect inertia altogether and return to a simple analytical model.

There are at least two reasons why highly simplified models are still of interest. First, they point to one possible modification of the velocity field through which the volumetric flow rate can grow beyond its critical value. Second, they provide a simple reference situation for future more complete simulations and a guideline for their interpretation.

Consider the induction equation

$$\partial_t \boldsymbol{B} + \nabla \times (\boldsymbol{B} \times \boldsymbol{u}_0) = \lambda \nabla^2 \boldsymbol{B} \tag{1}$$

for the velocity field

$$\boldsymbol{u}_0 = v_0 \begin{pmatrix} \sqrt{2}\sin(\pi\frac{x}{a})\cos(\pi\frac{y}{a}) \\ -\sqrt{2}\cos(\pi\frac{x}{a})\sin(\pi\frac{y}{a}) \\ w\sin(\pi\frac{x}{a})\sin(\pi\frac{y}{a}) \end{pmatrix}. \tag{2}$$

The treatment of this problem in an infinite domain (or for a periodicity length $2a$ much smaller than the dimension $L$ of the fluid volume) has been given in detail elsewhere (Roberts, 1972, see also Busse, 1978) so that only the main steps are repeated here. The magnetic field $\boldsymbol{B}$ is separated into a component $\boldsymbol{b}$ which fluctuates on the scale $a$ and a part $\bar{\boldsymbol{B}}$ whose average over a periodicity cell is zero: $\boldsymbol{B} = \bar{\boldsymbol{B}} + \boldsymbol{b}$. Assuming that the growth rate of the magnetic field is small and that $|\boldsymbol{b}| \ll |\bar{\boldsymbol{B}}|$, the following relations may be derived:

$$\partial_t \bar{\boldsymbol{B}} + < \nabla \times (\boldsymbol{b} \times \boldsymbol{u}_0) > = \lambda \nabla^2 \bar{\boldsymbol{B}} \tag{3}$$

$$\boldsymbol{b} = \frac{1}{2\lambda}(\frac{a}{\pi})^2 (\bar{\boldsymbol{B}}\nabla)\boldsymbol{u}_0 \tag{4}$$

$< .. >$ denotes the average over the $x, y$-plane. Growing modes have the form $\bar{\boldsymbol{B}} = B_0(\cos kz, \sin kz, 0)$. The critical magnetic Reynolds number depends on the wavenumber $k$. Direct numerical integration of the induction equation in finite domains confirms that the fastest growing mode has a shape similar to $(\cos kz, \sin kz, 0)$ in the inner region of the conducting volume, with half a wavelength fitting into the dynamo region (Tilgner, 1997).

Consider next the effect this magnetic field has on the velocity field. Neglecting inertia we write the total velocity field in the form $\boldsymbol{u} = \boldsymbol{u}_0 + \boldsymbol{u}_1$ where $\boldsymbol{u}_1$ is the solution of

$$0 = -\frac{1}{\rho}\nabla p + \nu \nabla^2 \boldsymbol{u}_1 + \frac{1}{\rho\mu_0}(\nabla \times \boldsymbol{B}) \times \boldsymbol{B} \tag{5}$$

with $\boldsymbol{B}$ taken from (3) and (4). $\mu_0$ is the magnetic permeability, $p$ is the pressure, $\rho$ is the density, and $\nu$ is the (turbulent) viscosity of the fluid. The Lorentz force is therefore proportional to

$$(\nabla\times\boldsymbol{B})\times\boldsymbol{B} = (\nabla\times\bar{\boldsymbol{B}})\times\bar{\boldsymbol{B}}+(\nabla\times\boldsymbol{b})\times\boldsymbol{b}+(\nabla\times\bar{\boldsymbol{B}})\times\boldsymbol{b}+(\nabla\times\boldsymbol{b})\times\bar{\boldsymbol{B}} \quad (6)$$

The first term of the right hand side is strictly zero in an infinite domain and is assumed to be small in the experimental geometry. The next three terms have orders of magnitude $|\boldsymbol{b}|^2/a$, $|\bar{\boldsymbol{B}}||\boldsymbol{b}|/L$, and $|\bar{\boldsymbol{B}}||\boldsymbol{b}|/a$ where $L$ is the typical dimension of the experiment. Since $|\boldsymbol{b}| \ll |\bar{\boldsymbol{B}}|$ and $a \ll L$ in accordance with the assumptions made to derive (3) and (4), only the last term in (6) needs to be retained. A solution $\boldsymbol{u}_1$ of (5) can then easily be obtained if $\partial_z \ll \partial_x, \partial_y$ and $\bar{B}_z \ll \bar{B}_x, \bar{B}_y$ are taken into account. The total velocity is found to be

$$\boldsymbol{u} = (1-\gamma)\boldsymbol{u}_0 + 2\gamma\frac{\bar{B}_x\bar{B}_y}{B_0^2}\tilde{\boldsymbol{u}} \quad (7)$$

with

$$\tilde{\boldsymbol{u}} = v_0 \begin{pmatrix} -\sqrt{2}\cos(\pi\frac{x}{a})\sin(\pi\frac{y}{a}) \\ \sqrt{2}\sin(\pi\frac{x}{a})\cos(\pi\frac{y}{a}) \\ w\cos(\pi\frac{x}{a})\cos(\pi\frac{y}{a}) \end{pmatrix} \quad (8)$$

and

$$\gamma = \frac{1}{4\lambda\nu}(\frac{a}{\pi})^2\frac{B_0^2}{\rho\mu_0}. \quad (9)$$

(7) shows that the magnetic field has two distinct effects. To sustain the amplitude of the original velocity distribution $\boldsymbol{u}_0$ the forcing must be increased by a factor $1+\gamma+\mathcal{O}(\gamma^2)$ as was assumed in the most naive model of saturation. But the velocity field is also modified in that a second component proportional to $|\boldsymbol{B}|^2$ is added. This new component is again a set of right handed helices identical to those making up $\boldsymbol{u}_0$ but shifted with respect to the original pattern (figure 1).

Equation (3) can now be revised by replacing the second term with $<\nabla\times[\boldsymbol{b}\times(\boldsymbol{u}_0+2\gamma\bar{B}_x\bar{B}_y/B_0^2\tilde{\boldsymbol{u}})]>$. One then obtains a nonlinear equation for $\bar{\boldsymbol{B}}$ which describes the saturation behavior of the magnetic field in terms of the imposed velocity field $\boldsymbol{u}_0$:

$$\partial_t\bar{\boldsymbol{B}} + \nabla\times v_0^2\frac{a}{\pi}\frac{w}{2\sqrt{2}\lambda}\begin{pmatrix}\bar{B}_x\\ \bar{B}_y\\ 0\end{pmatrix} - \nabla\times v_0^2\frac{a}{\pi}\frac{w}{\sqrt{2}\lambda}\gamma\frac{\bar{B}_x\bar{B}_y}{B_0^2}\begin{pmatrix}\bar{B}_y\\ \bar{B}_x\\ 0\end{pmatrix} = \lambda\nabla^2\bar{\boldsymbol{B}}. \quad (10)$$

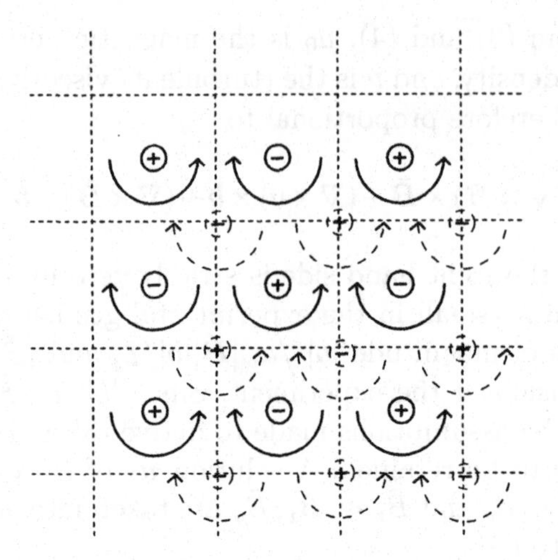

*Figure 1.* Sketch of the total velocity field (7) in the $x, y$-plane. Signs give the direction of the out of plane component of the flow. The solid lines indicate the original motion $\boldsymbol{u}_0$ in absence of a magnetic field. The cells of size $a$ are shown by dotted lines. The structure of $\tilde{\boldsymbol{u}}$ is indicated with dashed lines.

Multiplication of this equation with $(\cos kz, \sin kz, 0)$ and integration over $z$ yields the condition

$$v_0^2 = 2\sqrt{2}\frac{\pi}{a}\frac{\lambda^2 k}{w}\left(1 + \frac{1}{4\lambda\nu}\left(\frac{a}{\pi}\right)^2\frac{B_0^2}{\rho\mu_0} + ...\right) \tag{11}$$

where terms of higher order in $B_0^2 a^2/\lambda\nu\rho\mu_0$ have not been denoted explicitly. The kinematic dynamo onset is recovered in the limit $B_0 = 0$. In the Karlsruhe experiment, there is no control over the actual velocity distribution and only the flow rate through the pipes may be adjusted. Transferring the model to the experiment, the flow rate through a cell near the origin is given by

$$\int_0^a \int_0^a dx dy \boldsymbol{u}\hat{\boldsymbol{z}} \tag{12}$$

where $\hat{\boldsymbol{z}}$ is the unit vector in $z$-direction. The second term in (7) thus makes no contribution to the flow rate. The difference between the actual flow rate and the critical flow rate varies quadratically as a function of the saturation field. The fact that the flow rate increases beyond its critical value is due to the presence of $\tilde{\boldsymbol{u}}$ in (7).

The effect of $\tilde{\boldsymbol{u}}$ is next investigated with kinematic calculations. As in previous simulations of the Karlsruhe dynamo (Tilgner, 1997), a spherical volume of radius $r_o$ is assumed. The largest cylinder of aspect ratio 1 (height over radius) fitting into this sphere has the radius $R = r_o\sqrt{4/5}$ and represents the experimental dynamo cell. The symmetry axis of the

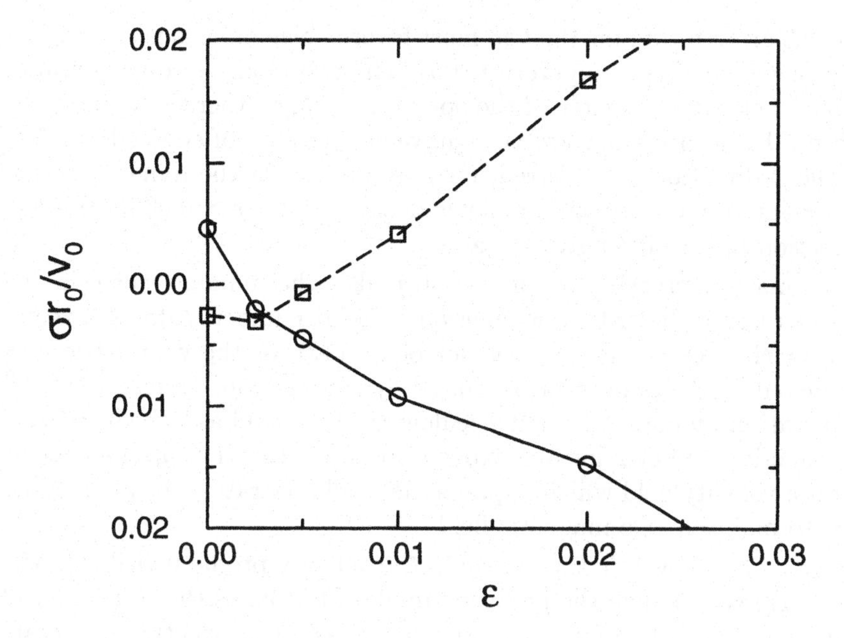


*Figure 2.* The growth rate $\sigma$ as a function of $\epsilon$ for eigenmodes with a magnetic field at the center of the computational volume pointing in the $x$- (circles) or $y$-direction (squares). The magnetic Reynolds number is approximately 0.2 % above critical.

cylinder is along $z$. The liquid in between the cylinder and the spherical boundary is at rest, whereas the motion of the conductor within the cylinder is prescribed. The velocity field inside the cylinder is constructed from the periodic field

$$\boldsymbol{u} = \boldsymbol{u}_0 + \frac{1}{\sqrt{2}}\epsilon \sin(2kz)\tilde{\boldsymbol{u}} \tag{13}$$

with $a = R \cdot 0.21/0.95$ and $k = \pi/R$. An $8 \times 8$ array of $\boldsymbol{u}$ is selected whose corners are removed so that 52 eddies remain in the cylinder. The velocity outside this region is set to zero. This velocity field is used in the following with different values of $\epsilon$. The factor $\sin 2kz$ is motivated by the observation that the fastest growing magnetic mode near the axis of the cylinder has a $z$-dependence in $(\cos kz, \sin kz, 0)$ with $k = \pi/R$, so that the factor in front of the last term in (7) varies in $\sin 2kz$. The important point to note is that the volumetric flow rate (12) does not depend on $\epsilon$. The critical magnetic Reynolds number does depend on $\epsilon$, however. This is demonstrated in figure

2 where the growth rate for the two most favored eigenmodes is shown as a function of $\epsilon$ at fixed and supercritical $v_0$. The two modes in question have a magnetic field at the center of the sphere pointing either in the $x$ or the $y$ direction. Both directions are not equivalent because of edge effects. For $\epsilon = 0$, the $x$-direction is preferred, for $\epsilon$ around 0.003 the dynamo action has disappeared and all magnetic fields decay, but for $\epsilon > 0.006$ the mode in $y$-direction has positive growth rates.

In summary, increasing $\epsilon$ from $\epsilon = 0$ makes the dynamo worse in the sense of increasing the critical magnetic Reynolds number. Adding a velocity field which does not modify the surrogate (12) for the volumetric flow rate delivered by the pumps has inhibited dynamo action. To restore it and obtain a stationary magnetic field, the flow rate (i.e. $v_0$) has to be increased. This is exactly the behavior one expects from a saturation mechanism: It generates a velocity field which requires the total flow rate to be augmented in order to maintain the magnetic field.

Figure 2 also shows that a change in the velocity profile invisible in the flow rate (12) can change the preferred mode. This opens the possibility of obtaining with the help of an exterior applied field two exactly degenerate modes. Transitions might occur between these two modes through nonlinear interactions.

The numerical results are in qualitative agreement with the analytical model leading to (11). However, the analytical approach relies on the existence of several small parameters which are not small in the actual experiment (such as the magnetic Reynolds number on the scale of the cell size) and expresses the results in terms of power series in these small parameters. Such a development can lead to qualitative discrepancies with direct computations, even if the series are pushed to higher orders than was done in order to obtain (11). As a cautionary tale, we report a numerical solution of a problem treated by Rädler et al. (1998). In the exact same geometry as used in the above kinematic calculation, the velocity field (13) was replaced by the profile specified in equation (10) of Rädler et al. (1998). Using the same notation as these authors, the kinematic growth rate shown in figure 3 has been determined for $V_1 = 0.0619m^3/s$ at different $V_2$. The critical $V_2$ is found to be $0.021m^3/s$. From figure 5 of Rädler et al. (1998) it follows that the growth rate $\sigma$ first increases as in figure 3 and then decreases with increasing $V_2$ such that it becomes negative for $V_2$ greater than approximately $0.04m^3/s$. This result is in stark contrast to the linear increase of $\sigma$ exhibited in figure 3 and shows that the truncated expansions employed in the mean field theory can lead to erroneous conclusions when applied to the parameter range pertaining to the experiment.

Better approximations of the real experiment are necessary in order to make quantitative predictions useful for the interpretation of the nonlinear

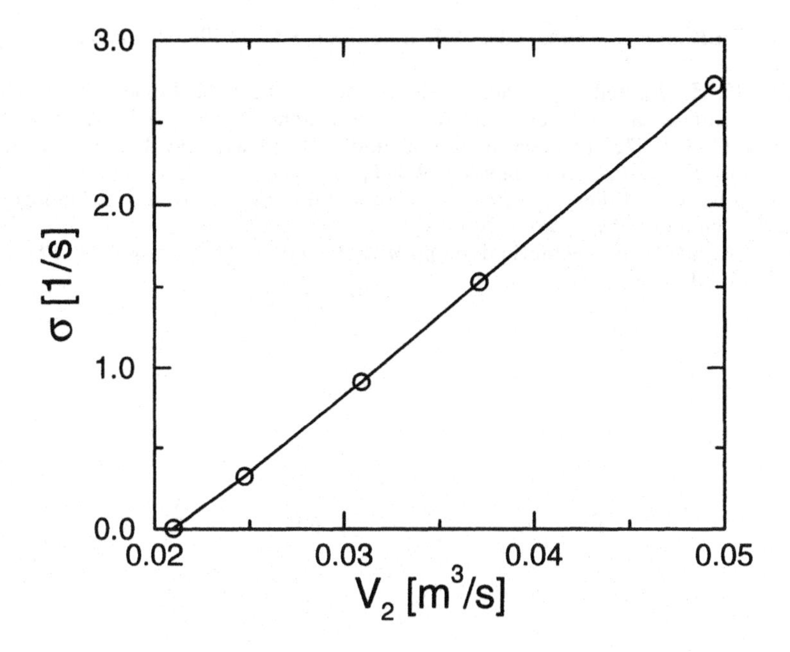


*Figure 3.* The growth rate $\sigma$ as a function of $V_2$.

properties of the laboratory dynamo. The idealized model presented at the beginning of this paper demonstrates the tendency of the Lorentz force to drive a flow which traverses the pipes present in the experiment. This is due to the phase shift between the term $(\nabla \times \boldsymbol{b}) \times \bar{\boldsymbol{B}}$ in (6) (which varies sinusoidally in space) and the velocity field $\boldsymbol{u}_0$. While a flow as depicted in figure 1 is impossible in the experiment, the spatial distribution of the Lorentz force must be very similar in the experiment and the model. One therefore expects that modifications of the real velocity field occur within each cell, such that the correction is zero at the center of the cells but strong in the corner regions. Future simulations and hopefully direct experimental measurements are expected to corroborate this picture.

## References

1. Busse, F.H. (1978) Magnetohydrodynamics in the Earth's dynamo, *Ann. Rev. Fluid Mech.*, **10**, 435–462.
2. Busse, F.H., Müller, U., Stieglitz, R. and Tilgner, A. (1998) Spontaneous Generation of Magnetic Fields in the Laboratory, In *Evolution of Spontaneous Structures in Dis-*

*sipative Continuous Systems*, eds. F.H. Busse and S.C. Müller, Springer Verlag, pp. 546–558.
3. Rädler, K.-H., Apstein, E., Rheinhardt, M. and Schüler, M. (1998) The Karlsruhe Dynamo Experiment. A Mean Field Approach, *Studia Geoph. Geod.*, **42**, 224–231.
4. Roberts, G.O. (1972) Dynamo action of fluid motions with twodimensional periodicities, *Phil. Trans. Roy. Soc. London*, **A271**, 411–454.
5. Stieglitz, R. and Müller, U. (2000) Experimental demonstration of a homogeneous two-scale dynamo, *Phys. Fluids*, in press.
6. Tilgner, A. (1998) A kinematic dynamo with a small scale velocity field, *Phys. Lett. A*, **226**, 75–79.

# SHEARED HELICAL TURBULENCE AND THE HELICITY CONSTRAINT IN LARGE-SCALE DYNAMOS

ALBERTO BIGAZZI
*Department of Mathematics, Politecnico di Milano, Piazza Leonardo da Vinci 32, I-20133 Milano, Italy*
AXEL BRANDENBURG
*Nordita, Blegdamsvej 17, DK-2100 Copenhagen Ø, Denmark*
*Mathematics Department, Univ. of Newcastle, NE1 7RU, UK*
AND
KANDASWAMY SUBRAMANIAN
*National Centre for Radio Astrophysics - TIFR, Poona University Campus, Ganeshkhind, Pune 411 007, India*

**Abstract.** The effect of shear on the growth of large scale magnetic fields in helical turbulence is investigated. The resulting large-scale magnetic field is also helical and continues to evolve, after saturation of the small scale field, on a slow resistive time scale. This is a consequence of magnetic helicity conservation. Because of shear, the time scale needed to reach an equipartition-strength large scale field is shortened proportionally to the ratio of the resulting toroidal to poloidal large scale fields.

## 1. Introduction

Magnetic helicity is conserved in ideal MHD. In non-ideal situations, when magnetic diffusivity is non-vanishing, it can only evolve on a long time scale governed by microscopic magnetic diffusivity. This is true in periodic or unbounded systems or in systems with perfectly conducting boundaries, where no flux of magnetic helicity through the boundaries is allowed. In systems with open boundaries, magnetic helicity can leak out and the evolution in time can thus be different.

The importance of magnetic helicity conservation for the evolution of large-scale magnetic fields in astrophysics has been recently discussed (Black-



*P. Chossat et al. (eds.), Dynamo and Dynamics, a Mathematical Challenge,* 117–124.

man & Field 2000, Brandenburg 2000, Brandenburg & Subramanian 2000, Kleeorin *et al.* 2000). Large-scale, helical magnetic fields where the outer scale of turbulent motions is much smaller than the scale of these fields, are observed in stars and galaxies. For the Sun significant amounts of magnetic helicity are observed at the solar surface (Berger & Ruzmaikin, 2000)

Boundary conditions are important in determining the overall dynamics of the large-scale field (Blackman & Field, 2000). In the case of a periodic homogeneous isotropic medium with no externally imposed magnetic field, recent numerical studies (Brandenburg 2000, hereafter referred to as B2000) show that, because of magnetic helicity conservation, the large scale magnetic field can only grow to its final (super-) equipartition field strength on a resistive time scale, which is usually many orders of magnitude longer than the dynamical time-scale determined by the turbulent eddy turnover time.

Besides allowing for a flux of magnetic helicity through the boundaries by imposing different boundary conditions, another way to allow for faster growth of the field is by means of shear which can amplify an existing field without changing its magnetic helicity. A regenerative mechanism for the cross-stream (poloidal) component of the field is also needed, because otherwise the sheared (toroidal) field would eventually decay (e.g. Moffatt 1978, Krause & Rädler 1980). Indeed a number of working dynamos which have both open boundaries and shear have been proposed (e.g., Glatzmaier & Roberts 1995, Brandenburg *et al.* 1995), but those models are rather complex and use sub-grid scale modelling, thus making it difficult to evaluate the role of magnetic helicity conservation.

Here we study the effect that shear alone can have on the dynamics of the large scale field, while keeping the system periodic. We find that the evolution of the large scale field is compatible with a mean-field model where the geometrical mean of the large-scale poloidal and toroidal fields evolves on a resistive time-scale. It is thus possible to have a larger toroidal field at the expense of the poloidal one without violating the helicity constraint. Equivalently, equipartition strength large scale fields can be attained in times shorter by the ratio of the resulting toroidal to poloidal field strength.

## 2. Equations and setup

The same set of MHD equation for an isothermal compressible gas as in B2000 is considered. The external forcing function $\boldsymbol{f}$ incorporates both the helical driving at intermediate scale $k = 5$ and the shear at $k = 1$.

$$\frac{\mathrm{D}\ln\rho}{\mathrm{D}t} = -\boldsymbol{\nabla}\cdot\boldsymbol{u}, \tag{1}$$

$$\frac{\mathrm{D}\boldsymbol{u}}{\mathrm{D}t} = -c_{\mathrm{s}}^{2}\boldsymbol{\nabla}\ln\rho + \frac{\boldsymbol{J}\times\boldsymbol{B}}{\rho} + \frac{\mu}{\rho}(\nabla^{2}\boldsymbol{u} + \tfrac{1}{3}\boldsymbol{\nabla}\boldsymbol{\nabla}\cdot\boldsymbol{u}) + \boldsymbol{f}, \tag{2}$$

$$\frac{\partial\boldsymbol{A}}{\partial t} = \boldsymbol{u}\times\boldsymbol{B} - \eta\mu_{0}\boldsymbol{J}, \tag{3}$$

where $\mathrm{D}/\mathrm{D}t = \partial/\partial t + \boldsymbol{u}\cdot\boldsymbol{\nabla}$ is the advective derivative, $\boldsymbol{u}$ is the velocity, $\rho$ is the density, $\boldsymbol{B} = \boldsymbol{\nabla}\times\boldsymbol{A}$ is the magnetic field, $\boldsymbol{A}$ is its vector potential, and $\boldsymbol{J} = \boldsymbol{\nabla}\times\boldsymbol{B}/\mu_0$ is the current density. The forcing function $\boldsymbol{f}$ takes the form $\boldsymbol{f} = \boldsymbol{f}_{\mathrm{turb}} + \boldsymbol{f}_{\mathrm{shear}}$, where

$$\boldsymbol{f}_{\mathrm{shear}} = C_{\mathrm{shear}}\frac{\mu}{\rho}\,\hat{\boldsymbol{y}}\sin x \tag{4}$$

balances the viscous stress once a sinusoidal shear flow has been established, and

$$\boldsymbol{f}_{\mathrm{turb}} = N\mathrm{Re}\{\boldsymbol{f}_{\boldsymbol{k}(t)}\exp[i\boldsymbol{k}(t)\cdot\boldsymbol{x} + i\phi(t)]\}, \tag{5}$$

is the small scale helical forcing with

$$\boldsymbol{f}_{\boldsymbol{k}} = \frac{\boldsymbol{k}\times(\boldsymbol{k}\times\boldsymbol{e}) - i|\boldsymbol{k}|(\boldsymbol{k}\times\boldsymbol{e})}{2\boldsymbol{k}^{2}\sqrt{1-(\boldsymbol{k}\cdot\boldsymbol{e})^{2}/\boldsymbol{k}^{2}}}, \tag{6}$$

where $\boldsymbol{e}$ is an arbitrary unit vector needed in order to generate a vector $\boldsymbol{k}\times\boldsymbol{e}$ that is perpendicular to $\boldsymbol{k}$, $\phi(t)$ is a random phase, and $N = f_0 c_{\mathrm{s}}(kc_{\mathrm{s}}/\delta t)^{1/2}$, where $f_0$ is a non-dimensional factor, $k = |\boldsymbol{k}|$, and $\delta t$ is the length of the time step. As in B2000 we choose the forcing wavenumbers such that $4.5 < |\boldsymbol{k}| < 5.5$. At each time step one of the 350 possible vectors is randomly chosen.

The equations are made non-dimensional with the choice $c_{\mathrm{s}} = k_1 = \rho_0 = \mu_0 = 1$, where $c_{\mathrm{s}}$ is the sound speed, $k_1$ is the smallest wavenumber in the box (so its size is $2\pi$), $\rho_0$ is the mean density (which is conserved), and $\mu_0$ is the vacuum permeability. The computational mesh is $120^3$ grid-points. Sixth order finite differences are used for spatial derivatives.

We consider the case when shear is strong compared to turbulence, but still subsonic. We choose for the shear parameter $C_{\mathrm{shear}} = 1$ and for the amplitude of the random forcing $f_0 = 0.01$. The resulting rms velocities in the meridional ($xz$) plane are around 0.015 and the toroidal rms velocities around 0.6.

The magnetic Prandtl number is ten for the simulations considered here, i.e. $\mu/(\rho_0\eta) = 10$, and $\eta = 5\times10^{-4}$. If calculated with respect to the box size ($=2\pi$), the Reynolds numbers for poloidal and toroidal velocities are $R_{\mathrm{m}}^{\mathrm{pol}} = 190$ and $R_{\mathrm{m}}^{\mathrm{tor}} = 7500$, respectively. By poloidal and toroidal components we mean those in the $xz$-plane and the $y$-direction, respectively. Based, instead, on the forcing scale, the poloidal magnetic Reynolds number

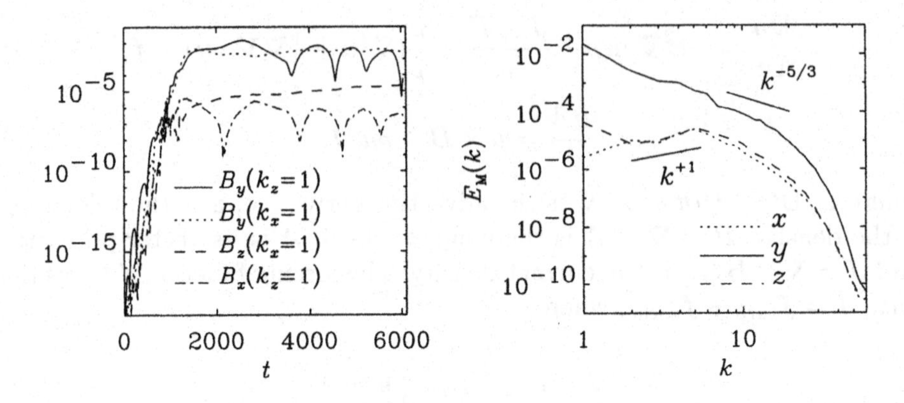


*Figure 1.* Evolution of the power, $|\hat{B}_i(k_j)|^2$, of a few selected Fourier modes (left panel). After $t = 1700$, most of the power is in the mode $|\hat{B}_y(k_z)|^2$, i.e. in the toroidal field component with variation in the $z$-direction. The three-dimensional power spectrum of the three field components is shown on the right. $120^3$ mesh-points, $t = 5000$.

is only about 40, and the kinetic Reynolds number is only 4, which is not enough to allow for a proper inertial range. The turnover time based on the forcing scale and the poloidal rms velocity is $\tau = 70$.

## 3. Time evolution of the field

A strong dynamo amplifies an initially weak random seed magnetic field exponentially on a dynamical time-scale up to equipartition. In Figure 1 we plot the evolution of the power, $|\hat{B}_i(k_j)|^2$, in a few selected modes. After $t = 1700$, most of the power is in the mode $|\hat{B}_y(k_z)|^2$, i.e. the toroidal field component with variation in the $z$-direction. The ratio of toroidal to poloidal field energies are around $10^4$, so $B_{\rm tor}/B_{\rm pol} \approx 100$.

A three dimensional power spectrum of the field components in Figure 1 shows the different behaviour of the poloidal and toroidal components. The dominating toroidal field has a $k^{-5/3}$-like spectrum from the largest scale to the dissipative cut-off. Poloidal fields, instead, are noisy and possess significant power near $k = 5$. The poloidal field saturates earlier than the toroidal, which is by then already dominated by large scales.

Longitudinal cross-sections show that the small scale contributions to the poloidal field result from variations in the toroidal direction. Whilst the toroidal field is relatively coherent in the toroidal direction, the poloidal field components are much less coherent and show significant fluctuations in the $y$-direction. We thus define mean fields $\overline{\boldsymbol{B}}$ to be the y-averaged fields. As can be seen from Figure 2 this is compatible with the definition of the mean

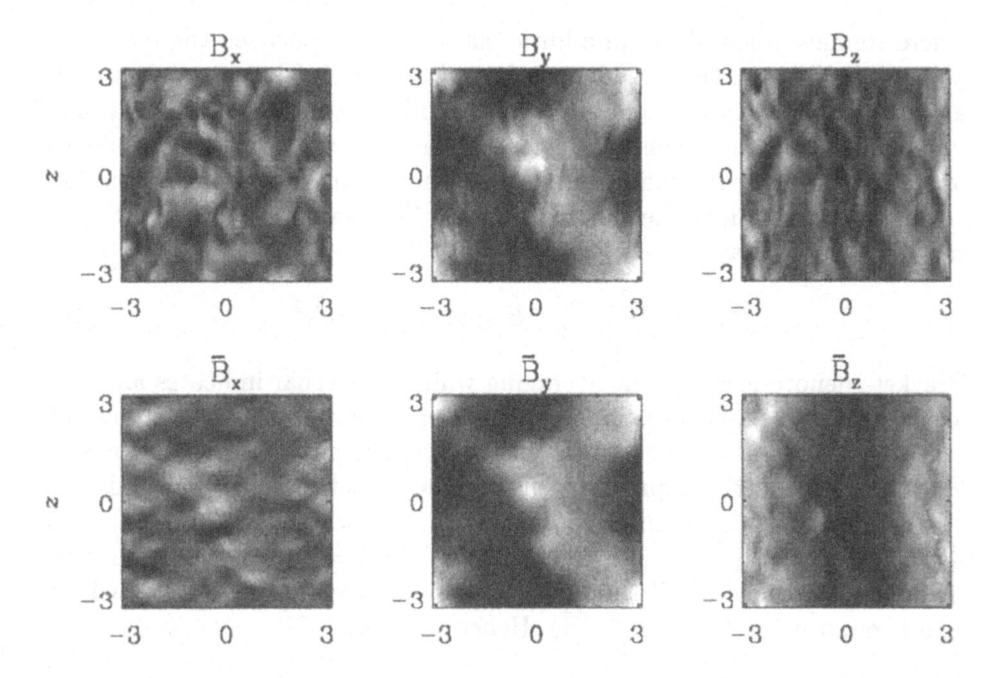


*Figure 2.* Images of the three components of $\boldsymbol{B}$ in an arbitrarily chosen $xz$ plane (first row), compared with the $y$-averaged fields (second row). $120^3$ mesh-points, $t = 5000$.

field as the large-scale Fourier expansion. Two-dimensional power-spectra of this averaged field show that poloidal mean field components also gain significant power at the largest scale (i.e. at $\boldsymbol{k}^2 < 2$) at later times.

## 4. Helicity constraint and the mean magnetic field

In an unbounded or periodic system the magnetic helicity, $\langle \boldsymbol{A} \cdot \boldsymbol{B} \rangle$, evolves according to

$$\frac{\mathrm{d}}{\mathrm{d}t}\langle \boldsymbol{A} \cdot \boldsymbol{B} \rangle = -2\eta \langle \boldsymbol{J} \cdot \boldsymbol{B} \rangle. \tag{7}$$

Taking into account the spectral properties of the above quantities, we may separate large-scale and small-scale contributions and write

$$\langle \overline{\boldsymbol{J}} \cdot \overline{\boldsymbol{B}} \rangle / k_1 \approx \mp B_{\mathrm{tor}} \, B_{\mathrm{pol}} \approx k_1 \langle \overline{\boldsymbol{A}} \cdot \overline{\boldsymbol{B}} \rangle. \tag{8}$$

The expression above would for instance be true for a field of the form

$$\overline{\boldsymbol{B}} = \begin{pmatrix} B_{\mathrm{pol}} \cos(k_1 z + \varphi_x) \\ B_{\mathrm{tor}} \sin(k_1 z + \varphi_y) \\ 0 \end{pmatrix}, \tag{9}$$

where we have allowed for an additional phase shift between the two components (relative to the already existing $\pi/2$ phase shift), $\varphi_y - \varphi_x$, but such a phase shift turned out to be small in our case. Furthermore, an additional $x$-dependence of the mean field, which is natural due to the $x$-dependence of the imposed shear profile could be accounted for. However, for the following argument all we need is relation (8). The amplitudes $B_{\rm pol}$ and $B_{\rm tor}$ can be calculated as

$$B_{\rm tor} \equiv \langle \overline{\boldsymbol{B}}_y^2 \rangle^{1/2}, \quad B_{\rm pol} \equiv \langle \overline{\boldsymbol{B}}_x^2 + \overline{\boldsymbol{B}}_z^2 \rangle^{1/2}. \tag{10}$$

Brackets denote here volume averaging while an overbar indicates an average over the toroidal direction ($y$). The upper sign applies to the present case where the kinetic helicity is positive (representative of the southern hemisphere), and the approximation becomes exact if Eq. (9) is valid.

Following B2000, in the steady case $\langle \boldsymbol{A} \cdot \boldsymbol{B} \rangle = \text{const}$, see Eq. (7), and so the r.h.s. of Eq. (7) must vanish, i.e. $\langle \boldsymbol{J} \cdot \boldsymbol{B} \rangle = 0$, which can only be consistent with Eq. (8) if there is a small scale component, $\langle \boldsymbol{j} \cdot \boldsymbol{b} \rangle$, whose sign is opposite to that of $\langle \overline{\boldsymbol{J}} \cdot \overline{\boldsymbol{B}} \rangle$. Hence we write

$$\langle \boldsymbol{J} \cdot \boldsymbol{B} \rangle = \langle \overline{\boldsymbol{J}} \cdot \overline{\boldsymbol{B}} \rangle + \langle \boldsymbol{j} \cdot \boldsymbol{b} \rangle \approx 0. \tag{11}$$

This yields, analogously to B2000,

$$-\frac{\mathrm{d}}{\mathrm{d}t} \left( B_{\rm tor} \, B_{\rm pol} \right) = +2\eta k_1^2 \left( B_{\rm tor} \, B_{\rm pol} \right) - 2\eta k_1 |\langle \boldsymbol{j} \cdot \boldsymbol{b} \rangle|, \tag{12}$$

which yields the solution

$$B_{\rm tor} \, B_{\rm pol} = \epsilon_0 B_{\rm eq}^2 \left[ 1 - e^{-2\eta k_1^2 (t - t_{\rm s})} \right], \tag{13}$$

where $\epsilon_0 = |\langle \boldsymbol{j} \cdot \boldsymbol{b} \rangle| / (k_1 B_{\rm eq}^2)$ is a prefactor, $B_{\rm eq}$ is the equipartition field strength with $B_{\rm eq}^2 = \mu_0 \langle \rho \boldsymbol{u}^2 \rangle$, and $t_{\rm s}$ is the time when the small scale field has saturated which is when Eq. (12) becomes applicable. All this is equivalent to B2000, except that $\langle \overline{\boldsymbol{B}}^2 \rangle$ is now replaced by the product $B_{\rm tor} \, B_{\rm pol}$. The significance of this expression is that large toroidal fields are now possible if the poloidal field is weak.

In Figure 3 we show the evolution of the product $B_{\rm tor} \, B_{\rm pol}$ as defined in (10) and compare with Eq. (13). There are different stages; for $1200 < t < 2200$ and $3000 < t < 3700$ the effective value of $k_1^2$ is 2 (because there are contributions from $k_x = 1$ and $k_z = 1$; see Figure 1), whilst at other times ($2500 < t < 2800$ and $t > 4000$) the contribution from $k_x = 1$ (for $2500 < t < 2800$) or $k_z = 1$ (for $t > 4000$) has become subdominant and we have effectively $k_1^2 = 1$. This is consistent with the change of field structure

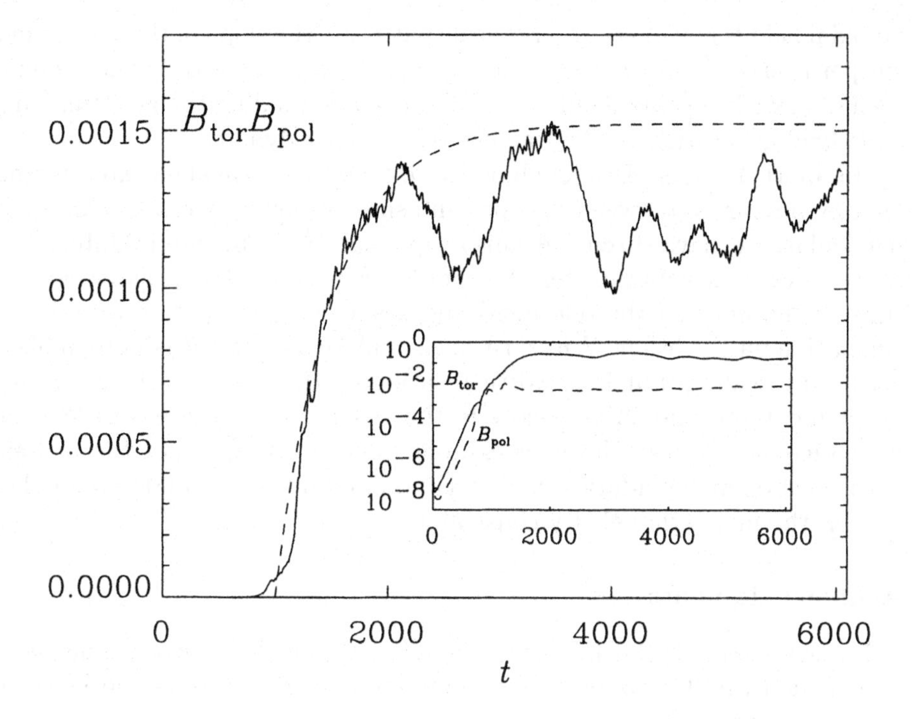


*Figure 3.* Growth of the product of poloidal and toroidal magnetic fields on a linear scale. The inset shows separately the evolution of poloidal and toroidal fields on a logarithmic scale. For the fit we have used $k_1^2 = 2$ and $\epsilon_0 = 3.8$.

discussed in the previous section: for $2000 < t < 3000$ and around $t = 4000$ the $B_y(k_x = 1)$ mode is less powerful than the $B_y(k_z = 1)$ mode.

## 5. Conclusions

The effects of the helicity constraint can clearly be identified in in our system even though much of the field amplification results from the shearing of a poloidal field. The constraint on the geometrical mean of the energies in the poloidal and toroidal field components is evident from Figure 3. The fit shows that the prefactor $\epsilon_0$ is about 3.8. Theoretically one may estimate $\epsilon_0$, which is proportional to $|\langle \boldsymbol{j} \cdot \boldsymbol{b} \rangle|$, by estimating $|\langle \boldsymbol{j} \cdot \boldsymbol{b} \rangle| \approx \rho_0 |\langle \boldsymbol{\omega} \cdot \boldsymbol{u} \rangle| \approx k_{\rm f} \rho_0 \langle \boldsymbol{u}^2 \rangle \approx k_{\rm f} B_{\rm eq}^2$. Since $\epsilon_0 = |\langle \boldsymbol{j} \cdot \boldsymbol{b} \rangle| / (k_1 B_{\rm eq}^2)$ this yields $\epsilon_0 \approx k_{\rm f}|/k_1 = 5$, in good agreement with the simulation.

Power spectra of the poloidal field show that most of the power is in the small scales, making the use of averages at first glance questionable. However, once the field is averaged over the toroidal direction the resulting poloidal field is governed by large scale patterns (the slope of the spectrum is steeper than $k^{-1}$, which is the critical slope for equipartition). The presence even of a weak poloidal field is crucial for understanding the resulting large scale field generation in the framework of an $\alpha\Omega$ dynamo.

In another paper (Brandenburg *et al.* 2000) we have elaborated further on the similarity between the present simulations and $\alpha\Omega$ dynamos. In particular, we have discussed anisotropic turbulent magnetic diffusivities as a possible explanation for the difference between the resistive growth time of the field on the one hand and a shorter cycle period seen in the simulation on the other. With just one simulations so far it is impossible to verify any scaling, but it is worth mentioning that the present cycle time of around 1000 time units is close to the geometrical mean of resistive and dynamical timescales. Nevertheless, one must not forget that the real sun does have open boundaries, and it is now important to understand their role on the magnetic helicity constraint.

## Acknowledgments

ABi and KS thank Nordita for hospitality during the course of this work. Use of the PPARC supported supercomputers in St Andrews and Leicester is acknowledged.

## References

1. Berger, M. A., & Ruzmaikin, A. (2000) Rate of helicity production by solar rotation. *J. Geophys. Res.* **105**, 10481–10490
2. Blackman, E. G., & Field, G. F. (2000) Constraints on the Magnitude of alpha in Dynamo Theory. *Astrophys. J.* **534**, 984–988
3. Brandenburg, A. (2000) The inverse cascade and nonlinear alpha-effect in simulations of isotropic helical hydromagnetic turbulence, *Astrophys. J.* (submitted) (astro-ph/0006186)
4. Brandenburg, A. & Subramanian, K. (2000) Large scale dynamos with ambipolar diffusion nonlinearity. *Astron. Astrophys.* **361**, L33–L36
5. Brandenburg, A., Bigazzi, A., & Subramanian, K., (2000) The helicity constraint in turbulent dynamos with shear, *Mon. Not. Roy. Astron. Soc.* (submitted), (astro-ph/0011081)
6. Brandenburg, A., Nordlund, Å., Stein, R. F., & Torkelsson, U. (1995) Dynamo-generated Turbulence and Large-Scale Magnetic Fields in a Keplerian Shear Flow. *Astrophys. J.* **446**, 741–754
7. Glatzmaier, G. A., & Roberts, P. H. (1995) A three-dimensional self-consistent computer simulation of a geomagnetic field reversal. *Nature* **377**, 203–209
8. Kleeorin, N. I, Moss, D., Rogachevskii, I., & Sokoloff, D. (2000) Helicity balance and steady-state strength of the dynamo generated galactic magnetic field. *Astron. Astrophys.* **361**, L5–L8
9. Krause, F., & Rädler, K.-H. (1980) *Mean-Field Magnetohydrodynamics and Dynamo Theory.* Pergamon Press, Oxford
10. Moffatt, H. K. (1978) *Magnetic Field Generation in Electrically Conducting Fluids.* CUP, Cambridge

# THE INVERSE CASCADE IN TURBULENT DYNAMOS

AXEL BRANDENBURG
*Nordita, Blegdamsvej 17, DK-2100 Copenhagen Ø, Denmark*
*Mathematics Department, Univ. of Newcastle, NE1 7RU, UK*

**Abstract.** The emergence of a large scale magnetic field from randomly forced isotropic strongly helical flows is discussed in terms of the inverse cascade of magnetic helicity and the $\alpha$-effect. In simulations of such flows the maximum field strength exceeds the equipartition field strength for large scale separation. However, helicity conservation controls the speed at which this final state is reached. In the presence of open boundaries magnetic helicity fluxes out of the domain are possible. This reduces the timescales of the field growth, but it also tends to reduce the maximum attainable field strength.

## 1. Introduction

It was since the mid-seventies when Frisch *et al.* (1975) and Pouquet *et al.* (1976) came up with the idea that the large scale magnetic fields seen in many astrophysical bodies could be caused by an inverse cascade-type phenomenon. Although there were close links with earlier results that helicity and lack of mirror symmetry are important (Steenbeck *et al.* 1966; see also Krause & Rädler 1980), the notion of an inverse cascade has put dynamo theory into a self-consistent framework of nonlinear turbulence theory.

Unfortunately the inverse cascade concept was not easily assimilated by the astrophysical community. The reasons are simple: the inverse cascade approach was developed in the framework of isotropic homogeneous turbulence, and was not readily applicable to astrophysical bodies that were stratified and enclosed in boundaries. Thus, people continued to use $\alpha^2$ and $\alpha\Omega$-dynamos (Moffatt 1978), which enabled modelling of a large variety of astrophysical bodies.

Here we want is to look more closely at inverse cascade dynamos. In particular, we want to know what kind of field they produce and how this



*P. Chossat et al. (eds.), Dynamo and Dynamics, a Mathematical Challenge,* 125–132.

relates to the fields generated by an $\alpha^2$-dynamo. The full results of this work are presented in a separate paper (Brandenburg 2000, hereafter referred to as B2000). In the present paper we also discuss the effects of open boundaries allowing magnetic helicity fluxes out of the domain into the exterior or across the equator. This is an important issue that has been raised recently in the context of dynamo theory (Blackman & Field 2000, Kleeorin *et al.* 2000).

In order to model isotropic random flows we adopt a forcing function that consists of randomly oriented Beltrami fields with a wavenumber, $k_{\rm f}$, that is larger than the smallest wavenumber, $k_1$, that fits into the box. For most of the calculations we use $k_{\rm f} = 5$ and $k_1 = 1$, leaving some margin for scale separation. In order to see more clearly the effects of scale separation we also have one run where $k_{\rm f} = 30$. In the wavenumber band $4.5 < |\boldsymbol{k}| < 5.5$ (for $k_{\rm f} = 5$) there are 350 wavevectors which are chosen randomly at each timestep, so the forcing is $\delta$-correlated in time, but the resulting velocity field is not. In fact, the velocity has a well defined correlation time that agrees well with the turnover time $\tau = \ell_{\rm f}/u_{\rm rms}$, where $\ell_{\rm f} = 2\pi/k_{\rm f}$ is the forcing scale and $u_{\rm rms}$ is the rms velocity.

The degree of turbulence that develops depends on the range of length scales left between the forcing scale and the dissipative cutoff scale. A reasonable range can only be obtained if the forcing wavenumber is not too high, so the run with $k_{\rm f} = 30$ (Run 6 in B2000) is an example where the flow is laminar on the forcing scale. The degree of mixing, as measured by the ratio of the turbulent to the microscopic diffusion coefficients for a passive scalar, $D_{\rm t}/D$, is here of order one. In a more turbulent run, Run 3 of B2000, this ratio is around 40. However, quite independently of how turbulent a run is, we find the emergence of a large scale magnetic field after some time. The resulting field resembles closely that obtained from an $\alpha^2$-dynamo with the same (periodic) boundary conditions. This analogy enables us to make contact with mean-field theory and to explain the resulting turbulent transport coefficients.

## 2. Emergence of a large scale field

The inverse cascade is traditionally described in terms of energy spectra. In Figure 1 we compare the spectral field evolution for two different forcing wavenumbers. Note that the envelope of the magnetic energy fits underneath a $k^{-1}$ slope. The peaks at $k = k_1$ and $k_{\rm f}$ also fit underneath the same slope. There are several features of the spectrum that are characteristic also of other cases investigated. For large enough scale separation one sees that the magnetic energy grows fast at two distinct wavenumbers, $k \approx 30$ and $k \approx 7$. However, when the energy at $k \sim 7$ reaches saturation the energy begins to be transferred to larger scales until much of the magnetic energy is at the largest scale possible. During this phase the magnetic energy at intermediate scales decreases to some minimum value which follows roughly a $k^{+3/2}$ spectrum. This effect may be referred to as 'self-cleaning', because by removing energy at intermediate scales the field at the largest scales appears less perturbed and hence cleaner. This self-cleaning effect is the result of nonlinearity, which suppresses the growth at intermediate scales. However, the type of nonlinearity does not seem to matter: even with ambipolar diffusion the same behaviour is found (Brandenburg & Subramanian 2000).

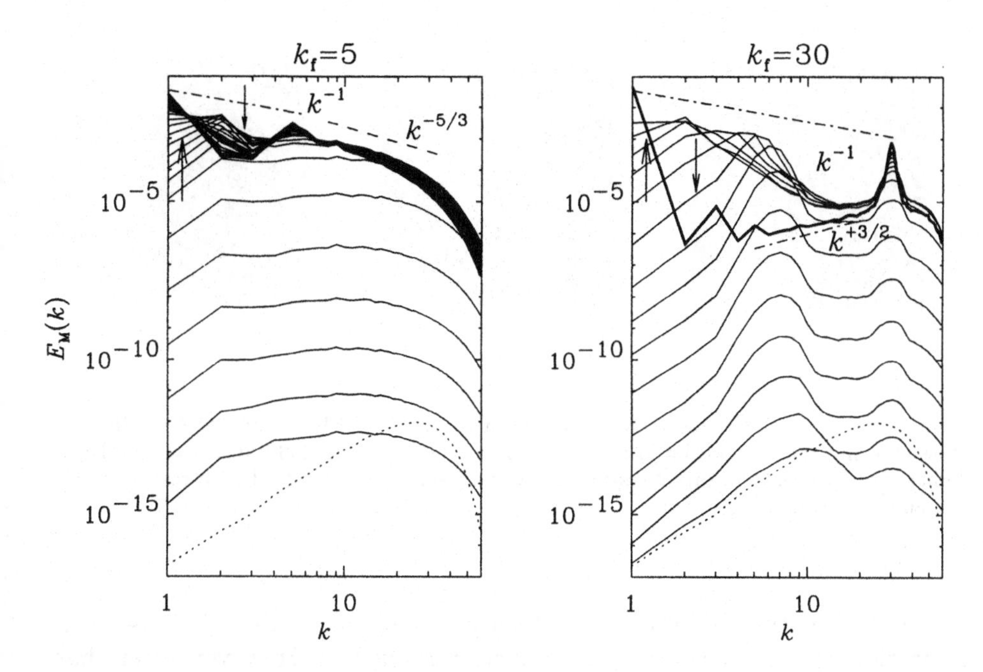


*Figure 1.* Magnetic energy spectra for runs with small and large scale separation, $k_f = 5$ and 30, respectively.

In Figure 2 we show cross-sections of $B_x$ at different times. Towards the end of the evolution the large scale magnetic field is essentially a Beltrami field which is here of the form $\overline{\boldsymbol{B}} \sim (\sin y, 0, \cos y)$, apart from some phase shift in $y$.

## 3. The final equilibrium field strength

Characteristic to all the runs reported in B2000 is the fact that super-equipartition field strengths are reached. In Figure 3 the evolution of magnetic and kinetic energies are shown for two runs with different forcing wavenumbers. Note also the extremely slow evolution past the moment

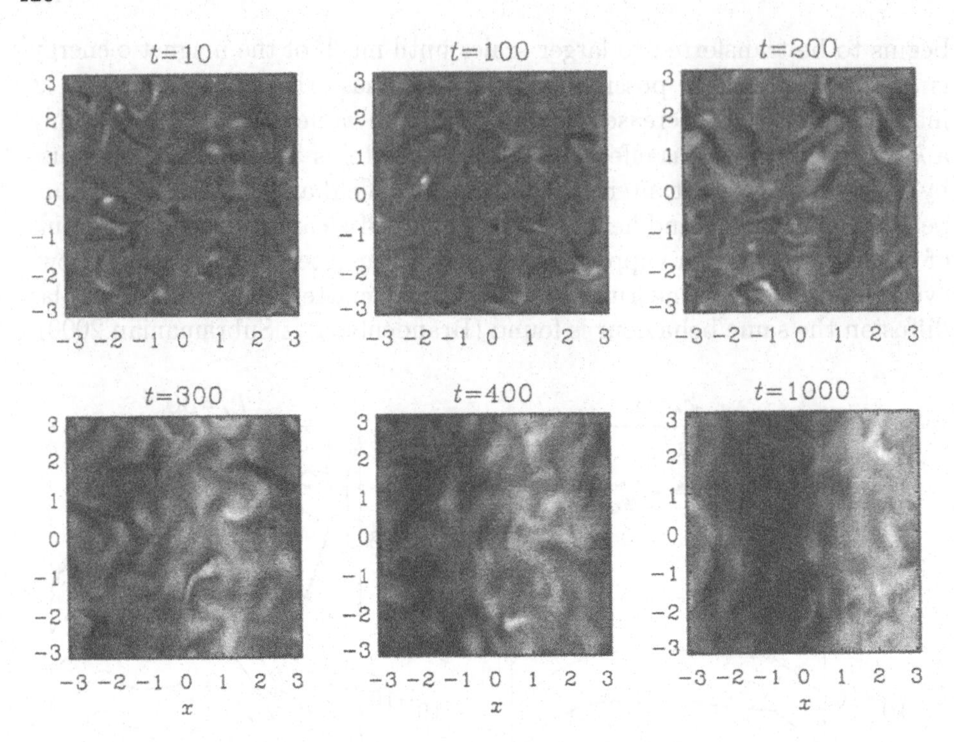


*Figure 2.* Gray-scale images of cross-sections of $B_x(x, y, 0)$ for Run 3 of B2000 at different times showing the gradual build-up of the large scale magnetic field after $t = 300$. Dark (light) corresponds to negative (positive) values. Each image is scaled with respect to its min and max values.

where the kinetic energy drops suddenly to a smaller value. This is when saturation of small scale magnetic energy is reached. However, after that moment the large scale magnetic energy continues to grow for some time, because the resulting large scale field is force-free and does hence not affect the velocity field directly.

The prolonged saturation behaviour found in the present simulations is at first glance unusual. Since at late times most of the magnetic energy is in the large scales, this slow evolution must have to do with the properties of the large scale field. An important property of this large scale field is that it possesses magnetic helicity. At the same time magnetic helicity is conserved by the nonlinear terms and can hence only change resistively on an ohmic timescale. (The case with open boundaries is different and will be discussed separately.) In order to demonstrate magnetic helicity conservation we consider a periodic box and write the magnetic field, $\boldsymbol{B}$, as the curl of a vector potential, $\boldsymbol{A}$, so $\boldsymbol{B} = \boldsymbol{\nabla} \cdot \boldsymbol{A}$. We use the uncurled

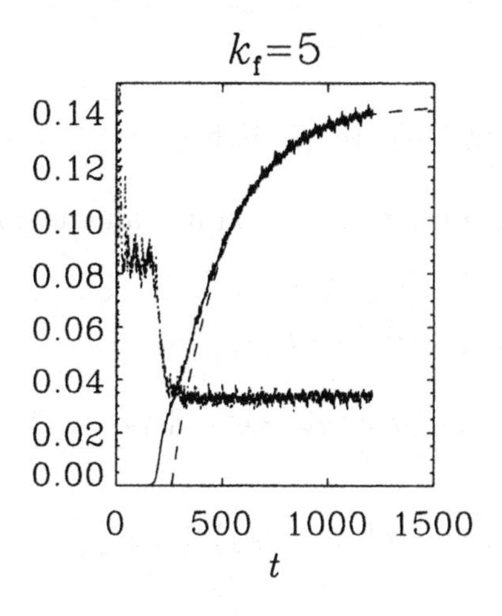


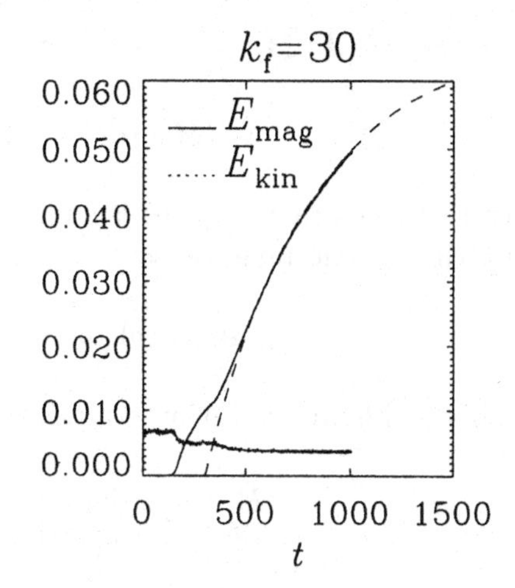


*Figure 3.* Evolution of kinetic (dotted) and magnetic (solid) energies. The slow evolution of the magnetic energy follows approximately the $1 - e^{-2\eta\Delta t}$ behaviour (dashed line) of Eq. (6) that results from helicity conservation.

induction equation,

$$\partial \boldsymbol{A}/\partial t = \boldsymbol{u} \times \boldsymbol{B} - \eta \boldsymbol{J} - \boldsymbol{\nabla}\phi, \tag{1}$$

where $\boldsymbol{J} = \boldsymbol{\nabla} \times \boldsymbol{B}$ is the current density, $\eta$ is the magnetic diffusivity, and $\phi$ is the electrostatic potential. The magnetic helicity, which can be defined as $\langle \boldsymbol{A} \cdot \boldsymbol{B} \rangle$, satisfies

$$\mathrm{d}\langle \boldsymbol{A} \cdot \boldsymbol{B} \rangle/\mathrm{d}t = -2\eta \langle \boldsymbol{J} \cdot \boldsymbol{B} \rangle, \tag{2}$$

where angular brackets denote volume averages over the periodic domain. The steady state solution must satisfy $\langle \boldsymbol{J} \cdot \boldsymbol{B} \rangle = 0$, so small and large scale current helicities must be comparable in magnitude, $|\langle \boldsymbol{j} \cdot \boldsymbol{b} \rangle| \approx |\langle \overline{\boldsymbol{J}} \cdot \overline{\boldsymbol{B}} \rangle|$, but of opposite sign. The magnetic helicity is however concentrated on large scales (its spectrum is $k^2$ times that of the current helicity), so its small scale contribution is negligible.

We measure the degree of magnetic helicity of the large scale field by the quantity $k_{AB}^{-1} = |\langle \overline{\boldsymbol{A}} \cdot \overline{\boldsymbol{B}} \rangle| / \langle \overline{\boldsymbol{B}}^2 \rangle$, which is a length scale characterizing the scale of the helical contribution. In a periodic domain of size $L$ the smallest wavenumber is $k_1 = 2\pi/L$, and so $k_{AB}$ is bounded from above by $k_{AB} \leq k_1$. The large scale current helicity is $k_1^2$ times the magnetic helicity, so we have

$$\langle \overline{\boldsymbol{A}} \cdot \overline{\boldsymbol{B}} \rangle = \mp \langle \overline{\boldsymbol{B}}^2 \rangle / k_{AB} = \langle \overline{\boldsymbol{J}} \cdot \overline{\boldsymbol{B}} \rangle / k_1^2, \tag{3}$$

where the upper sign applies to the case of positive kinetic helicity in the turbulence. Using this in Eq. (2) together with $\langle \boldsymbol{A} \cdot \boldsymbol{B} \rangle \approx \langle \overline{\boldsymbol{A}} \cdot \overline{\boldsymbol{B}} \rangle$ and

$\langle \boldsymbol{J} \cdot \boldsymbol{B} \rangle = \langle \overline{\boldsymbol{J}} \cdot \overline{\boldsymbol{B}} \rangle + \langle \boldsymbol{j} \cdot \boldsymbol{b} \rangle$ yields

$$\mp \mathrm{d}\langle \overline{\boldsymbol{B}}^2 \rangle / \mathrm{d}t = \pm 2\eta k_1^2 \langle \overline{\boldsymbol{B}}^2 \rangle - 2\eta k_{AB} \langle \boldsymbol{j} \cdot \boldsymbol{b} \rangle. \tag{4}$$

Prior to saturation $\langle \boldsymbol{j} \cdot \boldsymbol{b} \rangle$ is small, but during saturation its value is limited by the kinetic helicity, so

$$\langle \boldsymbol{j} \cdot \boldsymbol{b} \rangle \approx \langle \rho \rangle \langle \boldsymbol{\omega} \cdot \boldsymbol{u} \rangle \approx \pm k_{\mathrm{f}} \langle \rho \boldsymbol{u}^2 \rangle = \pm k_{\mathrm{f}} B_{\mathrm{eq}}^2 / \mu_0. \tag{5}$$

This yields an evolution equation for the mean magnetic field,

$$\frac{\langle \overline{\boldsymbol{B}}^2 \rangle}{B_{\mathrm{eq}}^2} \approx \frac{k_{\mathrm{f}} k_{AB}}{k_1^2} \left\{ 1 - \exp\left[ -2\eta k_1^2 (t - t_{\mathrm{sat}}) \right] \right\}, \tag{6}$$

which is only valid at late times when $\Delta t \equiv t - t_{\mathrm{sat}} > 0$. (Here, $t_{\mathrm{sat}}$ is the time when the small scale field saturates.) We emphasize that this relation is rather general and independent of the actual model of field amplification, because we used only the concept of magnetic helicity conservation. The important point here is that full saturation is only obtained after an ohmic diffusion time. In that sense Eq. (6) poses a constraint on the mean magnetic field at late times. It applies only as long as the magnetic field is helical. Indeed, lower degrees of helicity, i.e. smaller values of $k_{AB}^{-1}$, allow larger values of the final field strength; see Eq. (6). This has been verified in a model where differential rotation or shear contributed significantly to the field amplification (Brandenburg *et al.* 2000). However, this only relaxes the constraint by a certain factor which depends on the degree of helicity of the large scale field which, in turn, depends on the degree of linkage of poloidal and toroidal field. We found that to a good approximation this factor is given by the ratio $Q$ of toroidal to poloidal field strengths. In the sun this factor is less than a hundred, so this is a relatively minor effect compared with the value of the magnetic Reynolds number $(10^8 - 10^{10})$.

## 4. Helicity exchange across the equator

A different way of relaxing the slow growth problem is to allow for fluxes out of the dynamo volume either into the exterior of the dynamo (the corona or halo) or from one hemisphere to the other. In any case, there would be an extra surface term in Eq. (2). Here we want to discuss the latter alternative of a helicity flux between the two hemispheres. Such a flux could result from a turbulent exchange of magnetic helicity between the two hemispheres and should therefore be proportional to some turbulent magnetic diffusivity $\eta_{\mathrm{t}}$. Based on dimensional arguments, one may expect such a term to be of the form $\eta_{\mathrm{t}} k_{\mathrm{eff}}^2 H$, where $H$ is the gauge-invariant magnetic helicity for open volumes (Berger & Field 1984), which replaces $\langle \boldsymbol{A} \cdot \boldsymbol{B} \rangle$. The diffusion of magnetic helicity depends on the length scale $2\pi/k_{\text{eff}}$ over which the magnetic helicity varies (if evaluated over different volumes), so we expect $k_{\text{eff}} \leq k_1$. Equation (4) becomes then

$$\mp \mathrm{d}\langle \overline{\boldsymbol{B}}^2 \rangle / \mathrm{d}t = \pm 2\eta k_1^2 \langle \overline{\boldsymbol{B}}^2 \rangle - 2\eta k_{AB}^{-1} \langle \boldsymbol{j} \cdot \boldsymbol{b} \rangle \pm 2\eta_{\mathrm{t}} k_{\text{eff}}^2 \langle \overline{\boldsymbol{B}}^2 \rangle, \tag{7}$$

so the solution for the mean magnetic field is (assuming again $\langle \boldsymbol{j} \cdot \boldsymbol{b} \rangle \approx \pm k_{\mathrm{f}} B_{\text{eq}}^2 / \mu_0$) given by

$$\frac{\langle \overline{\boldsymbol{B}}^2 \rangle}{B_{\text{eq}}^2} \approx \frac{\eta k_{\mathrm{f}} k_{AB}}{\eta k_1^2 + \eta_{\mathrm{t}} k_{\text{eff}}^2} \left\{ 1 - \exp\left[ -2 \left( \eta k_1^2 + \eta_{\mathrm{t}} k_{\text{eff}}^2 \right) (t - t_{\text{sat}}) \right] \right\}. \tag{8}$$

Thus, the time dependence is no longer resistively dominated, because the microscopic diffusivity is now supplemented by an additional turbulent diffusivity. Unfortunately, however, the amplitude of the final field decreases in such a way that the initial linear growth in unchanged. This is simply because of the fact that the flux term, as modelled here, does not act as an effective driver, which is what the $\langle \boldsymbol{j} \cdot \boldsymbol{b} \rangle$-term did. This is also clearly seen in a simulation where we have included an equator by modulating the forcing function such that the kinetic helicity varied sinusoidally n the $z$-direction within the domain; see Figure 4.

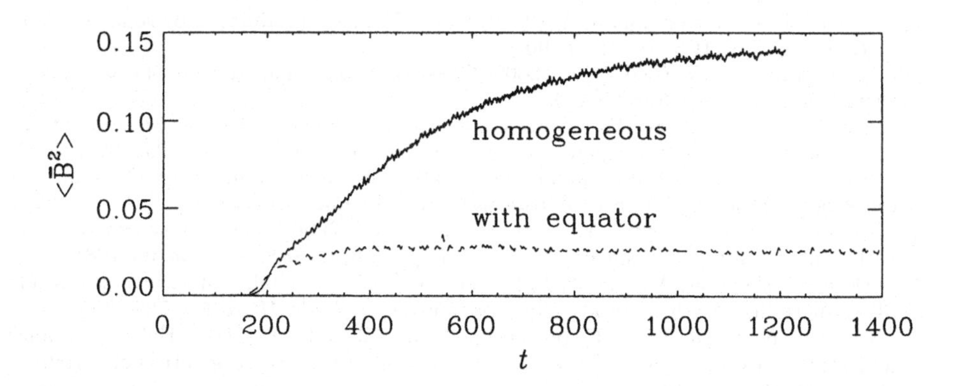


*Figure 4.* Evolution of the magnetic energy for a run with homogeneous forcing function (solid line) and a forcing function whose helicity varies sinusoidally throughout the domain (dotted line) simulating the effects of an equators.

## 5. Conclusions

The present work has shown that for helical velocity fields a large scale magnetic field is generated. There are strong parallels with the fields resulting from mean-field $\alpha^2$-dynamos (see B2000 for details). One aspect that has now begun to receive major attention is related to magnetic helicity conservation, which prevents rapid growth of magnetic helicity and hence helical large scale magnetic fields. Shear, which corresponds to differential rotation in a rotating system, relaxes this constraint only partially in that it lowers the fraction of the field that contributes to magnetic helicity. On the other hand, by allowing magnetic flux to escape through the boundaries, or allowing for mixing of magnetic helicity of opposite sign at the equator, the helicity constraint is modified such that the time scale is no longer resistively dominated. The problem however is that various attempts to model this effect result in significantly lower equilibrium amplitudes of the magnetic field. Part of the problem is that the loss of magnetic helicity implies at the same time a loss of magnetic energy. It would therefore be advantageous for the dynamo to lose preferentially small scale magnetic helicity and energy. This is something that the dynamo in the computer keep refusing to do. It may therefore be important to resort to more realistic simulations where the flows are driven naturally and not by some artificial stirring in space.

## References


1. Berger, M., & Field, G. B. (1984) The topological properties of magnetic helicity. *J. Fluid Mech.* **147**, 133–148
2. Berger, M. A., & Ruzmaikin, A. (2000) Rate of helicity production by solar rotation. *J. Geophys. Res.* **105**, 10481–10490
3. Blackman, E. G., & Field, G. F. (2000) Constraints on the magnitude of $\alpha$ in dynamo theory. *Astrophys. J.* **534**, 984–988
4. Brandenburg, A. (2000) The inverse cascade and nonlinear alpha-effect in simulations of isotropic helical hydromagnetic turbulence, *Astrophys. J.*, astro-ph/0006186 (B2000)
5. Brandenburg, A., & Subramanian, K. (2000) Large scale dynamos with ambipolar diffusion nonlinearity. *Astron. Astrophys.* **361**, L33–L36
6. Brandenburg, A., Bigazzi, A., & Subramanian, K. (2000) The helicity constraint in turbulent dynamos with shear, *Mon. Not. Roy. Astron. Soc.*, astro-ph/0011081
7. Frisch, U., Pouquet, A., Léorat, J., Mazure, A. (1975) Possibility of an inverse cascade of magnetic helicity in hydrodynamic turbulence. *J. Fluid Mech.* **68**, 769–778
8. Kleeorin, N. I, Moss, D., Rogachevskii, I., & Sokoloff, D. (2000) Helicity balance and steady-state strength of the dynamo generated galactic magnetic field. *Astron. Astrophys.* **361**, L5–L8
9. Krause, F., & Rädler, K.-H. (1980) *Mean-Field Magnetohydrodynamics and Dynamo Theory.* Pergamon Press, Oxford
10. Moffatt, H. K. (1978) *Magnetic Field Generation in Electrically Conducting Fluids.* CUP, Cambridge
11. Pouquet, A., Frisch, U., & Léorat, J. (1976) Strong MHD helical turbulence and the nonlinear dynamo effect. *J. Fluid Mech.* **77**, 321–354
12. Steenbeck, M., Krause, F., & Rädler, K.-H. (1966) Berechnung der mittleren Lorentz-Feldstärke $\overline{\boldsymbol{v} \times \boldsymbol{B}}$ für ein elektrisch leitendendes Medium in turbulenter, durch Coriolis-Kräfte beeinflußter Bewegung. *Z. Naturforsch.* **21a**, 369–376 See also the translation in Roberts & Stix (1971) The turbulent dynamo, Tech. Note 60, NCAR, Boulder, Colorado

# ROTATING MAGNETOCONVECTION IN DEPENDENCE ON STRATIFICATION, DIFFUSIVE PROCESSES AND BOUNDARY CONDITIONS.

J. BRESTENSKÝ, S. ŠEVČÍK AND J. ŠIMKANIN
*Dept of Geophysics, Faculty of Maths, Physics & Informatics, Comenius University, 842 48 Bratislava, Slovakia*

**Abstract.** Instabilities of MAC-waves type influenced by three diffusive processes in planar rapidly rotating stratified fluid layer permeated by the azimuthal magnetic field (1b) were investigated in dependence on Elsasser, Ekman, Roberts numbers and stratification parameter (7) for various mechanical and electrically conductive boundaries (8, 9). Among corresponding MC-waves the westward ones significantly determined by viscosity were revealed.
*Key words:* linear magnetoconvection, MAC waves, non-uniform stratification, boundary conditions.

## 1. Introduction

Despite the fact of succesful numerical experiments modelling the Geodynamo (see e.g. [7] or [8]), in frame of full theory described by non-linear partial differential equations in spherical geometry with complex boundary conditions, the simpler models, e.g. of rotating magnetoconvection – inherent ingredient of dynamo theory, are permanently important. Even studies of linear magnetoconvection in planar geometry can be heuristic due to many complexities, e.g. more coexisting diffusive processes in rotating thermally and electrically conductive fluid. Magnetoconvection studies are characterised by hunting for modes of convection, in a presence of magnetic field, with determining their properties.

Great amount of modes of our study are instabilities corresponding to Soward's model (1979) characterised by the almost magnetostrophic balance with the simplest boundary conditions (stress free perfect electric conductors). Developing modes propagate azimutally eastward or westward



*P. Chossat et al. (eds.), Dynamo and Dynamics, a Mathematical Challenge,* 133–144.

[10]. Their sensitivity to the mechanical and electromagnetic boundary conditions in the uniformly [3] and non-uniformly [12] stratified horizontal layer was investigated for either inviscid [4] or viscous fluids [11] of various Roberts numbers (7d), $q$ (= $\tau_\eta \,/\, \tau_\kappa$, ratio of magnetic and thermal diffusion time).

The cases of small $q \ll 1$ are characterised with periods (corresponding to $\sigma$ in (2)) of developing modes from $\approx 10^5$ to $10^8$ yr which are much greater than the characteristic Earth's core time $\tau_\eta \sim 10^4$ yr. The cases $q \gtrsim 1$ (e.g. due to diffusivities induced by turbulency) are characterised with periods of modes, which are of MAC waves type, more typical for geomagnetic secular variations. MAC waves are christianed by magnetic, Archimedean and Coriolis forces determining their dynamics [2].

Either modes of case $q \ll 1$ or of MAC waves type, strongly depend on Roberts number, $q$, and boundary conditions. On the contrary, MC waves (with no buoyancy), do not depend on $q$ but are significantly dependent on boundaries, because they do exist only for finitely electrically conducting boundaries [9, 4].

In this contribution our intention is to investigate how the non-uniform stratification of the layer influences the properties of modes formerly studied in uniformly stratified layer in large space of parameters [1]. Furthermore, to study whether this more complex stratification, which is more probable in the Earth's core conditions, can induce new modes or suppress some modes typical for the uniform stratification. The sensitivity of modes of MAC waves type to various electrical conductivities of boundaries is investigated, too.

In our triple diffusive system of rotating magnetoconvection we can expect many modes which were not so far revealed. Therefore the study of westwardly developing MC-waves, catalysed by viscosity, is thoroughly performed. Attention is focused also on eastward MC-waves, and in particular on the dependence of their onset value of magnetic field on various electrical conductivities of boundaries.

## 2. Governing equations

The models of uniformly and non-uniformly stratified layer of width $d$ with the following basic state is considered

$$\mathbf{U}_0 = \mathbf{0}\,, \qquad \mathbf{B}_0 = B_M \frac{s}{d}\,\hat{\varphi}\,, \qquad T_0 = T_l - \Delta T\,\frac{z + d/2}{d}\,(1 - \frac{z - d/2}{2\,z_d}) \qquad (1)$$

where $\mathbf{U}_0$ is velocity, $\mathbf{B}_0$ magnetic field and $T_0$ temperature; $z_d$ is the $z$-coordinate of the level dividing the layer into the stable and unstable sublayers (in uniformly stratified layers $z_d \to \pm\infty$). The linear stability study was performed. The governing equations for perturbations of basic state (1) can be derived from Navier-Stokes equation, induction and heat conduction equations for incompressible Boussinesq fluid. The simple basic magnetic field (1b) allows to search the solutions in separable form

$$\Re e\{f(z)J_m(ks)\exp[i\,(m\varphi+\sigma t)]\} \tag{2}$$

where $J_m(ks)$ is Bessel function, $k$ and $m$ are respectively radial and azimuthal wave numbers, and $\sigma$ is frequency of azimuthally travelling perturbations represented by $f(z)$, i.e. 5 functions, thus $w$ and $\omega$ for velocity, $b$ and $j$ for magnetic field, and $\vartheta$ for temperature. After linearisation, non-dimesionalisation and splitting the velocity and magnetic field into their poloidal and toroidal parts the set of governing equations for functions $f(z)$ is following

$$\begin{aligned} 0 &= Dw-\Lambda(2Db-imj)+E(D^2-k^2)\,\omega, & (3)\\ 0 &= D\omega-\Lambda[2\,Dj+im(D^2-k^2)\,b]+k^2R\vartheta-E(D^2-k^2)^2\omega, & (4)\\ i\sigma b &= (D^2-k^2)\,b+imw, \quad i\sigma j=(D^2-k^2)\,j+im\omega, & (5)\\ i\,(\sigma\,/\,q)\,\vartheta &= (1-z\,/\,z_M)\,w+(D^2-k^2)\,\vartheta, \quad \text{where } D=d\,/\,dz\,. & (6) \end{aligned}$$

The dimensionless parameters appearing in these equations are (7)

$$R=\frac{gd\Delta T\alpha}{2\Omega_0\kappa},\quad \Lambda=\frac{B_M^2}{2\Omega_0\rho_0\eta\mu},\quad E=\frac{\nu}{2d^2\Omega_0},\quad q=\frac{\kappa}{\eta},\quad z_M=-\frac{\rho_0\,c_p\kappa\,\Delta T}{d^2H}\,.$$

$R$, $\Lambda$, $E$, $q$ are modified Rayleigh, Elsasser, Ekman, and Roberts numbers, respectively, and $z_M$ is the stratification parameter $(=z_d\,/\,d)$.

Various boundary conditions were used. In all cases the boundaries were perfectly thermally conducting, i.e. $\vartheta=0$ for $z=\pm\frac{1}{2}$. The mechanical ones were for only impermeable (in the case of inviscid fluid), stress free, and rigid boundaries, i.e. for $\mathcal{I}$, $\mathcal{F}$, $\mathcal{R}$ cases of boundaries, respectively

$$w=0\ (\mathcal{I}),\qquad w=D^2w=D\omega=0\ (\mathcal{F}),\qquad \text{and } w=Dw=\omega=0\ (\mathcal{R}). \tag{8}$$

The electromagnetic boundary conditions for perfectly conducting boundaries are $b=Dj=0$, i.e. for $\xi\ \to\infty$, and for finitely conducting boundaries were used following ones

$$b\pm\tau Db=j\pm\xi\tau Dj=0,\ \text{ for } 0\neq\ \xi<\ \infty \tag{9}$$

with $\tau=(k^2+i\,\sigma\xi)^{-1/2}$, and $\pm$ is for upper/lower boundary. Parameter $\xi$ is the ratio of electrical conductivities of boundary and fluid layer.

Case $\mathcal{R}$ with values $\xi_+=10^{-3}$ and $\xi_-=1$ for upper and lower boundary, respectively, i.e. case $\mathcal{R}\frac{10^{-3}}{1}$, best fit Earth's core boundary conditions.

The method inspired by Chandrasekhar's (1961) book was developed (see e.g. [3], [11] or [12] for more details) for solution of governing equations (3 – 6) with boundary conditions (8, 9) for $w(z)$, $\omega(z)$, $b(z)$, $j(z)$, and $\vartheta(z)$ in the form $f(z) = \sum_l v_l f_l(z)$ where $f_l(z)$ are the proper linear combinations of trigonometric and hyperbolic functions.

The eigenvalue problem $\mathbf{A} \cdot \mathbf{v} = R\mathbf{v}$ with Rayleigh number $R$ as an eigenvalue and the set of coefficients $\{ v_l \}$ as an eigenvector was obtained.

## 3. Numerical results

The computation of the critical Rayleigh number $R_c$, the critical radial wave number $k_c$ and the critical frequency $\sigma_c$ is a main goal of numerical analysis. It was realized for $q = 5., 2., 1.$, $\Lambda \in \langle 0.05, 2500. \rangle$, and $m =$ 1, 2, 5. The values $\infty$, 0.45, 0.30, 0.10, $-0.10$, $-0.30$ were selected for $z_M$. The prime computational model is the case $\mathcal{R}\frac{10^{-3}}{1}$ which is provided for both types of stratification, for all $m$, and $E = 3 \cdot 10^{-7}$ (see Fig. 1, 2, and 3). The case $\mathcal{R}\frac{1}{1}$ was realized only for uniform stratification (Fig. 1). The computations relating to eastward MC waves were realized for wide range of conductivities of the boundaries, for $E = 3 \cdot 10^{-7}$, and for $q = 2$ (Fig. 4). On the other hand, for westward MC waves, the attention was focused on their dependence on $E = \mathcal{O}(10^{-3})$ in the model $\mathcal{R}\frac{10^{-3}}{1}$ (Fig. 5). Some following conclusions are also related to computations in other ranges of parameters and to other boundaries.

Westward modes (W) were found for the whole investigated range of $\Lambda$ and for each value of $q \gtrsim 1$. They exist for uniform as well as for non-uniform stratification (see left columns in the Fig. 1, 2, 3). The critical values $k_c$ and $\sigma_c$ are qualitatively similar for both kind of stratification. The computations of eigenfunctions again confirmed that W modes are suppressed in stably stratified sublayer as well as for $q \ll 1$. The decrease of the value $z_M$ for $z_M > 0$ ($z_M < 0$) has a destabilising (stabilising) influence on W modes (see Fig. 2, 3). On the other hand, the decrease of $z_M$ has stabilising influence on the E modes. Naturally, decrease of $q$ has stabilising influence for all investigated modes (6a, d).

Eastward modes (E) need sufficiently strong magnetic field for both kind of stratification. The lower $m$, the stronger field is necessary for marginal eastward modes (see right columns in Fig. 1, 2, 3).

Instabilities of MAC-waves type are slightly dependent on electromagnetic boundary conditions, i.e. the cases $\mathcal{R}\frac{10^{-3}}{1}$ and $\mathcal{R}\frac{1}{1}$ are very similar except for the eastward mode with $m = 2$.

"Viscous" modes, characterised by very high $k_c \geq \mathcal{O}(10^2)$, were found for small magnetic field. They propagate only westward. A range of $\Lambda$ of

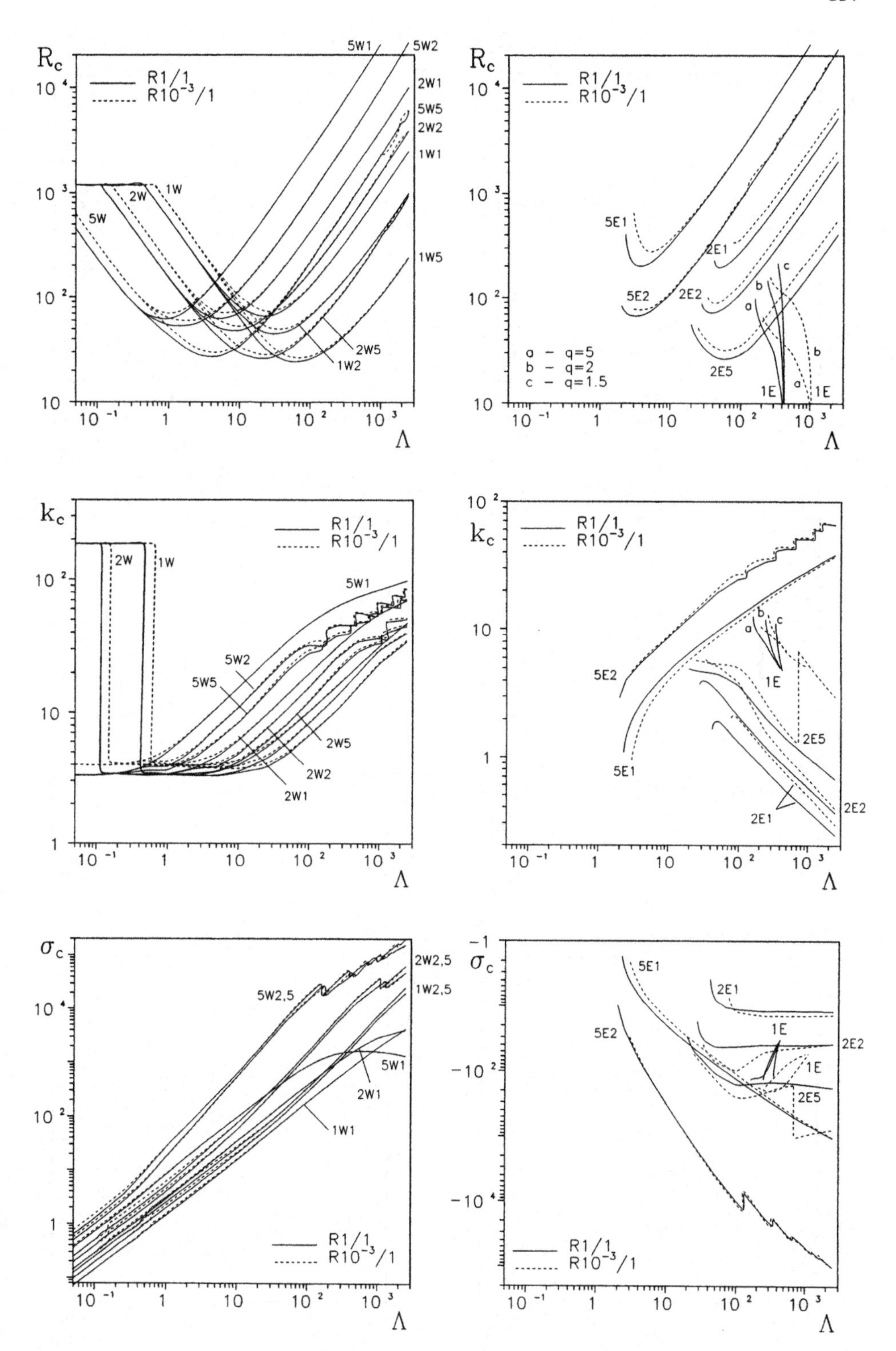


*Figure 1.* Dependence of critical numbers $R_c$, $k_c$, $\sigma_c$ on Elsasser number $\Lambda$, Roberts number $q$ and various electrical conductivities of boundaries for case of uniform stratification (in curves' labels, e.g. mWq means W mode with $m$ at $q$).

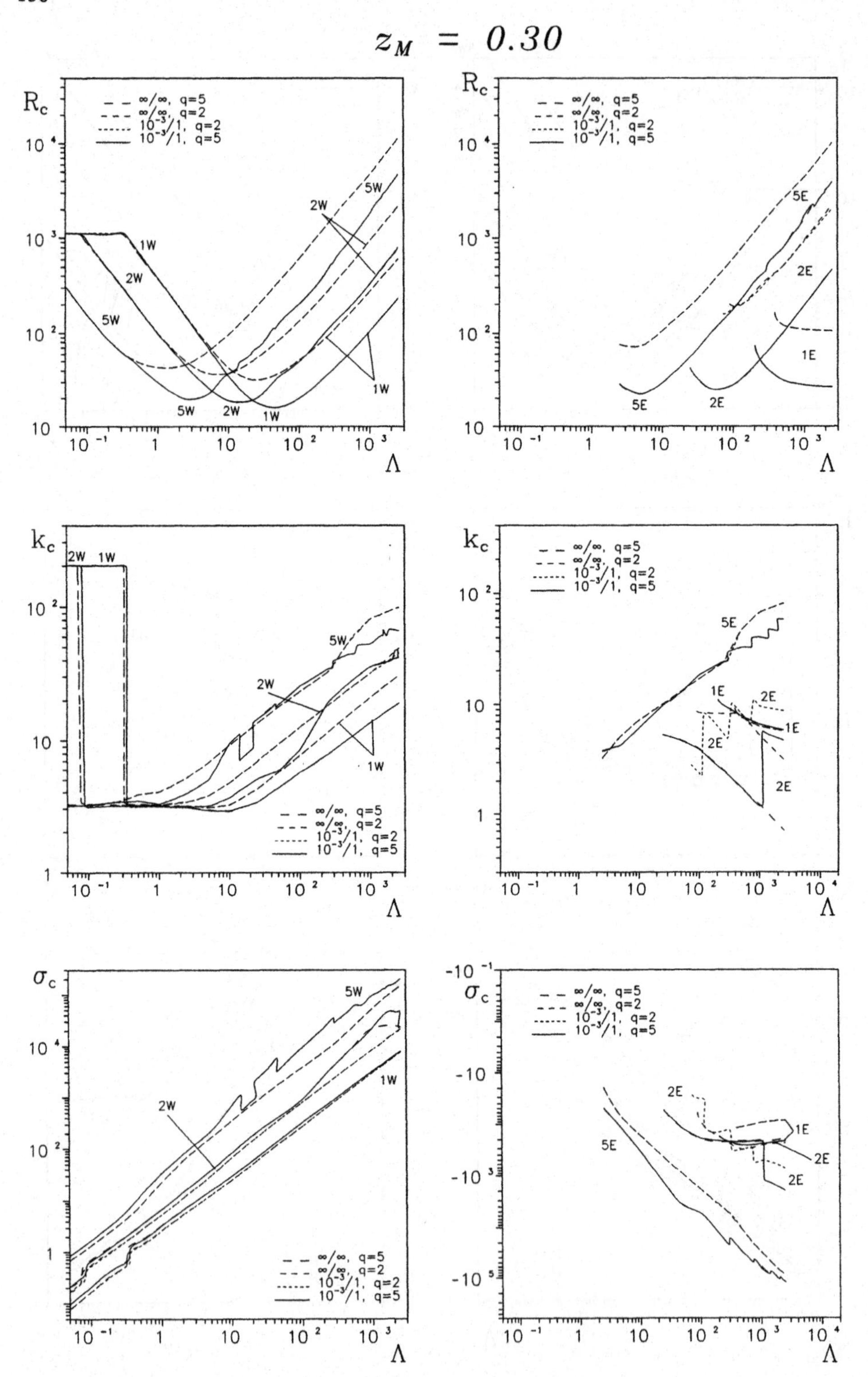


*Figure 2.* Dependence of critical numbers $R_c$, $k_c$, $\sigma_c$ on Elsasser number $\Lambda$, Roberts number $q$ and various electrical conductivities of boundaries for $z_M = 0.30$

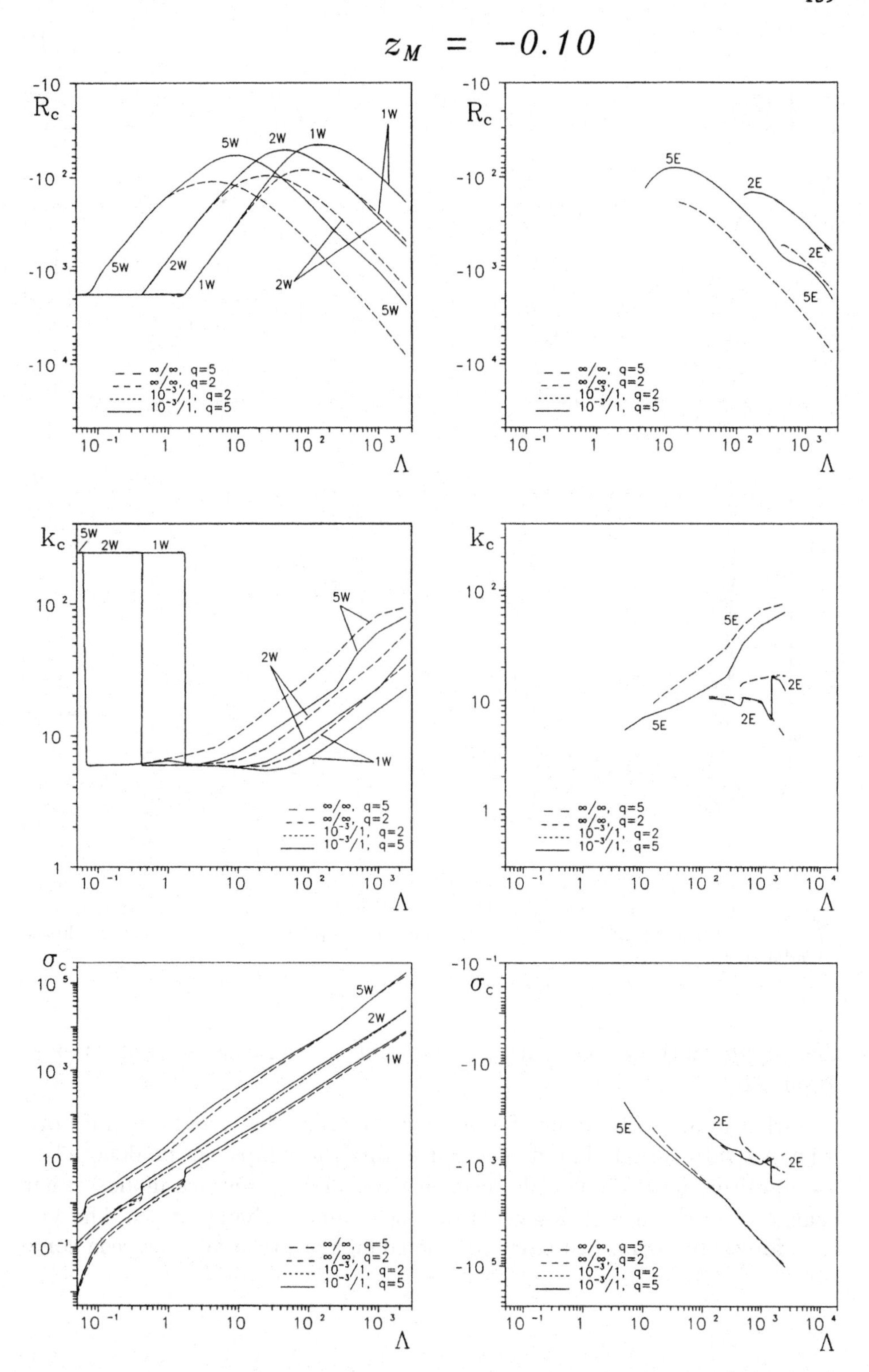


*Figure 3.* Dependence of critical numbers $R_c$, $k_c$, $\sigma_c$ on Elsasser number $\Lambda$, Roberts number $q$ and various electrical conductivities of boundaries for $z_M = -0.10$

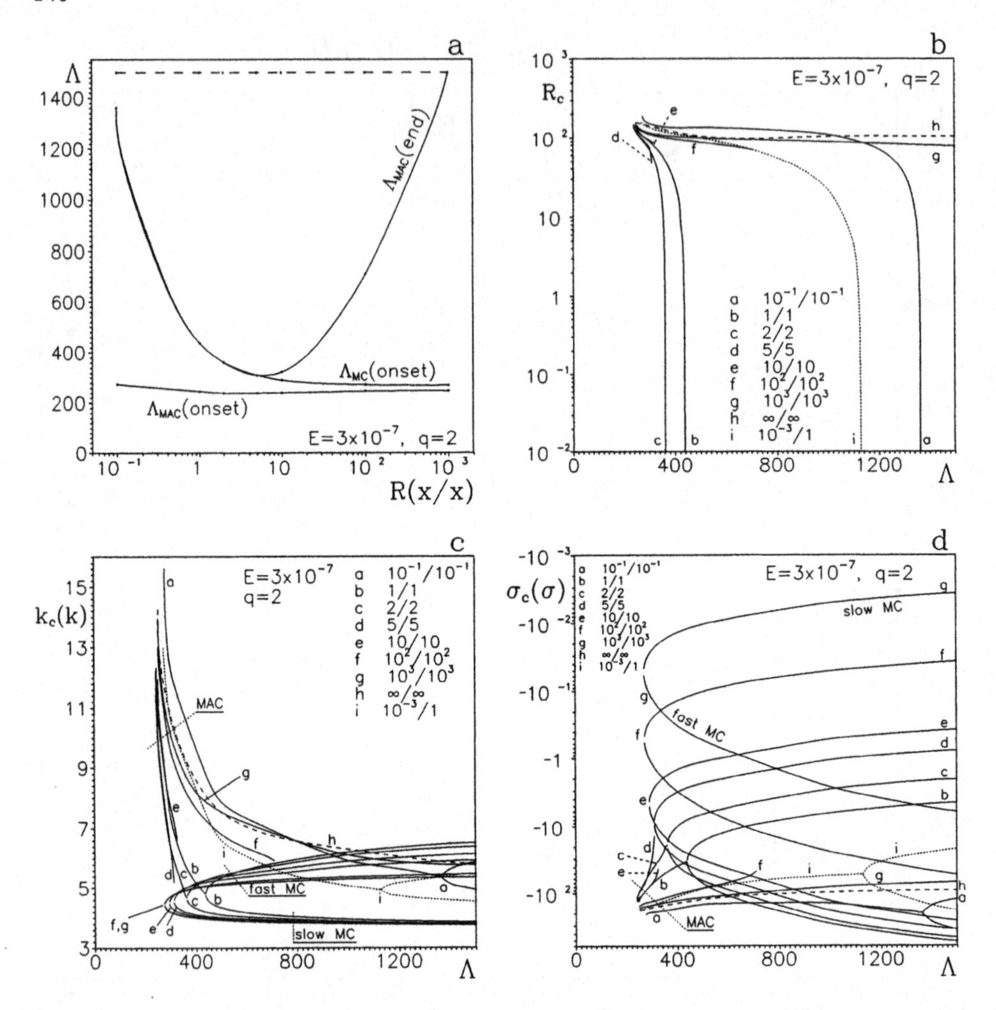


*Figure 4.* **Dependence of eastward MAC and MC waves on conductivity of boundaries. ($R(x/x)$ does mean rigid boundaries with the same conductivity $x$ of upper and lower boundary)**

their appearance depends on $z_M$. The jump into viscous mode does exist for all $z_M$.

Other jumps in $k_c$ and $\sigma_c$ for W as well as E-modes were found for both type of stratification. The modes concerning the jumps can be denoted as "competitive modes" [11]. However, in contrary to "viscous modes" their change of preference corresponds to much smaller change in $k_c$. The viscous forces are again necessary for the change of preference between some

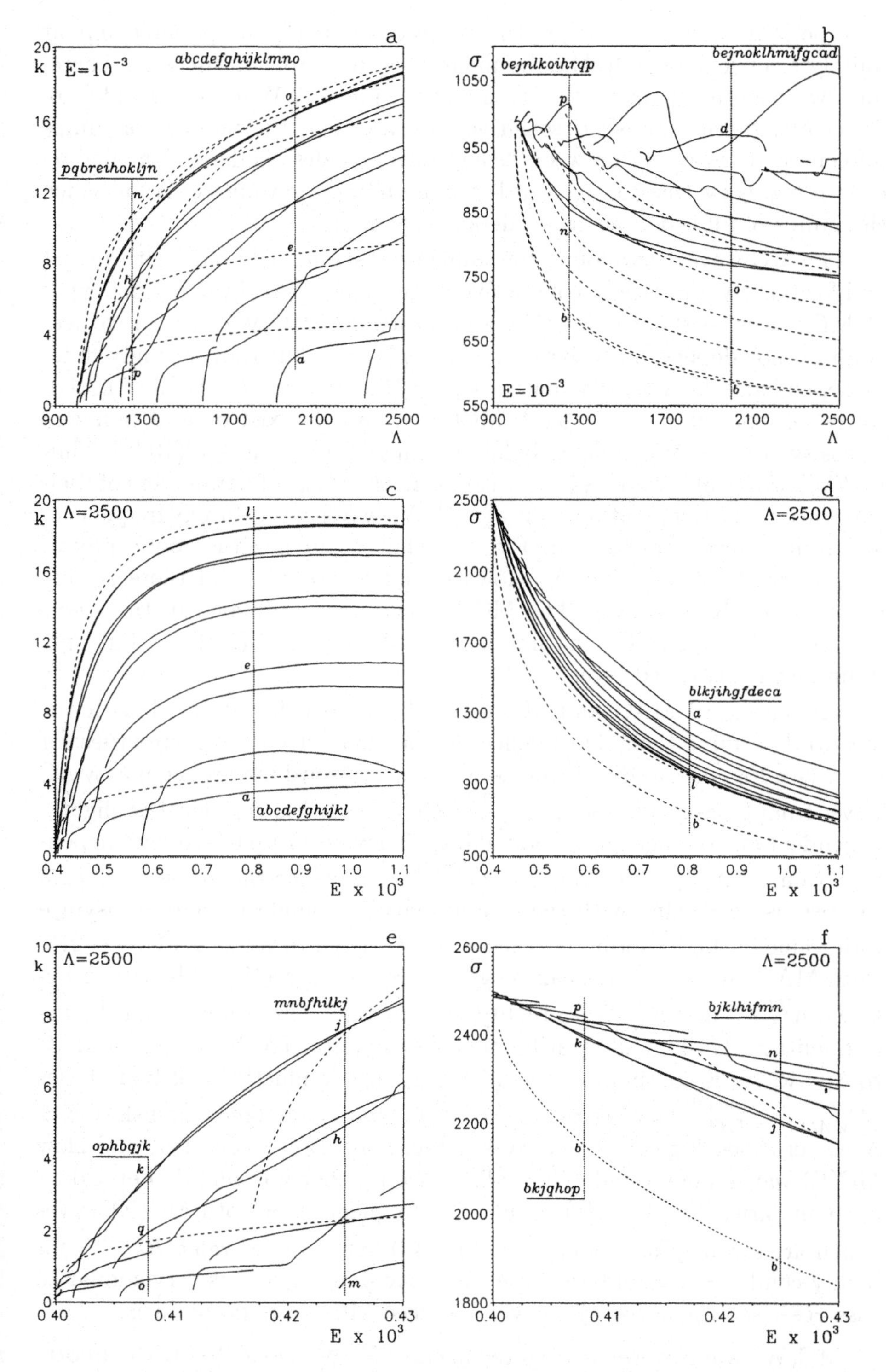


*Figure 5.* Radial wave numbers, $k$, and frequencies, $\sigma$, of MC(W)-waves in dependence on Elsasser, $\Lambda$, and Ekman, $E$, numbers. Each wave is characterised by couple of curves $k(\Lambda)$ and $\sigma(\Lambda)$ or $k(E)$ and $\sigma(E)$ denoted by the same small letters: $a$, $b$, $c$, ...

"competitive modes". However, the jumps at $m = 2(\mathrm{E})$ are possible only at finitely electrically conducting boundaries with no role of viscosity. Therefore, we denoted them as "electromagnetic jumps". While the number of "electromagnetic jumps" is sensitive to $q$ and $z_M$, the number of jumps influenced by viscosity between "competitive modes" is dependent mainly on $z_M$, e.g. they miss for $z_M \leq 0.1$. Furthermore, the values of $\Lambda$ for which the jumps of all types occur are dependent on $z_M$.

The MC waves were found for sufficiently strong field only with $m = 1$ and for finitely electrically conductive boundaries cases. Two distinct types of MC-waves, eastward, MC(E), and westward, MC(W), were revealed. Both do not depend on Roberts number $q$ [4, 11], but they strongly differ in dependence on viscosity, measured by Ekman number. While MC(E)-waves do exist at $E = \mathcal{O}(10^{-7})$, MC(W)-waves do exist only at non-zero viscosity at small but sufficiently large Ekman number, $E \geq \mathcal{O}(10^{-3})$. Thus MC(W)-waves are catalysed by viscosity. In the range of parameters of their existence there is a great amount of MC(W)-waves with various frequencies and radial wave numbers (see Fig. 5). They occur in either uniformly or non-uniformly stratified layer. However, on the other hand there is only a couple of MC(E)-waves. The MC(E)-waves were revealed in the whole investigated range of Ekman number, $\mathcal{O}(10^{-7}) \leq E \leq \mathcal{O}(10^{-3})$, but only in uniformly stratified layer.

The eastward MAC and MC waves have been investigated with the aim to determine their dependence on the conductivity of boundaries for $q = 2$ (see Fig. 4, also Fig. 1). Generally, the eastward modes begin as MAC waves from Elsasser number $\Lambda_{MAC}^{(onset)} = \mathcal{O}(10^2)$ (see Fig. 4a) which is slightly dependent on conductivity of boundaries. They occur up to strongly dependent value of $\Lambda_{MAC}^{(end)}$. The value of $\Lambda_{MC}^{(onset)}$, from which the MC waves begin to arise, is decreasing with the conductivity of boundaries and an asymptotic value for high conductivities can be predicted to value $\Lambda_{MC}^{(onset)}) \sim 270$. Both MAC and MC waves can be generated for magnetic fields with $\Lambda$ between $\Lambda_{MC}^{(onset)}$ and $\Lambda_{MAC}^{(end)}$. In reality, they can not arise simultaneously due to qualitatively different stratification they require. The transition of MAC to MC waves is continuous for sufficiently low conductivity of boundaries ($\Lambda_{MAC}^{(end)} = \Lambda_{MC}^{(onset)}$) and critical $R_c$ of MAC waves strongly diminishes with $\Lambda$ to zero (see Fig. 4b, the curves $a, b, c$, and also $i$). The fast and slow MC(E) waves were found. The radial wave number $k$ is slightly dependent on boundaries (see Fig. 4c), unlike the frequencies of both MC(E)-waves which are strongly dependent (see Fig. 4d). The critical value $R_c^{(MAC)}$ for highly conductive boundaries (see the curve $g$ in Fig. 4b) is very similar to case $R\frac{\infty}{\infty}$ (see the curve $h$) for which indeed MC waves do not arise.

MC(W)-waves were investigated in cases $\mathcal{R}\frac{10^{-3}}{1}$ and $\mathcal{F}\frac{10^{-3}}{1}$ with uniform as well as non-uniform stratification. The results in Fig. 5 are presented for case $\mathcal{R}\frac{10^{-3}}{1}$ of uniformly stratified layer. We can distinguish two groups of MC(W)-waves. While the 1st one is almost independent on mechanical boundary conditions, i.e. stress free or no-slippery ones, the 2nd one is very sensitive to them. In Fig. 5 simpler, i.e. dotted (more complex, i.e. solid) curves correspond to the 1st (2nd) group of MC(W)-waves. Monotonous dependence of radial wave numbers and frequencies on $\Lambda$ and $E$ is not valid, first, for the 2nd group where frequencies of some modes do not decrease with $\Lambda$ in some of its ranges, and, second, for curves $k(E)$ which have maxima at $E \sim 10^{-3}$.

## 4. Conclusions

In this study many known properties of developing instabilities were confirmed [1, 3, 4, 9, 10, 11, 12]. The most interesting new results are following.

The changes of preference between various modes jointed with jumps in values of critical radial wave numbers and frequencies depend on stratification parameter $z_M$. Jumps into "viscous modes" do exist for all considered $z_M$. Number of other jumps between "competitive modes" of either viscous or electromagnetic origin strongly depends on $z_M$.

MC-waves are an example of modes strongly dependent on boundary conditions or/and stratification. Great deviations in the MC(E) waves onset values of $\Lambda$ are among various cases of finite electrical conductivities of boundaries, e.g. in sequence of cases $\frac{10^{-3}}{1}$, $\frac{1}{1}$, and $\frac{10^{3}}{10^{3}}$ the onset $\Lambda$ are 1127., 437., and 271., respectively. Thus, there is indication, that the role of electrical conductivity of boundaries is not negligible in onset of magnetic instabilities which in spherical geometry with magnetic field of dipole symmetry arise at $\Lambda = \mathcal{O}(10)$ (see e.g. [6]). Despite this fact, MC(E)-waves do not arise at perfectly conductive boundaries. Furthermore, MC(E)-waves are suppressed in non-uniformly stratified layer also for finitely conductive boundaries.

MC(W)-waves necessarily need viscosity for their existence. Their great amount in investigated ranges of parameters with many examples of close values of either radial wave numbers or frequencies indicates necessity to study their mutual non-linear interactions. It seems that the role of the MC(W)-waves, catalysed by viscosity, in the Earth's core is very important.

## Acknowledgements

We would like to thank the organisers of the meeting "Dynamo and Dynamics, a mathematical challenge", in particular Prof Pascal Chossat and

Dr Juliana Oprea, for all their work in producing such a stimulating and enjoyable conference.

This work has been supported by the Scientific Grant Agency VEGA (No 1/7177/20).

## References

1. Boďa J. (1988) Thermal and magnetically driven instabilities in a non-constantly stratified fluid layer, *Geophys. Astrophys. Fluid dynamics*, **44**, pp. 77–90
2. Braginsky S.I. (1964) Magnetohydrodynamics of the Earth's Core, *Geomagn. Aeron.*, **4**, pp. 898–916 (Engl. Transl. 698–712)
3. Brestenský J. and Ševčík S. (1994) Mean electromotive force due to magnetoconvection in rotating horizontal layer with rigid boundaries, in: M.R.E. Proctor, ed., Geodynamo Modelling, *Geophys. Astrophys. Fluid dynamics*, **77 (1-4)**, pp. 191–208
4. Brestenský J., Ševčík S. and Šimkanin J. (1998) Magnetoconvection in dependence on Roberts number, *Studia geoph. et geod.*, **42**, pp. 280–288
5. Chandrasekhar S. (1961) Hydrodynamic and hydromagnetic stability, *Clarendon press, Oxford*
6. Fearn D.R. (1998) Hydromagnetic flow in planetary cores, *Rep. Prog. Phys.*, **61**, pp. 175–235
7. Glatzmaier G. A. and Roberts P. H. (1995) A three-dimensional self-consistent computer simulation of geomagnetic field reversal, *Nature*, **377**, pp. 203-209
8. Glatzmaier G. A. and Roberts P. H. (1997) Numerical Simulations of the Geodynamo, in J. Brestenský and S. Ševčík, eds., Stellar and Planetary Magnetoconvection (Bratislava 1996), *Acta Astron. et Geophys. Univ. Comenianae*, **XIX**, pp. 125–143
9. Roberts P.H. and Loper D. (1979) On the diffusive instability of some simple steady magnetohydrodynamic flows. *J. Fluid Mech.*, **90**, pp. 641–668
10. Soward A.M. (1979) Thermal and magnetically driven convection in a rapidly rotating fluid layer, *J. Fluid Mech.*, **90**, pp. 669–684
11. Ševčík S., Brestenský J. and Šimkanin J. (2000) MAC waves and related instabilities influenced by viscosity in dependence on boundary conditions, *Phys. Earth Planet. Inter.*, **122 (3-4)**, pp. 161–174
12. Šimkanin J., Brestenský J. and Ševčík S. (1997) Dependence of rotating magnetoconvection in horizontal layer on boundary conditions and stratification, in: J. Brestenský and S. Ševčík, eds., Stellar and Planetary Magnetoconvection (Bratislava 1996), *Acta Astron. et Geophys. Univ. Comenianae*, **19**, pp. 195–220

# SMALL- AND LARGE-SCALE DYNAMO ACTION IN SIMPLE FORCED FLOWS

N.H. BRUMMELL
*JILA, University of Colorado, Boulder, CO, U.S.A.*
F. CATTANEO
*Dept. of Mathematics, University of Chicago, Chicago, U.S.A.*
AND
S.M. TOBIAS
*JILA, University of Colorado, Boulder, CO, U.S.A.* †

**Abstract.** We study the fully nonlinear magnetohydrodynamic (MHD) equations in 3–D driven by a specific time-dependent forcing function, characterised by a frequency, $\Omega$. Stable hydrodynamic solutions exist that are or are not dynamos depending on $\Omega$. We examine the effect of such flows on an imposed large-scale (mean) field and measure the associated $\alpha$-effect. We find that a significant non-zero $\alpha$-effect can exist, indicating possible dynamo action on scales larger than the velocity scale, when sometimes no dynamo action exists at the velocity scale itself. Flows exist which are not small-scale dynamos, yet may be large-scale dynamos.

## 1. Introduction

Dynamo processes are commonly invoked to explain the origin of cosmical magnetic fields [11]. An outstanding problem in the study of astrophysical magnetic fields is the origin of our sun's magnetic activity. There is much evidence to lead us to believe that the sun's magnetic field is derived from a dynamo (see e.g. [14], [14]). However, solar magnetic activity manifests itself over a wide variety of spatio-temporal scales, ranging from the current limit of observational resolution (few hundred km) and a few minutes, to active regions extending over much of the solar surface emerging with a

†Present address: Dept. of Applied Mathematics, University of Leeds, Leeds, UK



*P. Chossat et al. (eds.), Dynamo and Dynamics, a Mathematical Challenge,* 145–152.

cyclic pattern with a period of eleven years. The question of the origin of this solar magnetic activity has two components: firstly, why does the sun have a magnetic field at all, and secondly, why is there an activity cycle? The first question is related to the production and maintenance of magnetic fields on scales similar to those of the driving velocity (granulation, supergranulation, possibly giant cells) – a small-scale dynamo; the second refers instead to the creation of a magnetic field on a scale much larger than any velocity field present – a large-scale dynamo.

Fast dynamo theory, the study of magnetic field generation in the limit of vanishing magnetic diffusivity, has shed light on the conditions under which small-scale dynamo action is likely (see e.g. [8], [8]). In particular, it is found that chaotic trajectories of fluid elements are necessary (but not sufficient) for fast dynamo action [9]. Provided that the magnetic diffusivity is not enhanced by the contraction and packing of field lines sufficiently to overcome the exponential stretching, then the flow will be a dynamo. Such favourable conditions are likely to be prevalent throughout the convection zone of the sun, and therefore all scales of the convection might be expected to produce and sustain magnetic field on similar (or generally smaller and highly intermittent) scales. Observations appear to concur, exhibiting magnetic fields on scales related to the velocity patterns present. However, fields are disordered in both space and time with very little net large-scale, ordered field.

The generation of magnetic fields on scales larger than the driving velocity and with considerable order (such as the butterfly diagram, Hale's polarity law, Joy's inclination law etc.) is the problem more traditionally associated with a solar dynamo, and yet possibly the tougher of our two questions. To achieve such fields, some symmetries of the system must be broken. To this end, rotation is typically invoked to break the reflectional symmetry of the motions, and magnetic reconnection can further provide an irreversibility to the processes. The most widely studied mechanism for large-scale magnetic field generation is the $\alpha$-effect derived from mean-field theory ([13], [13]; [10] [10]). This mechanism describes the generation of mean magnetic fields through the interaction of small-scale velocity and magnetic fluctuations, where here "mean" identifies a field that varies slowly compared to the scale of variation of the fluctuations.

Thus, efforts under the auspices of dynamo theory, where we seek to learn about the types of flows that can amplify and sustain magnetic field, could be loosely grouped into two categories: the search for those flows that exhibit small-scale dynamo action (scales comparable with the velocity field), and those that can generate and sustain a field on a much larger scale. Here, we examine a familiar class of flows for both small-scale and large-scale dynamo action in the non-linear regime via numerical simulations.

## 2. Formulation

We follow the ealier work of [3] ([3], [4]; hereinafter termed BCT1 and BCT2) and consider dynamo action in an incompressible fluid with finite viscosity and electrical conductivity, in a periodic domain with $2\pi$ periodicity in all three directions. The evolution of this system is described by the following dimensionless equations:

$$(\partial_t - R_m^{-1}\nabla^2)\mathbf{B} = \nabla \times (\mathbf{U} \times \mathbf{B}), \tag{1}$$

$$(\partial_t - R_e^{-1}\nabla^2)\mathbf{U} + \mathbf{U} \cdot \nabla\mathbf{U} = -\nabla p + \mathbf{J} \times \mathbf{B} + \mathbf{F}, \tag{2}$$

$$\nabla \cdot \mathbf{B} = \nabla \cdot \mathbf{U} = 0. \tag{3}$$

Here $\mathbf{U}$ is the velocity, $\mathbf{B}$ is the magnetic field, $p$ is the pressure, $\mathbf{J} = \nabla \times \mathbf{B}$ is the electric current and $R_e$ and $R_m$ are the kinetic and magnetic Reynolds numbers respectively. The Reynolds numbers, $R_e = ul/\nu$ and $R_m = ul/\eta$ where $u$ and $l$ are typical velocities and lengthscales and $\nu$ and $\eta$ are the viscous and magnetic diffusivities, are set to $R_e = R_m = 100$ here.

The system is driven by a forcing function $\mathbf{F}$, chosen to induce (at least initially, when $\mathbf{J} \times \mathbf{B}$ is negligible) a velocity $\mathbf{U}_0$,

$$\mathbf{F}_0(\mathbf{x}, t) = (\partial_t - R_e^{-1}\nabla^2)\mathbf{U}_0(\mathbf{x}, t). \tag{4}$$

We choose $\mathbf{U}_0$ to be a modified version of an ABC flow ([1] [1], [2] [2], [7] [7])

$$\begin{aligned}\mathbf{U}_0(\mathbf{x}, t) = (&\sin(z + \epsilon \sin \Omega t) + \cos(y + \epsilon \sin \Omega t),\\ &\sin(x + \epsilon \sin \Omega t) + \cos(z + \epsilon \sin \Omega t),\\ &\sin(y + \epsilon \sin \Omega t) + \cos(x + \epsilon \sin \Omega t)).\end{aligned} \tag{5}$$

Since the nonlinear term $\mathbf{U}_0 \cdot \nabla\mathbf{U}_0$ can be exactly balanced by the pressure gradient, with no contribution to the driving of a solenoidal flow, $\mathbf{U}_0(\mathbf{x}, t)$ is an exact solution of equation (2) with the forcing (4). The magnetic initial conditions consist of a weak random seed field with amplitude (in units of the velocity) of less than $10^{-5}$. The system is then evolved from this initial state of weak magnetisation so that the linear and nonlinear phases of dynamo action can be clearly identified. We solve the equations (1-3) using standard pseudospectral methods ([5] [5]; for details, see BCT1 and BCT2).

## 3. Small-scale dynamo action

The small-scale dynamo action of this class of flows has been investigated in detail in BCT1 and BCT2. Here we briefly summarise some of the findings of those papers.

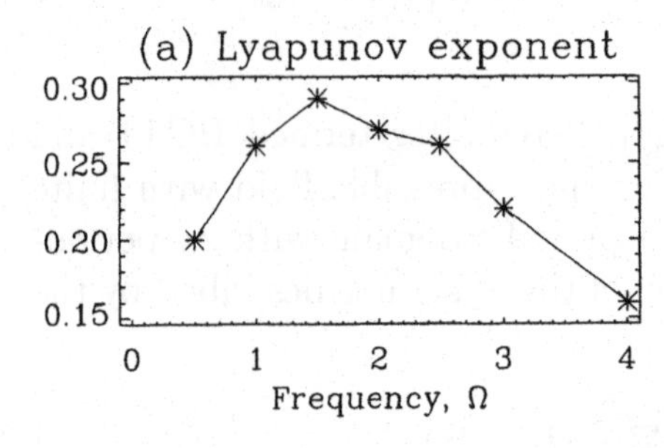


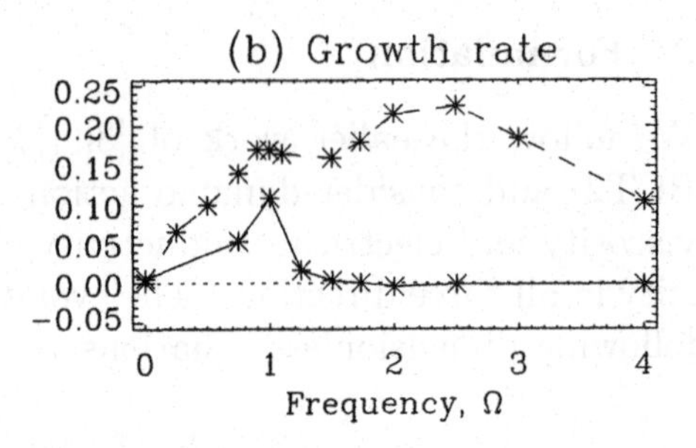


*Figure 1.* (*a*) Plot of the average finite-time Lyapunov exponent calculated for $\mathbf{U}_0$ for various forcing frequencies $\Omega$. (*b*) Plots of the kinematic growth rate for small-scale dynamo action for the same forcing at various frequencies $\Omega$ but with different initial velocity conditions: $\mathbf{U}_0$ dashed line, $\mathbf{U}_1$ solid line.

The target velocity field has been chosen since its Lagrangian properties indicate that kinematic dynamo action is likely. The modified ABC flows $\mathbf{U}_0$ all contain chaotic trajectories, with varying coverage depending on $\epsilon$ and $\Omega$. Finite-time Lyapunov exponents indicate that the averge degree of chaoticity increases monotonically with $\epsilon$ but exhibits a maximum when $\Omega$ is varied (figure 1*a*). We consider here simulations at a fixed value of $\epsilon = 1$, but investigate the variation with the forcing frequency, $\Omega$, using $\mathbf{U}_0$ and a weak magnetic field as initial conditions. Even though these calculations are fully nonlinear at all times, there exists a regime early on in the simulations where the magnetic field is still small enough not to affect the velocity via the Lorenz force significantly. In this pseudo-linear state (which we term the *kinematic* regime), the velocity field remains close to the target velocity, $\mathbf{U}_0$, and the magnetic field either amplifies or decays exponentially. A growth rate can be measured and the flow can be deemed a small-scale *kinematic* dynamo or not accordingly. Figure 1*b* (dashed line) shows the kinematic growth rates for these simulations. The behaviour with $\Omega$ is not monotonic, exhibiting a maximum growth rate at $\Omega = 2.5$. The behaviour agrees qualitatively with the information from the Lyapunov exponents but not quantitavely, as has been anticipated previously [12]. All the flows $\mathbf{U}_0(\Omega)$ (for the range of $\Omega$ examined here) are kinematic dynamos, although some are more efficient than others.

The kinematic growth cannot continue indefinitely. Eventually, the amplification results in magnetic fluctuations that are strong enough to influence the velocity field, and a fully nonlinear MHD state is achieved. It appears initially that two behaviours are observed. First, the velocity field may be modified by the magnetic fluctuations such that the hydrodynamic flow becomes a statistically marginal dynamo, i.e. the magnetic field is neither amplified nor reduced on average, but sustained at a finite level. This

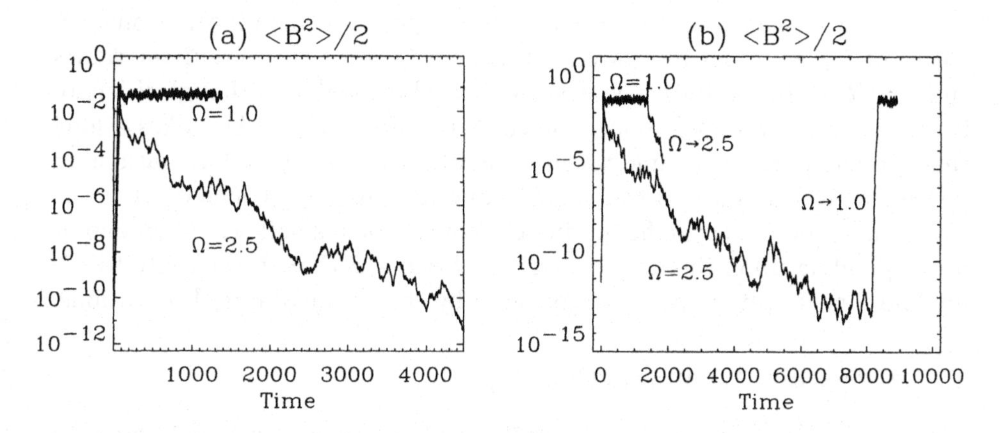


*Figure 2.* Plots of the magnetic energy versus time for the wto forcing frequencies considered. (*a*) Nonlinear evolution of the two cases. (*b*) Continuation after the forcing frequency is switched in each case after a period of nonlinear evolution.

may be considered the normal nonlinear saturation of a dynamo process. An example is shown in figure 2*a* for $\Omega = 1$. In the second scenario, the magnetic fluctuations appear to perturb the underlying hydrodynamic state sufficiently that it evolves to a different velocity state altogether. The dynamo properties of this new state are such that it is incapable of sustaining the magnetic field, and so it decays away. An example of this behaviour is shown in figure 2*a* for $\Omega = 2.5$.

These two seemingly disparate behaviours can both be understood in terms of the existence of another hydrodynamic state, which we shall call $\mathbf{U}_1$, that is another solution of the equations with the same forcing. That is, the $\mathbf{U}_0$ state is not a unique solution of the nonlinear equations at these parameters. The magnetic perturbations that are induced mediate the transition to this other state, and the ultimate fate of the nonlinear MHD system depends on the dynamo properties of $\mathbf{U}_1$. Using the hydrodynamic solution obtained at the end of a decaying nonliner simulation (e.g. $\Omega = 2.5$) as initial conditions, the kinematic dynamo properties of $\mathbf{U}_1$ can be investigated, varying $\Omega$ in the forcing, as before. The resultant growth rates are shown in figure 1*b* (solid line). These show that the kinematic growth rates for this new initial state are in general lower and indeed sometimes negative whereas the growth rates for the initial condition $\mathbf{U}_0$ were always larger and positive. Whether the nonlinear MHD system sustains magnetic field indefinitely (i.e. is a *nonlinear* dynamo) depends on the kinematic dynamo properties of $\mathbf{U}_1$ since this flow ultimately governs the hydrodynamics underlying the nonlinear MHD flow. Thus, $\Omega = 1$ is a nonlinear dynamo, whereas $\Omega = 2.5$ is not. This is further demonstrated in figure 2*b* where the forcing frequencies are switched after a period of nonlinear evolution. The flow $\mathbf{U}_0$ is in fact hydrodynamically unstable at the parameter values chosen ($R_e = R_m = 100$) whereas the flow $\mathbf{U}_1$ is stable, although if left to hydrodynamic processes alone, the transition from $\mathbf{U}_0$ to $\mathbf{U}_1$ takes a long time (in comparison to time taken to amplify the magnetic fluctuations to a strength comparable to the velocity fluctuations). We conclude that the equations under the specific forcing chosen can be a small-scale dynamo or not depending upon the frequency of the forcing, although the route to the final nonlinear state may be complicated by the choice of initial conditions.

## 4. Large-scale dynamo action

We now examine the ability of the flows used above to act as large-scale dynamos. The domain size chosen for the simulations so far has been comparable to the velocity correlation length of the flows. To study the generation and maintenance of flows on scales larger than this length scale, it would be necessary to increase the domain size substantially. Since this is numerically costly, we instead examine the effect of the flows on an imposed mean magnetic field in a domain of size as before. The mean field is uniform in this context and therefore varies on a large (infinite) scale compared to that of the domain. Thus we may consider the associated $\alpha$ formalism to search for any possibilities of large-scale field generation (following [6] [6]). The $\alpha$-effect expresses the relationship between the average electromotive force $\mathcal{E}$ and the average magnetic field $\langle \mathbf{B} \rangle$, i.e.

$$\mathcal{E} = \langle \mathbf{u} \times \mathbf{b} \rangle_i = \alpha_{ij} \langle \mathbf{B} \rangle_j - \beta_{ijk} \partial_j \langle \mathbf{B} \rangle_k + \cdots \tag{6}$$

where $\mathbf{u}$ ($= \mathbf{U}$ in this case) and $\mathbf{b}$ are the fluctuating velocity and magnetic field and averages are taken over volumes large compared to the velocity correlation length. Since we are confined to a domain size close to the correlation length, we may expect the components of $\alpha$ to be highly fluctuating. In order to perform a meaningful average recovering the the correct statistical properties of $\alpha$, we must therefore average over long times as well as the domain.

We simulate as before, except that a mean field component in one spatial direction is added to the weak initial magnetic field,

$$\mathbf{B}(\mathbf{x}, t = 0) = (B_0, 0, 0) + \mathbf{b}, \tag{7}$$

where $B_0$ is a constant and $\mathbf{b}$ are the magnetic perturbations with zero mean (so that $\langle B \rangle = B_0 \hat{\mathbf{x}}$). The $\alpha$ tensor defined in equation (6) then only has a measurable component $\alpha_{i1}$, which may be calculated (knowing $B_0$) from $\langle \mathbf{u} \times \mathbf{b} \rangle_i$. Figure 3 shows results from two simulations at different forcing

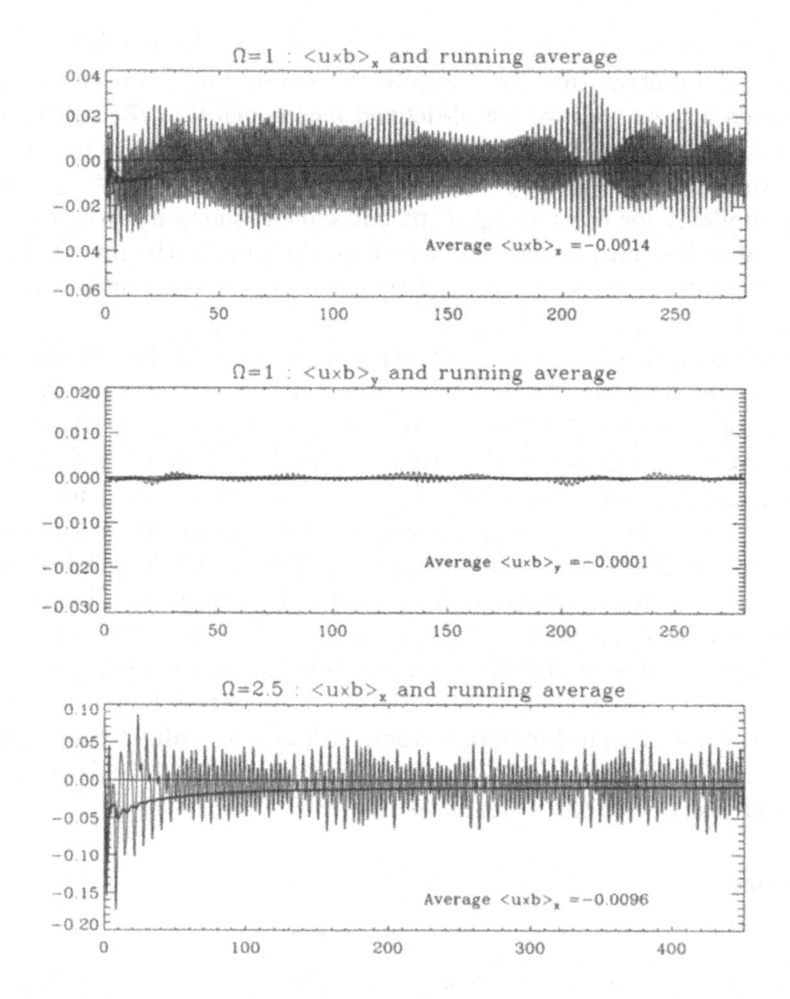


*Figure 3.* Plots of the components of $\langle \mathbf{u} \times \mathbf{b} \rangle$ and their running time average (thick line) for $\Omega = 1$ ($a$ is the $x$-component, and $b$ the y-component) and $\Omega = 2.5$ ($c$, the $x$-component).

frequencies $\Omega$. Both have an imposed mean magnetic field of strength $B_0^2 = 0.01$ and start from the stable hydrodynamic configuration $\mathbf{U}_1$.

Figure 3$a$ and $b$ show the $x$- and $y$-components of the $\langle \mathbf{u} \times \mathbf{b} \rangle$ vector from the calculation with $\Omega = 1$. The superposed thick line shows the running time average of $\langle \mathbf{u} \times \mathbf{b} \rangle_i$ (the average over the domain) up to time, $t$, i.e.

$$\frac{1}{t} \int_0^t \langle \mathbf{u} \times \mathbf{b} \rangle_i (t) dt = \frac{B_0}{t} \int_0^t \alpha_{i1}(t) dt = B_0 \, \overline{\alpha_{i1}}, \tag{8}$$

which defines our time-average approximation, $\overline{\alpha_{i1}}$, to the volume-average measure that is the $\alpha$-effect. Figures 3 show that $\langle \mathbf{u} \times \mathbf{b} \rangle$ is indeed a highly intermittent quantity. However, despite the strong fluctuating nature, the components appear to have a well-defined mean such that $\overline{\alpha_{i1}}$ converges in the long time average. For the $x$-component (figure 3$a$), $\overline{\alpha_{11}}$ converges to a finite value, $\overline{\alpha_{11}} = -0.0014/B_0 = -0.014$. The $y$-component $\overline{\alpha_{21}}$ (figure 3$b$, and similarly for the $z$-component not shown) shows a zero mean with much smaller fluctuations, as might be expected due to the imposed mean field configuration. A significant $\alpha$-effect appears to be present in this class of flows.

A more remarkable case is demonstrated in figure 3$c$. This figure shows the $x$-component of $\langle \mathbf{u} \times \mathbf{b} \rangle$ for a simulation with forcing frequency $\Omega = 2.5$. Again, a significant $\overline{\alpha_{11}}$ results, with a final value close to $\overline{\alpha_{11}} = -0.1$ demonstrating the presence of a *stronger* $\alpha$-effect than for the lower $\Omega$. This is intriguing, since we recall that, without the imposed mean field, the $\Omega = 1$ case is a small-scale dynamo, whereas the $\Omega = 2.5$ case is not. It appears that, under the forcing at frequency $\Omega = 2.5$, the resultant nonlinear flow is **not a small-scale dynamo, but may be a large-scale dynamo**, due to the presence of a significant $\alpha$-effect. In other words, flows exist that are unable to amplify magnetic fields on the scale of the velocity coherence length, but which may be able to amplify much larger scale fields. In the study of the solar dynamo, it would certainly be enlightening to know the classes of flows that were both small-scale and large-scale dynmos, and those that lost one or the other property.

## References


1. Arnold, V.I., (1965), *C.R. Acad. Sci. Paris*, **261**, 17
2. Beltrami, E., (1889), *Opera Matematiche*, **4**, 304 (1889)
3. Brummell, N.H., Cattaneo, F. & Tobias, S.M., (1998) *Phys. Lett. A* **249**, 437 (BCT1)
4. Brummell, N.H., Cattaneo, F. & Tobias, S.M., (2000) *Fluid Dyn. Res.* in press (BCT2)
5. Canuto, C., Hussaini, M.Y., Quarteroni, A. & Zang, T.A., (1988) *Spectral Methods in Fluid Dynamics*, Springer-Verlag
6. Cattaneo, F. & Hughes, D.W., (1996), *Phys. Rev. E*, **54**, 4532
7. Childress, S., (1970), *J. Math. Phys.*, **11**, 3063
8. Childress, S. & Gilbert, A.D., (1995) *Stretch, Twist, Fold: The Fast Dynamo*, Springer.
9. Klapper, I. & Young, L.S., (1995) *Comm. Math. Phys.*, **173**, 623
10. Moffatt, H.K., (1978), *Magnetic Field Generation in Electrically Conducting Fluids*, CUP
11. Parker, E.N. (1979), Cosmical Magnetic Fields, Clarendon, Oxford
12. Ponty, Y., Pouquet, A., Rom-Kedar, V. & Sulem, P.L., (1993) In *Solar and Planetary Dynamos* (ed. M.R.E. Proctor, P.C. Matthews & A.M. Rucklidge), CUP, 241
13. Steenbeck, M., Krause, F. & Rädler, K-H, (1966), *Z. Naturforsch*, **21a**, 369
14. Weiss, N. O., (1994), in *Lectures on Solar and Planetary Dynamos*, (ed. M.R.E. Proctor & A.D. Gilbert), CUP, 59

# CONVECTION DRIVEN DYNAMOS IN ROTATING SPHERICAL FLUID SHELLS

F.H. BUSSE AND E. GROTE
*Institute of Physics, University of Bayreuth, D-95440 Bayreuth*

Many dynamical processes occuring on large scales in planets and in stars can be modeled surprisingly well on a laboratory scale, and experimental measurements and observations have made major contributions to our understanding of flows in the atmosphere, in the oceans or on the surface of the sun. The similarity of dynamical processes on the laboratory and on planetary or even stellar scales has been rationalized by the concept of eddy diffusivities which is based on the idea that small scale turbulence acts on large scale fluid motions in approximately the same way as molecular diffusivities affect the dynamics of fluids in a laboratory experiment. It is generally expected that the concept of eddy diffusivity also holds for the dynamo process of the generation of planetary and stellar magnetic fields. In fact, in the case of the Earth's core there is no need to use the eddy version of the magnetic diffusivity since the magnetic Reynolds number may not be larger than a few hundreds which can be easily accommodated in numerical simulations. Since dynamical processes often depend sensitively on the ratio of diffusivities, the fact that numerical simulations of the geodynamo have not been able to reach magnetic Prandtl numbers $P_m$ much less than unity, is more worrying. In this review we shall consider some typical results for convection driven dynamos in rotating spherical shells down to values $P_m = 0.4$.

The mathematical formulation of the problem is the same as that used in earlier papers [2, 5, 6, 7]. A minimum of four parameters is used representing the essential dynamical dependences of the problem. The Rayleigh number$R$, the Coriolis parameter $\tau$, the Prandtl number $P$ and the magnetic Prandtl number $P_m$ are defined by

$$R = \frac{\alpha\gamma\beta d^6}{\nu\kappa}, \quad \tau = \frac{2\Omega d^2}{\nu}, \quad P = \frac{\nu}{\kappa}, \quad P_m = \frac{\nu}{\lambda} \tag{1}$$



*P. Chossat et al. (eds.), Dynamo and Dynamics, a Mathematical Challenge,* 153–162.

where $\nu, \kappa, \lambda$ are the kinematic viscosity, the thermal and the magnetic diffusivity, respectively, $\alpha$ is the coefficient of thermal expansion, $d$ is the thickness of the spherical fluid shell and $\Omega$ is its angular velocity of rotation. The radial components of gravity and of the basic temperature gradient are given by $-\gamma r d$ and $-\beta r d^2$ where $r$ is the distance from the center measured in multiples of $d$. In principle the ratio $\eta$ between inner and outer radius of the shell is another important parameter. But it will be fixed here at $\eta = 0.4$.

It is convenient to use $\nu/d$ and $\nu(\mu\varrho)^{1/2}/d$ as scales for the velocity field $\boldsymbol{u}$ and for the magnetic flux density $\boldsymbol{B}$, respectively, where $\mu$ and $\varrho$ are the magnetic permeability and the density of the fluid. For the solenoidal fields $\boldsymbol{u}$ and $\boldsymbol{B}$ we introduce the general representations in terms of poloidal and toroidal components,

$$\boldsymbol{u} = \nabla \times (\nabla v \times \boldsymbol{r}) + \nabla w \times \boldsymbol{r} \ , \qquad \boldsymbol{B} = \nabla \times (\nabla h \times \boldsymbol{r}) + \nabla g \times \boldsymbol{r} \qquad (2)$$

where the poloidal parts are given by $v$ and $h$ and the toroidal parts by $w$ and $g$. $\boldsymbol{r}$ is the dimensionless position vector with respect to the center. At the boundaries $r = r_i \equiv \eta(1-\eta)^{-1}$ and $r = r_o \equiv (1-\eta)^{-1}$ stress-free conditions are assumed and the outside is regarded as electrically insulating.

The basic state of hydrostatic equilibrium becomes unstable to convective overturning when the Rayleigh number $R$ exceeds a critical value $R_c$ dependent on $\tau$ and $P$. The convection arranges itself in the form of stationarily drifting columns as shown in the upper left of figure 1. Further bifurcations follow either without magnetic field to vacillating convection and chaotic convection as discussed by Zhang and Busse (1988), Sun *et al.* (1993), Ardes *et al.* (1997), Tilgner and Busse (1997) and Grote and Busse (2001), or bifurcations involving magnetic fields occur in which case we speak of a dynamo. Dependent on the number of symmetries which the state of convection has retained various kinds of broken or retained symmetries characterize the bifurcating magnetic state as indicated in the figure. Various examples for these bifurcating solutions have been explored in several papers [18, 9, 16]. The most important distinction of symmetry is that between magnetic fields that are either symmetric (quadrupolar) or antisymmetric (dipolar) with respect to the equatorial plane as has been indicated in the upper right corner of the figure. These symmetries of dynamos persist far into the turbulent regime because even turbulent convection retains to a high degree of accuracy its symmetry with respect to the equatorial plane as long as $\tau$ is large enough. Both, dipolar and quadrupolar magnetic fields are compatible with this symmetry since the Lorentz force is quadratic in $\boldsymbol{B}$.

In the following we shall discuss recent systematic computations of spherical dynamos first as a function of $P_m$ at fixed values of $P$ and $\tau$

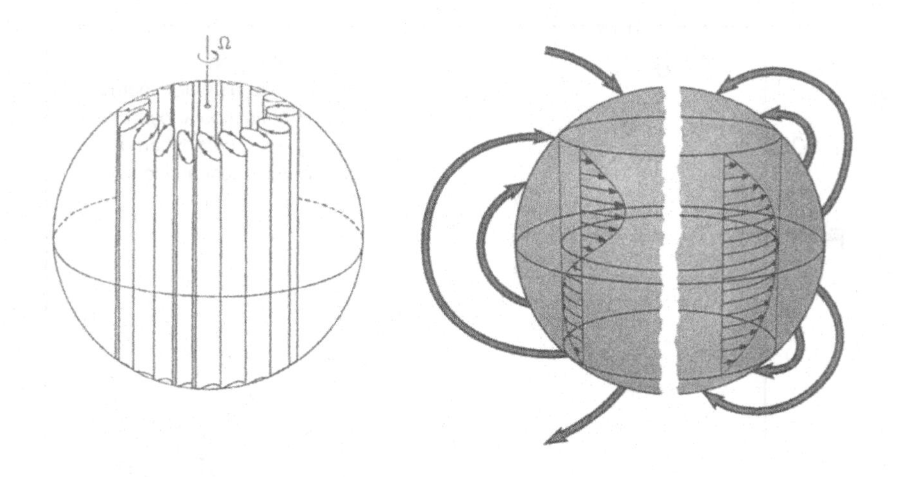


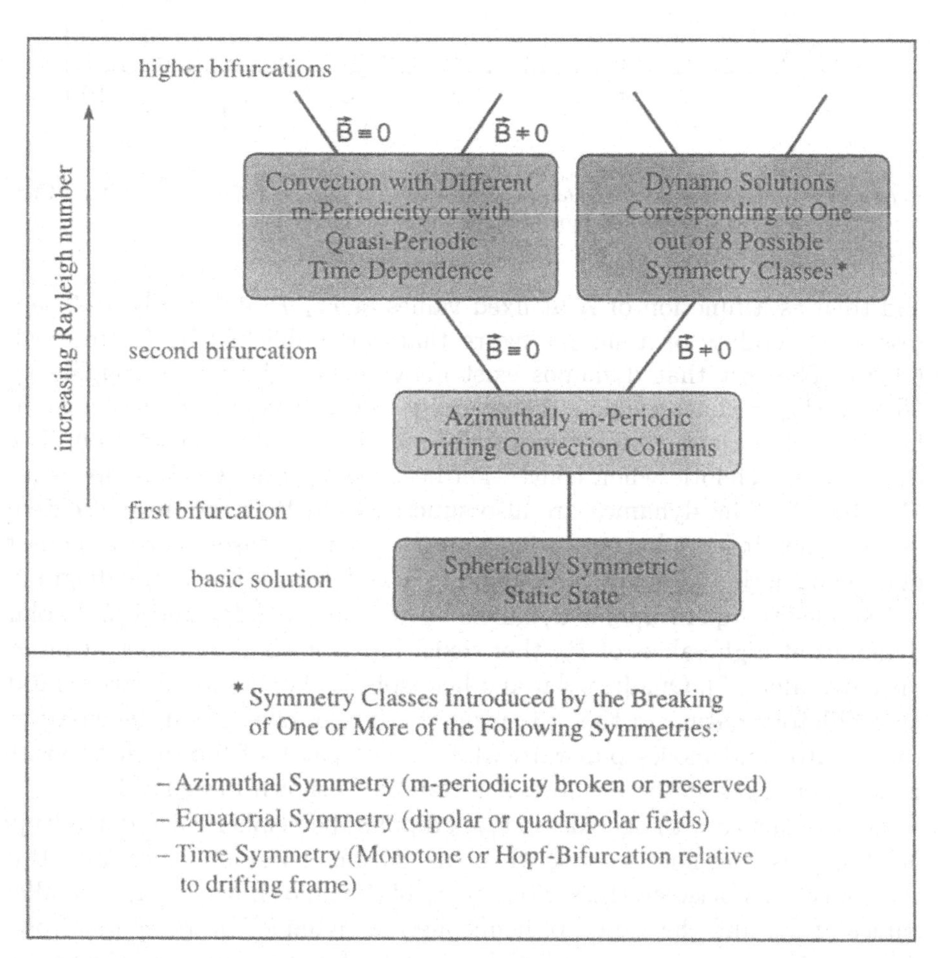


*Figure 1.* Bifurcation properties

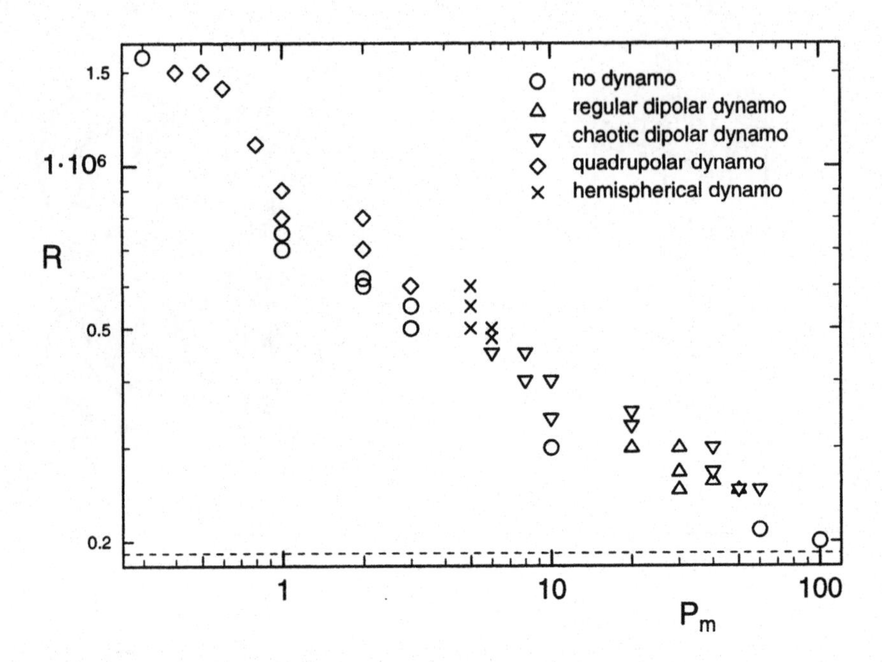


*Figure 2.* Existence of Dynamo Solutions in the $R$-$P_m$-Plane in the Case $\tau = 10^4, P = 1$.The dashed line indicates the onset of convection.

and then as a function of $R$ at fixed values of $P_m, P$ and $\tau$. Figure 2 represents an update of a similar figure that was published by Grote et al. (2000). The fact that dynamos exist above a line given very roughly by $(R - R_c)P_m = \text{const.}$ reflects the property that a magnetic Reynolds number $Re_m$ of the order of $10^2$ is usually required for dynamos action. The dynamos are chaotic when convection becomes chaotic which happens for $R > 3 \cdot 10^5$. The dynamos are also subcritical in that the magnetic field decays when it is no longer strong enough. For this reason alone the onset of dynamo action does not correspond to a well defined line in the diagram.

Besides the quadrupolar dynamos at low values of $P_m$ and the dipolar dynamos at high values of $P_m$ there exist hemispherical dynamos at intermediate values [7]. Quadrupolar and hemispherical dynamos always exhibit an oscillatory nature in that the magnetic flux of one polarity emerges at the equator and moves poleward while a new patch of flux with opposite polarity emerges at the equator. An example is shown in figure 3.

The sequence with decreasing $P_m$ of dipolar, hemispherical and quadrupolar dynamos continues to hold when the Coriolis parameter increases. But the transitions between the various types of dynamos move towards smaller values of $P_m$ and the region of hemispherical dynamos becomes more predominant. An increase of $\tau$ also seems to be the best way to achieve dynamos at smaller values of $P_m$. The decrease of $P_m$ with increasing $\tau$ is

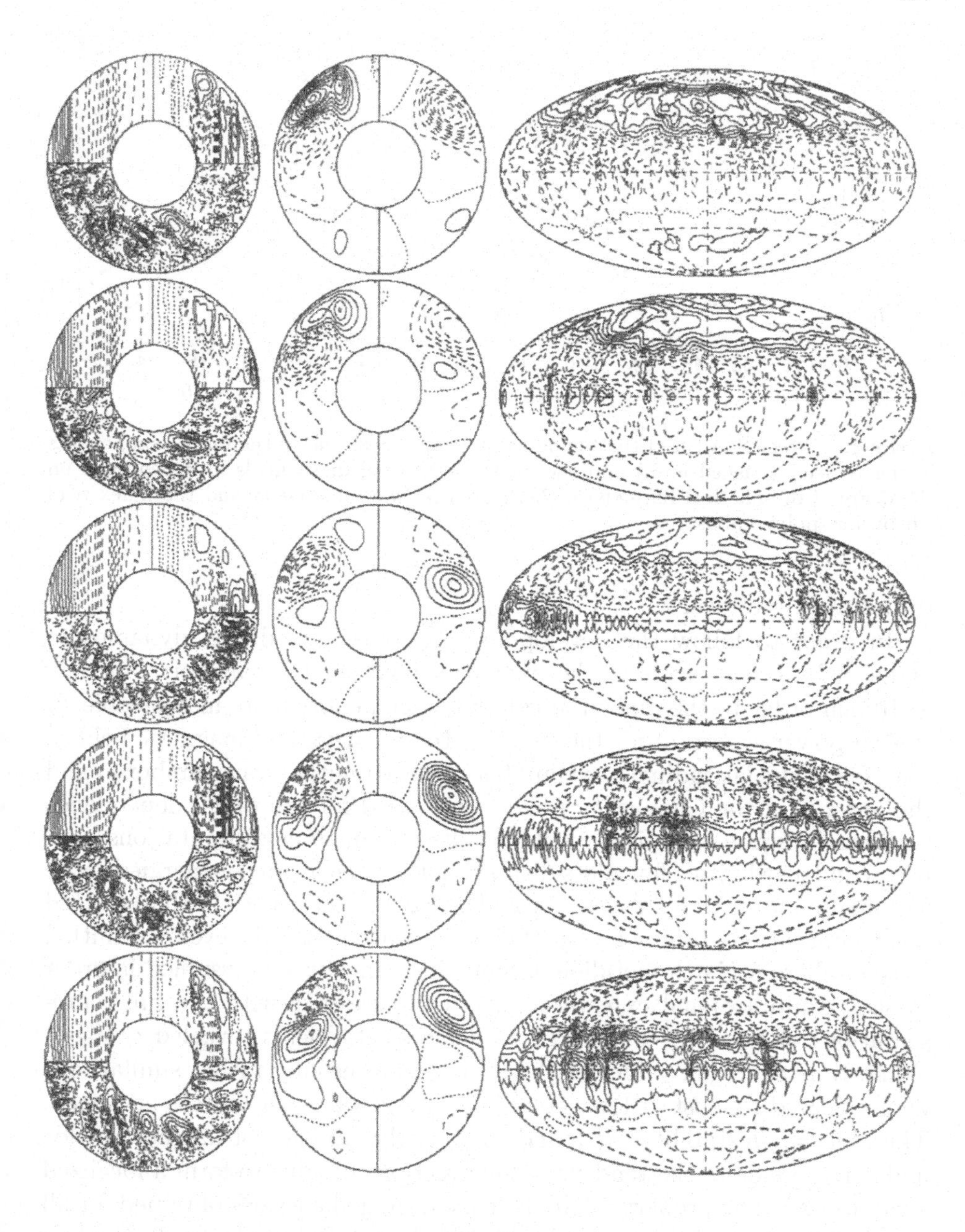

*Figure 3.* Oscillating Hemispherical Dynamo with $R = 2.7 \cdot 10^6$, $\tau = 3 \cdot 10^4$, $P = 1$, $P_m = 0.5$. A time sequence of 5 plots, $\Delta t = 0.012$ apart, is shown from top to bottom in each column. The first column shows lines of constant $\bar{u}_\varphi$ in the upper left quarter, meridional streamlines, $r \sin\theta \partial\bar{v}/\partial\theta =$ const., in the upper right quarter and streamlines in the equatorial plane, $r\partial v/\partial\varphi =$ const., in the lower half. The middle column shows lines of constant $\bar{B}_\varphi$ in the left half and meridional field lines, $r \sin\theta \partial\bar{h}/\partial\theta =$ const., in the right half. The right column shows lines of constant $B_r$ on the surface of the sphere. Solid (dashed) lines indicate positive (negative) values. The $^-$ indicates the azimuthal average.

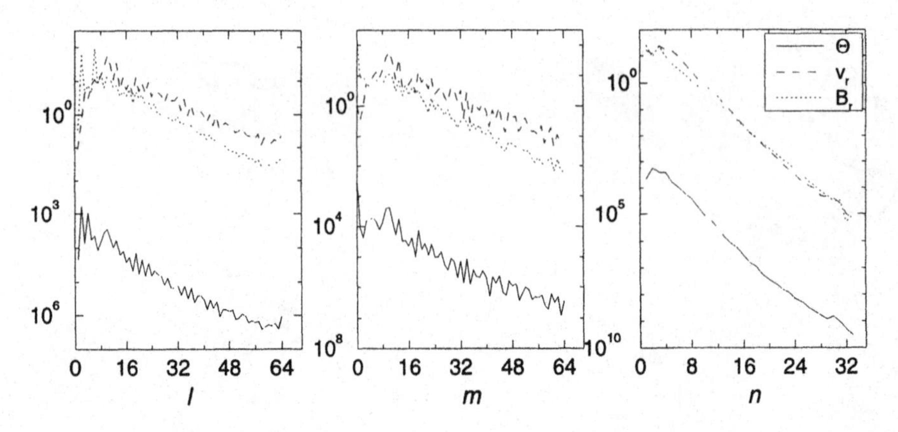


*Figure 4.* Power Spectra of the temperature (solid lines) and of the radial components of the velocity (dashed lines) and the magnetic (dotted lines) fields in dependence on the degree $l$ of spherical harmonics, the azimuthal wavenumber $m$ and the index $n$ of Chebychev polynomials.

rather weak, however.

In order to follow the development of convection driven dynamos to higher values of $R$, we have chosen a somewhat smaller value of $\tau$, namely $5 \cdot 10^3$, in order to the make the computational effort more manageable. A typical power spectrum for the velocity, temperature and magnetic field as function of the latitudinal wavenumber $l$, the azimuthal wavenumber $m$ and the order $n$ of the Chebychev polynomials used for the radial dependence is shown in figure 4. In figure 5 the results of numerous computations have been condensed into a few plots displaying the time averaged energies of the various components of the velocity and of the magnetic field. We had started the computations originally by taking into account only even azimuthal wavenumbers $m$, i.e. a meridional plane of symmetry was assumed in order to save computational expenses. Later computations with the full representations of velocity, temperature and magnetic fields revealed changes. While time averaged values of the energies are usually rather similar, dynamos ceased to exist at low values of $R$ if odd values of $m$ were admitted. This "$m = 1$-instability" is found even in the absence of magnetic fields and corresponds to the tendency of convection columns to form a localized structure [8]. The property that the time averaged energies obtained for all $m$ are not as smooth as those obtained with even $m$ only is due to the fact that the computations have not run yet for a sufficiently long time. Some remarkable features can be noticed. At $R = 5 \cdot 10^5$ there exists a purely quadrupolar dynamo as well as a dynamo of mixed parity with a relatively small dipolar component of the magnetic field corresponding to about 1/20 of the energy of the quadrupolar component in the time average. Both dynamos are chaotic and smaller perturbations will not change them. Another

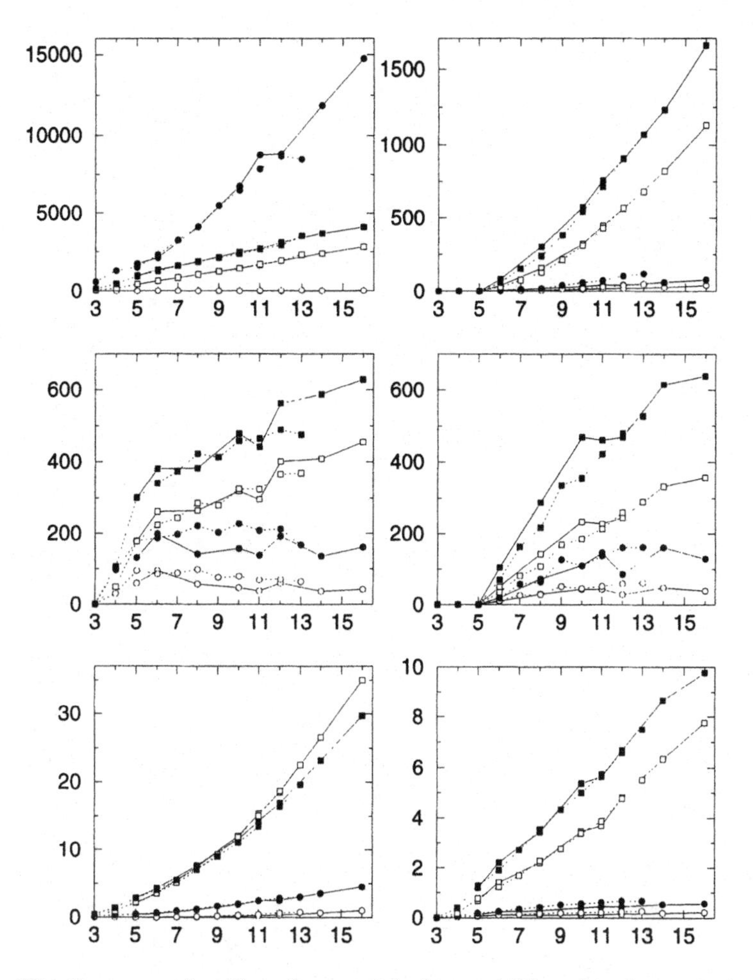


*Figure 5.* Kinetic energy densities of symmetric (upper left) and antisymmetric (upper right) components, magnetic energy densities of quadrupolar (middle left) and dipolar (middle right) components, and viscous (lower left) and Ohmic (lower right) dissipations are plotted as a function of $R \cdot 10^{-5}$ for convection driven dynamos in the case $\tau = 5 \cdot 10^3, P = P_m = 1$. Filled (open) symbols indicate toroidal (poloidal) components of the energies and dissipations, circles (squares) indicate axisymmetric (non-axisymmetric) components. In the case of the dissipations the contributions have not been separated with respect to their equatorial symmetry. The scales of the two lower plots must be multiplied by the factor $10^5$. Symbols connected by solid (dotted) lines correspond to computations carried out with all azimuthal wavenumbers $m$ (with even $m$ only).

feature is the appearance of oscillatory predominantly dipolar dynamos at higher values of $R$. While quadrupolar and hemispherical dynamos have always been found to possess an oscillatory nature, this has not been the case for the dipolar dynamo found at lower values of $\tau$ or at higher values of $P_m$. In the case of the dynamos of figure 5 the dipolar component increases in importance with increasing $R$ relative to the quadrupolar component,

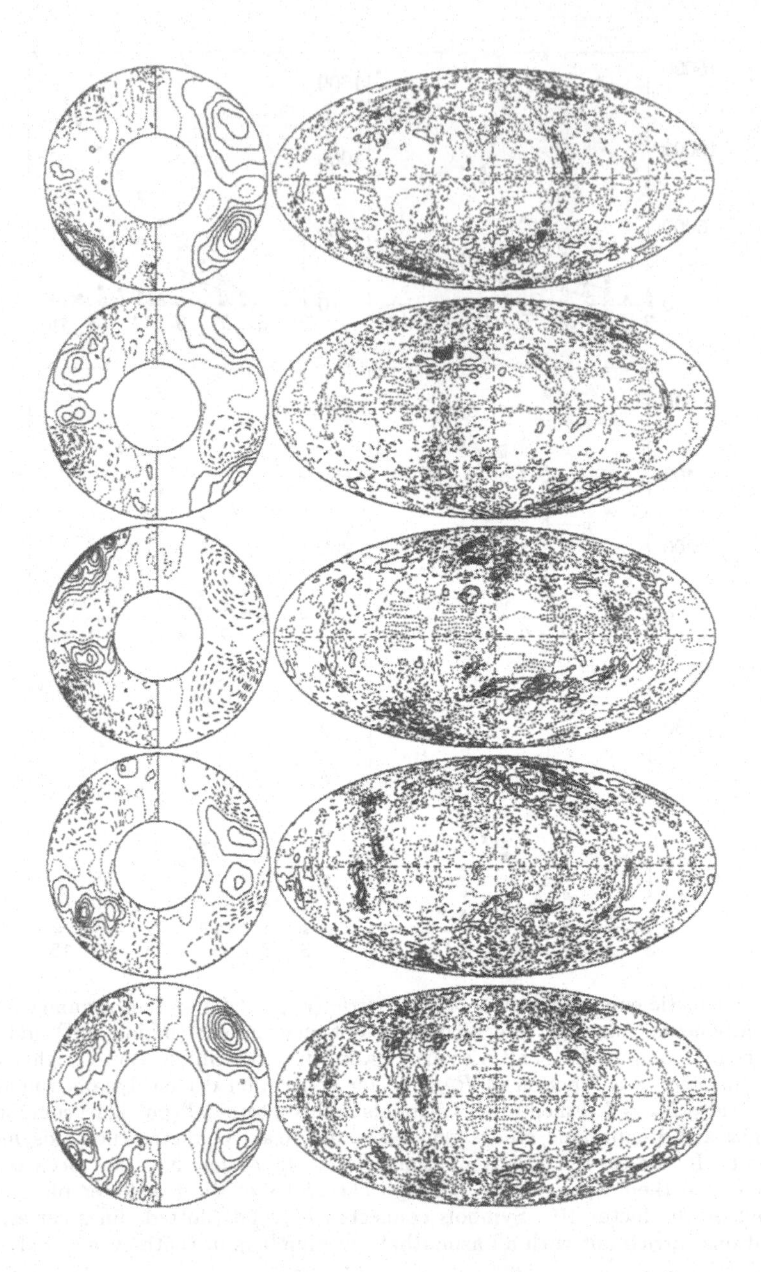

*Figure 6.* Oscillatory Dipolar Dynamo for $\tau = 5 \cdot 10^3, R = 1.6 \cdot 10^6, P = P_m = 1$. A sequence of 5 plots, $\Delta t = 0.01$ apart, is shown from top to bottom covering about one period. The left column shows lines of constant $\bar{B}_\varphi$ in the left half of the circle and meridional field lines, $r \sin \theta \partial h / \partial \theta =$ const., in the right half. The right column shows lines of constant $B_r$ at the surface of the spherical shell.

but retains the tendency of the latter towards oscillations. An example of an oscillatory dipolar dynamo is displayed in figure 6. It is remarkable how coherently the oscillation of the axisymmetric components of the magnetic field occurs even though the fluctuating components of the magnetic field dominate. The latter are visible in the form of filamentary structures at the surface of the sphere and indicate the flux expulsion that governs the interaction between magnetic field and convection flow at these high magnetic Reynolds numbers. The frequency of oscillations is also proportional to the magnetic Reynolds number and thus grows with increasing Rayleigh number as it does in the case of hemispherical and quadrupolar dynamo oscillations.

The onset of the dipolar component of the magnetic field in the case of figure 5 is connected with the onset of convection in the polar regions. Since these regions inside the tangent cylinder touching the equator of the inner boundary are not connected by geostrophic contours, the polar convection flows do not exhibit a symmetry with respect to the equatorial plane. This property contributes to the enhancement of the dipolar component of the magnetic field. At values of $R$ of the order $10^6$ or above, the polar convection rolls have become far more vigorous than the convection columns outside the tangent cylinder and the former carry most of the heat transport.

In this paper only some typical examples have been discussed of numerous simulations of convection driven dynamos in rotating spherical shells that have been performed during the past years. Similar results have been obtained by other groups [3, 10, 12] based on analogous treatments of the basic equations. Certain differences arise from the use of different basic temperature distributions and boundary conditions for the velocity field. The use of a basic temperature gradient proportional to $r^{-1}$ instead of one proportional to $r$, for instance, appears to support a tendency towards dipolar rather than of quadrupolar dynamos [11]. Similar tendencies can be expected if effects of compressibility or of partial stable stratification are taken into account. Besides the simulations based on the fundamental equations of the problem, convection driven spherical dynamos have been simulated through the use of additional assumptions such as hyperdiffusivity concepts or the mean field approximation. There is no place here to survey the numerous papers on the subject and we refer for further details to the review of Fearn (1998).

## References


1. Ardes, M., Busse, F.H. and Wicht, J. (1997) Thermal Convection in Rotating Spherical Shells, *Phys. Earth Planet. Int.* **99**, pp. 55–67
2. Busse, F.H., Grote, E. and Tilgner, A. (1998) On convection driven dynamos in rotating spherical shells, *Studia geoph. et geod.* **42**, pp. 211–223

3. Christensen, U., Olson, P. and Glatzmaier, G.A. (1999) Numerical modeling of the geodynamo: A systematic parameter study, *Geophys. J. Int.* **38**, pp. 393–409
4. Fearn, D.R. (1998) Hydromagnetic flow in planetary cores, *Rep. Prog. Phys.* **61**, pp. 175–235
5. Grote, E., Busse, F.H. and Tilgner, A. (1999) Convection driven quadrupolar dynamos in rotating spherical shells, *Phys. Rev. E* **60**, pp. R5025–R5028
6. Grote, E., Busse, F.H. and Tilgner, A. (2000) Regular and Chaotic Spherical Dynamos, *Phys. Earth Planet. Int.* **117**, pp. 259–272
7. Grote, E. and Busse, F.H. (2000) Hemsipherical dynamos generated by convection in rotating spherical shells, *Phys. Rev. E* **62**, pp. 4457–4460
8. Grote, E. and Busse, F.H. (2001) Dynamics of Convection and Dynamos in Rotating Spherical Fluid Shells, to be published in *Fluid Dyn. Res.*
9. Hirsching, W. and Busse, F.H. (1995) Stationary and chaotic dynamos in rotating spherical shells, *Phys. Earth Planet. Int.* **90**, pp. 243–254
10. Katayama, J.S., Matsushima, M. and Honkura, Y. (1999) Some characteristics of magnetic field behavior in a model of MHD dynamo thermally driven in a rotating spherical shell, *Phys. Earth Planet. Int.* **111**, pp. 141–159
11. Kutzner, C. and Christensen, U. (2000) Effects of driving mechanisms in geodynamo models, *Geophys. Res. Lett.* **27**, 29–32
12. Olson, P., Christensen, U. and Glatzmaier, G.A. (1999) Numerical modeling of the geodynamo: Mechanism of field generation and equilibration, *J. Geophys. Res.* **104**, pp. 10383–10404
13. Sun, Z.-P., Schubert, G. and Glatzmaier, G.A. (1993) Transitions to chaotic thermal convection in a rapidly rotating spherical fluid shell, *Geophys. Astrophys. Fluid Dyn.* **69**, pp. 95-131
14. Tilgner, A., Ardes, M. and Busse, F.H. (1997) Convection in Rotating Spherical Fluid Shells, *Acta Astron. Geophys. Univ. Comeniae* **19**, pp. 337–358
15. Tilgner A. and Busse, F.H. (1997) Finite amplitude convection in rotating spherical fluid shells, *J. Fluid Mech.* **332**, pp. 359–376
16. Wicht, J. and Busse, F.H. (1997) Magnetohydrodynamic dynamos in rotating spherical shells, *Geophys. Astrophys. Fluid Dyn.* **86**, pp. 103-129
17. Zhang, K.-K. and Busse, F.H. (1988) Finite amplitude convection and magnetic field generation in a rotating spherical shell, *Geophys. Astrophys. Fluid Dyn.* **44**, pp. 33-53
18. Zhang, K.-K. and Busse, F.H. (1989) Convection Driven Magnetohydrodynamic Dynamos in Rotating Spherical Shells, *Geophys. Astrophys. Fluid Dyn.* **49**, pp. 97–116

# DYNAMOS IN ROTATING AND NONROTATING CONVECTION IN THE FORM OF ASYMMETRIC SQUARES

A. DEMIRCAN AND N. SEEHAFER
*Institut für Physik, Universität Potsdam*
*PF 601553, 14415 Potsdam, Germany*

**Abstract.**
We study the dynamo properties of asymmetric square patterns in Boussinesq Rayleigh-Bénard convection in a plane horizontal layer. Cases without rotation and with weak rotation about a vertical axis are considered. There exist different types of solutions distinguished by their symmetry, among them such with flows possessing a net helicity and being capable of kinematic dynamo action in the presence as well as in the absence of rotation. In the nonrotating case these flows are, however, always only kinematic, not nonlinear dynamos. Nonlinearly the back-reaction of the magnetic field then forces the solution into the basin of attraction of a roll pattern incapable of dynamo action. But with rotation added parameter regions are found where the Coriolis force counteracts the Lorentz force in such a way that the asymmetric squares are also nonlinear dynamos.

## 1. Introduction

Most cosmic dynamos are driven by convection. Studies of convection-driven dynamos have concentrated either on turbulent convection [2] or on convection near onset, where simple steady flows can be obtained [6, 16]. For Rayleigh-Bénard Boussinesq convection with symmetric top and bottom boundary conditions, i.e. for convection with up-down reflection symmetry, the preferred convection pattern near onset is rolls. However, recent experimental and theoretical investigations show other possible attractors in this kind of convection. Resonant square and hexagon patterns appear in a range where only rolls were previously known to be stable [1, 5, 7, 9]. Usually these *asymmetric* hexagons and squares, with rising or with descending motion in the center (and descending or rising motion near the



*P. Chossat et al. (eds.), Dynamo and Dynamics, a Mathematical Challenge,* 163–171.

boundary) are observed in convection lacking up-down reflection symmetry, namely in fluids with strongly temperature dependent viscosity or in Bénard-Marangoni convection [10, 15, 18], the asymmetric square pattern representing the dominating pattern over a wide range of the control parameters. Details about this pattern can be found in [5, 9, 19]. In the present paper we report on the dynamo properties of the asymmetric square pattern.

## 2. Equations and Parameters

We consider buoyancy-driven rotating convection in an electrically conducting plane fluid layer of thickness $d$ heated from below. Using the Oberbeck-Boussinesq approximation, the governing system of partial differential equations reads as follows:

$$\nabla \cdot \mathbf{v} = \nabla \cdot \mathbf{B} = 0 \tag{1}$$

$$\frac{\partial \mathbf{v}}{\partial t} + (\mathbf{v} \cdot \nabla)\mathbf{v} = -\nabla p + \mathcal{P}\Delta \mathbf{v} + \mathcal{P}\mathcal{R}\theta \mathbf{e}_z + (\nabla \times \mathbf{B}) \times \mathbf{B} + \mathcal{P}\sqrt{\mathcal{T}}\, \mathbf{v} \times \mathbf{e} \tag{2}$$

$$\frac{\partial \mathbf{B}}{\partial t} + (\mathbf{v} \cdot \nabla)\mathbf{B} = \mathcal{P}\mathcal{P}_m^{-1}\Delta \mathbf{B} + (\mathbf{B} \cdot \nabla)\mathbf{v} \tag{3}$$

$$\frac{\partial \theta}{\partial t} + \mathbf{v} \cdot \nabla \theta = v_z + \Delta \theta \,. \tag{4}$$

Here $\mathbf{v}$ is the fluid velocity, $\mathbf{B}$ the magnetic field, and $p$ and $\theta$ represent the deviations of pressure and temperature from their values in the pure conduction state. We use Cartesian coordinates $x$, $y$ and $z$ with the $z$ axis in the vertical direction parallel to the gravitational force. $\mathbf{e}_z$ is the unit vector in the vertical direction whereas the vector $\mathbf{e}$ is the general notation for the unit vector in the direction of the rotation axis. For our special choice $\mathbf{e} = \mathbf{e}_z$ one has $\mathbf{v} \times \mathbf{e} = (v_y, -v_x, 0)$ in Eq. (2). Equations (1)–(4) are given in dimensionless form where the units of length and time are $d$ and $d^2/\kappa$, respectively, with $\kappa$ being the thermal diffusivity. $\theta$ is measured in units of the temperature difference $\delta T$ between lower and upper boundaries of the fluid layer and $p$ and $\mathbf{B}$ in units of $\rho\kappa^2/d^2$ and $\sqrt{\mu_0 \rho}\kappa/d$, respectively, where $\rho$ is the mass density and $\mu_0$ the vacuum magnetic permeability. There are four dimensionless parameters, the Prandtl number $\mathcal{P}$, the magnetic Prandtl number $\mathcal{P}_m$, the Rayleigh number $\mathcal{R}$ and the Taylor number $\mathcal{T}$, defined by

$$\mathcal{P} = \frac{\nu}{\kappa}, \qquad \mathcal{P}_m = \frac{\nu}{\eta}, \qquad \mathcal{R} = \frac{\alpha g d^3}{\nu\kappa}\delta T, \qquad \mathcal{T} = \left(\frac{2\Omega d^2}{\nu}\right)^2, \tag{5}$$

where $\nu$ is the kinematic viscosity, $\eta$ the magnetic diffusivity, $\alpha$ the volumetric expansion coefficient, $\Omega$ the angular velocity of the rotation and $g$ the gravitational acceleration. $\mathcal{R}$ measures the strength of the buoyancy forces. We apply periodic boundary conditions with period $L$ in the horizontal directions $x$ and $y$. The top and bottom planes are assumed to be stress-free, isothermal and impenetrable for matter and electromagnetic energy:

$$\frac{\partial v_x}{\partial z} = \frac{\partial v_y}{\partial z} = v_z = \theta = \frac{\partial B_x}{\partial z} = \frac{\partial B_y}{\partial z} = B_z = 0 \quad \text{at } z = 0, 1 \,. \tag{6}$$

As in [8, 20] we restrict ourselves to the case of a vanishing mean horizontal flow since such a flow can be removed by a Galilean transformation. In our numerics we used a pseudospectral method with a spatial resolution of $32^3$ points for simulations and $16^3$ points for non-simulative eigenvalue and eigenvector calculations. Time integration was performed using a Runge-Kutta scheme. The aspect ratio is kept fixed at a value of $L = 4$ and the Prandtl number is 6.8. The Taylor number, measuring the rotation rate, is restricted to values below the critical one for the Küppers-Lortz [14] instability.

## 3. Convection in the Form of Asymmetric Squares

An example of steady up-square convection in the absence of rotation is shown in Fig. 1(a). The square pattern appears via the skewed-varicose [4] instability of primary rolls (details are described in [9]). Cells with rising or descending motion in the center can appear. The spectrum of the excited Fourier modes shows that the asymmetric squares can be represented to lowest order by

$$\left(A_1 e^{i\mathbf{k_1 x}} + A_2 e^{i\mathbf{k_2 x}}\right) + \left(B_1 e^{i(\mathbf{k_1}+\mathbf{k_2})\mathbf{x}} + B_2 e^{i(\mathbf{k_1}-\mathbf{k_2})\mathbf{x}}\right) + c.c. \tag{7}$$

where $\mathbf{k_1}$ and $\mathbf{k_2}$ are horizontal wave numbers with $\mathbf{k_1} \perp \mathbf{k_2}$, $|\mathbf{k_1}| = |\mathbf{k_2}| = k$ being the fundamental wave number of the square pattern ($2\pi/k$ the side length of the asymmetric squares). (7) represents a superposition of two checkerboard or symmetric square patterns as shown in Fig. 1(b), one with the fundamental wave number and the other, rotated by an angle of $\pi/4$, with the wave number $q = |\mathbf{k_1} + \mathbf{k_2}| = \sqrt{2}k$, which is the wave number of the skewed varicose unstable rolls. We find the two checkerboard pattern solutions to be always unstable. However, rolls with the smaller wave number $k$ are stable in the region where we observe stable asymmetric squares. The wave numbers $k$ and $q$ are in resonance through triadic interactions of the associated wave vectors [21]. A representation like (7) was used in [19] to study square cells in non-Boussinesq convection near onset and is

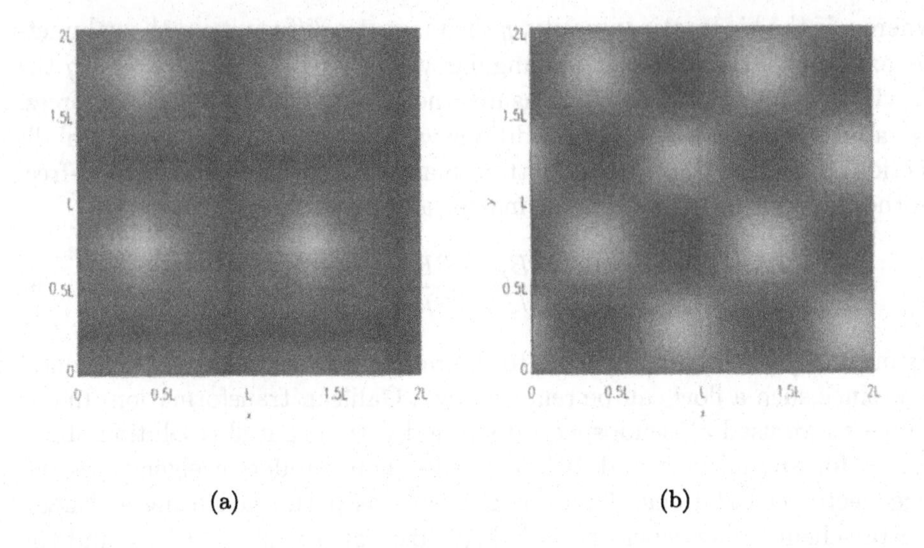


(a) (b)

*Figure 1.* Shadowgraph images of the vertical velocity component $v_z$ in the horizontal midplane for $\mathcal{T} = 0$. Bright areas refer to positive values where the motion of the fluid is upwards. (a) Asymmetric square pattern for $\mathcal{R} = 7000$, (b) unstable checkerboard pattern for $\mathcal{R} = 1010$.

contained in a more general Galerkin ansatz used in [5] to study asymmetric squares in Boussinesq convection. Asymmetric squares were also found numerically in compressible magnetoconvection near onset [17].

A nonvanishing kinetic helicity, for a given volume $V$ defined by $H = \int_V \mathbf{v} \cdot \nabla \times \mathbf{v} \, d^3\mathbf{x}$, is known to be favourable (though not necessary [12]) for dynamo action. Without rotation, the checkerboard pattern solutions are symmetric to reflections in vertical planes parallel to one of the sides of a square and to up-down reflections combined with horizontal translations by one square. The symmetry to reflections in vertical planes then implies zero net helicity (since helicity is a pseudoscalar and thus changes sign under reflections).

The bifurcation of the asymmetric squares from the rolls with wave number $q$ being apparently subcritical [9], simulations starting from a superposition of the unstable rolls with a small perturbation lead, depending on the perturbation added to the rolls, either to asymmetric squares or to rolls of wave number $k$ as final states. In the case of $\mathcal{T} = 0$, the square solutions obtained may or may not possess horizontal $D_4$ symmetry (the dihedral group $D_4$ contains all rotations and reflections which transform a square into itself). For the $D_4$ symmetric solutions one has $A_1 = A_2$ and $B_1 = B_2$ in (7). Like for the checkerboard pattern solutions, the symmetry to reflections in vertical planes then implies zero net helicity.

We now turn to the solutions that lack the horizontal $D_4$ symmetry at $\mathcal{T} = 0$. For these we always find the $D_4$ symmetry to be completely broken. In particular, all reflection symmetries are absent, allowing a nonzero net helicity even in the absence of rotation. Fig. 2(a) shows the helicity of a

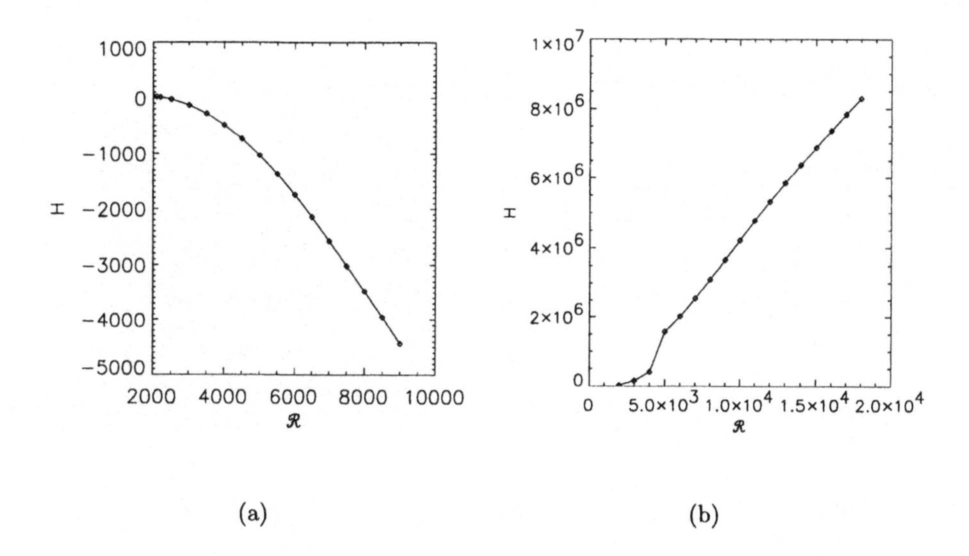


*Figure 2.* The helicity of an upflow square as a function of the Rayleigh number for (a) $\mathcal{T} = 0$ and (b) $\mathcal{T} = 100$.

nonrotating upflow square as a function of $\mathcal{R}$ in the range where the flow is stationary. Rotation at low rates about the vertical axis leaves the asymmetric square pattern stable. The stability boundary towards higher values of $\mathcal{R}$, where the pattern loses stability to different kinds of oscillations, is shifted upwards compared to the case without rotation. The helicity due to rotation [see Fig. 2(b)] is much larger than the "self helicity" of the nonrotating squares already for small $\mathcal{T}$.

## 4. Dynamo Action of Asymmetric Square Convection

We always find the primary convection states to be incapable of kinematic dynamo action. This obviously results from our choice of the parameters, with in particular the rotation rate being restricted to values below the critical one for the Küppers-Lortz instability — for studies of dynamos in *rapidly* rotating plane layers see [13, 22]. We also find the asymmetric square solutions with horizontal $D_4$ symmetry (obtainable only in the absence of rotation), as well as the (unstable) checkerboard pattern solutions in the case of no rotation, to be incapable of dynamo action. The convection flows in the form of asymmetric squares without horizontal $D_4$ symmetry, however, can act as dynamos.

Through solving the magnetic induction equation, first the kinematic dynamo properties of the asymmetric squares were determined. In Fig. 3(a) results for the nonrotating case and for $\mathcal{T} = 100$ are given. The two curves

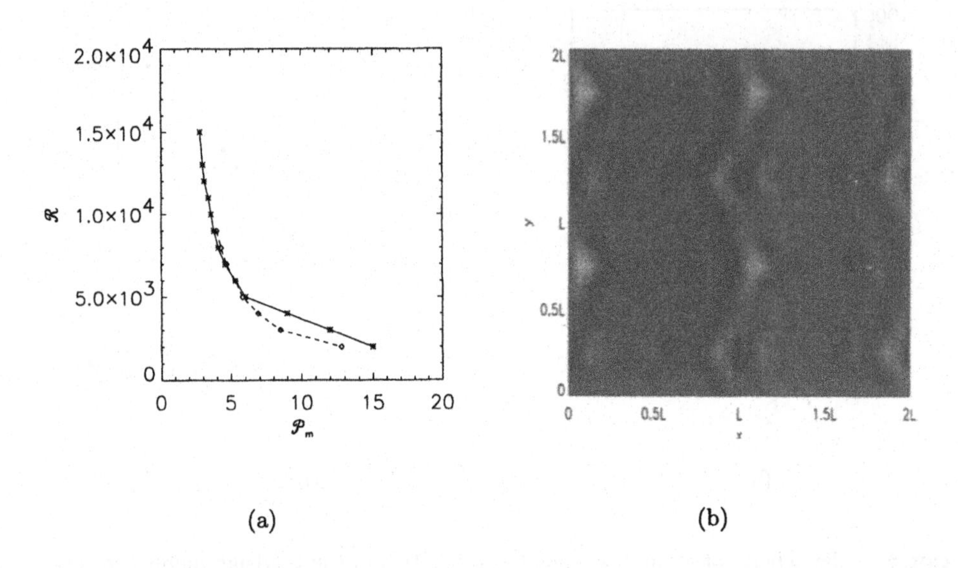


*Figure 3.* **(a) Stability boundary for the kinematic dynamo instability in the $\mathcal{P}_m$-$\mathcal{R}$ plane; the dashed line corresponds to the nonrotating case and the continuous line to $\mathcal{T} = 100$. (b) Shadowgraph image of the vertical component $B_z$ of the unstable magnetic eigenmode for the velocity field shown in Fig. 1(a) and $\mathcal{P}_m = 5.5$; the values in the horizontal midplane are shown, bright areas indicating positive values.**

in the $\mathcal{P}_m$-$\mathcal{R}$ plane are stability boundaries where a single real eigenvalue becomes positive and the dynamo starts. The magnitude of the helicity does not seem to be the most crucial factor for the onset of the kinematic dynamo, though after onset the dynamo growth rates increase much faster with $\mathcal{R}$ if rotation is present. An example of the generated magnetic field in the nonrotating case is depicted in Fig. 3(b). The flow stretches and folds the magnetic field lines and concentrates magnetic flux near cell boundaries.

Traditionally, it is assumed that the introduction of nonlinearity leads to a saturation of the exponential growth of the magnetic field and to a modified velocity field that maintains the magnetic field at a finite amplitude. However, recent investigations show that the back-reaction of the magnetic field can also extinguish the dynamo. This was observed for flow in triply periodic Cartesian geometry driven by an explicit forcing [3], spherical dynamo models with rotation and explicit forcing [11] and two-dimensional convection rolls in a plane layer rotating about an oblique axis [16]. Fig. 4(a) shows the time evolutions of magnetic and kinetic energies starting

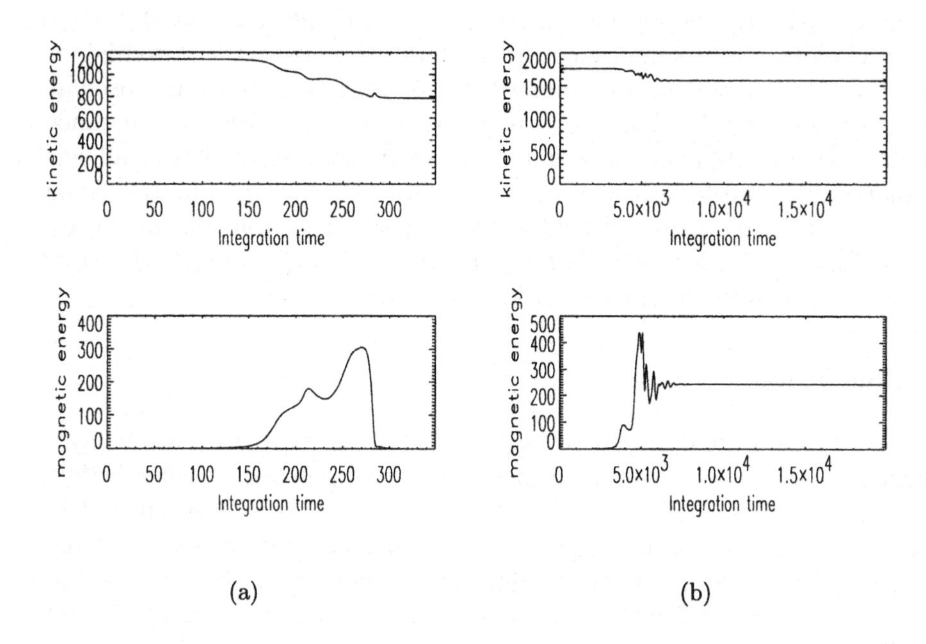


*Figure 4.* Time evolutions of kinetic and magnetic energies for (a) $\mathcal{T} = 0$, $\mathcal{R} = 5000$ and $\mathcal{P}_m = 6$ and (b) $\mathcal{T} = 10$, $\mathcal{R} = 7000$ and $\mathcal{P}_m = 4.65$.

from a square pattern and a small seed magnetic field for the case without rotation. Initially the magnetic field grows exponentially with a well defined growth rate. In this kinematic phase the Lorentz force is negligible and the square pattern remains undisturbed. After reaching a sufficient strength the magnetic perturbation forces the solution into the basin of attraction of the two-dimensional roll state with wave number $k$. The roll solution is incapable of dynamo action and the magnetic field decays to zero.

In order to achieve nonlinear dynamo action, additional effects have to be included. We consider background rotation at very low rates ($0 \leq \mathcal{T} \leq 150$) where the asymmetric square pattern is not affected by hydrodynamical instabilities. Although the mechanism underlying the self-extinguishing is still acting, there are parameter ranges where a nonlinear dynamo is found. Time evolutions of kinetic and magnetic energies in such a case are shown in Fig. 4(b). After the initial kinematic phase, a back reaction of the magnetic field is clearly visible. But though the velocity field is modified, it still corresponds to an asymmetric square pattern. The magnetic field saturates and is maintained for all time. A necessary condition for nonlinear dynamo action is a certain balance between the Coriolis and Lorentz forces. A similar balance between these forces characterizes the strong-field branch of the Childress-Soward [6] dynamo where in a *rapidly* rotating layer the Lorentz force counteracts the Coriolis force such as to facilitate convection [22]. By increasing $\mathcal{P}_m$ (that is, the Lorentz force) and $\mathcal{T}$ (that is, the Coriolis force) simultaneously we find magnetic attractors different from the stationary squares, among them oscillating squares and oscillating or stationary rolls. For $\mathcal{P}_m = 6$, $\mathcal{T} = 27$ and $\mathcal{R} = 7000$ for instance, simulations starting from an asymmetric square solution with a small seed magnetic field lead to stationary magnetic rolls showing modulation along the roll axis; if the magnetic field is switched off these rolls disappear and the solution falls back on the simple roll state (without modulation) of wave number $k$, which is not capable of kinematic dynamo action.

## 5. Conclusion

We have studied the dynamo properties of convection in the form of asymmetric squares. There exist solutions with flows possessing a net helicity and being capable of kinematic dynamo action even in the absence of rotation. Nonlinear dynamos require the presence of rotation and are found in parameter regions where the Coriolis force counteracts the Lorentz force in such a way that self-extinguishing of the dynamo by the Lorentz force is prohibited.

## References


1. Assenheimer, M. and V. Steinberg: 1996, 'Observation of coexisting upflow and downflow hexagons in Boussinesq Rayleigh-Bénard convection'. *Phys. Rev. Lett.* **76**, 756–759.
2. Brandenburg, A., R. L. Jennings, Å. Nordlund, M. Rieutord, R. F. Stein, and I. Tuominen: 1996, 'Magnetic structures in a dynamo simulation'. *J. Fluid Mech.* **306**, 325–352.
3. Brummell, N. H., F. Cattaneo, and S. M. Tobias: 1998, 'Linear and nonlinear dynamo action'. *Phys. Lett. A* **249**, 437–442.
4. Busse, F. H. and R. M. Clever: 1979, 'Instabilities of convection rolls in a fluid of moderate Prandtl number'. *J. Fluid Mech.* **91**, 319–335.
5. Busse, F. H. and R. M. Clever: 1998, 'Asymmetric squares as an attracting set in Rayleigh-Bénard convection'. *Phys. Rev. Lett.* **81**, 341–344.
6. Childress, S. and A. M. Soward: 1972, 'Convection driven hydromagnetic dynamo'. *Phys. Rev. Lett.* **29**, 837–839.
7. Clever, R. M. and F. H. Busse: 1996, 'Hexagonal convection cells under conditions of vertical symmetry'. *Phys. Rev. E* **53**, R2037–R2040.
8. Demircan, A., S. Scheel, and N. Seehafer: 2000, 'Heteroclinic behavior in rotating Rayleigh-Bénard convection'. *Eur. Phys. J. B* **13**, 765–775.
9. Demircan, A. and N. Seehafer: 2001, 'Nonlinear square patterns in Rayleigh-Bénard convection'. *Europhys. Lett.* **53**, 202–208.

10. Eckert, K., M. Bestehorn, and A. Thess: 1998, 'Square cells in surface-tension-driven Bénard convection: experiment and theory'. *J. Fluid Mech.* **356**, 155–197.
11. Fuchs, H., K.-H. Rädler, and M. Rheinhardt: 1999, 'On self-killing and self-creating dynamos'. *Astron. Nachr.* **320**, 129–133.
12. Hughes, D. W., F. Cattaneo, and E. Kim: 1996, 'Kinetic helicity, magnetic helicity and fast dynamo action'. *Phys. Lett. A* **223**, 167–172.
13. Jones, C. A. and P. H. Roberts: 2000, 'Convection-driven dynamos in a rotating plane layer'. *J. Fluid Mech.* **404**, 311–343.
14. Küppers, G. and D. Lortz: 1969, 'Transition from laminar convection to thermal turbulence in a rotating fluid layer'. *J. Fluid Mech.* **35**, 609–620.
15. M. F. Schatz, S. J. VanHook, W. D. McCormick, J. B. Swift, and H. L. Swinney: 1999, 'Time-independent square patterns in surface-tension-driven Bénard convection'. *Phys. Fluids* **11**, 2577–2582.
16. Matthews, P. C.: 1999, 'Dynamo action in simple convective flows'. *Proc. R. Soc. London A* **455**, 1829–1840.
17. Matthews, P. C., M. R. E. Proctor, and N. O. Weiss: 1995, 'Compressible magnetoconvection in three dimensions: planforms and nonlinear behaviour'. *J. Fluid Mech.* **305**, 281–305.
18. Oliver, D. S. and J. R. Booker: 1983, 'Planform of convection with strongly temperature-dependent viscosity'. *Geophys. Astrophys. Fluid Dyn.* **27**, 73–85.
19. Proctor, M. R. E. and P. C. Matthews: 1996, '$\sqrt{2}:1$ resonance in non-Boussinesq convection'. *Physica D* **97**, 229–241.
20. Scheel, S. and N. Seehafer: 1997, 'Bifurcation to oscillations in three-dimensional Rayleigh-Bénard convection'. *Phys. Rev. E* **56**, 5511–5516.
21. Silber, M. and A. C. Skeldon: 1999, 'Parametrically excited surface waves: Two-frequency forcing, normal form symmetries, and pattern selection'. *Phys. Rev. E* **59**, 5446–5456.
22. St. Pierre, M. G.: 1993, 'The strong field branch of the Childress-Soward dynamo'. In: M. R. E. Proctor, P. C. Matthews, and A. M. Rucklidge (eds.): *Theory of Solar and Planetary Dynamos*. Cambridge: Cambridge University Press, pp. 295–302.

# MAGNETOCONVECTION

T. EMONET
*Astronomy and Astrophysics, University of Chicago,*
*5640 South Ellis, Chicago, IL 60637, USA;*
*emonet@flash.uchicago.edu*

F. CATTANEO
*Department of Mathematics, University of Chicago,*
*5640 South Ellis, Chicago, IL 60637, USA;*
*cattaneo@mhd2.uchicago.edu*

AND

N. O. WEISS
*Department of Applied Mathematics and Theoretical Physics,*
*University of Cambridge, Cambridge CB3 9EW, UK;*
*N.O.Weiss@damtp.cam.ac.uk*

**Abstract.** We present results from a systematic numerical survey of the interaction between turbulent convection and magnetic fields, as the net vertical magnetic flux through a large aspect ratio computational box is increased from low to high values. Different regimes are identified and discussed.

## 1. Introduction

The white light granulation of the solar photosphere displays morphological differences between quiet Sun regions, plages and active regions. Magnetograms reveal that the amount of net magnetic flux through these regions varies from zero in the quiet Sun, to a substantial fraction of the equipartition value in plages. The common denominator between the corresponding convective patterns is that they all result from the interaction between turbulent convection and magnetic fields.

In the present paper, we report results from a numerical survey designed to study these interactions for different values of the total vertical magnetic



*P. Chossat et al. (eds.), Dynamo and Dynamics, a Mathematical Challenge,* 173–180.

flux, $\mathcal{F}$, through the layer. The simulations have been carried out for a Boussinesq fluid within a computational box of large aspect ratio ($10 \times 10 \times 1$) and at large Rayleigh number (760 times critical). These numbers require high resolution ($512 \times 512 \times 97$ collocation points) but ensure that the solutions are within the turbulent regime, and that the convective flows are only weakly affected by the boundaries. Several cases have been considered, with $\mathcal{F}$ increasing from zero to values high enough to halt the convection. Here we only present the magnetoconvective cases ($\mathcal{F} \neq 0$). The dynamo calculation ($\mathcal{F} = 0$) is described elsewhere in this volume by Cattaneo (2001, called Paper I in the following; see also [1], [1]).

Three-dimensional interactions between magnetic fields and convectively driven turbulent flows have been investigated by many authors during the past decades. Of particular interest for the present work are the papers by [4] [4] for the dynamo case and those by [6] [6] and [5] [5] for magnetoconvection. The simulations presented here are distinguished from previous calculations by the large values of the Rayleigh number and aspect ratio that we use. Also, our work is the first systematic survey of these interactions as $\mathcal{F}$ is increased. For a more complete description see [3] [3]. The corresponding computer animations are available on the web at `http://flash.uchicago.edu/~mhd`.

The numerical model is the same as in Paper I (we refer to it for further details of the parameters). The initial conditions are described in the next section, where we also provide an overview of the different asymptotic regimes. Details about how the convective flow is affected by the increase of the magnetic flux are given in § 3. The corresponding changes in the structure of the magnetic field are described in § 4. Conclusions are contained in § 5.

## 2. Initial conditions and overview of the asymptotic regimes

We start all the simulations by adding a uniform vertical magnetic field of intensity $B_0$ (expressed as a dimensionless Alfvén speed) to a fully developed unmagnetized convective solution. The only difference from the initial condition used in Paper I is that, for the dynamo case, a random seed field was added instead of a uniform one. In the following we compare the output of seven runs, numbered 1 to 7, obtained by setting $B_0 = 25$, $50$, $50\sqrt{2}$, $100$, $100\sqrt{2}$, $200$ and $200\sqrt{2}$.

The uniform field added to the system is rapidly deformed by the turbulent flow. The volume averaged magnetic energy $\langle B^2 \rangle/2$ is amplified, whereas the kinetic energy $\langle u^2 \rangle/2$ diminishes. Eventually, each energy settles down to some asymptotic value. The different regimes obtained in this way are represented schematically in figure 1. They can be conveniently

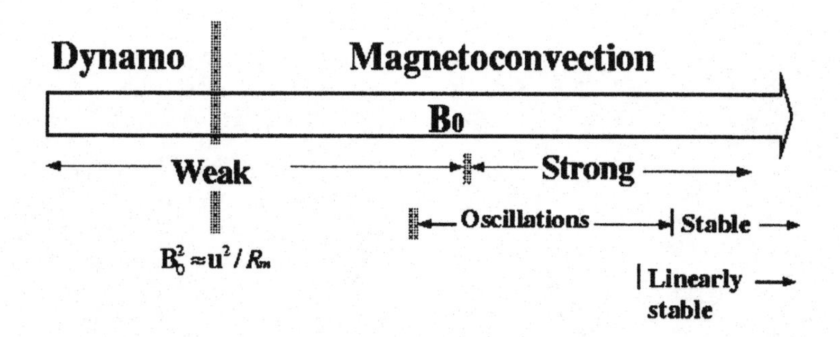


*Figure 1.* Schematic representation of the different regimes obtained for increasing values of the net magnetic flux through the box.

classified into weak and strong magnetic regimes. With very weak fields (e.g. case 1) the pattern is barely distinguishable from that in the dynamo calculation with $B_0 = 0$ described in Paper I. The weak field regime is characterized by a vigorous flow able to concentrate the magnetic field on the border of the convective cells, whereas in the strong field regime flux concentration has ceased. This is best illustrated by looking at the volume rendering of the magnetic intensity (figure 2). The asymptotic regimes can also be classified as non-oscillatory or oscillatory: at some point between the intermediate and strong magnetic field regimes, $\langle u^2 \rangle$ and $\langle B^2 \rangle$ begin to oscillate around their asymptotic values with a definite frequency and a long-time phase coherence. Finally, we note that convection is possible for values of $B_0$ higher than the critical value for linear stability: case 7 is a subcritical regime (see [3], [3] for more details).

In the next two sections we describe in more detail how the structures of the convective flow and of the magnetic field are affected by the systematic enhancement of the curvature force, due to magnetic tension, in the box.

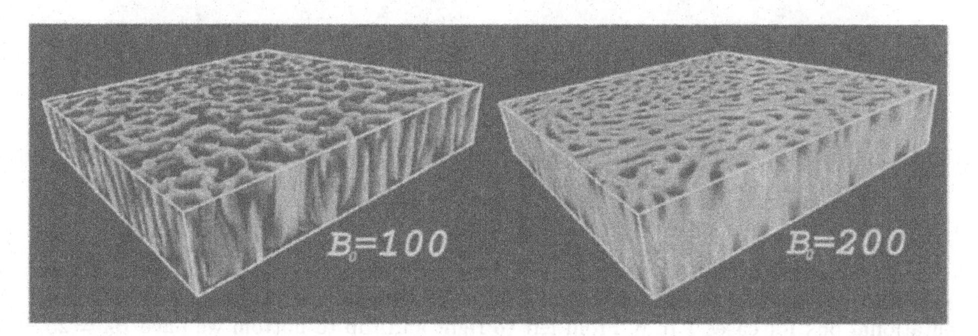


*Figure 2.* 3D structure of the magnetic intensity in the weak and strong magnetic regimes ($B_0 = 100$ and 200). The magnetic intensity is represented by variations in color and transparency. For $B_0 = 100$ the magnetic field is concentrated around the convective cells. No field-free regions are left when $B_0 = 200$.

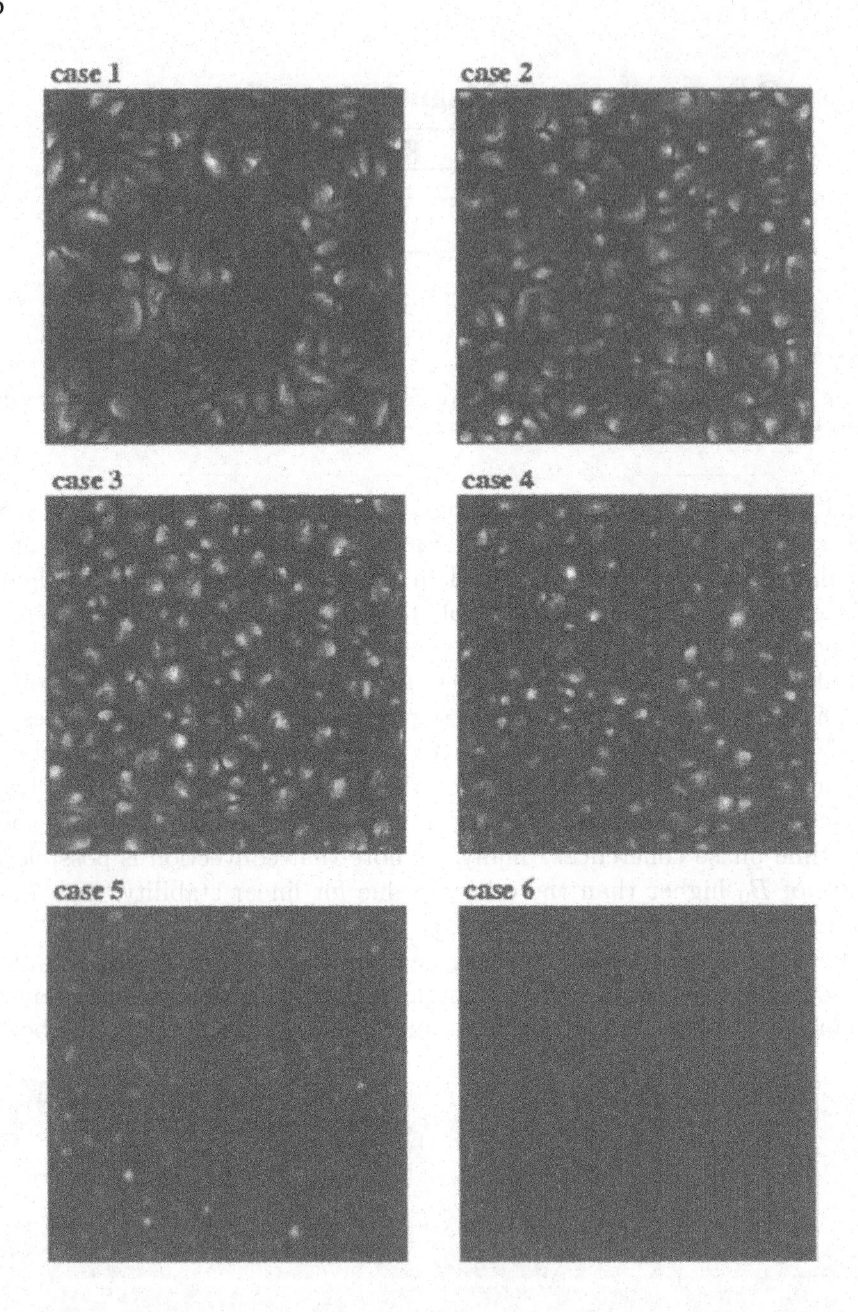


*Figure 3.* Temperature fluctuations in a horizontal plane near the surface of the computational box for cases 1 to 6. From left to right and top to bottom we have $B_0 = 25$, 50, $50\sqrt{2}$, 100, $100\sqrt{2}$ and 200.

## 3. The structure of the convective flow

Near the surface, the structure of convection is best revealed by the temperature fluctuations in a horizontal plane located close to the top of the box. These are shown for cases 1 to 6 in figure 3. Bright regions correspond to hot rising fluid whereas the dark lanes correspond to the cooler downflows. There is an evident similarity between these convective patterns and those observed at the surface of the Sun. For increasing values of $\mathcal{F}$ the horizontal size of the convection cells diminishes, reproducing observed differences between quiet Sun granulation and abnormal granulation in a plage. The general darkening of the temperature map indicates that convective transport becomes less efficient. Finally, notice that the transition from the very weak to the weak and then to the strong magnetic field regime is smooth.

For very small values of the imposed uniform field, the structure of the convective flow is similar to that obtained in the dynamo case. For larger values of $B_0$, the Lorentz force becomes increasingly effective. A consequence is that $\langle u^2 \rangle$ begins (cases 4 and higher) to oscillate with a frequency proportional to $1/B_0$ (the inverse of the mean Alfvén speed) around its asymptotic value.

As $\mathcal{F}$ increases, the flow becomes more uniform along $z$. This can be quantified by measuring the amount of horizontal kinetic energy stored into the different Fourier modes as function of $B_0$ (figure 4). Beginning with case 3, a growing part of the horizontal kinetic energy ends up in the lowest vertical wavenumbers ($k_z = 0$); for case 7 almost all of the energy is in the $k_z = 0$ mode. Similarly, the amount of energy stored in the low

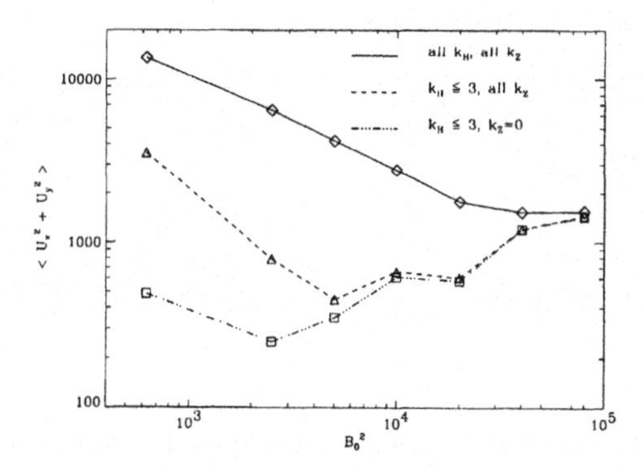


*Figure 4.* Volume average of the horizontal kinetic energy (full line). The dashed and dotted-dashed lines represent the part of the energy which is stored in the modes $k_H = \sqrt{k_x^2 + k_y^2} \leq 3$, and $k_H \leq 3$, $k_z = 0$, respectively.

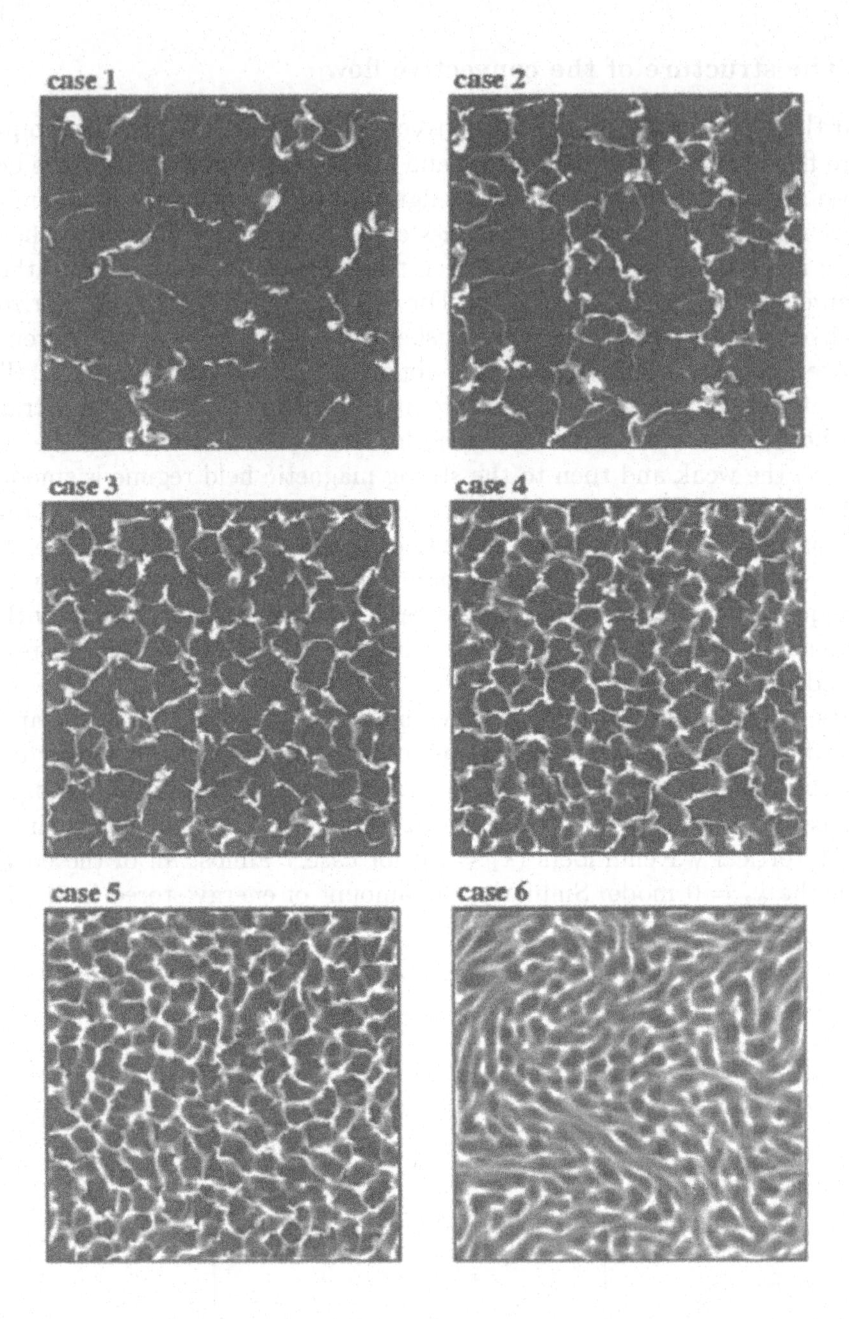


*Figure 5.* Vertical component of the magnetic field near the surface of the computational box for $B_0 = 25$, 50, $50\sqrt{2}$, 100, $100\sqrt{2}$ and 200. Grey indicates a region of zero magnetic field, white means upwards (positive) and black downwards (negative) oriented field. In the weak and intermediate magnetic regimes (cases 1 to 4) the magnetic flux is expelled from the centers of the cells. Flux concentration has ceased in the strong magnetic regime (case 6).

*horizontal* wavenumbers increases. This indicates that a 2D inverse cascade may be at work, transferring energy from small to large horizontal scales.

## 4. The structure of the magnetic field

The intensity of the vertical component of the magnetic field in a horizontal plane near the surface of the box is shown for cases 1 to 6 in figure 5. In all cases, the magnetic field is clearly concentrated into the downflowing regions (compare with fig. 3). In the weak magnetic regime (cases 1 to 4) the flux is concentrated into isolated structures away from the centers of the cells. Case 5 seems to be in an intermediate state and in the strong magnetic regime (case 6) the field appears to be everywhere. In case 6 the distribution of the cells reveals some large scale structure in the horizontal flow. More insights about the structure of the magnetic field are obtained by studying its probability density functions (PDF). As shown in figure 6, apart for the hump due to the added uniform magnetic field, the shape of the PDF in the weak magnetic regime is a stretched exponential, similar to that obtained in the dynamo case (see Paper I); for larger values of $B_0$ the hump grows but

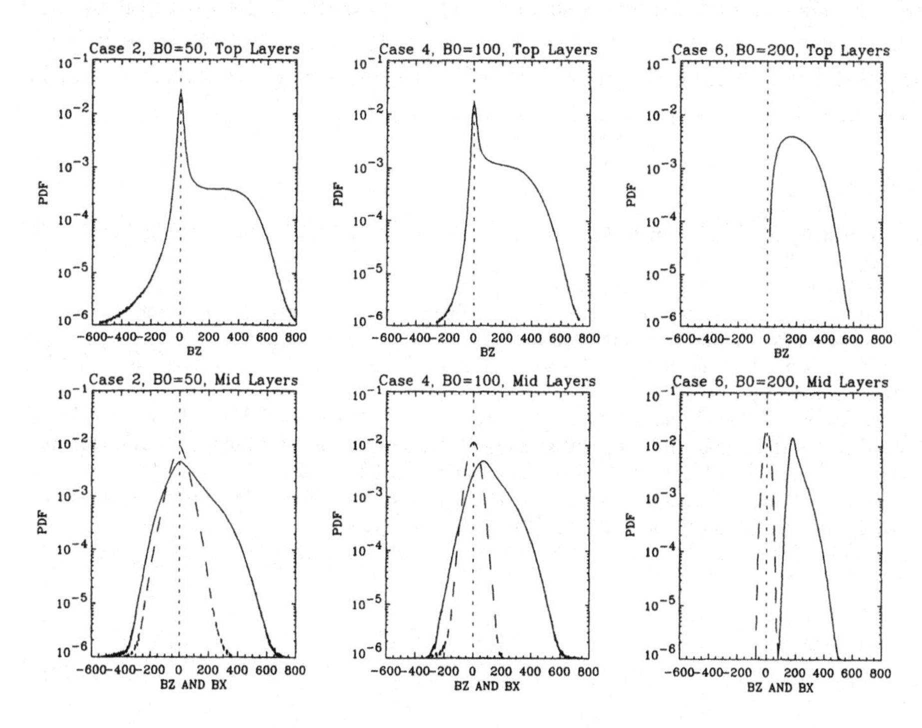


*Figure 6.* Probability density functions of the magnetic field near the top of the box (upper row) and in the middle. The PDFs of $B_z$ and $B_x$ are represented by full and dashed lines respectively. Three cases are considered, corresponding to the weak (case 1), intermediate (case 4) and strong (case 6) magnetic field regimes.

in all cases $B_z = 0$ remains the most probable magnetic intensity, indicating that flux concentration is effective. This picture changes radically for case 6: in the strong magnetic regime, the most probable intensity is close to the intensity of the mean field.

## 5. Conclusion

As the imposed flux $\mathcal{F}$ is increased, two distinctive regimes are observed, *weak* and *strong.* The very weak field regime differs only slightly from the small-scale dynamo discussed in Paper I, while the weak regime itself is characterized by vigorous turbulent convective flows and magnetic fields concentrated into individual structures, mostly aligned with the downflows. In the strong field regime, there is no flux concentration and convection is strongly affected by Lorentz forces. The flow remains turbulent in the horizontal direction but is mostly uniform in the vertical direction. We find that the transition from the weakest magnetic regime to the strongest occurs in a very gradual manner.

The resemblance of the convective patterns and associated magnetic fields in the weak regime (cases 1 to 4) with white light granulation and magnetograms of quiet Sun and plages is gratifying, given the simplicity of the model which includes no effects of compressibility, ionization or radiative transfer.

## References

1. Cattaneo, F. (1999) On the origin of magnetic fields in the quiet photosphere, *ApJ.*, **515**, pp. L39–L42.
2. Cattaneo, F. (2001) These Proceedings, Paper I
3. Cattaneo, F., Emonet, T., Weiss, N.O. (2001) On the interaction between convection and magnetic Fields, in preparation.
4. Meneguzzi, M., Pouquet, A. (1989) Turbulent dynamos driven by convection, *J. Fluid Mech.*, **205**, pp. 297–318.
5. Rucklidge, A.M., Weiss, N.O., Brownjohn, D.P., Matthews, P.C., Proctor, M.R.E. (2000) Compressible magnetoconvection in three dimensions: pattern formation in a strongly stratified layer, *J. Fluid Mech.*, **419**, pp. 283–323.
6. Tao, L., Weiss, N.O., Brownjohn, D.P., Proctor, M.R.E. (1998) Flux Separation in Stellar Magnetoconvection, *ApJ.*, **496**, pp. L39–L42.

# ON THE APPLICATION OF GRID-SPECTRAL METHOD TO THE SOLUTION OF GEODYNAMO EQUATIONS

P. HEJDA AND I. CUPAL
*Geophysical Institute, Acad. Sci, 141 31 Prague, Czech Republic*

AND

M. RESHETNYAK
*Institute of the Physics of the Earth, Russian Acad. Sci, 123810 Moscow, Russia*

## 1. Introduction

The geodynamo process of magnetic field generation occurs in the outer core of the Earth and is described by 3D MHD-equations. The focus of this paper is on a numerical solution and thus the equations will be considered, without the loss of generality, in a simple Boussinesq approximation. Denoting $\mathbf{B}$ the magnetic field, $\mathbf{v}$ the velocity, $p$ the pressure and $T$ the temperature, the dimensionless equations read

$$\frac{\partial \mathbf{B}}{\partial t} = \nabla\times(\mathbf{v}\times\mathbf{B}) + \nabla^2\mathbf{B}, \tag{1}$$

$$R_o(\frac{\partial \mathbf{v}}{\partial t} + \mathbf{v}\cdot\nabla\mathbf{v}) = -\nabla p + \mathbf{F} + E\nabla^2\mathbf{v}, \tag{2}$$

$$\frac{\partial T}{\partial t} + \mathbf{v}\cdot\nabla(T + T_o) = q\nabla^2 T, \tag{3}$$

$$\nabla\cdot\mathbf{B} = 0, \tag{4}$$

$$\nabla\cdot\mathbf{v} = 0, \tag{5}$$

where $\mathbf{F}$ is the sum of Archimedean, Corioliss and Lorenz forces

$$\mathbf{F} = qR_aTr\mathbf{1}_r - (\mathbf{v}\times\mathbf{1}_z) + \Lambda(\nabla\times\mathbf{B})\times\mathbf{B},$$



*P. Chossat et al. (eds.), Dynamo and Dynamics, a Mathematical Challenge,* 181–187.

and where the following dimensionless numbers are introduced

$$E = \frac{\nu}{2\Omega L^2} \text{ [Ekman number]}, \qquad \mathrm{R_o} = \frac{\eta}{2\Omega \mathrm{L}^2} \text{ [Rossby number]},$$

$$\Lambda = \frac{B_0^2}{2\Omega\eta\mu\rho} \text{ [The Elsasser number]}, \qquad \mathrm{q} = \frac{\kappa}{\eta} \text{ [Roberts number]},$$

$$R_a = \frac{\alpha g_o \Theta L}{2\Omega\kappa} \text{[Rayleigh number]}.$$

Here $\eta = (\mu\sigma)^{-1}$ is the magnetic diffusivity, $\mu$ the permeability, $\sigma$ the electrical conductivity, $\rho$ the mean density, $\nu$ the kinematic viscosity, $\Omega$ is the Earth's rotation rate, $\alpha$ is the coefficient of volume expansion, $\Theta$ is the unit of temperature and $\kappa$ represents the eddy thermal diffusivity. The Earth's core radius is the unit of length, L, the velocity is measured by $\eta/L$, the unit of time is $L^2/\eta$ and $B_0$ is the mean magnetic field within the Earth's outer core.

The most frequently used method of solution was outlined in the 1950's by Bullard and Gellman (1954). It consists of

- decomposition of vector fields into toroidal and poloidal parts
- spectral transformation with spherical harmonics in angular coordinates
- special functions (Chebysheff polynomials, Bessel functions) or finite differences in radial direction.

In spite of the general decrease of popularity of spectral methods in the 1970's in favour of finite differences or finite elements, the above method has kept its privileged position in the planetary magnetohydrodynamics because of its indisputable advantages:

- the condition of non-divergence is automatically satisfied
- 'boundary conditions' on the axis of rotation are involved in the properties of spherical harmonic functions
- magnetic field can be easily fitted to an outer source-free potential field.

A different approach was used only for some 2D models (Jepps, 1975; Braginsky, 1978; Braginsky and Roberts, 1987; Cupal and Hejda, 1989; Anufriev *et al.*, 1995; Jault, 1995; Anufriev and Hejda, 1998) In this approach, the resolution into toroidal and poloidal parts is retained but the resulting equations are approximated by finite differences scheme on a 2D spherical grid in the meridional plane.

In our new method for 3D hydromagnetic dynamo we build on our previous positive experience with finite differences approximation in the meridional plane. It is combined with the spectral method in $\phi$-component.

Similar method, except in cylindrical coordinates, was developed by Nakajima and Roberts (1995) for kinematic dynamo.

## 2. Grid-spectral approach

Vector fields are represented in the form

$$\mathbf{F}(r,\,\theta,\,\varphi,\,t) = \begin{pmatrix} F_r \\ F_\theta \\ F_\varphi \end{pmatrix} = \sum_{m=0}^{M} \begin{pmatrix} F_r^{c\,m} \\ F_\theta^{c\,m} \\ F_\varphi^{c\,m} \end{pmatrix} \cos m\varphi + \begin{pmatrix} F_r^{s\,m} \\ F_\theta^{s\,m} \\ F_\varphi^{s\,m} \end{pmatrix} \sin m\varphi,$$

and scalars functions analogously as

$$F(r,\,\theta,\,\varphi,\,t) = \sum_{m=0}^{M} F^{c\,m} \cos m\varphi + F^{s\,m} \sin m\varphi.$$

Since spherical coordinates are used, not only boundary conditions on the physical boundaries must be formulated (the inner-outer core boundary, ICB, and the core-mantle boundary, CMB), but also conditions along the axis of rotation and (for magnetic field) in the center of the sphere. Using the properties of analytical scalar or vector functions for $\theta \to 0$, $\theta \to \pi$ we obtain

$$\frac{\partial F_r^m}{\partial \theta} = F_\theta^m = F_\varphi^m = 0, \qquad \text{for} \quad m = 0,$$

$$F_r^m = \frac{\partial F_\theta^m}{\partial \theta} = \frac{\partial F_\varphi^m}{\partial \theta} = 0, \qquad \text{for} \quad m = 1,$$

$$F_r^m = F_\theta^m = F_\varphi^m = 0, \qquad \text{for} \quad m > 1.$$

The boundary conditions for scalar functions coincide with those for component $r$. The spherical components of the dipolar magnetic field are not defined in the center of the sphere. It would cause problems in the numerical solutions and thus the new variable $\mathbf{b} = r\mathbf{B}$ is introduced. This variable is equal to zero for all spherical components at the center of the sphere. The toroidal-poloidal decomposition is not used in this method and the validity of eqs.(4) and (5) is not automatically satisfied. Their respective roles in the magnetic induction equation (1) and in the fluid momentum equations (2) are different. If the initial magnetic field were non-divergent, the exact solution of (1) would preserve the same property. However, there is a danger that the non-divergence of the solution will deteriorate due to rounding errors during the numerical process. Therefore, eq. (1) is integrated only for components $\theta$ and $\varphi$ and the $r$-component is computed using the condition (4). The detailed description of the numerical solution of the magnetic induction equation, including numerical tests, can be found in Hejda and Reshetnyak, 1999 and 2000.

## 3. Fractional step method

With regards to the fluid momentum equation (2), the incompressibility condition (5) is a constraint in the space of possible solutions and it is related in the implicit form to the pressure field. The solution must further satisfy the boundary conditions. We will consider non-slip boundary conditions with possible rotation of the solid inner core, i.e.

$$v_r = v_\vartheta = 0; \quad v_\varphi = \Omega_i r_i \sin\vartheta \quad \text{at ICB,}$$
$$v_r = v_\vartheta = v_\varphi = 0 \quad \text{at CMB.} \tag{6}$$

The angular velocity of the inner core rotation $\Omega_i$ can be obtained from the impuls momentum equation

$$R_o = I\frac{\partial \Omega_i}{\partial t} = 2\pi r_i^3 \int_0^\pi \left[ Er\frac{\partial}{\partial r}\left(\frac{\overline{v_\varphi}}{r}\right)_{r=r_i} + \overline{B_r B_\varphi}\Big|_{r=r_i} \right] \sin^2\theta d\theta \tag{7}$$

where $I = \frac{8}{15}\pi r_i^5$ is the moment of inertia of the inner core. The equation (7) is reduced to the balance between viscous and magnetic torque when $I$ is ignored. The velocity boundary conditions will be further denoted simply as $\hat{\mathbf{v}}$ for clear explanation of the algorithm, which is a variation of the Fractional step method (Heinrich and Pepper,1999). The velocity is sought at each time step $n$ in the form of a sum of two parts:

$$\mathbf{v}^n = \mathbf{U}^n + \mathbf{u}^n,$$

where $\mathbf{U}^{n+1}$ is obtained in the first substep as the solution of eq.(2) without pressure term

$$R_o \frac{\mathbf{U}^{n+1} - \mathbf{v}^n}{\delta t} = -R_o \mathbf{v}^n \cdot \nabla \mathbf{v}^n + \mathbf{F}^n + E\nabla^2 \mathbf{v}^n$$

with the boundary conditions $U_\tau^{n+1} = \delta t\ \partial p^n/\partial\tau + \hat{v}_\tau$ and $U_r^{n+1} = \hat{v}_r = 0$ at ICB and CMB. Here $\tau$ denotes tangential components $\vartheta$ or $\varphi$.

In the second substep the correction of the velocity is computed using the pressure term. Applying $\nabla\cdot$ to (2) the Poisson equation

$$\nabla^2 p^{n+1} = \nabla\cdot(-R_o \mathbf{v}^n \cdot \nabla \mathbf{v}^n + \mathbf{F}^n + E\nabla^2 \mathbf{v}^n) = (\delta t)^{-1} \nabla\cdot \mathbf{U}^{n+1}$$

is obtained. It is solved with boundary condition $\partial p^{n+1}/\partial r = 0$ at ICB and CMB. The correction of the velocity is then

$$\mathbf{u}^{n+1} = -\delta t \nabla p^{n+1}.$$

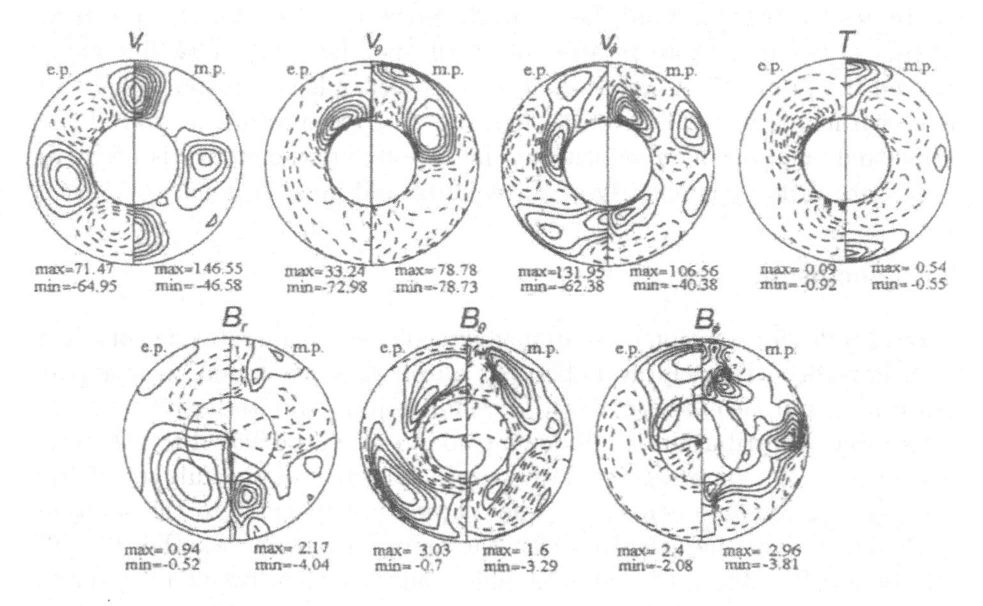


**Figure 1.** The snapshot of the solution of the dynamo equation with an eastward rotating inner core $\Omega_i = 36.4$. The dashed contours relate to the negative values.

It can be easily verified that $\mathbf{v}^{n+1}$ is incompressible and satisfies (2) and (6).

## 4. Computer code and numerical tests

The form of the equation of temperature conductivity (3) is similar to the components of the magnetic induction equation (1) and its solution does not represent any additional difficulties. The computer code was written in Fortran 77 and implemented on PC Pentium III. The power of the computer has allowed us to carry out the tests in the so-called 2.5D approximation, i.e. just for a few modes in $\varphi$. As an example of the solution of the geodynamo model the equation are solved for $m \leq 2$ using the following parameters:

$$\Lambda = 1,\ q = 1.82,\ R_o = 2 \times 10^{-3},\ E = 2 \times 10^{-3},\ R_a = 3 \times 10^2.$$

The temperature profile within the outer core is

$$T_0 = \frac{r_i/r - 1}{1 - r_i}.$$

An interesting feature of the solution is that by solving the equation(7) with non-zero $I$ we obtain an eastward rotation of the inner core $\Omega_i$ that fluctuates (in time) around the value 20. However, when $I = 0$ is assumed in eq. (7) the inner core rotates westward and the value of $\Omega_i$ fluctuates around the value -15. A snapshot of the time dependent solution with $I \neq 0$ is shown in the figure 1. Even though the inner core rotates eastward the azimuthal component of velocity at the core-mantle boundary is negative (westward drift), particularly at the equator and polar regions.

## 5. Conclusion

Several codes for solution of hydromagnetic dynamos in a rotating spherical shell have been recently available, but all of them are based on a similar (spectral) approach. Therefore, it is useful to have an alternative method. This recent contribution is a follow-up to our studies devoted to the kinematic dynamo and presents the first test solutions of the full set of hydromagnetic dynamo equations. The method is applicable to the solution of the geodynamo models. However, the question remains whether we will still be able to obtain the solution when choosing all parameters closer to the real parameters of the Earth's core, particularly the Ekman number, whose real values are assumed to be under $E = 10^{-12}$.

**Acknowledgements**: This work was supported by INTAS foundation (grant 99-00348), Russian Foundation of Basic Research (grant 00-05-65258) and Grant Agency of the Academy of Sciences of the Czech Republic (grant A3012006).

## References

1. Anufriev, A.P., Cupal, I. and Hejda, P. (1995) The weak Taylor state in an $\alpha\omega$-dynamo. *Geophys. Astrophys. Fluid. Dyn.*, **79**, 125–145.
2. Anufriev, A.P. and Hejda, P. (1998) Effect of the magnetic field at the inner core boundary on the flow in the Earth's core. *Phys. Earth Planet. Inter.*, **106**, 19–30.
3. Braginsky, S.I. (1978) Nearly axially symmetric model of the hydrodynamic dynamo of the Earth II. *Geomag. Aeron.*, **18**, 225–231.
4. Braginsky, S.I. and Roberts, P.H. (1987) A model-Z geodynamo. *Geophys. Astrophys. Fluid. Dyn.*, **38**, 327–349.
5. Bullard, E.C. and Gellman, H. (1954) Homogeneous dynamos and terrestrial magnetism. *Phil. Trans. R. Soc. Lond.*, **A 247**, 213–278
6. Cupal, I. and Hejda, P. (1989) On the computation of a model-Z with electromagnetic core-mantle coupling. *Geophys. Astrophys. Fluid. Dyn.*, **49**, 161–172.
7. Heinrich, C.J., Pepper, D.W. (1999) Intermediate finite element method, Taylor & Francis, New York, pp.1-585
8. Hejda, P., Reshetnyak, M. (1999) A grid-spectral method of the solution of the 3D kinematic geodynamo with the inner core. *Studia geoph. et. geod.*, **43**, 319–325
9. Hejda, P., Reshetnyak, M. (2000) The grid-spectral approach to 3-D geodynamo modelling. *Computers & Geosciences*, **26**, 167–175
10. Jault, D. (1995) Model Z by computation and Taylor's condition. *Geophys. Astrophys. Fluid. Dyn.*, **79**, 99–124.

11. Jepps., S.A. (1975) Numerical models of hydromagnetic dynamos. *J. Fluid. Mech.*, **67**, 625–646.
12. Nakajima, T. and Roberts, P.H. (1995) An application of mapping method to asymmetric kinematic dynamos. *Phys. Earth Planet. Inter.*, **91**, 53–61.

# SUPER– AND COUNTER–ROTATING JETS AND VORTICES IN STRONGLY MAGNETIC SPHERICAL COUETTE FLOW

RAINER HOLLERBACH
*Department of Mathematics*
*University of Glasgow*
*Glasgow, G12 8QW, U.K.*

## 1. Introduction

Spherical Couette flow is the flow induced in a spherical shell by fixing the outer sphere and rotating the inner one. Magnetic Couette flow is then the natural magnetohydrodynamic extension in which the fluid is taken to be electrically conducting, and a magnetic field is imposed. For the very strong fields we will consider here, the topology of this imposed field — which field lines thread both boundaries, and which only one or the other — turns out to be crucial. In this work we will therefore present a systematic survey of the possible field configurations, and in each case study the effect on the resulting flow.

The reason the topology of the field lines is so important is due to the anisotropic nature of the magnetic tension force, coupling different regions only along the field lines. There is thus a natural, and dynamically extremely important division of the fluid into regions magnetically coupled to both boundaries, or only to one or the other. Starchenko (1997) was the first to appreciate this point, and to explore some of its logical consequences. In particular, he addressed the question of what behaviour should be expected on those field lines separating one region from another, and showed that one obtains a shear layer across which the fluid adjusts from one regime to the other. Similarly (and independently), Dormy *et al.* (1998) discovered that this shear layer can in fact contain within it a super-rotating jet, in which the fluid essentially over-compensates, and ends up rotating faster than the imposed rotation of the inner sphere.

Aside from the fact that Starchenko's work is asymptotic whereas Dormy *et al.*'s is numeric, the only other difference is that Dormy *et al.* took the inner region $r < r_i$ to be a conductor and the outer region $r > r_o$ to be



*P. Chossat et al. (eds.), Dynamo and Dynamics, a Mathematical Challenge,* 189–197.

an insulator, whereas initially Starchenko took them both to be insulators. Although it is not immediately obvious why, switching the inner boundary from insulating to conducting must therefore be the underlying cause of the jet. And indeed, when Starchenko redid his asymptotic analysis with a conducting inner boundary, he too obtained a jet.

Motivated by this finding that the precise choice of electromagnetic boundary conditions can have such a strong, and entirely unexpected effect, Hollerbach (2000) systematically considered all four possible choices, namely taking both inner and outer boundaries independently to be either insulating or conducting. He found that taking both boundaries to be conducting yielded more dramatic results still. In particular, the strength of the jet then appears to increase indefinitely with increasing imposed field strength, whereas before it had ultimately leveled off, at around 30% of the imposed rotation. In this work we will therefore also take both boundaries to be conducting.

## 2. Equations

As in Hollerbach and Skinner (2001), in the limit of small magnetic Reynolds number $Rm$, the governing equations become

$$\frac{\partial \mathbf{U}}{\partial t} + Re\, \mathbf{U} \cdot \nabla \mathbf{U} = -\nabla p + \nabla^2 \mathbf{U} + M^2 \left(\nabla \times \mathbf{b}\right) \times \mathbf{B}_0,$$

$$0 = \nabla^2 \mathbf{b} + \nabla \times (\mathbf{U} \times \mathbf{B}_0),$$

where $\mathbf{B}_0$ is the imposed field and $Rm\, \mathbf{b}$ the induced field — and we note that $Rm$ only enters here, in the meaning we ascribe to $\mathbf{b}$, but not in the actual equations to be solved. The two non-dimensional parameters that do appear in the equations are the Hartmann number

$$M = \frac{B_0 r_i}{(\mu \rho \nu \eta)^{1/2}}$$

measuring the strength of the imposed field, and the Reynolds number

$$Re = \frac{\Omega_0 r_i^2}{\nu}$$

measuring the imposed rotation.

Hollerbach and Skinner took the axial field $\mathbf{B}_0 = \hat{\mathbf{e}}_z$ for their imposed field, and computed not only the axisymmetric basic state as it depends on $M$ and $Re$, but also the non-axisymmetric instabilities that arise for sufficiently large $Re$. In this work we will also insist that $\mathbf{B}_0$ be purely axisymmetric, to ensure that we still have an axisymmetric basic state. However, we will then consider only these axisymmetric solutions, and only in the $Re = 0$ limit of infinitesimally small imposed rotation. We therefore have a two-dimensional, linear problem that depends only on the spatial structure of $\mathbf{B}_0$ and the Hartmann number $M$, enabling us to do a systematic survey of these two dependencies. For the details of the conducting boundary conditions and the rest of the numerical implementation the reader is again referred to Hollerbach and Skinner.

## 3. Choice of $\mathbf{B}_0$

Let us begin by considering the question of how many fundamentally different field topologies there are. We agreed above that there are three different types of field line, namely ones threading both boundaries, and ones threading only one or the other. (Requiring $\mathbf{B}_0$ to be a potential field excludes the possibility of field lines threading neither boundary, that is, of field lines forming closed loops entirely within the shell.) There are thus eight fundamentally different types of field, corresponding to the three types of field line being independently either present or absent. Of course, one of these eight, namely when all three types of field line are absent, is just non-magnetic Couette flow again, which is certainly an interesting problem, but not the one we want to consider here. Also, one can show easily enough that having all field lines threading both boundaries, and none only one or the other, would require $\mathbf{B}_0$ to have a monopole component. We are thus left with the six possible configurations

$$bi, \quad bo, \quad bio, \quad i, \quad o, \quad io,$$

where the $b/i/o$ indicates the presence of field lines threading both boundaries/inner only/outer only. Of these six, only *bi* and *bo* have been considered so far, with Hollerbach (2000) finding that a dipolar field (*bi*) yields a super-rotating jet, whereas an axial field (*bo*) yields a counter-rotating jet, both occurring on the particular field line separating the $b$ region from the $i/o$ region (as expected), and both increasing with increasing field strength roughly as $M^{0.6}$. So, our goal in this work is to come up with simple examples of the other four configurations as well, and see whether they yield super- or counter-rotating jets, or something else entirely, and also whether whatever structures emerge again scale as $M^{0.6}$, or something else.

Very conveniently, we can produce examples of all six configurations simply by taking appropriate linear combinations of the dipolar and axial fields

$$\mathbf{B}_d = 8r^{-3}\cos\theta\,\hat{\mathbf{e}}_r + 4r^{-3}\sin\theta\,\hat{\mathbf{e}}_\theta,$$

$$\mathbf{B}_a = \cos\theta\,\hat{\mathbf{e}}_r - \sin\theta\,\hat{\mathbf{e}}_\theta.$$

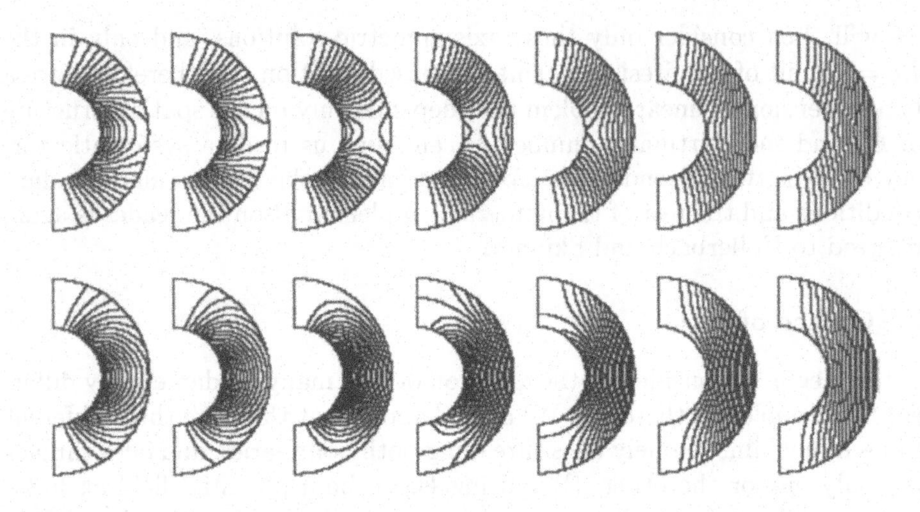

*Figure 1.* The field lines of $\mathbf{B}_0$, with the top row showing $\mathbf{B}_+$ and the bottom row $\mathbf{B}_-$. In both cases $\epsilon = 1/4$, 1/2, 1, 2, 4, 8 and 16, going from left to right.

Figure 1 shows the result of taking

$$\mathbf{B}_0 = \mathbf{B}_d \pm \epsilon \mathbf{B}_a \equiv \mathbf{B}_\pm,$$

and then of course renormalizing so that the volume-averaged $|\mathbf{B}_0|^2 = 1$.

Considering $\mathbf{B}_+$ first, we see that for $\epsilon < 1/2$ the field is qualitatively like a purely dipolar field, that is, it is of *bi* type. Between $\epsilon = 1/2$ and $\epsilon = 4$, however, an X-type neutral point first appears at $r_o$ (=2), steadily moves inward, and finally disappears at $r_i$ (=1). In this range of $\epsilon$, the field is seen to be of *bio* type. And finally, for $\epsilon > 4$ it is of *bo* type, and is qualitatively like a purely axial field.

Considering $\mathbf{B}_-$ next, we now find that for $\epsilon < 1$ it is of *bi* type, and for $\epsilon > 8$ of *bo* type. At $\epsilon = 1$ and 8, though, it is *i* and *o*, respectively, and in the intermediate range $1 < \epsilon < 8$ it is *io*. We thus see that these simple linear combinations of $\mathbf{B}_d$ and $\mathbf{B}_a$ do indeed yield examples of all six configurations, and are thus ideally suited for our purposes here.

## 4. Jets

For $\mathbf{B}_+$ we will take $\epsilon = 1/2$, $4/1.5^3$ and 4, corresponding to the X-type neutral point being at $r = 2$, 1.5 and 1, that is, just appearing, exactly half-way, and just disappearing. The reason we wish to consider $\epsilon = 1/2$ and 4 as well, even though technically they are just the *bi* and *bo* cases

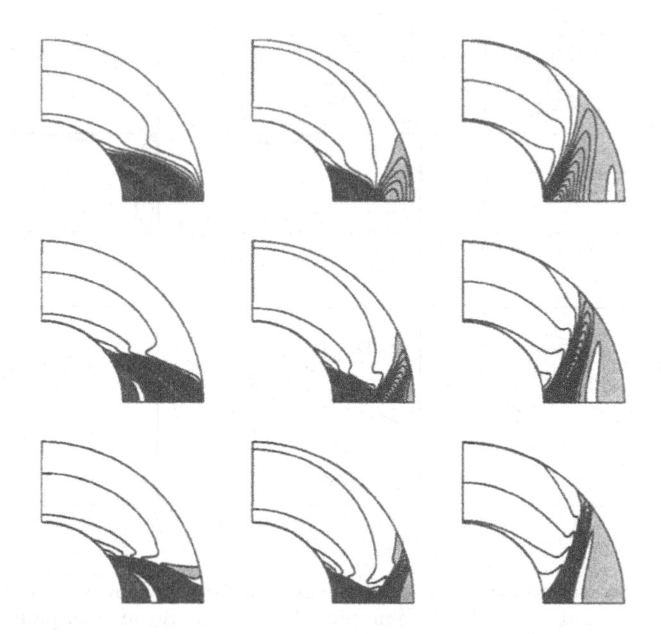

*Figure 2.* Contour plots of the angular velocity for $\mathbf{B}_+$, with a contour interval of 0.2. From left to right $\epsilon = 1/2$, $4/1.5^3$ and 4, and from top to bottom $M = 10^2$, $10^{2.5}$ and $10^3$. Regions of super-rotation ($\Omega > 1$) are dark-shaded, regions of counter-rotation ($\Omega < 0$) light-shaded.

already considered, is that they are somewhat degenerate cases. In particular, Hollerbach and Skinner conjectured that this $M^{0.6}$ scaling must have something to do with the scalings of the boundary layer at precisely these points where the field line separating $b$ from $i/o$ just grazes the outer/inner boundary, that is, where $B_r = 0$, so if we choose degenerate fields for which $B_\theta = 0$ there as well, it might perhaps yield a different scaling. And of course $\epsilon = 4/1.5^3$ is the genuinely new *bio* case.

Figures 2 and 3 show these results, and we note that *bi*/*bo* again yield super-/counter-rotating jets, and that *bio* yields super- and counter-rotating jets simultaneously. Furthermore, the strengths of all of these jets once again increase with increasing $M$, and even with roughly the same 0.6 exponent as before (although it is perhaps worth noting that all four curves are still becoming slightly flatter, so presumably the true asymptotic exponent is somewhat less than 0.6). It would thus appear that this degeneracy mentioned above has no effect on the scalings after all. Finally, we note that as $M$ increases, the width of all of these jets decreases, at a rate consistent with the expected asymptotic scaling $M^{-1/2}$ (Starchenko, 1997).

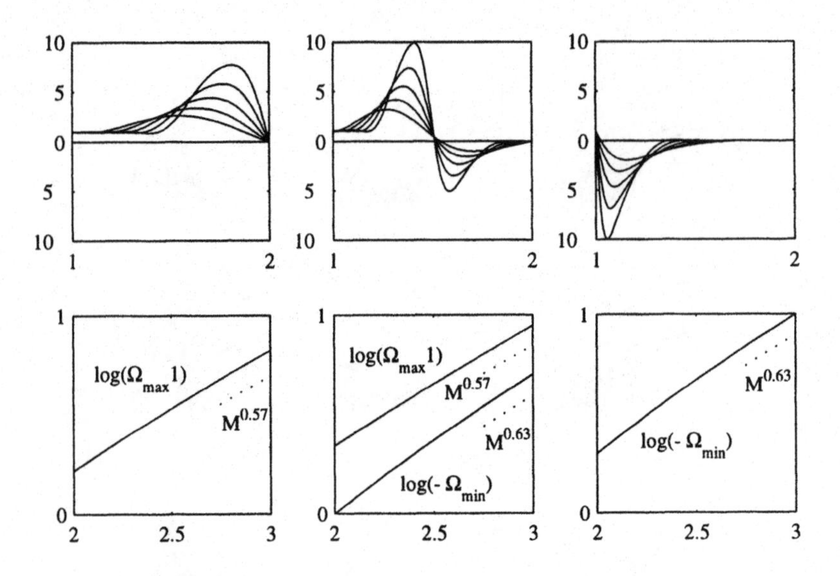


*Figure 3.* The top row shows cross-sections through the jets shown in figure 2, that is, $\Omega$ as a function of $r$, at $\theta = \pi/2$ (the equator). The five curves in each panel correspond to $\log(M) = 2$, 2.25, 2.5, 2.75 and 3. The bottom row shows the strengths of the jets as functions of $\log(M)$, and indicates the scaling roughly as $M^{0.6}$.

## 5. Vortices

For $\mathbf{B}_-$ we take $\epsilon = 1$, $8/1.5^3$ and 8, corresponding to the dividing line between $i$ and $o$ being at $r = 2$, 1.5 and 1. These are thus the three remaining cases $i$, $io$, and $o$. The results, shown in figures 4 and 5, look rather different now. In particular, whereas before we had increasingly strong but narrow jets, here there is remarkably little variation with $M$. Focussing attention on the equatorial regions first, shown in cross-section in the top row of figure 5, we note that there are again regions of super-rotation and other regions of counter-rotation, which again become increasingly thin with increasing $M$. Now, however, the degree of super- or counter-rotation never exceeds 10%, and does not increase with $M$, but appears to saturate instead. These are thus more like the jets obtained by Dormy *et al.* and Starchenko.

The most interesting dynamics in these cases occur not at the equator, but at the pole, shown in cross-section in the bottom row of figure 5. We clearly see a counter-rotating vortex in the $i$ case, and a super-rotating vortex in the $o$ case. And like the jets above, the strengths of these vortices too appear to saturate with increasing $M$, at around 70% for the counter-rotating one, and around 100% for the super-rotating one. Unlike any of the previous jets though, these vortices do not become thinner and thinner with increasing $M$; they seem to become essentially independent of $M$ instead.

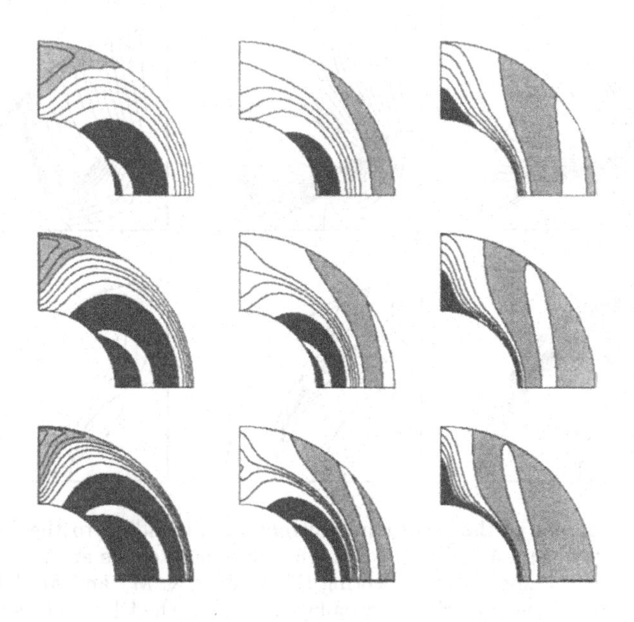

*Figure 4.* As in figure 2, but now for $\mathbf{B}_-$, and $\epsilon = 1$, $8/1.5^3$ and 8.

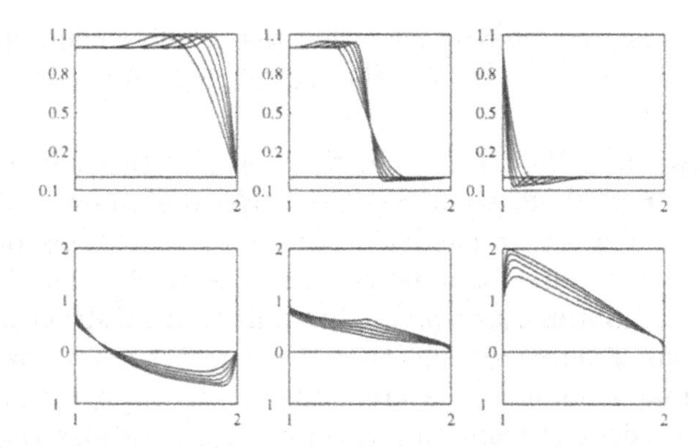


*Figure 5.* As in figure 3, the top row shows $\Omega$ as a function of $r$, at $\theta = \pi/2$. The bottom row shows $\Omega$ as a function of $r$, at $\theta = 0$.

## 6. Discussion

In order to even begin to understand the results presented here, one must consider not just the angular velocity, but also the induced electric currents $\mathbf{j} = \nabla \times \mathbf{b}$, shown in figure 6. Perhaps not surprisingly, the currents follow the imposed field lines very closely, so that the Lorentz force $\mathbf{j} \times \mathbf{B}_0$ is less than one might have expected. However, if one looks carefully one finds that $\mathbf{j}$ is not parallel to $\mathbf{B}_0$ everywhere; it is in fact just at the regions of

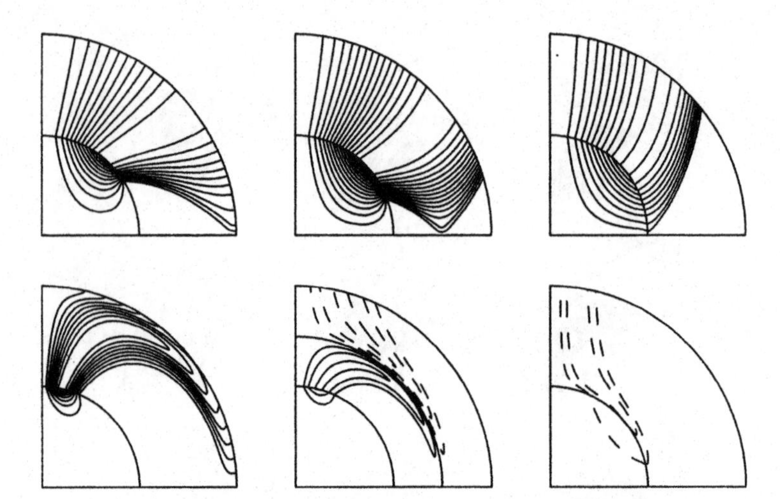

*Figure 6.* Streamlines of the meridional current corresponding to the $M = 10^3$ cases in figures 2 (top row) and 4 (bottom row). The contour intervals are $10^{-2}$ and $10^{-4}$ for the top/bottom rows respectively, reflecting the scalings as $M^0$ and $M^{-1}$ (see figure 7). Solid lines indicate counter-clockwise circulation, and dashed lines clockwise, with the recirculation through the outer boundary not shown.

strongest super- and counter-rotation that the misalignment, and hence the Lorentz force, is greatest — and of the right sign to explain the anomalous rotation.

Nevertheless, to fully understand the origin of these anomalous rotations, it is not enough to compare $\mathbf{j}$ with $\mathbf{B}_0$ and simply see where the Lorentz force is greatest. In particular, in the corresponding problem with insulating inner and outer boundaries (not presented in detail here due to lack of space), one obtains a surprisingly similar spatial structure for $\mathbf{j}$, with the only obvious difference being that the currents must now recirculate through Hartmann boundary layers rather than through the boundaries themselves. One does not obtain any anomalous rotations though, so the spatial pattern alone cannot explain the conducting boundary results.

In order to explain these results, one must therefore consider not just the spatial structure, but the magnitude of $\mathbf{j}$ as well, where one does find a striking difference between the conducting and insulating cases. Figure 7 shows these results, and one notes that for all choices of $\mathbf{B}_0$ the current is much weaker in the insulating case, and even appears to have a fundamentally different scaling with Hartmann number, $M^{-1}$ versus $M^0$ for one set, and $M^{-1.3}$ versus $M^{-1}$ for the other. Why this should be is easy to understand, at least qualitatively: it is simply much easier to recirculate current

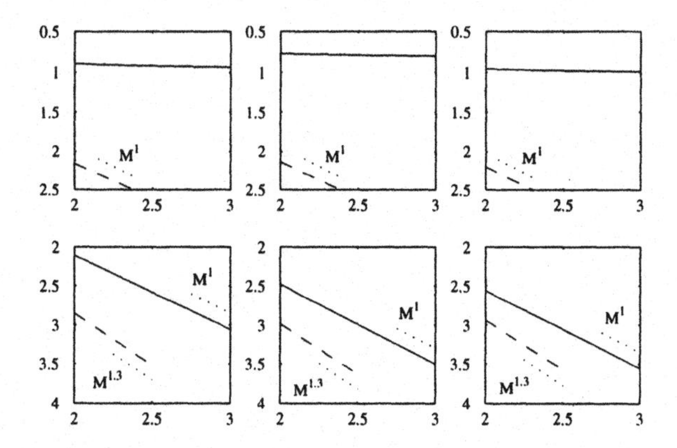


*Figure 7.* The magnitudes of the currents, $\log(j_{tot})$, as functions of $\log(M)$, with the top and bottom rows corresponding to the solutions in figures 2 and 4, respectively. The solid lines in each case correspond to the conducting boundary results, the dashed lines to the insulating boundary results. The dotted lines indicate the presumed asymptotic scalings.

through the boundaries rather than through increasingly thin boundary layers. It is not immediately obvious though how that leads to the specific scalings obtained here; that can only emerge from a detailed asymptotic analysis of this problem, which clearly deserves further study.

## References

1. Dormy, E., Cardin, P. and Jault, D. (1998) MHD flow in a slightly differentially rotating spherical shell, with conducting inner core, in a dipolar magnetic field, *Earth Planet. Sci. Lett.*, **160**, 15–30.
2. Hollerbach, R. (2000) Magnetohydrodynamic flows in spherical shells, to appear in Egbers, C. and Pfister, G. (eds.) *Physics of Rotating Fluids*, Lecture Notes in Physics, vol. **549**. Springer, Heidelberg.
3. Hollerbach, R. and Skinner, S. (2001) Instabilities of magnetically induced shear layers and jets, *Proc. Roy. Soc.* A, in press.
4. Starchenko, S.V. (1997) Magnetohydrodynamics of a viscous spherical layer rotating in a strong potential field, *J. Exper. Theor. Phys.*, **85**, 1125–1137.

# LARGE- AND SMALL-SCALE DYNAMO ACTION

D.W. HUGHES
*Department of Applied Mathematics,*
*University of Leeds, Leeds LS2 9JT, UK*

F. CATTANEO
*Department of Mathematics,*
*University of Chicago, Chicago, IL 60637, USA*

AND

J.-C. THELEN
*Department of Astronomy and Astrophysics,*
*University of Chicago, Chicago, IL 60637, USA*

**Abstract**

For an idealized model of helically forced flow in an extended domain, we have investigated the interaction between dynamo action on different spatial scales. The evolution of the magnetic field is studied numerically; from an initial state of weak magnetization, through the kinematic and into the dynamic regime. We show how the choice of initial conditions is a crucial factor in determining the structure of the magnetic field at subsequent times. Furthermore, with initial conditions chosen to favour the growth of the small-scale field, the evolution of the large-scale magnetic field can be described in terms of the $\alpha$-effect of mean field magnetohydrodynamics.

## 1. Introduction

The evolution of cosmical magnetic fields is most naturally explained in terms of dynamo action, in which the fields are amplified and maintained by the motions of an electrically conducting fluid. Traditionally, astrophysical dynamo theory has concentrated on describing the evolution of *large-scale* magnetic fields, such as the component of the Sun's magnetic field that is responsible for the solar cycle. In particular, *mean field electrodynamics* was developed specifically to study this kind of process, in which the magnetic



*P. Chossat et al. (eds.), Dynamo and Dynamics, a Mathematical Challenge,* 199–206.

field evolves on a scale large compared to a typical scale of the velocity; indeed, practically all studies of astrophysical dynamos have been cast within its framework. One of the assumptions of mean field theory, possibly one that is not always explicitly stated (though see the discussions in [5] [5] and [4] [4]), is that the small-scale magnetic fields — of scales comparable to or smaller than that of the velocity — are stable to dynamo growth. Indeed, for low magnetic Reynolds number turbulence lacking reflectional symmetry (e.g. helical turbulence), it is indeed the case that there will be dynamo amplification of the large-, but not the small-scale field. However, at high or even moderate magnetic Reynolds numbers the situation is likely to be more complex. Recent advances in dynamo theory suggest that at high magnetic Reynolds numbers (the case of astrophysical interest) small-scale dynamo action is prevalent even in the absence of helicity (see, for example, [3] [3]). Thus, turbulence in highly (electrically) conducting fluids is always likely to act as a small-scale dynamo, and in the presence of helicity, as a large-scale dynamo also.

The aim of the work described here is to go beyond studies of large- or small-scale dynamos in isolation, and to investigate the important issue of the interplay between magnetic field generation on different spatial scales. We base our approach on moderately high magnetic Reynolds number simulations of helically forced flows, in which we follow in detail the magnetic field evolution in both the kinematic and dynamic regimes.

## 2. Formulation of the model

We consider dynamo action in an incompressible fluid with constant electrical conductivity and viscosity. The flow is driven by the forcing function $\mathbf{F}$, related to the velocity $\mathbf{U}_0$ through

$$\mathbf{U}_0 = (\partial_y\psi, -\partial_x\psi, \psi), \qquad \psi = \sqrt{3/2}[\cos(x+\cos t) + \sin(y + \sin t)], \quad (1)$$

and

$$\mathbf{F} = (\partial_t - Re^{-1}\nabla^2)\mathbf{U}_0. \quad (2)$$

This construction guarantees that in the absence of magnetic field, $\mathbf{U} = \mathbf{U}_0$ is a solution of the momentum equation. The flow $\mathbf{U}_0$ is $z$-independent and hence, for this flow, the induction equation has separable solutions of the form

$$\mathbf{B}(\mathbf{x}, t) = \mathbf{B}_p(x, y, t)\exp(st + ik_z z), \qquad s = \sigma + i\omega, \quad (3)$$

where $\mathbf{B}_p$ is periodic in time with the same period as $\mathbf{U}_0$. This form of the solution affords a formal reduction of the induction equation from three to two spatial dimensions and allows a numerical investigation of the kinematic problem up to reasonably high values of the magnetic Reynolds number

($Rm \approx 10^5$). The dynamo growth rate $\sigma$ depends, in general, on $Rm$ and $k_z$. In the simplest case where $\mathbf{B}_p$ has the same spatial periodicity as $\mathbf{U}_0$, the maximum growth rate (taken over all values of $k_z$) increases with $Rm$ before levelling off at a value of approximately 0.3 at $Rm \approx 80$. Significantly, increases in $Rm$ up to the level of computational feasibility ($Rm \approx 10^5$) leave the growth rate unaffected, suggesting that the flow is indeed a fast dynamo [2]. For $Rm$ sufficiently large ($\gtrsim 50$), the mode of maximum growth rate becomes independent of $Rm$, assuming the value $k_z = k_{max} \approx 0.57$; i.e. the mode of maximum growth rate attains asymptotically a fixed "size".

In the present paper our interests are chiefly in the nonlinear regime, where we anticipate that the problem will become three-dimensional (with $\mathbf{U} \neq \mathbf{U}_0$), and that therefore we will lose the computational advantage afforded by two-dimensionality. The motivation for our choice of forcing function is thus somewhat different. We wish to study the development of nonlinear dynamo states from initial conditions of weak magnetization and so as a first consideration it makes sense to build upon a well-studied kinematic regime. As a second factor, the forcing function $\mathbf{F}$ given by (2) gives rise to a spatially periodic problem that can be efficiently implemented numerically. Third, even at moderate $Rm$ the velocity $\mathbf{U}_0$ is a healthy dynamo, and thus the nonlinear regime can be attained in a reasonably short computational time. Fourth, and crucially for the work described in the following section, the dynamo growth rate remains positive as $k_z \to 0$, thus implying that the velocity $\mathbf{U}_0$ can amplify magnetic fields of arbitrarily large extent in the $z$-direction. We conclude this section by noting that since, by computational necessity, nonlinear studies are restricted to moderate values of $Rm$, we do not explicitly invoke the fast dynamo property of $\mathbf{U}_0$. Nonetheless, it gives us hope that the results derived here may apply even in the large $Rm$ regime.

## 3. Magnetic Field Evolution in a Tall Box

We explore the interplay between the evolution of magnetic structures of different sizes, with particular interest in the conditions that lead to the generation of large-scale magnetic fields. Since we are mostly concerned with the interaction between large- and small-scale magnetic structures it is appropriate to specify precisely what we mean by these terms in the present context. Typically in dynamo theory, scales are defined relative to the characteristic scale of the velocity. In the case of turbulent flows this scale naturally corresponds to the correlation length. For the flow $\mathbf{U}_0$, which is $z$-independent, the only meaningful scale is that of the horizontal periodicity. (For convenience we shall refer to $x$ and $y$ as the horizontal directions and $z$ as the vertical, even though there is no gravity in this problem.) However, it would be misleading to equate this particular scale with a correlation length, since for periodic flows all periodic subdomains are in phase and hence, in some sense, the correlation length is infinite. Instead, for this problem it is more meaningful to shift the emphasis from the velocity to the magnetic field. The horizontal structure of the (kinematic) magnetic eigenfunction is not appropriate since it turns out that its horizontal average is non-zero, and therefore the magnetic field can be thought of as having infinite extent. On the other hand, the vertical extent of the mode of maximum growth rate is a well-defined, meaningful scale, and we shall adopt it here as the scale characterizing this problem. Thus, a large-scale field is to be understood as one whose *vertical* extent is large compared to that of the mode of maximum growth rate.

For our numerical experiments we consider a computational domain of length $2\pi$ in the $x$ and $y$ directions, and $8 \times 2\pi/k_{max}$ in the $z$-direction, this choice being motivated by a compromise between computational feasibility and the desire for a genuine scale separation. It is convenient to define a new vertical wavenumber $K$ measured in units of $k_{max}/8$; with this notation the mode of maximum growth rate has $K = 8$. As initial conditions we impose

$$\mathbf{U}(\mathbf{x}, 0) = \mathbf{U}_0(\mathbf{x}, 0), \qquad \mathbf{B}(\mathbf{x}, 0) = A_1 \mathbf{B}_p^{(1)} \mathrm{e}^{iz} + A_8 \mathbf{B}_p^{(8)} \mathrm{e}^{8iz}, \tag{4}$$

where $\mathbf{B}_p^{(K)}$ describes the horizontal structure of the kinematic eigenmode of wavenumber $K$. If the constants $A_1, A_8 \ll 1$ then, initially, the Lorentz force will be negligible and the velocity will satisfy $\mathbf{U}(\mathbf{x}, t) \approx \mathbf{U}_0(\mathbf{x}, t)$. Initially the magnetic field will evolve kinematically, before eventually attaining sufficient strength that the Lorentz force becomes dynamically significant. Clearly, the structure of the magnetic field as it enters into this nonlinear phase will depend on the initial choice of the amplitudes $A_1$ and $A_8$; we explore this dependence by considering two distinct cases.

### 3.1. LARGE-SCALE SATURATION

We consider first a case in which the initial amplitudes $A_1$ and $A_8$ are chosen such that the energy in the large-scale component exceeds that in the small-scale component throughout the kinematic regime. The transition to the dynamical regime and the subsequent development can be followed by the time evolution of the magnetic energy density binned by vertical wavenumber. Thus we define

$$M_K(t) = \frac{1}{2} \sum_{k_x, k_y} |\widehat{\mathbf{B}}(k_x, k_y, K)|^2, \tag{5}$$

where $\widehat{\mathbf{B}}$ is the Fourier transform of $\mathbf{B}$; no distinction is made here between positive and negative values of $K$. Analogous quantities may be defined for

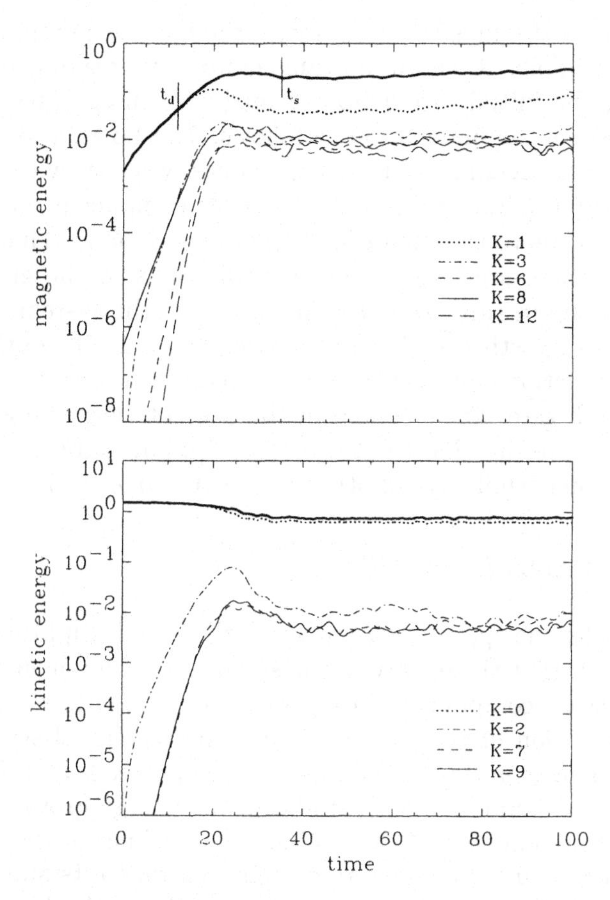


*Figure 1.* Time evolution of the magnetic and kinetic energies as functions of time for a case with $Rm = 100$. The magnetic initial conditions were such that $A_1(0) \gg A_8(0)$. Also shown are the time histories of the energies binned by vertical wavenumbers for specific choices of wavenumbers.

the velocity. Figure 1 shows the time evolution of the total magnetic and kinetic energies, together with the energies of some specific wavenumbers. Two important epochs can be identified; $t = t_d \approx 12$, corresponding to the beginning of the dynamical regime and characterized by substantial departures from an exponential behaviour; $t = t_s \approx 35$, corresponding to the beginning of the stationary state. The interval $t_d < t < t_s$ is a period of dynamical readjustment in which the system evolves markedly from the kinematic state. The ratio $M_1/M_8$ is approximately 100 at $t = t_d$ implying that, at least initially, the modification to the velocity is due entirely to the Lorentz force associated with the large-scale field component. Interestingly, this ratio decreases only slightly during the period $t_d < t < t_s$; thus, even after the nonlinear readjustment, the energy in the large-scale component substantially exceeds that of the small-scale component.

Note that in the kinematic regime, although the Lorentz force is dynamically negligible, it has important symmetry-breaking consequences. For instance, the initial production of magnetic energy in wavenumber 3 (see figure 1a) can be understood as follows. The Lorentz force associated with the $K = 1$ eigenfunction drives a weak velocity with wavenumber $K = 2$; the interaction between this velocity component and the $K = 1$ eigenfunction leads to the excitation of the $K = 3$ mode. Thus in the kinematic regime there are two distinct contributions to the growth of each mode. One is the intrinsic dynamo instability of the system and proceeds at the dynamo growth rate for that particular mode. The other is due to nonlinear interactions and can in many circumstances exceed the natural dynamo growth rate. For example in the present case, the growth of the $K = 1$ and 8 modes in what we define to be the kinematic regime is mostly due to the former, while for all other modes it is due to the latter.

### 3.2. SMALL-SCALE SATURATION

We now consider the opposite case in which the initial amplitudes $A_1$ and $A_8$ are chosen such that the energy in the small-scale component exceeds that in the large-scale component throughout the kinematic regime. Figure 2 shows the evolution of the energies binned by wavenumber; here $t_d \approx 20$ and $t_s \approx 65$. The evolution of the $K = 8$ mode proceeds in a similar way to that of the $K = 1$ mode described above; the energy grows exponentially in the kinematic regime, overshoots and then declines in the readjustment phase between $t_d$ and $t_s$, eventually settling down to its stationary value. Likewise, the $K = 1$ mode grows exponentially in the kinematic phase, is further amplified in the readjustment phase, and finally saturates at a level that is substantially less than that of the $K = 8$ mode. Superficially, it would appear that the present case is analogous to that of §3.1 above, only with the roles of the $K = 1$ and $K = 8$ modes reversed. However, we should note that there is a fundamental difference between them; whereas the nonlinear amplification of modes with $K > 1$ in the previous case corresponds to a *forward* cascade, the amplification of modes with $K < 8$ in the present case corresponds to a *backward* cascade. This can be exemplified by consideration of the evolution of the $K = 1$ mode (see inset in figure 2a). For times less then $t = t_1 \approx 20$ the mode grows exponentially at its natural dynamo growth rate. At $t = t_1$ the contributions to the mode energy due to nonlinear interactions exceed that of the dynamo instability and the energy begins to grow at a higher rate. This accelerated growth continues up to $t = t_2 \approx 40$, after which the mode gradually approaches its stationary value. During the period $t_1 < t < t_2$ the dominant nonlinear contributions to the

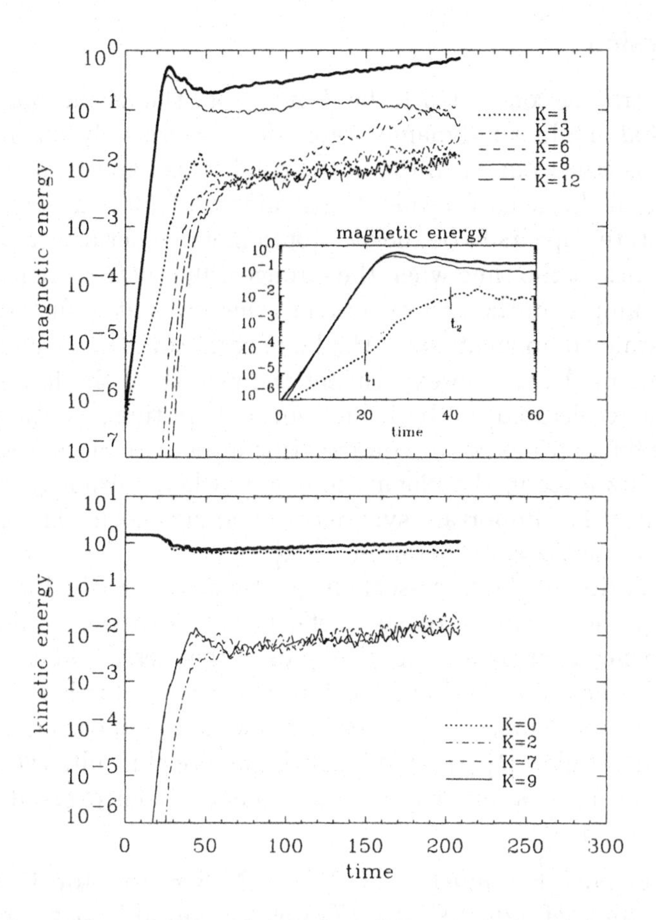


*Figure 2.* As figure 1 but with initial conditions such that $A_1(0) \approx A_8(0)$.

energy growth of the $K = 1$ mode are due to the interaction between magnetic fluctuations with $K = 8$ and velocity fluctuations with $K = 7$ and 9, the latter being driven by the Lorentz force associated with the interaction between the $K = 1$ and $K = 8$ modes. (We believe that the equality between $t_d$ and $t_1$ is coincidental.) The significant feature to note is that the nonlinear contributions to the *large-scale* field ($K = 1$ mode) can be represented as spatial averages of products of *small-scale* fluctuations $(K = 7, 8, 9)$. These have the form of the familiar $\alpha$–effect term of mean field electrodynamics, and it is therefore tempting to interpret them within the framework of this theory. Of particular significance from an astrophysical perspective is the issue of when the $\alpha$-effect ceases to be effective. From figure 2a it can be seen that the $\alpha$-effect "shuts down" when the energy in the large-scale ($K = 1$) mode is still small, a result that is consistent with an $\alpha$-prescription of the form $\alpha = \alpha_0/(1 + RmB_1^2)$.

## 4. Discussion

The most striking conclusion to be drawn from the calculations described above is that nonlinear dynamo states depend crucially on initial conditions. This is true even when the initial conditions describe a state of weak magnetization. In general terms, the significant factor in determining the saturated state appears to be the distribution of magnetic energy among the different scales at the time when the strongest fluctuations reach equipartition. For example, in the first experiment there was little further growth of the small-scale components once the large-scale component had saturated. By contrast, in the second experiment the growth of the large-scale component was accelerated, all be it only for a short time, by the presence of small-scale fluctuations at the equipartition level. Also of interest is the role of the Lorentz force in the kinematic regime where, although dynamically insignificant, it has important symmetry-breaking effects. These results are discussed in much greater detail in [1] [1].

The evolution of the large-scale magnetic field described in §3.2 is consistent with the existence of an $\alpha$–effect that is driven nonlinearly, and whose efficiency decreases dramatically once the mean field energy exceeds a critical *Rm-dependent* value. This interpretation is further supported by a further series of "small box" experiments with an imposed magnetic field, described in detail in [1] [1]. Despite such numerical results the issue of the nature of the suppression of the $\alpha$-effect remains controversial; we discuss it at length in [1] [1].

We thank N.H. Brummell, S.M. Tobias, M.R.E. Proctor, R. Rosner and N.O. Weiss for useful discussions. FC was supported in part by the NASA SR&T program, the DOE Flash initiative at the University of Chicago, and PPARC, DWH was supported in part by PPARC, and JCT was supported by the NASA SR&T and TRACE programs at the University of Chicago.

## References

1. Cattaneo, F., Hughes, D.W. and Thelen, J.-C. (2001) The nonlinear properties of a large-scale dynamo driven by helical forcing, *J. Fluid Mech.*, submitted.
2. Galloway, D.J. and Proctor, M.R.E. (1992) Numerical calculations of fast dynamos in smooth velocity fields with realistic diffusion, *Nature* **356**, pp. 691–693.
3. Hughes, D.W., Cattaneo, F. and Kim, E. (1996) Kinetic helicity, magnetic helicity and fast dynamo action, *Phys. Lett. A*, **223**, pp. 167–172.
4. Krause, F. and Rädler, K.-H. (1980) *Mean-field magnetohydrodynamics and dynamo theory*, Pergamon.
5. Moffatt, H.K. (1978) *Magnetic field generation in electrically conducting fluids*, Cambridge University Press.

# DYNAMO PROBLEMS IN SPHERICAL AND NEARLY SPHERICAL GEOMETRIES

D. J. IVERS
*School of Mathematics and Statistics*
*The University of Sydney, NSW 2006, Australia.*

AND

C. G. PHILLIPS
*Mathematics Learning Centre*
*The University of Sydney, NSW 2006, Australia.*

**Abstract.** Hybrid vector spherical harmonic / poloidal-toroidal spherical spectral forms of the linearised magnetohydroynamic equations are described. The equations are highly structured with relatively few terms and form the basis of computer codes, which implement a wide range of dynamo problems in spherical and nearly spherical geometries.

## 1. Introduction

The numerical solution of various forms of the rotating Boussinesq/anelastic non-linear and linearised magnetohydrodynamic (mass, momentum, induction and heat) equations in spherical and nearly-spherical geometries is outlined. Linearisation of the equations may be made about axisymmetric or three-dimensional steady basic states. The prototype physical model consists of uniformly electrically-conducting spherical solid-inner and spherical-shell fluid-outer cores, and an insulating exterior. Various extensions have been considered, including no inner-core or an insulating inner-core, a non-uniformly electrically-conducting mantle, anistropic turbulent viscous or thermal diffusion, pressure buoyancy, non-spherically-symmetric gravitation and departures from spherical boundaries. The linearised equations are discretised using scalar, vector and tensor spherical harmonic expansions and toroidal-poloidal representations in angle, and second- or fourth-order finite-differences on a variable grid or Chebychev collocation in radius. The equations are solved using eigenvalue, critical-value or time-stepping techniques. These codes have been extensively tested against many known



*P. Chossat et al. (eds.), Dynamo and Dynamics, a Mathematical Challenge,* 207–215.

analytical and numerical solutions, including kinematic dynamo problems and thermal-/magneto-convection problems. The relevant topics of interest include topographic effects on the convection, locking of convection to a non-spherically-symmetric gravitation varying in colatitude and/or azimuth and turbulent viscous and thermal diffusivities due to the effects of the magnetic field and rotation. More restricted use of vector and tensor spherical harmonic techniques has been made by various authors ([9], 1970; [7], 1974; [11], 1974; [13], 1975; [15], 1985; [12], 1997).

## 2. The Non-Linear Equations

For an electrically-conducting Boussinesq fluid in a frame rotating with angular velocity $\boldsymbol{\Omega} = \mathbf{1}_z$ the non-dimensional governing equations are

$$\begin{aligned} Ro(\partial_\tau \mathbf{v} + \boldsymbol{\omega} \times \mathbf{v}) + 2\boldsymbol{\Omega} \times \mathbf{v} + \frac{d\boldsymbol{\Omega}}{d\tau} \times \mathbf{r} &= -\nabla P + \Lambda \mathbf{J} \times \mathbf{B} - Ra\,\Theta \mathbf{g} + E\nabla^2 \mathbf{v} \\ \tau_\eta \partial_\tau \mathbf{B} &= \nabla^2 \mathbf{B} + R_{\mathrm{m}} \nabla \times (\mathbf{v} \times \mathbf{B}) \\ \partial_\tau \Theta + \mathbf{v} \cdot \nabla \Theta &= Pe^{-1} \nabla^2 \Theta + Q + \epsilon \mathbf{J}^2 , \end{aligned}$$

where $\nabla \cdot \mathbf{v} = 0$, $\nabla \cdot \mathbf{B} = 0$ $P = p + \frac{1}{2}\mathbf{v}^2$, $\boldsymbol{\omega} = \nabla\times \mathbf{v}$, $\mathbf{J} = \nabla\times \mathbf{B}$, $\mathbf{v}$, $\mathbf{B}$ and $\Theta$ are the velocity, magnetic induction and temperature, $Q$ is the rate of heat production per unit volume, $\tau_\eta$ dimensionless magnetic diffusion time, $R_{\mathrm{m}}$ is the magnetic Reynolds number, $Ro$ is the Rossby number, $Ra$ is a modified Rayleigh number, $\Lambda$ is a dimensionless measure of the magnetic field, $E$ is the Ekman number, $Pe$ is the Peclet number, $\epsilon$ is a dimensionless measure of the Ohmic dissipation. The choice of independent dimensionless parameters depends on the problem.

Poloidal-toroidal representations are used for the magnetic field and the velocity, $\mathbf{B} = \mathbf{S}\{S\} + \mathbf{T}\{T\}$ and $\mathbf{v} = \mathbf{S}\{s\} + \mathbf{T}\{t\}$, where $\mathbf{T}\{T\} := \nabla\times \{T\mathbf{r}\}$ and $\mathbf{S}\{S\} := \nabla\times \mathbf{T}\{S\}$. Representations of the current and vorticity are induced by the identity $\nabla\times \mathbf{S}\{S\} = \mathbf{T}\{-\nabla^2 S\}$, $\mathbf{J} = \mathbf{T}\{-\nabla^2 S\} + \mathbf{S}\{T\}$ and $\boldsymbol{\omega} = \mathbf{T}\{-\nabla^2 s\} + \mathbf{S}\{t\}$.

## 3. Vector Spherical Harmonics

Vector spherical harmonics play a key role in the derivation of our structurally simple angle-spectral equations and codes. Spherical harmonics in colatitude $\theta$ and east-longitude $\phi$ are defined for $m \geq 0$ by,

$$\begin{aligned} Y_n^m(\theta, \phi) &:= (-)^m \sqrt{\frac{(2n+1)(n-m)!}{(n+m)!}} P_{n,m}(\cos\theta) e^{im\phi} \\ P_{n,m}(z) &= (-)^n \frac{(1-z^2)^{m/2}}{2^n n!} \frac{d^{m+n}(1-z^2)}{dz^{m+n}} . \end{aligned}$$

For $m < 0$, $Y_n^m = (-)^m (Y_n^{-m})^*$, where the star denotes complex conjugation. The spherical harmonics are orthonormal with respect to the inner-product, $(f, g) := \frac{1}{4\pi} \oint f g^* \, d\Omega$. The poloidal and toroidal potentials, temperature and pressure are expanded in spherical harmonics, $S = \sum_\alpha S_\alpha Y_\alpha$, $T = \sum_\alpha T_\alpha Y_\alpha$, $\Theta = \sum_\alpha \Theta_\alpha Y_\alpha$ $s = \sum_\alpha s_\alpha Y_\alpha$ $t = \sum_\alpha t_\alpha Y_\alpha$ $P = \sum_\alpha P_\alpha Y_\alpha$, where a lowercase Greek subscript denotes the 2-index of a spherical harmonic and $S_\alpha = (S, Y_\alpha)$, etc.

The vector spherical harmonics used are defined by [6]

$$\mathbf{Y}_{n,n_1}^m := (-)^{n-m} \sqrt{2n+1} \sum_{m_1,\mu} \begin{pmatrix} n & n_1 & 1 \\ m & -m_1 & -\mu \end{pmatrix} Y_{n_1}^{m_1} \mathbf{e}_\mu ,$$

where $\mathbf{e}_0 := \mathbf{1}_z$ and $\mathbf{e}_{\pm 1} := \mp(\mathbf{1}_x \pm i\mathbf{1}_y)/\sqrt{2}$. In particular, $\mathbf{Y}_{n,0}^\mu = \delta_n^1 \mathbf{e}_\mu$, $\mathbf{Y}_{0,1}^0 = -\mathbf{1}_r$. Other definitions have been used ([1], [1]; [10], [10]; [4], [4]; [3], [3]). All have useful advantages: the gradient, curl, divergence and Laplace operators take simple forms for $\mathbf{Y}_{n,n_1}^m$. The $\mathbf{Y}_{n,n_1}^m$ are orthonormal with respect to the inner-product $(\mathbf{F}, \mathbf{G}) := \frac{1}{4\pi} \oint \mathbf{F} \cdot \mathbf{G}^* \, d\Omega$ of complex vector functions $\mathbf{F}$ and $\mathbf{G}$. The magnetic induction, electric current, velocity, vorticity and gravity are expanded in vector spherical harmonics, $\mathbf{B} = \sum_\alpha B_\alpha \mathbf{Y}_\alpha$, $\mathbf{J} = \sum_\alpha J_\alpha \mathbf{Y}_\alpha$, $\mathbf{v} = \sum_\alpha v_\alpha \mathbf{Y}_\alpha$, $\boldsymbol{\omega} = \sum_\alpha \omega_\alpha \mathbf{Y}_\alpha$, $\mathbf{g} = \sum_\alpha g_\alpha \mathbf{Y}_\alpha$, where a lower-case Greek subscript on a vector quantity denotes the 3-index of a vector spherical harmonic; e.g. $\mathbf{Y}_\alpha$ denotes $\mathbf{Y}_{n_\alpha, n_{\alpha 1}}^{m_\alpha}$. The summations are over $n_\alpha = 0, 1, 2, \ldots$, $n_{1\alpha} = n_\alpha - 1, n_\alpha, n_\alpha + 1$ and $m_\alpha = -n_\alpha, -n_\alpha + 1, \ldots, n_\alpha - 1, n_\alpha$, and $B_\alpha = (\mathbf{B}, \mathbf{Y}_\alpha)$, etc.

The following properties of vector spherical harmonics are important.

$$\nabla(f Y_n^m) = \sqrt{\frac{n}{2n+1}} \mathbf{Y}_{n,n-1}^m \partial_n^{n-1} f - \sqrt{\frac{n+1}{2n+1}} \mathbf{Y}_{n,n+1}^m \partial_n^{n+1} f ,$$

where $f$ is a radial function, $\partial_n^{n_1} := \partial_r + (n+1)/r$, if $n_1 = n-1$, and $\partial_n^{n_1} := \partial_r - n/r$, if $n_1 = n+1$. In 3-index notation, $\partial_\gamma := \partial_{n_\gamma}^{n_{1\gamma}}$ and $\partial^\gamma := \partial_{n_{1\gamma}}^{n_\gamma}$.

$$\begin{aligned}
\nabla \times (f \mathbf{Y}_{n,n-1}^m) &= i\sqrt{\frac{n+1}{2n+1}} \mathbf{Y}_{n,n}^m \partial_{n-1}^n f \\
\nabla \times (f \mathbf{Y}_{n,n}^m) &= \frac{i}{\sqrt{2n+1}} \left\{ \sqrt{n} \mathbf{Y}_{n,n+1}^m \partial_n^{n+1} f + \sqrt{n+1} \mathbf{Y}_{n,n-1}^m \partial_n^{n-1} f \right\} \\
\nabla \times (f \mathbf{Y}_{n,n+1}^m) &= i\sqrt{\frac{n}{2n+1}} \mathbf{Y}_{n,n}^m \partial_{n+1}^n f .
\end{aligned}$$

The Laplacian satisfies $\nabla^2 f Y_\alpha = Y_\alpha D_{n_\alpha} f$ and $\nabla^2 f \mathbf{Y}_\alpha = \mathbf{Y}_\alpha D_{n_{1\alpha}} f$, where $D_{n_\alpha} := \partial_r^2 + 2r^{-1}\partial_r - n_\alpha(n_\alpha + 1) r^{-2}$. Thus, $D_\alpha = D_{n_\alpha}$, if $\alpha$ is a 2-index, and $D_\alpha = D_{n_{1\alpha}}$, if $\alpha$ is a 3-index.

The $Y$-coefficients of $s$ and $t$ are related to the $\mathbf{Y}$-coefficients of $\mathbf{v}$ by

$$v_{n,n_1}^m = f_v(n,n_1)\begin{cases} \partial_n^{n_1} s_n^m, & n_1 = n \pm 1; \\ t_n^m, & n_1 = n; \end{cases}$$

where

$$f_v(n,n_1) := \begin{cases} (n+1)\sqrt{n/(2n+1)}, & n_1 = n-1; \\ -i\sqrt{n(n+1)}, & n_1 = n; \\ n\sqrt{(n+1)/(2n+1)}, & n_1 = n+1. \end{cases}$$

The coefficients $B_{n,n_1}^m$ are similarly related to $S_n^m$ and $T_n^m$. To extract the toroidal and poloidal parts of expressions the inverse relation is used,

$$t_n^m = e_v(n,n)v_{n,n}^m, \quad D_n s_n^m = e_v(n,n-1)\partial_{n-1}^n v_{n,n-1}^m + e_v(n,n+1)\partial_{n+1}^n v_{n,n+1}^m, \tag{1}$$

where the equation factor

$$e_v(n,n_1) := \begin{cases} 1/\sqrt{n(2n+1)}, & n_1 = n-1; \\ i/\sqrt{n(n+1)}, & n_1 = n; \\ 1/\sqrt{(n+1)(2n+1)}, & n_1 = n+1. \end{cases}$$

The $\mathbf{Y}$-components of the vorticity are given by

$$\omega_{n,n_1}^m = f_\omega(n,n_1)\begin{cases} \partial_n^{n_1} t_n^m, & n_1 = n \pm 1; \\ D_n s_n^m, & n_1 = n; \end{cases}$$

where $f_\omega(n, n\pm 1) = f_{\mathbf{v}}(n, n\pm 1)$ and $f_\omega(n,n) = -f_v(n,n)$. The coefficients of the electric current $J_{n,n_1}^m$ are similarly related to the magnetic potentials $T_n^m$ and $S_n^m$.

## 4. The Non-Linear Spectral Equations

In the $\mathbf{Y}_\gamma$-spectral forms of the non-linear equations in §2 only two coupling integrals arise from the product terms. Moreover, the terms are simple in structure and limited in number, which greatly simplify coding. The $\mathbf{Y}_\gamma$-spectral momentum equation is obtained by simplifying the inner-product with $\mathbf{Y}_\gamma$. In particular, the vector-product of the $\mathbf{Y}$-expansions of two vector fields $\mathbf{F}$ and $\mathbf{G}$ yields $(\mathbf{F}\times\mathbf{G})_\gamma = \sum_{\alpha,\beta} F_\alpha G_\beta(\mathbf{Y}_\alpha \times \mathbf{Y}_\beta, \mathbf{Y}_\gamma)$, and hence the $\mathbf{Y}$-forms of $\omega \times \mathbf{v}$, $2\boldsymbol{\Omega} \times \mathbf{v}$, $d\boldsymbol{\Omega}/dt \times \mathbf{r}$ and $\mathbf{J} \times \mathbf{B}$ can be expressed in terms of the coupling integral $(\mathbf{Y}_\alpha \times \mathbf{Y}_\beta, \mathbf{Y}_\gamma)$. In a similar way the $\mathbf{Y}$-spectral form of the buoyancy force can be expressed in terms of the coupling integral $(Y_\alpha \mathbf{Y}_\beta, \mathbf{Y}_\gamma)$, $(\Theta\mathbf{g}, \mathbf{Y}_\gamma) = \sum_{\alpha,\beta} \Theta_\alpha g_\beta (Y_\alpha \mathbf{Y}_\beta, \mathbf{Y}_\gamma)$. The time derivative, viscous and pressure terms are $(\partial_\tau \mathbf{v}, \mathbf{Y}_\gamma) = \partial_\tau v_\gamma$, $(\nabla^2\mathbf{v}, \mathbf{Y}_\gamma) = D_\gamma v_\gamma$

and, from the gradient formula, $(\nabla P, \mathbf{Y}_\gamma) = f_P(\gamma)\partial_\gamma P_\gamma$, where $f_P(\gamma) = \sqrt{n_\gamma/(2n_\gamma+1)}$, if $n_{1\gamma} = n_\gamma - 1$, and $f_P(\gamma) = -\sqrt{(n_\gamma+1)/(2n_\gamma+1)}$, if $n_{1\gamma} = n_\gamma + 1$. Combining the $\mathbf{Y}$-form of each term yields

$$\begin{aligned}(Ro\ \partial_\tau - ED_\gamma)\, v_\gamma &= -f_P(\gamma)\partial_\gamma P_\gamma - \sum_{\alpha,\beta}\{Ra\Theta_\alpha g_\beta(Y_\alpha \mathbf{Y}_\beta, \mathbf{Y}_\gamma) + \\ &\qquad (-Ro\ \omega_\alpha v_\beta - 2\Omega_\alpha v_\beta - \dot{\Omega}_\alpha r_\beta + \\ &\qquad \Lambda J_\alpha B_\beta)(\mathbf{Y}_\alpha \times \mathbf{Y}_\beta, \mathbf{Y}_\gamma)\}\,.\end{aligned}$$

The coupling integrals of three harmonics, $(\mathbf{Y}_\alpha \times \mathbf{Y}_\beta, \mathbf{Y}_\gamma)$ and $(Y_\alpha \mathbf{Y}_\beta, \mathbf{Y}_\gamma)$, have been evaluated in closed form ([6] [6], [8]) in terms of $3j$-, $6j$- and $9j$-symbols.

It is easier to implement the spectral magnetic vector potential equation with $\mathbf{v}\times\mathbf{B}$, and then take the curl, rather than the induction equation with $\nabla\times(\mathbf{v}\times\mathbf{B})$. Uncurling the induction equation gives $\tau_\eta\partial_\tau\mathbf{A} = \nabla^2\mathbf{A} + R_\mathrm{m}\mathbf{v}\times\mathbf{B} + \nabla V$, where $\mathbf{B} = \nabla\times\mathbf{A}$. From $\mathbf{v}\times\mathbf{B} = \sum_{\alpha,\beta} v_\alpha B_\beta(\mathbf{Y}_\alpha \times \mathbf{Y}_\beta, \mathbf{Y}_\gamma)\mathbf{Y}_\gamma$,

$$(\tau_\eta\partial_\tau - D_\gamma)A_\gamma = R_\mathrm{m}\sum_{\alpha,\beta} v_\alpha B_\beta(\mathbf{Y}_\alpha \times \mathbf{Y}_\beta, \mathbf{Y}_\gamma) + (\nabla V)_\gamma\,.$$

The spectral temperature equation is given by the inner-product of the temperature equation with $Y_\gamma$,

$$(\partial_\tau - Pe^{-1}D_\gamma)\Theta_\gamma = Q_\gamma + \sum_{\alpha,\beta}\{-v_\alpha(\nabla\Theta)_\beta + \epsilon J_\alpha J_\beta\}(\mathbf{Y}_\alpha \cdot \mathbf{Y}_\beta, Y_\gamma)\,.$$

## 5. The Linearised Equations

The MHD equations linearised about the steady basic state $(\mathbf{v}_0, \mathbf{B}_0, \Theta_0, P_0)$ are

$$\begin{aligned} Ro\ \partial_\tau\mathbf{v}' &= -Ro(\boldsymbol{\omega}_0 \times \mathbf{v}' + \boldsymbol{\omega}' \times \mathbf{v}_0) - 2\boldsymbol{\Omega}\times\mathbf{v}' - \nabla P' + \\ &\qquad \Lambda(\mathbf{J}_0 \times \mathbf{B}' + \mathbf{J}' \times \mathbf{B}_0) - Ra\ \Theta'\mathbf{g} + E\nabla^2\mathbf{v}' \\ \tau_\eta\partial_\tau\mathbf{B}' &= \nabla^2\mathbf{B}' + R_\mathrm{m}[\nabla\times(\mathbf{v}_0\times\mathbf{B}') + \nabla\times(\mathbf{v}'\times\mathbf{B}_0)] \\ \partial_\tau\Theta' &= -\mathbf{v}_0\cdot\nabla\Theta' - \mathbf{v}'\cdot\nabla\Theta_0 + Pe^{-1}\nabla^2\Theta' + Q' + 2\epsilon\mathbf{J}_0\cdot\mathbf{J}' \end{aligned}$$

We have derived hybrid spectral forms in angle [5] for these equations, in which the basic state is expressed by the vector fields $\mathbf{v}_0$, $\boldsymbol{\omega}_0$, $\mathbf{B}_0$, $\mathbf{J}_0$, $\nabla\Theta_0$ and $\mathbf{g}$, but the perturbation state by the scalar fields $s'$, $t'$, $S'$, $T'$ and $\Theta'$. The basic state fields are expanded in vector spherical harmonics and the perturbation fields in scalar spherical harmonics. Spectral forms of the interaction terms are a hybrid of $\mathbf{Y}$-coefficients of the basic state and $Y$-coefficients of the perturbation toroidal-poloidal potentials and temperature. The notation $v^0_\alpha = v^{0,m_\alpha}_{n_\alpha,n_{1\alpha}}$, etc, is used: $\mathbf{v}_0 = \sum_\alpha v^0_\alpha\mathbf{Y}_\alpha$, etc.

The hybrid spectral forms of the linearised $t'_\gamma$- and $s'_\gamma$-equations are obtained by linearising the **Y**-spectral momentum equation, substituting toroidal-poloidal representations only for the perturbation magnetic, velocity, electric current and vorticity fields, and extracting the toroidal and poloidal parts using (1),

$$\begin{aligned}(Ro\,\partial_\tau - ED_\gamma)t'_\gamma &= \sum_{\alpha,\beta}\{-Ro\,(\omega^0_\alpha v'_\beta)_\gamma + Ro\,(v^0_\alpha \omega'_\beta)_\gamma + \Lambda(J^0_\alpha B'_\beta)_\gamma - \\ &\qquad \Lambda(B^0_\alpha J'_\beta)_\gamma - Ra\,(g_\alpha \Theta'_\beta)_\gamma - 2(\Omega_\alpha v'_\beta)_\gamma\}\,, \\ n_{1\gamma} &= n_\gamma;\end{aligned}$$

$$\begin{aligned}(Ro\,\partial_\tau - ED_\gamma)D_\gamma s'_\gamma &= \sum_{\alpha,\beta,n_{1\gamma}}\{-Ro\,(\omega^0_\alpha v'_\beta)_\gamma + Ro\,(v^0_\alpha \omega'_\beta)_\gamma + \Lambda(J^0_\alpha B'_\beta)_\gamma - \\ &\qquad \Lambda(B^0_\alpha J'_\beta)_\gamma - Ra\,(g_\alpha \Theta'_\beta)_\gamma - 2(\Omega_\alpha v'_\beta)_\gamma)\}\,, \\ n_{1\gamma} &\neq n_\gamma.\end{aligned}$$

The Lorentz and Coriolis interaction terms are

$$(J^0_\alpha B'_\beta)_\gamma := e_v(\gamma) f_B(\beta)(\mathbf{Y}_\alpha \times \mathbf{Y}_\beta, \mathbf{Y}_\gamma)\begin{cases}\partial^\gamma(J^0_\alpha \partial_\beta S'_\beta), & n_{1\gamma}\neq n_\gamma,\ n_{1\beta}\neq n_\beta;\\ \partial^\gamma(J^0_\alpha T'_\beta), & n_{1\gamma}\neq n_\gamma,\ n_{1\beta}= n_\beta;\\ J^0_\alpha \partial_\beta S'_\beta, & n_{1\gamma}= n_\gamma,\ n_{1\beta}\neq n_\beta;\\ J^0_\alpha T'_\beta, & n_{1\gamma}= n_\gamma,\ n_{1\beta}= n_\beta;\end{cases}$$

$$(B^0_\alpha J'_\beta)_\gamma := e_v(\gamma) f_J(\beta)(\mathbf{Y}_\alpha \times \mathbf{Y}_\beta, \mathbf{Y}_\gamma)\begin{cases}\partial^\gamma(B^0_\alpha \partial_\beta T'_\beta), & n_{1\gamma}\neq n_\gamma,\ n_{1\beta}\neq n_\beta;\\ \partial^\gamma(B^0_\alpha D_\beta S'_\beta), & n_{1\gamma}\neq n_\gamma,\ n_{1\beta}= n_\beta;\\ B^0_\alpha \partial_\beta T'_\beta, & n_{1\gamma}= n_\gamma,\ n_{1\beta}\neq n_\beta;\\ B^0_\alpha D_\beta S'_\beta, & n_{1\gamma}= n_\gamma,\ n_{1\beta}= n_\beta;\end{cases}$$

$$(\Omega^0_\alpha v'_\beta)_\gamma := e_v(\gamma) f_v(\beta)(\mathbf{Y}^0_{1,0} \times \mathbf{Y}_\beta, \mathbf{Y}_\gamma)\begin{cases}\partial^\gamma \partial_\beta s'_\beta, & n_{1\gamma}\neq n_\gamma,\ n_{1\beta}\neq n_\beta;\\ \partial^\gamma t'_\beta, & n_{1\gamma}\neq n_\gamma,\ n_{1\beta}= n_\beta;\\ \partial_\beta s'_\beta, & n_{1\gamma}= n_\gamma,\ n_{1\beta}\neq n_\beta;\\ t'_\beta, & n_{1\gamma}= n_\gamma,\ n_{1\beta}= n_\beta.\end{cases}$$

The advective terms, $(\omega^0_\alpha v'_\beta)_\gamma$ and $(v^0_\alpha \omega'_\beta)_\gamma$, are similar to the Lorentz force terms. The general buoyancy term is

$$(g_\alpha \Theta'_\beta)_\gamma := e_v(\gamma)(\mathbf{Y}_\alpha Y_\beta, \mathbf{Y}_\gamma)\begin{cases}\partial^\gamma(g_\alpha \Theta'_\beta), & n_{1\gamma}\neq n_\gamma;\\ g_\alpha \Theta'_\beta, & n_{1\gamma}= n_\gamma.\end{cases}$$

The hybrid spectral linearised $S'_\gamma$- and $T'_\gamma$-equations are obtained similarly from the magnetic vector potential equation,

$$(\tau_\eta \partial_\tau - D_\gamma)S'_\gamma = R_{\mathrm{m}}\sum_{\alpha,\beta}\{(v^0_\alpha B'_\beta)_\gamma - (B^0_\alpha v'_\beta)_\gamma\}\,, \quad n_{1\gamma} = n_\gamma\,;$$

$$(\tau_\eta \partial_\tau - D_\gamma) T'_\gamma = R_\mathrm{m} \sum_{\alpha,\beta,n_{1\gamma}} \{(v^0_\alpha B'_\beta)_\gamma - (B^0_\alpha v'_\beta)_\gamma\}, \quad n_{1\gamma} \neq n_\gamma .$$

The first interaction term is

$$(v^0_\alpha B'_\beta)_\gamma := e_B(\gamma) f_B(\beta) (\mathbf{Y}_\alpha \times \mathbf{Y}_\beta, \mathbf{Y}_\gamma) \begin{cases} \partial^\gamma (v^0_\alpha \partial_\beta S'_\beta), & n_{1\gamma} \neq n_\gamma,\ n_{1\beta} \neq n_\beta; \\ \partial^\gamma (v^0_\alpha T'_\beta), & n_{1\gamma} \neq n_\gamma,\ n_{1\beta} = n_\beta; \\ v^0_\alpha \partial_\beta S'_\beta, & n_{1\gamma} = n_\gamma,\ n_{1\beta} \neq n_\beta; \\ v^0_\alpha T'_\beta, & n_{1\gamma} = n_\gamma,\ n_{1\beta} = n_\beta . \end{cases}$$

The term $(B^0_\alpha v'_\beta)_\gamma$ is similar. The hybrid spectral linearised heat equation with $\mathbf{q}' = -\nabla \Theta'$ is

$$(\partial_t - Pe^{-1} D_\gamma) \Theta'_\gamma = \sum_{\alpha,\beta} \{(v^0_\alpha q'_\beta) - (q^0_\alpha v'_\beta) + \epsilon (J^0_\alpha J_\beta)\},$$

where the interaction terms $(v^0_\alpha q'_\beta)_\gamma$, $(q^0_\alpha v'_\beta)_\gamma$, $(J^0_\alpha J'_\beta)_\gamma$, depend on the one coupling integral, $(\mathbf{Y}_\alpha \cdot \mathbf{Y}_\beta, Y_\gamma)$.

The highly structured form of the interaction terms in these hybrid spectral equations and the limited number of radial function types which arise, namely $D_\gamma D_\gamma f$, $\partial^\gamma (f_0 D_\beta f)$, $\partial^\gamma (f_0 \partial_\beta f)$, $\partial^\gamma (f_0 f)$, $f_0 D_\beta f$, $f_0 \partial_\beta f$, $f_0 f$, produce compact code, which is easier to implement than Bullard-Gellman [2] type spectral equations.

## 6. Extensions of the Model

We have extended the model of §2 to non-linear anelastic and possibly turbulent mass, momentum and heat equations of the form $\nabla \cdot (\rho_a \mathbf{v}) = 0$ and

$$\begin{aligned} Ro\ \rho_a (\partial_\tau \mathbf{v} + \boldsymbol{\omega} \times \mathbf{v}) + \rho_a 2 \boldsymbol{\Omega} \times \mathbf{v} &= -\nabla P + (-Ra\ \Theta + Ra_p p) \mathbf{g}_a + \\ &\quad \Lambda \mathbf{J} \times \mathbf{B} + E \nabla \cdot \boldsymbol{\tau} \\ \rho_a (\partial_\tau \Theta + \mathbf{v} \cdot \nabla \Theta) - \epsilon_a \Theta \rho_a \mathbf{v} \cdot \mathbf{g}_a &= -Pe^{-1} \nabla \cdot \mathbf{q} + \epsilon \mathbf{J}^2 + Q , \end{aligned}$$

where $Ra_p$ and $\epsilon_a$ are dimensionless parameters. In anelastic conditions the density $\rho_a$ is spherically-symmetric, the deviatoric stress tensor $\boldsymbol{\tau} = \rho_a \left\{ \nabla \mathbf{v} + (\nabla \mathbf{v})^T - \frac{2}{3} \mathbf{I} \nabla \cdot \mathbf{v} \right\}$ and heat flux $\mathbf{q} = -\nabla \Theta$. In turbulent conditions, where $\mathbf{v}$, $\mathbf{B}$ and $\Theta$ are to be interpreted as mean fields, $\boldsymbol{\tau}$ takes one of the forms $\boldsymbol{\tau} = \{\mathbf{D}_\nu \cdot (\nabla \mathbf{v})_S\}_S$ or $\boldsymbol{\tau} = \mathbf{D}_\nu \cdot (\nabla \mathbf{v})_S$, $\mathbf{q} = -\mathbf{D}_\kappa \cdot \nabla \Theta$ and we introduce an isotropic $\alpha$-effect. The subscript $S$ denotes the trace-free symmetric part and invariance arguments have been used to impose preferred directions on the viscous diffusion tensor,

$$\mathbf{D}_\nu = (\nu_{00} + \nu_{01} \boldsymbol{\Omega} \cdot \mathbf{B}) \mathbf{I} + \nu_1 \mathbf{B}\,\mathbf{B} + \nu_2 \mathbf{B}\,\boldsymbol{\Omega} + \nu_3 \boldsymbol{\Omega}\,\mathbf{B} + \nu_4 \boldsymbol{\Omega}\,\boldsymbol{\Omega} .$$

The turbulent thermal diffusion tensor $\mathbf{D}_\kappa$ is of similar form to $\mathbf{D}_\nu$. Second-rank tensor spherical harmonics [8] are required in the derivation of the spectral momentum equation [14].

In problems with nearly spherical boundary topography, $r = a + h(\theta, \phi)$, where $\oint h\, d\Omega = 0$, $|h| << a$, scalar and (solenoidal) vector fields are analytically continued (solenoidally) using Taylor expansions. Thus the boundary conditions at $\mathbf{r} = (a+h)\mathbf{1}_r$ are imposed at $\mathbf{r}_0 = a\mathbf{1}_r$ using the expansions,

$$
\begin{aligned}
f(a+h(\theta,\phi),\theta,\phi) &= f(r,\theta,\phi) + h(\theta,\phi)\partial_r f(r,\theta,\phi) + \mathcal{O}(h^2) \\
\mathbf{F}(a+h(\theta,\phi),\theta,\phi) &= \mathbf{F}(r,\theta,\phi) + h(\theta,\phi)\partial_r \mathbf{F}(r,\theta,\phi) + \mathcal{O}(h^2)\,.
\end{aligned}
$$

If $h = \sum_\alpha h_\alpha Y_\alpha$, the spectral forms to $\mathcal{O}(h^2)$ are given by inner-products with $Y_\gamma$ and $\mathbf{Y}_\gamma$, $f_\gamma + \sum_{\alpha,\beta} h_\alpha \partial_r f_\beta (Y_\alpha Y_\beta, Y_\gamma)$ and $F_\gamma + \sum_{\alpha,\beta} h_\alpha \partial_r F_\beta (Y_\alpha \mathbf{Y}_\beta, \mathbf{Y}_\gamma)$ The coupling integral $(Y_\alpha Y_\beta, Y_\gamma)$ does not occur previously.

The computer codes have been extended to the linearised forms of the above equations. The codes can be further extended to the adjoint equations for weakly non-linear studies.

## References


1. Brink, D.M. and Satchler, G.R. (1968) *Angular Momentum*, 2nd ed., OUP, Oxford
2. Bullard, E.C. and Gellman, H. (1954) Homogeneous Dynamos and Terrestrial Magnetism, *Phil. Trans. R. Soc. Lond. A*, **Vol. no. 247**, pp. 213–278
3. Burridge, R. (1969) Spherically Symmetric Differential Equations, the Rotation Group, and Tensor Spherical Functions", *Proc. Camb. Phil. Soc.* **Vol. no. 65**, pp. 157–175
4. Gelfand, I.M. and Shapiro, Z.Ya. (1956) Representations of the Group of Rotations in Three-Dimensional Space and their Applications, *Amer. Math. Soc. Transl.*, **Vol. no. 2**, pp. 207–316
5. Ivers, D.J. and Phillips, C.G. (2000), Scalar and Vector Spherical Harmonic Spectral Equations of Rotating Non-Linear and Linearised Magnetohydrodynamics, in preparation
6. James, R.W. (1973), The Adams and Elsasser Dynamo Integrals, *Proc. R. Soc. Lond. A*, **Vol. no. 331**, pp. 469–478
7. James, R.W. (1974), The Spectral Form of the Magnetic Induction Equation, *Proc. R. Soc. Lond. A*, **Vol. no. 340**, pp. 287–299
8. James, R.W. (1976), New Tensor Spherical Harmonics for Application to the Partial Differential Equations of Mathematical Physics, *Phil. Trans. R. Soc. Lond. A*, **Vol. no. 281**, pp. 195–221
9. Jones, M.N. (1970) Atmospheric Oscillations–I, *Planet. Space Sci.*, **Vol. no. 18**, pp. 1393–1416
10. Morse, P.M. and Feshbach, H. (1953) *Methods of Theoretical Physics, Part 2*, McGraw-Hill, New York
11. Moses, H.E. (1974) The Use of Vector Spherical Harmonics in Global Meteorology and Aeronomy, *J. Atmos. Sci.*, **Vol. no. 31**, pp. 1490–1499
12. Oprea, I., Chossat, P. & Armbruster, D. (1997) Simulating the Kinematic Dynamo Forced by Heteroclinic Convective Velocity Fields, *Theor. Comp. Fluid Dynam.*, **Vol. no. 9**, pp. 293–309
13. Pekeris, C.L. and Accad, Y. (1975) Theory of Homogeneous Dynamos in a Rotating Liquid Sphere, *Proc. Nat. Acad. Sci. USA*, **Vol. no. 72**, pp. 1496–1500

14. Phillips, C.G. and Ivers, D.J. (2000), Spherical Anisotropic Diffusion Models for the Earth's Core, *Phys. Earth Planet. Int.*, **Vol. no. 117**, pp. 209–223
15. Swarztrauber, P.N. and Kasahara, A. (1985), The Vector Analysis of Laplace's Tidal Equations, *SIAM J. Sci. Stat. Comput.*, **Vol. no. 6**, pp. 464–491

# ANELASTIC PLANETARY MAGNETOHYDRODYNAMICS

S.V. STARCHENKO
*GFO Borok 42-4, Yaroslavskaya obl.152742, Russia*

**Abstract.** A self-consistent anelastic planetary/satellite MHD system is optimally scaled. This scaling identifies key properties of MHD generators. Those are primarily located in thin ($\sim r/R^n$) buoyancy layers at the liquid core boundary. Here $n = 1/3$ at the onset of convection, $n = 1/2$ for the developed magneto-convection and $R$, which is defined via the preliminary Reference State of the planet/satellite, is about the 'turbulent' Reynolds or/and magnetic Reynolds number. Simple diffusion and heat equations together with non-inertia state for magnetic and velocity equations are proposed in order to solve the real 3D MHD problems in planet or satellite. Boussinesq and anelastic approaches are compared.

## 1. Introduction

The preliminary Reference State (RS in abbreviated notation, corresponding values being barred hereafter) of the fluid in planetary core is no-flow ($\overline{\mathbf{V}} = 0$) and non-magnetic ($\overline{\mathbf{B}} = 0$). This adiabatic and well-mixed state has spatially invariable specific entropy ($\nabla \overline{S} = 0$) and mass fraction of the light constituent ($\nabla \overline{\xi} = 0$). The last is named concentration here. Density $\overline{\rho}$, modified pressure $\overline{p}$, temperature $\overline{T}$ and chemical potential $\overline{\mu}$ are only dependent on the spherical radius $r$. Prime " $'$ " will denote the radial derivative for them. For example, $\overline{\mu}' = d\overline{\mu}/dr$ that is proportional to the gravitational acceleration and $\overline{\mu}' \approx -2.5(r/r_i)m/s^2$ in the Earth ($r_i$ is the internal rigid core radius).

Convection is further described by non-barred variables those are represented as disturbances of RS. Thus, $0 + \mathbf{B}$, $0 + \mathbf{V}$, $\overline{\xi} + \xi$, $\overline{S} + S$ etc. are the actual physical values. They obey the following fully self-consistent anelastic system of equations that consists of the magnetic (1), velocity (2),



*P. Chossat et al. (eds.), Dynamo and Dynamics, a Mathematical Challenge,* 217–224.

diffusion (3) and heat (4) equations [2]:

$$\partial \mathbf{B}/\partial t = \nabla\times(\mathbf{V}\times\mathbf{B}) + \kappa_B \nabla^2 \mathbf{B}, \qquad \nabla\cdot\mathbf{B} = 0; \tag{1}$$

$$\frac{D\mathbf{V}}{Dt} + 2\mathbf{\Omega}\times\mathbf{V} - \kappa_V \nabla^2 \mathbf{V} = \mathbf{1}_r A + \frac{(\nabla\times\mathbf{B})\times\mathbf{B}}{\mu_0 \overline{\rho}} - \nabla\frac{p}{\overline{\rho}}, \ \nabla\cdot(\overline{\rho}\mathbf{V}) = 0; \tag{2}$$

$$\overline{\rho}\partial(\overline{\xi} + \xi)/\partial t + \nabla\cdot(\overline{\rho}\xi\mathbf{V}) = \nabla\cdot(\overline{\rho}\kappa_\xi \nabla\xi); \tag{3}$$

$$\overline{T}\overline{\rho}\frac{\partial(\overline{S} + S)}{\partial t} + \overline{T}\nabla\cdot(\overline{\rho}S\mathbf{V}) = \nabla\cdot(\overline{\rho}\overline{T}\kappa_S \nabla S) + \frac{\kappa_B}{\mu_0}(\nabla\times\mathbf{B})^2 + \overline{\mu}'\overline{\rho}\kappa_\xi \frac{\partial\xi}{\partial r} + Si. \tag{4}$$

Here: $A = -\overline{T}'S - \overline{\mu}'\xi$ is the Archimedian buoyancy acceleration in (2) that $A$, the only force driving the magneto-convection, and its direction is parallel to the radial unit vector $\mathbf{1}_r$. Sink $Si = c_p \kappa_T (r^2 \overline{\rho}\overline{T}')'/r^2$ is a negative heat source in (4). Specific heat $c_p \approx 800 J/kg\ K$ and thermal diffusivity $\kappa_T \approx .5\cdot 10^{-5} m^2/s$ in the Earth. This sink appears because the temperature $\overline{T}$ in RS is not a solution to the Fourier equation for thermal diffusion. So, the magneto-convection supports RS compensating negative $Si$ by positive magnetic and other heat sources in (4) where viscosity heating is omitted because it's usually too small in the planetary and satellite's outer cores.

'Turbulent' transport coefficients $\kappa_{V,\xi,S}$ are about the magnetic diffusivity $\kappa_B$ [2] that's order is between 1 and 10 $m^2/s$ in the planets/satellites with own magnetic field. Thus, order one 'turbulent' Prandtl-type numbers can be introduced for the velocity $V$, concentration $\xi$ and entropy $S$ as

$$P_V = \kappa_V/\kappa_B \sim 1, \qquad P_\xi = \kappa_\xi/\kappa_B \sim 1, \qquad P_S = \kappa_S/\kappa_B \sim 1. \tag{5}$$

Corresponding molecular Prandtl numbers ($P_T = \kappa_T/\kappa_B$ instead of $P_S$ because entropy molecular diffusion is impossible) are less or of order $10^{-5} \geq P_{V,\xi,T}^{molecular}$.

No-slip boundary conditions for the velocity $\mathbf{V}(r = r_i) = \mathbf{V}_{IC}$ in (2) should be imposed at the inner rigid core (IC - hereafter). The same is applied at the outer ($r = r_o$) - core-mantle (CM) boundary for the Earth's type planets, while stress-free conditions should be used instead in the major planets. The sum of angular momenta of IC, of the outer core (OC) and of the mantle is usually constrained to be zero in our reference frame that is rotating with an angular velocity $\Omega \approx 7\cdot 10^{-5}/s$ for the Earth. Magnetic coupling is due to thin conducting layer in the mantle at CM. The rest of the mantle can be regarded as an insulator where magnetic field is potential and decay as $1/r^3$. The magnetic field $\mathbf{B}$ and normal electric current $\mathbf{1}_r\cdot\nabla\times\mathbf{B}/\mu_0$ from equation (1) are continuous at IC and CM.

The convection is mainly driven by the buoyancy release during freezing OC to IC. That imposes boundary conditions on (3) and (4) as

$$\rho_{IC}\Delta S \partial r_c/\partial t = -\overline{\rho}\kappa_S \partial S/\partial r, \ \ \rho_{IC}\Delta\xi \partial r_c/\partial t = -\overline{\rho}\kappa_\xi \partial\xi/\partial r \text{ at } r = r_i, \tag{6}$$

where IC density $\rho_{IC}$ is typically a few percents more than OC density $\overline{\rho}(r = r_i) = \overline{\rho}_i$. The growth of the local IC radius

$$\partial r_c(t, \theta, \varphi)/\partial t = -r_i[(f_S/c_p)\partial S/\partial t + f_\xi \partial\xi/\partial t]$$

and depends on the correspondent freezing numbers those are $f_S \approx 33,\ f_\xi \approx 70$ in the Earth in accordance with [2], while the specific latent heat over the temperature $\Delta S$ and the concentration release $\Delta\xi$ satisfy $(100 \le \Delta S \le 300) J/(kg\ K)$ and $(.06 \le \Delta\xi \le .08)$ there.

No concentration flux should be towards impermeable CM from OC, while entropy boundary condition at CM is set up by

$$\partial\xi/\partial r = 0, \qquad -\overline{T}\overline{\rho}\kappa_S \partial S/\partial r = q_S \qquad \text{at } r = Cr_i. \qquad (7)$$

The non-adiabatic heat flux satisfies $|q_S(t)| \approx 10^{-2} W/m^2$ at the Earth's CM [2], while the conductive (RS) heat flux $q_T = -c_p\overline{\rho}\kappa_T\overline{T}'$ is $Q = (2 \div 4)$ times large. Total CM heat flux from OC to mantle is $4\pi(Cr_i)^2(q_S + q_T)$ that's larger (lower) than $4\pi(Cr_i)^2 q_T$ for superadiabatic (subadiabatic) state. Buoyancy acceleration $A$ in (2) has concentration $-\overline{\mu}'\xi$ and entropy $-\overline{T}'S$ parts. Their ratio in OC is rather close to $q_T/|q_S| = Q$. This could be due to domination of the concentration buoyancy that eventually supports the RS heat flux $q_T$ via magneto-convection, while the entropy is connected with smaller non-adiabatic $|q_S|$.

## 2. Optimal Scaling

Let's choose optimal units for the planetary dynamo system (1-7). The unit of length and time could optimally be the radius of the inner rigid core $r_i$ and the corresponding magnetic diffusion time $t_i = r_i^2/\kappa_B$, those $r_i$ =1222 km and $t_i \approx 23\ 000$ years in the Earth's core where $\kappa_B \approx 2m/s^2$. This magnetic time $t_i$ is rather close to 'turbulent' diffusion times for the other $(S, \xi, \mathbf{V})$ variables due to (5).

Density and temperature units could optimally be their maximal values $\max(\overline{T}) = T_i$ and $\max(\overline{\rho}) = \rho_i$. Those maxima are reached at $r = r_i$, $T_i = 5300K$, $\rho_i = 1.2 \cdot 10^4 kg/m^3$ in the Earth. The minimal values at CM are just ~15% less. The heat flux optimal unit $q_i = |q_S|$ is described after the boundary condition for entropy in (7). The latter gives the entropy unit $S_i = q_i r_i/(T_i \rho_i \kappa_S)$ if we estimate $\partial S/\partial r$ as $S_i/r_i$. This optimal $S_i$ is close to $1.4 \cdot 10^{-4} J/kg\ K$ in the Earth's core. The scaling of entropy was suggested from [1].

The radial derivative to the RS chemical potential and temperature could well be estimated as $\overline{\mu}_i' = -\mu_i'' r$ and $\overline{T}_i' = -T_i'' r$. Constant $\mu_i'' = 2 \cdot 10^{-6}/s^2$ and $T_i'' = 2.5 \cdot 10^{-10} K/m^2$ in the Earth. Those derivatives determine the value of the buoyancy acceleration $A$ from (2) that drives the magneto-convection. The Coriolis force in (2) should be of the order of the buoyancy force. Thus, we would estimate $\Omega V_i = \mu_i'' r_i \xi_i$ because the concentration release at $r = r_i$ is the main buoyancy drive. The mean entropy contribution to the buoyancy $A$ from (2) is about $Q = q_T/q_i$ times smaller (see the end of the previous section). So, our second estimation is $(\mu_i'' r)\xi_i = Q(T_i'' r)S_i$. Both estimations and $S_i$ from above give us the optimal velocity $V_i = q_T r_i^2 T_i''/(T_i \rho_i \Omega \kappa_S)$ and concentration $\xi_i = q_T r_i T_i''/(T_i \mu_i'' \rho_i \kappa_S)$ units. Earth's $V_i \approx 1.4 \cdot 10^{-3} m/s$ and $\xi_i \approx 4 \cdot 10^{-8}$.

We complete our optimal 'i'-units by the standard unit of the magnetic field $B_i = \sqrt{\mu_0 \rho_i \kappa_B \Omega}$, that $B_i \approx 1mT$ in the Earth [2]. To verify our optimal units one could compare them with typical values in 3D solution to (1-7). Such solutions exist for only one set of the principal parameters and our 'i'-units are very close to the typical results of those solutions presented in [3]. Unfortunately, this particular set is not so suitable for planetary or satellite MHD models because it has totally unrealistic viscosity ($P_V \approx 10^3$, while it should be $\leq 1$ as in (5)) and artificial damping in small scales for the numerical stability.

To investigate the real planetary MHD let's first scale $r, t$, $\mathbf{B}, \mathbf{V}$, $\xi, S$, $\overline{\rho}, \overline{T}$, $\overline{\mu}', \overline{T}'$ in (1-7) by the above 'i'-units. Then we get the following non-dimensional system of equations. All the convective variables are of order one $(S, \xi, \mathbf{V}) \sim 1$ in it. Its non-dimensional magnetic equations are

$$\partial \mathbf{B}/\partial t = R \nabla \times (\mathbf{V} \times \mathbf{B}) + \nabla^2 \mathbf{B}, \nabla \cdot \mathbf{B} = 0, \tag{8}$$

where $R$ is the main number controlling the problem and is defined via the given RS values and diffusion coefficients as $R = q_T r_i^3 T_i''/(T_i \Omega \rho_i \kappa_S \kappa_B)$. That is about the magnetic Reynolds number because $R$ is also equal to $V_i r_i/\kappa_B (\approx 890$ in the Earth). Non-dimensional magnetic and velocity boundary conditions have the same form as dimensional ones (see previous section).

The non-dimensional velocity equations include the anelastic condition $(\nabla \cdot \mathbf{V})/V_r = -\overline{\rho}'/\overline{\rho} (= r/19$ in the Earth) and the force balance:

$$\frac{\mathbf{B} \times \nabla \times \mathbf{B}}{R \overline{\rho}} + \epsilon \mathbf{V} \nabla \mathbf{V} + \frac{\epsilon}{R}\left(\frac{\partial \mathbf{V}}{\partial t} - P_V \nabla^2 \mathbf{V}\right) + 2 \mathbf{1}_z \times \mathbf{V} + \nabla \frac{p}{\overline{\rho}} = A \mathbf{1}_r \tag{9}$$

where the buoyancy acceleration $A/r = S/Q + \xi$, while the Rossby number $\epsilon = V_i/(r_i \Omega)$ is as small as $10^{-5}$. The Ekman number $E = \kappa_V/(r_i^2 \Omega)$ is equivalent to ours $\epsilon P_V/R = E$ that is about $2 \cdot 10^{-8}$ in the Earth.

Balancing magnetic and buoyancy terms in (9) we conclude that the magnetic field $B$ is of order 1 (in $B_i$), if it's typical scale is as small as $1/R$. That is rather hard to treat by direct simulation. MHD simulations were so far based on viscous-magnetic balance. See that in this volume or in [4, 5].

Using unrealistically large viscosity with $P_V \geq 10^3$ in contrast to (5) they usually have $B \sim E^{1/4}R^{1/2}$ in all the known 2÷3D models. Their Ekman number $E = P_V\epsilon/R \geq 10^{-4}$ is many orders more than real $E \ll 10^{-7}$. To handle this, we will later (section 4) consider omitting the viscous and all the other inertia terms from (9) of order $\epsilon$. This gives magnetic field between $R^{1/2}$ and 1 when the excitation condition $R \geq 10^2$ holds. Let's now finish with diffusion and heat equations:

$$\partial\xi/\partial t + R\mathbf{V}\cdot\nabla\xi = P_\xi\nabla^2\xi - \dot{\overline{\xi}}, \tag{10}$$

$$\frac{\partial S}{\partial t} + R\mathbf{V}\cdot\nabla S = P_S\nabla^2 S - 3P_S Q - \dot{\overline{S}} - P_\xi Q\epsilon_T r\frac{\partial\xi}{\partial r} + Q\frac{\epsilon_T}{R}|\nabla\times\mathbf{B}|^2, \tag{11}$$

where $\epsilon_T = r_i^2 T_i''/T_i \approx .07$ in the Earth, while some $r\overline{\rho}', r\overline{T}' \leq 1/10$ effects are neglected in (10-11) because the corresponding terms ($\sim \kappa_{T,\xi}$ etc.) are known even with lower accuracy in the planets and satellites. Given $\dot{\overline{S}} = d\overline{S}/dt$ is negative due to cooling of the liquid core, while $\dot{\overline{\xi}} = d\overline{\xi}/dt > 0$ because $\overline{\xi}$ is growing there. At the present epoch, Earth's $\dot{\overline{S}} \approx -.5$ and $\dot{\overline{\xi}} \approx +.9$ following [2, 3]. The equations (10-11) are subjected to four boundary conditions:

$$\frac{\partial\xi}{\partial r} = 0 = \frac{\partial S}{\partial r} \pm 1 \text{ at } r = C, \qquad \delta_S\frac{\partial S}{\partial r} = \frac{\partial S}{\partial t} + P_t\frac{\partial\xi}{\partial t} = \delta_\xi\frac{\partial\xi}{\partial r} \text{ at } r = 1 \tag{12}$$

where $+1/-1$ is for super/sub-adiabatic state, $P_t = f_\xi\xi_i c_p/(f_S S_i) \sim 1$, while small $\delta_S = P_S c_p\rho_i/(f_S\Delta S\rho_{IC})$ and $\delta_\xi = P_\xi c_p\rho_i\xi_i/(S_i f_S\rho_{IC}\Delta\xi)$ give us the buoyancy layers thickness ($\sim r_i/10$). In the Earth $P_t \approx .48$, $\delta_S \approx .12$ and $\delta_\xi \approx .10$ at the rigid core boundary. Its radius $r_c \approx 1$ is growing slowly on the same time-scale ($\sim 10^9$ years) as for $\overline{\xi}$ following $R_c\partial r_c/\partial t = -\partial S/\partial r$ where $R_c = \rho_{IC}\Delta S/(S_i\rho_i P_S) \approx 1.5\cdot 10^6$.

## 3. Onset and Buoyancy Layers

A usually accepted Boussinesq system in planetary hydrodynamics could be formally obtained from equations (8-11), if one neglects by the small $V_r\overline{\rho}'/\overline{\rho}$ ($= \nabla\cdot\mathbf{V}$) in the anelastic condition and omits some suitable small terms in (11) in order to get such a linear combination of (10) and (11) that gives one equation for $A/r = S/Q + \xi$ [4] instead of two. However, we don't know any attempt in Boussinesq frame to model boundary conditions (12). Those are time-dependent and directly provide us with buoyancy boundary layers at $r = 1$ via $\delta_{S,\xi} \ll 1$. This is so in contrast to the usual Boussinesq approach! Thus, anelastic onset/maintain of the convection should be investigated and seems to be for the first time.

Equations (10-11) allow $r$–stratified solution without convection:

$$\dot{\xi}_o = P_\xi \frac{(r^2 \xi_o')'}{r^2} - \dot{\bar{\xi}},\ \dot{S}_o = P_S \frac{(r^2 S_o')'}{r^2} - 3P_S Q - \dot{\bar{S}} - P_\xi Q \epsilon_T r \xi_o'. \quad (13)$$

Those as the subject to ($\xi_o' = 0,\ S_o' = -(\pm 1)$ at $r = C$) give

$$\xi_o' = -P_\xi^{-1} r^{-2} \int_r^C (\dot{\xi}_o + \dot{\bar{\xi}}) r^2 dr = P_\xi^{-1} \frac{\dot{\xi}_o + \dot{\bar{\xi}}}{3}(r - \frac{C^3}{r^2}), \quad (14)$$

$$S_o' = Q\epsilon_T \frac{\dot{\xi}_o + \dot{\bar{\xi}}}{30 P_S}(2r^3 + \frac{3C^5}{r^2} - 5C^3) + (Q + \frac{\dot{S}_o + \dot{\bar{S}}}{3P_S})(r - \frac{C^3}{r^2}) - (\pm 1) \quad (15)$$

if we suppose $(S_o, \xi_o) = f(t) + F(r)$. Using the other boundary conditions ($\delta_\xi \xi_o' = \dot{S}_o + P_t \xi_o = \delta_S S_o'$ at $r = 1$) we get (16) that give us $\xi_o', S_o'$ via (14-15). Necessary condition for convection onset $\xi_o' < 0$ are hold when

$$\dot{\xi}_o + \dot{\bar{\xi}} = \frac{3P_S P_\xi[(\pm 1) + (Q + P_S^{-1}(\dot{\bar{S}} + \dot{\bar{\xi}} P_t)/3)(C^3 - 1)]}{Q\epsilon_T(.3C^5 - .5C^3 + .2)P_\xi + (C^3 - 1)[\delta_\xi(C^3 - 1 + \delta_S^{-1} P_S) + P_\xi P_t]} \quad (16)$$

is positive. The lower part in (16) is always positive. Thus, the excitation is primarily determined by the balance between RS heat flux ($Q > 0$) and heat/mixing history ($\dot{\bar{S}} + P_t \dot{\bar{\xi}} \le 0$), while the influence of super/sub-adiabatic state ($\pm 1$) is not so crucial because $C^3 Q \gg 1$. The driving term for linear magneto- convection onset is $V_r(S_o', \xi_o')$ in (10,11), both$\sim QCV_r$. Following similar arguments as in [5] we prove that magneto- convection onsets in layers with thickness about $1/R^{1/3}$ at the liquid boundaries.

In the developed convection, non-linear term influence will result in flattening the entropy and concentration outside the boundaries due to domination of $R\mathbf{V}\cdot\nabla(\xi, S)$ terms of (10,11) in the large scales, while $(S, \xi)$ are sharply changed in the buoyancy layer at $r = 1$. The heat and diffusion equations are there reduced to

$$\frac{V_\theta}{r}\frac{\partial(S_m, \xi_m)}{\partial\theta} + \omega\frac{\partial(S_m, \xi_m)}{\partial\varphi} + v'x\frac{\partial(S_m, \xi_m)}{\partial x} = P_{S,\xi}\frac{\partial^2(S_m, \xi_m)}{\partial x^2} \quad (17)$$

and are subjected to the boundary conditions at $r = 1$ in (12) and to the flattening conditions ($\xi_m \approx -\xi_o, S_m \approx -S_o$) well outside the boundary. We also use onset-state (14-16) and separate the onset (subscript - 'o') from maintain ('m') variables as $\xi = \xi_o + \xi_m$ and $S = S_o + S_m$. Angular velocity $\omega = V_\varphi/s$ ($s = r\sin\theta$) and radial velocity derivative $v' = \partial V_r/\partial r$ provide the major contribution in (17) being actually independent on $x = R^{1/2}(r - 1) \sim 1$ in the layer.

The buoyancy layer equations at $r = C$ are similar to (17), but they obey to homogeneous boundary condition $\xi'_m = 0 = S'_m$ because non-homogeneous conditions in (12) at $r = C$ are already held by $\xi_o, S_o$. This results in a passive role of this boundary layer where $x = R^{1/2}(r - C) < 0$.

## 4. Non-inertia State

In non-inertia approximation, terms of order $\epsilon$ should be omitted in (9). Let's first integrate those anelastic equations in cylindrical co-ordinates $(z, s, \varphi)$ when the magnetic force $-\mathbf{B}\times\nabla\times\mathbf{B}/R$ can be omitted as well:

$$V_s = -\frac{1}{2s}\frac{\partial P}{\partial \varphi}, \quad V_\varphi = \frac{1}{2}\frac{\partial P}{\partial s} - \frac{s}{2r}A, \quad V_z = \frac{1}{2\overline{\rho}}\int^z \left(\frac{\partial A}{\partial \varphi} + \overline{\rho}'\frac{\partial P}{\partial \varphi}\right)\frac{dz}{r}. \qquad (18)$$

The lower integration limit in $P = p/\overline{\rho} = \int^z Azdz/r$ and in $V_z$ include free $z$-independent functions of $(t, s, \varphi)$. Those two functions are uniquely defined for non-axisymmetric $(\mathbf{V} - \mathbf{V}^0)$ when we satisfy non-penetrating conditions $V_r = 0$ at $r = 1, C$. Axisymmetric $\mathbf{V}^0 = \int_0^{2\pi} \mathbf{V} d\varphi/2\pi$ has $V_s^0 = 0$ and non-unique $V_\varphi^0$ that is defined up to an arbitrary $(z, \varphi)$-independent geo-strophic velocity $V_g^0(t, s)$.

To remove this non-uniqueness one has to return somehow the magnetic or/and viscosity effects. Following [6] we multiply by $\overline{\rho}$ the $\varphi$-component of (9) and integrate it over the cylinder of constant $s$ that is rigidly bounded by $z = Z_t(s)$ on its top and by $z = Z_b(s)$ on its bottom. Now using anelastic condition $\nabla \cdot (\overline{\rho}\mathbf{V}) = 0$ we obtain the famous Taylor constrain:

$$\mathcal{T}(s) = \int_{Z_b}^{Z_t} \int_0^{2\pi} \mathbf{B}\cdot\mathbf{1}_\varphi\times\nabla\times\mathbf{B}\, dz\, d\varphi = 0 \qquad (19)$$

still neglecting terms of the order of or smaller than $\epsilon$ in (9). The expression (19) defines geo-strophic $V_g^0$ via (8) and remains valid even when we completely return the magnetic field in (18). Besides, (19) naturally divide the liquid planetary core on three parts.

Those are separated by the rigid boundaries and by the cylinder parallel to z-axis and tangent to the inner rigid core equator. Separation and concentration of MHD generators located in those parts should explain basic symmetry, value and variations of the magnetism and hydrodynamics in the deep planetary interiors.

Returning the inertia and viscous terms to (19) we, finally, get

$$\frac{\partial V_g^0}{\partial t} = \frac{\mathcal{T}}{\overline{\rho}\epsilon} - \int_{Z_b}^{Z_t}\int_0^{2\pi} \mathbf{1}_\varphi\cdot(R\mathbf{V}\nabla\mathbf{V} - P_V\nabla^2\mathbf{V})\, dz\, d\varphi, \qquad (20)$$

that'd be used in creating non-inertia 3D codes. This'd give us the real solutions to magneto-convection problems in the planetary liquid cores.

## 5. Conclusions

A self-consistent anelastic planetary magneto-hydrodynamic system was optimally scaled. Simplifying the system this scaling identified key properties of the MHD generators. Those are primarily located in non-homogeneous thin and numerically accessible buoyancy (Archimedian) boundary layers at the liquid planetary core. A simple buoyancy layer system was obtained instead of the diffusion and heat equations. Non-inertia state is proposed in order to solve real 3D problems in magnetism and hydrodynamics of deep interiors in the planets or satellites.

The main control number of the problem $R$ (see its definition after (8)) is similar to the modified Rayleigh number, but its value is well determined and rather moderate, being about the magnetic Reynolds or 'turbulent' Reynolds number because $R \sim 10^2 \div 10^3$ in planet/satellite with own magnetic field.

The usual homogenous Boussinesq approach is shown to be not so suitable for the planetary MHD. The special on-set state should be derived for the anelastic magneto-convection because anelastic boundary condition are time-dependent and testify the appearance of the buoyancy boundary layers with thickness of order $r/R^{1/3}$. In the developed magneto-convection, similar layers have thickness of order $r/R^{1/2}$ and the external layer becomes passive depending on the internal layer.

**Acknowledgements**: The author appreciated discusssion with Prof. Soward which contributed in his understanding of the anelastic planetary MHD system, and he also thanks Prof. Cupal and Dr Anufriev who gave him their preprint to appear in PEPI 2001.
This work was supported by the Royal Society in the Exeter University (2000 June-August) and by INTAS-99-00348 grant.

## References

1. Anufriev, A.P. and Cupal, I. (2001) Characteristic amplitudes in the solution of anelastic geodynamo model, *Phys. Earth Planet. Inter.*, in print
2. Braginsky, S.I. and Roberts, P.H. (1995) Equations governing convection in the Earth's core and the geodynamo, *Geoph. Astroph. Fluid Dyn.*, **Vol. 79**, pp. 1-97
3. Glatzmaier, G.A. and Roberts, P.H. (1997) Simulating the geodynamo, *Contemporary Physics*, **Vol. 38** (4), pp. 269-288
4. Starchenko, S. V. (1999) Supercritical convection associated with ultra fast MHD rotation, *JETP (Zh. Eksp. Teor. Fiz.)*, **Vol. 115** (5), pp. 1708-1720
5. Starchenko, S. V. (2000) Supercritical magneto-convection in rapidly rotating planetary cores, *Phys. Earth Planet. Inter.*, **Vol. 117** (1-4), pp. 225-235
6. Taylor, J. B. (1963) The magneto-hydrodynamics of a rotating fluid and the Earth's dynamo problem, *Proc. R. Soc. London Ser.*, **Vol. A 274**, pp. 274-283

# THE GALACTIC DYNAMO

K.M. FERRIERE
*Observatoire Midi-Pyrénées*
*14 avenue E. Belin*
*31400 Toulouse*
*France*

**Abstract.** The purpose of this paper is to present a model of Galactic dynamo driven by supernova explosions. I first describe, in physical and mathematical terms, the threefold impact of supernova-driven turbulence on the large-scale Galactic magnetic field, namely, the alpha-effect, the vertical advection, and magnetic diffusion. I then present recent numerical solutions of the Galactic dynamo equation, which support the idea that the large-scale magnetic field can be amplified through a combination of large-scale differential rotation and supernova-driven turbulence.

## 1. Introduction

Various types of observations – synchrotron radiation, Faraday rotation, Zeeman splitting – converge to show that our Galaxy, as well as other spiral galaxies, is pervaded by strong interstellar magnetic fields, whose pressure is comparable to both the interstellar gas pressure and the cosmic ray pressure [7].

The origin and temporal evolution of the large-scale component of these galactic magnetic fields is still a matter of debate. The most likely scenario to date involves a hydromagnetic dynamo operating through the combined action of the large-scale differential rotation and small-scale turbulent motions [10, 9, 11]. The differential rotation stretches magnetic field lines in the azimuthal direction about the galactic center. The role played by turbulent motions is threefold. First, since they occur in a rotating medium, they have a net helicity, whereby they can generate magnetic field in the direction perpendicular to the prevailing field (alpha-effect). Second, they



*P. Chossat et al. (eds.), Dynamo and Dynamics, a Mathematical Challenge,* 225–232.

contribute to the effective advection of the large-scale magnetic field. Third, by mixing field lines together, they lead to magnetic field diffusion.

Whereas the rotation curve of our Galaxy is reasonably well established observationally, mainly thanks to H I and CO velocity measurements (e.g., [6], [6]), the properties of turbulent motions are still poorly understood. Mathematically, their impact on the large-scale magnetic field can be parameterized with the help of two tensors: the alpha-tensor, $\alpha_{ij}$, which embodies both the alpha-effect and the effective magnetic advection, and the diffusivity tensor, $\beta_{ijk}$, which represents turbulent magnetic diffusion [8].

In this paper, I present a calculation of the tensors $\alpha_{ij}$ and $\beta_{ijk}$ in our Galaxy, based on the assumption that the turbulent motions responsible for dynamo action are primarily driven by supernova explosions. I distinguish between isolated supernovae, which create individual supernova remnants, and clustered supernovae, which, together with the powerful winds from their progenitor stars, produce giant cavities known as superbubbles. I then present numerical solutions of the Galactic dynamo equation in which the alpha- and diffusivity tensors take on the values given by the supernova-driven calculation.

## 2. The Dynamo Equation

The time evolution of the large-scale magnetic field in our Galaxy is governed by the dynamo equation,

$$\frac{\partial\langle\mathbf{B}\rangle}{\partial t} = \nabla \times \Big(\langle\mathbf{v}\rangle \times \langle\mathbf{B}\rangle\Big) + \nabla \times \mathcal{E} \, , \tag{1}$$

where $\mathbf{B}$ is the magnetic field, $\mathbf{v}$ is the velocity field,

$$\mathcal{E} \equiv \langle\delta\mathbf{v} \times \mathbf{B}\rangle \tag{2}$$

is the electromotive force due to turbulent motions, angle brackets denote large-scale (or ensemble-averaged) quantities, and the symbol $\delta$ denotes small-scale turbulent quantities (e.g., [10], [10]), In general, $\mathcal{E}$ can be expressed as a linear function of $\mathbf{B}$ and its first-order spatial derivatives, with the help of the tensors $\alpha_{ij}$ and $\beta_{ijk}$ mentioned in the introduction:

$$\mathcal{E}_i = \alpha_{ij}\, B_j + \beta_{ijk}\, \frac{\partial B_j}{\partial x_k} \tag{3}$$

[8].

In the case of the Galactic disk, where the parameters of the interstellar medium and the sources of turbulence vary essentially in the vertical direction, the alpha-tensor takes on the form

$$\alpha_{ij} = \begin{pmatrix} \alpha_R & -v_{\rm esc} & 0 \\ v_{\rm esc} & \alpha_\Phi & 0 \\ 0 & 0 & \alpha_Z \end{pmatrix} \tag{4}$$

in a cylindrical reference frame $(\hat{e}_R, \hat{e}_\Phi, \hat{e}_Z)$ with origin at the Galactic center and $\hat{e}_Z$ perpendicular to the Galactic plane. $\alpha_R$, $\alpha_\Phi$, and $\alpha_Z$ give the effective rotational velocity associated with the alpha-effect when $\mathbf{B}$ is radial, azimuthal, and vertical, respectively, and $v_{\rm esc}$ is the effective vertical velocity at which $\mathbf{B}$ is advected by turbulent motions [1]. The diffusivity tensor can be written as

$$\beta_{ijk} = \beta_h \left(\epsilon_{ijR}\,\delta_{kR} + \epsilon_{ij\Phi}\,\delta_{k\Phi}\right) + \beta_v\,\epsilon_{ijZ}\,\delta_{kZ} \ , \tag{5}$$

where $\beta_h$ and $\beta_v$ are the horizontal and vertical magnetic diffusivities, respectively, $\delta_{ij}$ is the unit tensor, and $\epsilon_{ijk}$ is the three-dimensional permutation tensor [2].

## 3. The Alpha- and Diffusivity Tensors

To calculate the tensors $\alpha_{ij}$ and $\beta_{ijk}$ in our Galaxy, I proceeded in two successive steps.

In the first step, I derived a formal analytical expression for the electromotive force, and hence for the different components of $\alpha_{ij}$ and $\beta_{ijk}$, due to a collection of explosions in a differentially rotating, stratified medium, with rotation and stratification axes aligned. I first calculated the contribution from a given explosion, $\delta\mathbf{v} \times \mathbf{B}$, by assuming that the explosion sweeps up the ambient gas into an expanding shell of negligible thickness (thin-shell approximation), that magnetic field lines remain tied to the thermal gas (frozen-in approximation), and that the swept field lines reconnect with the ambient magnetic field at the time when the shell merges with the background. I then averaged individual contributions over a vertical distribution of explosions. The resulting analytical expressions for the different components of $\alpha_{ij}$ and $\beta_{ijk}$ can all be written in the form

$$\int dZ_0 \ \ \mathcal{R}(Z_0) \ \ \ldots \ , \tag{6}$$

where $\mathcal{R}(Z_0)$ is the explosion rate per unit volume at a height $Z_0$ above the equatorial plane and the suspension dots represent a complicated function depending, on the one hand, on the parameters of the large-scale differential rotation (more specifically, the large-scale rotation rate, $\Omega$, and shear rate, $G$) and, on the other hand, on a number of parameters characteristic of

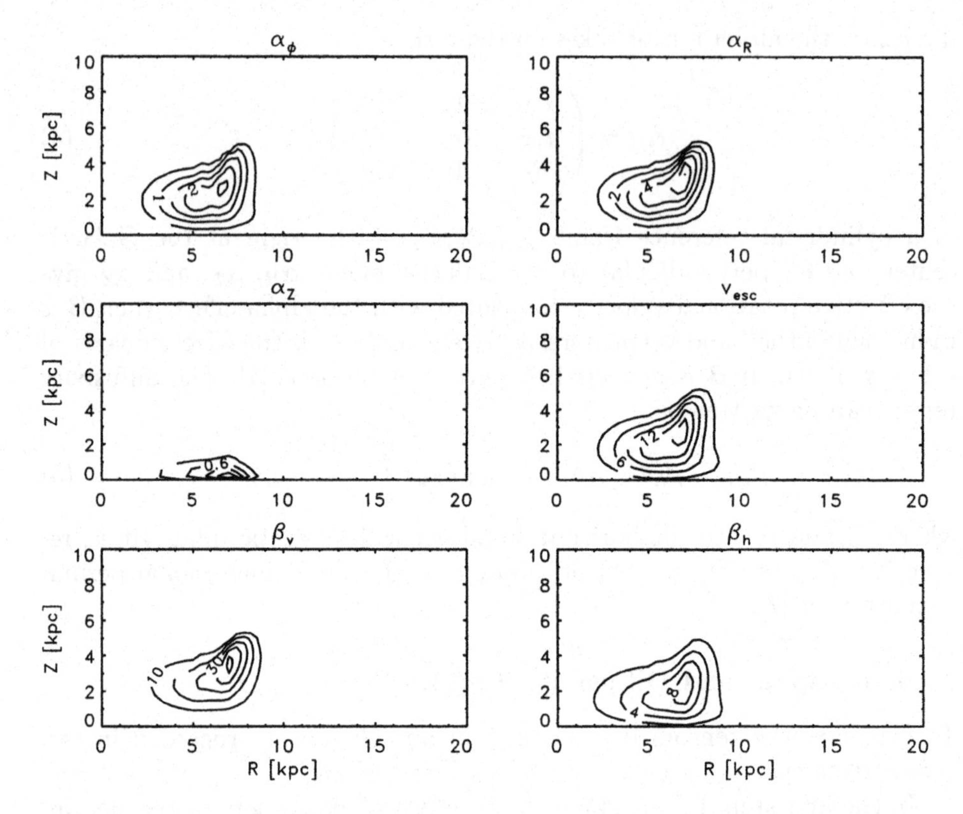


*Figure 1.* Contour lines of the four different components of the alpha-tensor (expressed in km $s^{-1}$) and of the two different components of the diffusivity tensor (expressed in $10^{26}$ $cm^2$ $s^{-1}$) in a given meridional plane of our Galaxy.

the considered explosion (like the lifetime of the shell and its final cross-section).

In the second step, I applied the analytical expressions obtained in the first step to supernova explosions in the Galactic disk. I started by constructing an axisymmetric model of the interstellar medium of our Galaxy, based on the best-to-date observational data on the spatial distribution of the different interstellar constituents, namely, the thermal gas, cosmic rays, magnetic fields, the gravitational field, the rotation curve, and supernovae [3]. I then wrote a numerical code able to simultaneously follow the expansion of a large number of shells, associated either with individual supernova remnants or with superbubbles, and originating at different points in my model interstellar medium. For each shell, the code computes the characteristic parameters relevant to dynamo action and, from this, deduces its contribution to the alpha- and diffusivity tensors. Averaging individual contributions over the adopted supernova distribution then yields a numerical estimate for each component of $\alpha_{ij}$ and $\beta_{ijk}$ as a function of Galactic radius, $R$, and height, $Z$ [4]. The corresponding contour lines in a given meridional plane are plotted in Figure 1.

It turns out that superbubbles give a much larger contribution to $\alpha_{ij}$ and $\beta_{ijk}$ than individual supernova remnants, despite the fact that isolated and clustered supernovae occur at about the same frequency ($\simeq 1/100$ yr). The different functions reach their maximum value at a Galactic radius $\simeq 7$ kpc, i.e., somewhat farther out than the peak in supernova rate (at $\simeq 5$ kpc), while still inside the solar circle (at 8.5 kpc). The peak values are $\alpha_\Phi \simeq 2.6$ km s$^{-1}$, $\alpha_R \simeq 6.0$ km s$^{-1}$, $\alpha_Z \simeq 1.7$ km s$^{-1}$, $v_{\rm esc} \simeq 16$ km s$^{-1}$, $\beta_v \simeq 54 \times 10^{26}$ cm$^2$ s$^{-1}$, and $\beta_h \simeq 9 \times 10^{26}$ cm$^2$ s$^{-1}$. It is interesting to note that, with these values, the dynamo number, defined by

$$D = \frac{|G|\,\alpha_\Phi\,|Z|^3}{\beta_v^2}\,, \tag{7}$$

is $\simeq 8$ in the peak region. This is a comfortable value, large enough to support the idea of a supernova-driven Galactic dynamo and, at the same time, small enough to rule out the existence of very high dynamo modes.

## 4. Numerical Solutions

The rest of the work described in this paper was done in collaboration with D. Schmitt (see [5], [5]). We introduced the components of $\alpha_{ij}$ and $\beta_{ijk}$ given by Figure 1, together with the Galactic rotation curve of [6] ([6]), into the Galactic dynamo equation, and we solved the latter numerically on a $(R, Z)$-mesh extending radially from 0 to 20 kpc and vertically from 0 to 10 kpc (for linear solutions) and from $-10$ kpc to 10 kpc (for nonlinear solutions). The external medium was assumed to be a perfect conductor, except in a few trial runs where a vacuum was assumed.

We first studied linear modes, which were allowed to be either axisymmetric (azimuthal wavenumber $m = 0$) or bisymmetric ($m = 1$) with respect to the rotation axis, and either symmetric (symbol S) or antisymmetric (symbol A) with respect to the equatorial plane. For each mode, we started by computing a reference solution corresponding to the case when all the input parameters take on their default values. Then, we systematically varied each of the input parameters, so as to determine its impact on the temporal evolution and on the spatial structure of the large-scale magnetic field.

We found that axisymmetric modes are always easier to excite than bisymmetric modes. Under the reference conditions, the S0 and A0 modes have very similar properties. Both grow monotonously with time at a slow exponential rate $\simeq 0.45$ Gyr$^{-1}$, which suggests that the Galactic magnetic

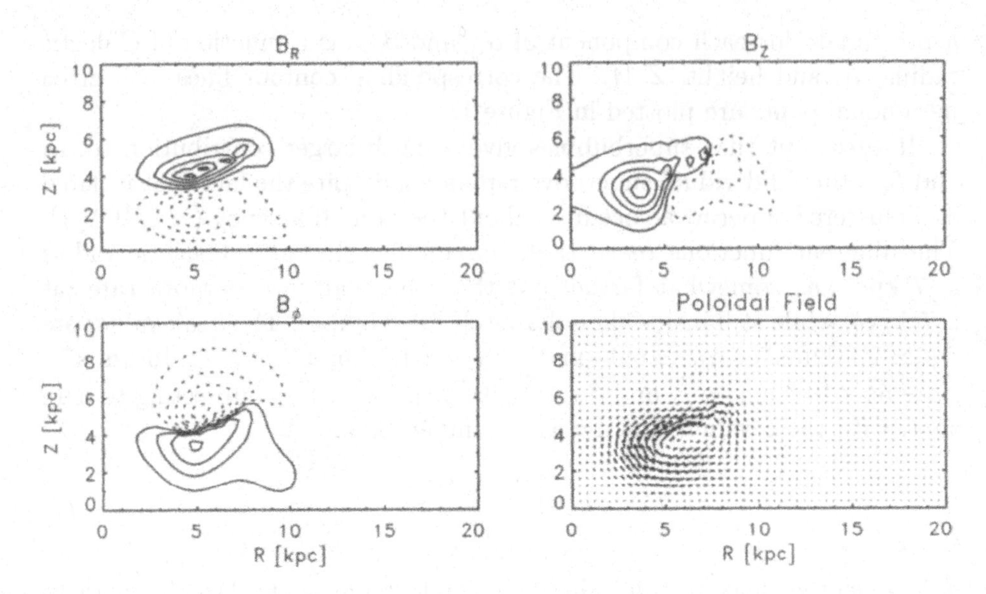


*Figure 2.* **Contour lines of the three cylindrical components of the large-scale magnetic field and lines of force of its poloidal component, in a given meridional plane of our Galaxy, for the S0 mode under the reference conditions.**

field has presently reached a state close to saturation. The azimuthal field component dominates by more than one order of magnitude; as indicated by Figure 2, it exhibits, on each side of the midplane, two well-defined polarities peaking at $R \simeq 5$ kpc and separated by a node at $|Z| \simeq 4$ kpc. The poloidal field peaks at a slightly smaller radius, and its lines of force rotate about a point located close to the maximum of $\alpha_\Phi$.

When the effective escape velocity, $v_{\rm esc}$, is set to zero, the exponential growth rate decreases and even turns slightly negative for the S0 mode. The magnetic field becomes oscillatory, and its overall pattern propagates toward the midplane. During most of a dynamo cycle, $B_\Phi$, which is again the dominant field component, possesses four alternating polarities along the vertical (see Figure 3). Oscillatory behaviors are also obtained, albeit with an increased growth rate, when the diagonal components of $\alpha_{ij}$ are enhanced by a factor $> 3$ or when the magnetic diffusivities are reduced by a factor $> 1.7$ with respect to their reference values.

The parameters of the Galactic differential rotation, too, have a considerable influence on the solutions. In particular, when in addition to being a decreasing function of $R$, $\Omega$ is assigned a vertical dependence in $\exp\left[-(Z/5\,\mathrm{kpc})^2\right]$, both the S0 and A0 modes decay oscillatorily with time. As before, the magnetic field is predominantly azimuthal, but now, the alternating polarities of $B_\Phi$ (displayed in Figure 4) migrate along an oblique

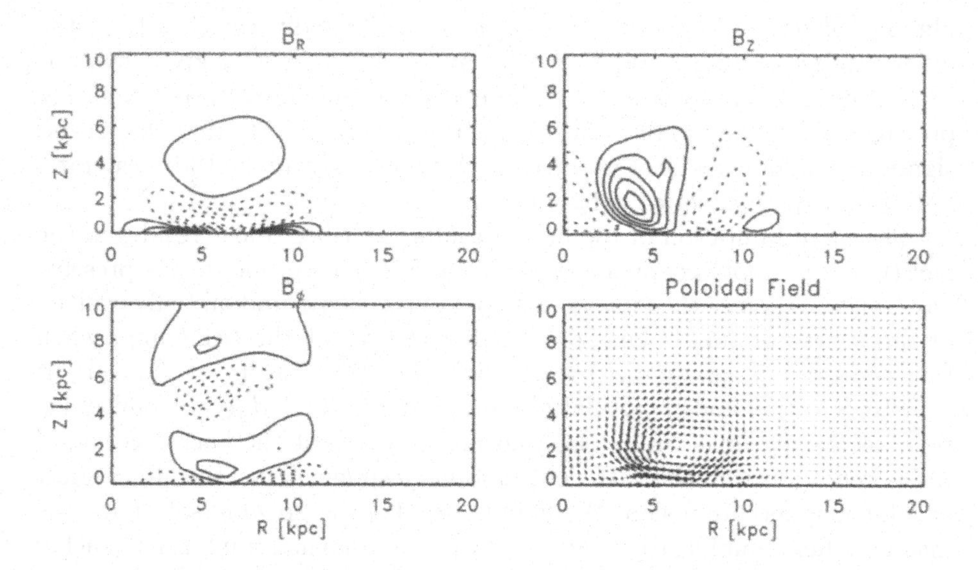


*Figure 3.* Same as Figure 2, with $v_{\rm esc} = 0$.

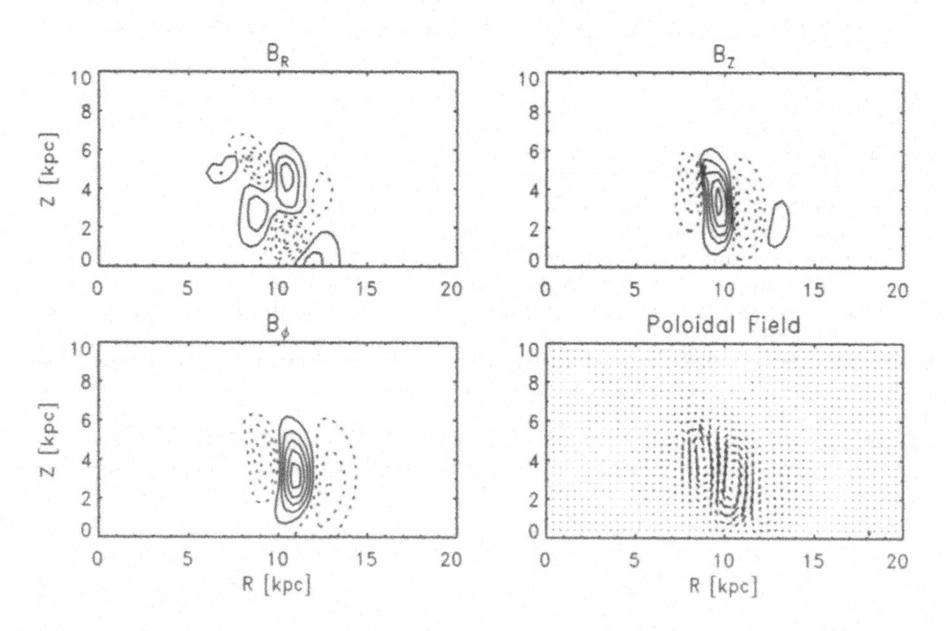


*Figure 4.* Same as Figure 2, with $\Omega \propto \exp\left[-(Z/5\,\mathrm{kpc})^2\right]$.

axis, which can be verified as roughly parallel to lines of constant $\Omega$.

For bisymmetric modes, the azimuthal stretching of magnetic field lines by the large-scale differential rotation has a destructive, rather than amplifying, effect. As a result, the magnetic field rapidly vanishes from the differentially-rotating parts of the Galaxy (outside $R \simeq 4$ kpc), while it undergoes a much slower exponential decay, accompanied by an azimuthal propagation, in the rigidly-rotating innermost region. Note that there exist significant differences between S1 and A1 modes, not only in the decay rate, but also in the magnetic configuration.

The main limitation of the linear results described above resides in the fact that our adopted expressions for $\alpha_{ij}$ and $\beta_{ijk}$ are proper to the present-day Galaxy, whereas the true dynamo parameters have undoubtedly evolved in the course of time, being presumably greater at the beginning, when the large-scale magnetic field was weaker. For this reason, we carried out a number of nonlinear calculations in which $\alpha_{ij}$ and $\beta_{ijk}$ are multiplied by a decreasing function of $|\langle \mathbf{B} \rangle|$, meant to represent the back-reaction of the Lorentz force on the turbulent motions responsible for the alpha-effect and for magnetic diffusion. We found that, for the growing solutions, the magnetic field amplification saturates when its maximum intensity reaches $\sim 20\ \mu$G, corresponding to a magnetic pressure roughly equal to four times the local gas pressure. The time to saturation, which depends on the seed field strength adopted, is typically on the order of a few 10 Gyr.

## References


1. Ferrière, K.M. (1993a) *ApJ*, **404**, p. 162
2. Ferrière, K.M. (1993b) *ApJ*, **409**, p. 248
3. Ferrière, K.M. (1998a) *ApJ*, **497**, p. 759
4. Ferrière, K.M. (1998b) *A&A*, **335**, p. 488
5. Ferrière, K.M. and Schmitt, D. (2000) *A&A*, **358**, p. 125
6. Fich, M., Blitz, L. and Stark, A.A. (1989) *ApJ*, **342**, p. 272
7. Heiles, C. (1987) in *Interstellar Processes*, eds. D.J. Hollenbach and H.A. Thronson (Dordrecht: Reidel), p. 171
8. Moffatt, H.K. (1978) Magnetic Field Generation in Electrically Conducting Fluids (Cambridge: Cambridge Univ. Press), p. 145
9. Parker, E.N. (1971) *ApJ*, **163**, p. 255
10. Steenbeck, M., Krause, F. and Rädler, K.-H. (1966) *Z. Nat.*, **A21**, p. 369
11. Vainshtein, S.I. and Ruzmaikin, A.A. (1971) *AZh.*, **48**, p. 902; English trans. *Soviet Astron.*, **15**, p. 714 (1972)

# TURBULENT DIAMAGNETISM AND GALACTIC DYNAMO

A.S. GABOV

*Scientific Research Computing Center,*
*Moscow State University, Moscow, 119899 Russia*

D.D. SOKOLOFF

*Department of Physics,*
*Moscow State University, Moscow, 119899 Russia*

AND

A. SHUKUROV

*Department of Mathematics,*
*University of Newcastle upon Tyne, NE1 7RU, UK*

## 1. Introduction

Turbulent diamagnetism results in an expulsion of large-scale magnetic field from regions with a high intensity of turbulent motions (see, e.g. [1]): a turbulent conductive fluid behaves like a diamagnetic with an effective magnetic permeability $\mu \propto \beta^{-1/2}$, with $\beta$ being the turbulent magnetic diffusivity. This phenomenon can be described as a transport of the large-scale magnetic field at a velocity which is proportional to $-\nabla\beta$.

The role of turbulent diamagnetism in galactic dynamo is nontrivial because large-scale magnetic fields could be generated in gaseous halos of galaxies, where the turbulent magnetic diffusivity is larger than in the disc (see [2]). It means that turbulent diamagnetism leads to a transport of the magnetic field to the galactic mid-plane where the generation is more efficient. This can enhance the dynamos action. We restrict our attention to turbulent diamagnetism in galactic dynamos with a fixed velocity field, so the dynamo is described by linear equations (a one-dimensional kinematic dynamo model).



*P. Chossat et al. (eds.), Dynamo and Dynamics, a Mathematical Challenge,* 233–237.

## 2. Basic equations

The generation of the regular magnetic field in the case of inhomogeneous diffusion is governed by [3], [4]

$$\frac{\partial \mathbf{B}}{\partial t} = \nabla \times (\alpha \mathbf{B}) + \nabla \times (\mathbf{V} \times \mathbf{B}) + \nabla \times \left( -\frac{1}{2} \nabla \beta \times \mathbf{B} \right) - \nabla \times (\beta \nabla \times \mathbf{B}), \quad (1)$$

where $\mathbf{B}$ is large-scale magnetic field, $\mathbf{V}$ is mean velocity field, and $\alpha$ and $\beta$ describe helicity and magnetic viscosity of turbulence.

The transport of magnetic field caused by inhomogeneous diffusion is given by $\nabla \times \left( -\frac{1}{2} \nabla \beta \times \mathbf{B} \right)$ [3]. We consider a turbulent disc which rotates at an angular velocity $\omega$ directed along the $z$-axis of a cylindrical polar reference frame. The half-thickness of the disc $h$ is much smaller than the disc radius $R$. We discuss only axially symmetric solutions of Eq. (1).

With $B_r = -\partial A_\varphi / \partial z$, $B_z = r^{-1} \partial (r A_\varphi)/\partial r$ (where $\nabla \times \mathbf{A} = \mathbf{B}$), $A \equiv A_\varphi$, $B \equiv B_\varphi$, the local dynamo equations reduce to

$$\frac{\partial A}{\partial t} = \alpha B + \frac{1}{2} \frac{\partial \beta}{\partial z} \frac{\partial A}{\partial z} + \beta \frac{\partial^2 A}{\partial z^2},$$

$$\frac{\partial B}{\partial t} = -G \frac{\partial A}{\partial z} + \frac{1}{2} \frac{\partial}{\partial z} \left( \frac{\partial \beta}{\partial z} B \right) + \frac{\partial}{\partial z} \left( \beta \frac{\partial B}{\partial z} \right),$$

where $G = r\, \partial \omega / \partial r$ is the measure of differential rotation. We add the vacuum boundary conditions $B(\pm h) = 0, \partial A/\partial z(\pm h) \simeq 0$.

We seek solutions in the form $A(z,t) = a(z) e^{\gamma t}, B(z,t) = (Gh/\beta_0) b(z) e^{\gamma t}$. Making a transformation to the dimensionless variables

$$z \to z/h, t \to t/(h^2/\beta_0), \gamma \to \gamma/(\beta_0/h^2), \alpha \to \alpha_0 \alpha(z)$$

($\beta_0$ is the value of $\beta$ in the symmetry plane of the disk) we obtain the local dynamo equations in dimensionless form

$$2\beta a'' + \beta' a' - 2\gamma a = -2D\alpha b, \quad (2)$$

$$2(\beta b')' + (\beta' b)' - 2\gamma b = 2a', \quad (3)$$

$$b(\pm 1) = a'(\pm 1) = 0, \quad (4)$$

where $D = G\alpha_0 h^3/\beta_0^2$ is the dynamo number.

Since helicity is described by an antisymmetric function of $z$, the solutions of the system (2)–(4) can be divided into even (quadrupolar) ones, with the symmetry conditions $b'(0) = a(0) = 0$, and odd (dipolar) ones with $b(0) = a'(0) = 0$. We analyze quadrupole solutions since quadrupole

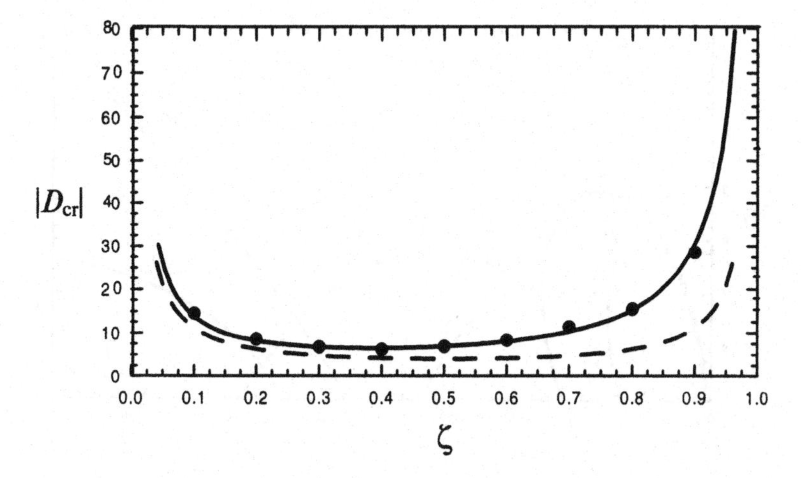


*Figure 1.* Dependence of the absolute value of the critical dynamo number on the source position $\zeta$. The dashed line shows the dependence for $\beta = \text{const}$. The solid line corresponds to the exact solution for $\beta(z) = 1 + z^2$. The points show the results obtained with the Galerkin expansion.

modes are preferred in disc geometry, and therefore consider just a half of the disk, $0 \le z \le 1$.

## 3. Concentrated Helicity

We assume here that $G$ does not depend on $z$, and $\alpha(z) = \delta(z-\zeta) - \delta(z+\zeta)$. The singularity in $\alpha$ results in a discontinuity in $a'$ and $b'$ [6, 5], such that

$$[a] = 0, \quad [a'] = -D\frac{b(\zeta)}{\beta(\zeta)}, \tag{5}$$

$$[b] = 0, \quad [b'] = 0. \tag{6}$$

where square brackets denote the magnitude of the discontinuity. We adopt $\beta(z) = 1 + f(z)$, where $f(z)$ is a growing even function of $z$. We take $f(z) = Cz^2$ and consider two cases, $C = 1$ and $C = 9$.

Below we use Galerkin expansion choosing, as the basic functions, the eigenfunctions of the problem (2)–(6) with $\beta = \text{const}$. This set of functions is complete, however we include only the first mode, which is the only mode to grow under typical galactic conditions, and the next two (decaying) modes (see details in [5]). The dependence of the critical dynamo number $D_{cr}$ on the source position $\zeta$ is shown in Fig. 1. For $\beta = \text{const}$, $D_{cr}$ has a minimum at $\zeta = 0.5$ [6]. When $\beta$ varies with $z$, the value of $\zeta$ minimizing $D_{cr}(\zeta)$ is reduced. For $\beta = \text{const}$ the $D_{cr}(\zeta)$ profile is symmetric with respect to $\zeta = 0.5$, but inhomogeneous $\beta(z)$ destroys this symmetry.

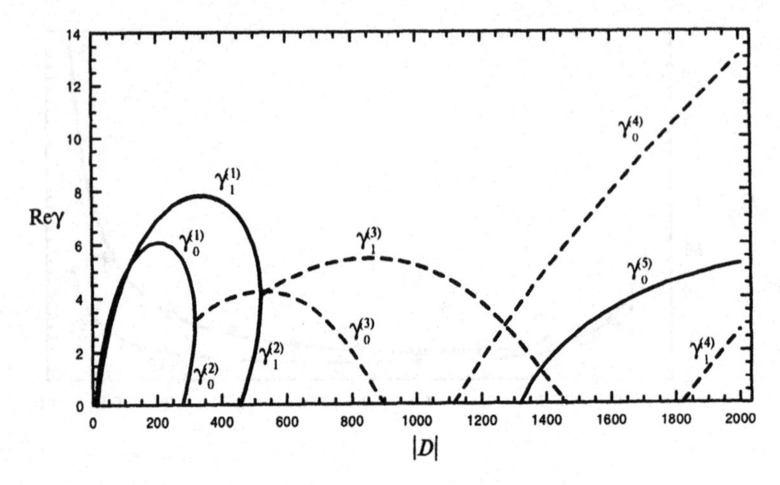


*Figure 2.* The dependence of the growth rate of quadrupole magnetic modes $\mathrm{Re}\,\gamma_C^{(i)}$ on $|D|$ for $\alpha(z) = \sin(\pi z)$. Subscript refers to the value of $C$, $C = 0$ (thin line, $\beta$ independent of $z$) and $C = 1$ (thick line); and superscript $i$ refers to the mode order. Solid lines show real eigenvalues of the system (2)–(4) and dashed lines show the real parts of complex eigenvalues.

At first glance, the term $\nabla \times \left(-\frac{1}{2}\nabla\beta \times \mathbf{B}\right)$ in the mean field equation (1) should lead to a shift in the field maximum toward the disc midplane and to a reduction of the field magnitude near the disc surface. However, our results demonstrate the opposite effect: the field becomes weaker at the disk center. Comparing diffusion flux density $\Pi_1 = -\beta b'$ and advection flux density $\Pi_2 = -\beta' b$ one can demonstrate that diffusion plays the dominant role near the disc midplane. Correspondingly, turbulent diamagnetism is noticeably weaker than diffusion processes and cannot substantially influence the distribution of azimuthal field.

## 4. A Smooth Distribution of Helicity

Consider now numerical solutions of equations (2)–(4) for a smooth distributions of helicity $\alpha(z)$. The case when the turbulent magnetic diffusivity is taken to be constant and the helicity is described by a smooth antisymmetric function $\alpha(z) = \sin(\pi z)$ has been well explored [1].

Figure 2 presents the dependence of the growth rates of the magnetic modes $\mathrm{Re}\,\gamma$ on $|D|$. We see that turbulent diamagnetism weakens the generation of a magnetic field at modest dynamo numbers and substantially enhances it at large dynamo numbers. There is an interval of dynamo number values where all modes decay, i.e. our calculations indicate the existence of a 'window' of decaying solutions for $\alpha(z) = \sin(\pi z)$. The 'window' expands when turbulent magnetic diffusivity varies with $z$. For a continuous

$\alpha(z)$, turbulent diamagnetism leads to a displacement of the magnetic field toward the $z = 0$ plane.

## 5. Conclusions

In a thin disk surrounded by vacuum, turbulent diamagnetism affects the distribution of the large-scale magnetic field only weakly, but significantly increases the efficiency of the dynamo action for large dynamo numbers and leads to an increase in the critical dynamo numbers.

## References


1. Zel'dovich, Ya.B., Ruzmaikin, A.A., Sokoloff, D.D. (1983) *Magnetic Fields in Astrophysics.* Gordon and Breach, NY
2. Sokoloff, D.D., Shukurov, A.M. (1990) Regular magnetic fields in coronae of spiral galaxies, *Nature*, **Vol.347**, pp. 51–53
3. Roberts, P., Soward, A.M. (1975) A Unified Approach to Mean Field Electrodynamics, *Astron. Nachr.*, **296**, pp. 49–64
4. Krause, F., Rädler, K.-H. (1980) *Mean-Field Magnetohydrodynamics and Dynamo Theory.* Akademie-Verlag, Berlin.
5. Gabov, A.S., Sokoloff, D.D., Shukurov, A.M. (1996) Turbulent diamagnetism in a galactic disk, *Astronomy Reports*, **Vol.73, No.4**, pp. 511–519
6. Moffatt, H.K. (1978) *Magnetic Field Generation in Electrically Conducting Fluids.* Cambridge University Press, Cambridge.

# PARKER INSTABILITY WITH CORIOLIS FORCE AND MAGNETIC RECONNECTION AS A PART OF THE GALACTIC FAST DYNAMO ACTION

M. HANASZ

*Toruń Centre for Astronomy, Nicolaus Copernicus University PL-87148 Piwnice/Toruń, Poland*

K. OTMIANOWSKA-MAZUR

*Astronomical Observatory, Jagiellonian University, Orla 171, 20-244 Kraków, Poland*

AND

H. LESCH

*University Observatory, München University, Scheinerstr. 1, D-81679 Germany*

## Abstract

In the present paper, we investigate the influence of the Coriolis force and magnetic reconnection on the evolution of the Parker instability in galactic disks. We apply a model of a local gas cube, permeated by an azimuthal large-scale magnetic field and solve numerically resistive 3D MHD equations with the contribution of Coriolis force. We introduce a current dependent resistivity which switches on the magnetic reconnection above a certain critical current density. Our main goal is to study the magnetic field topology and the formation of large scale poloidal magnetic fields from the initial azimuthal field. Our simulations demonstrate that the Parker instability leads to the formation of helically twisted magnetic flux tubes which are next agglomerated by reconnection forming significant poloidal magnetic field component on the scale of the whole cube. Such an evolution represents a kind of the fast dynamo process as proposed by Parker (1992).



*P. Chossat et al. (eds.), Dynamo and Dynamics, a Mathematical Challenge,* 239–246.

## 1. Introduction

Recent observations of magnetic fields in the irregular galaxies LMC (Klein et al. 1996) and NGC4449 (Chyży et al. 2000) show that in very slowly and nearly rigidly rotating dwarfs the well-organized large-scale magnetic field structures are present. The well-known processes of a MHD turbulent dynamo are ineffective in disks rotating with velocities an order of magnitude smaller than those found in normal spirals. The outstanding problem is what kind of physical mechanism could be responsible for producing the observed large-scale magnetic configuration in dwarf galaxies. One needs a process which is faster and much more efficient than the 'classical' turbulent dynamo because the dynamical age of NGC 4449 is only of the order of $10^8$ yr (Hartmann et al. 1986).

The amplification timescale of the dynamo results from timescales of several independent processes underlying the production of the poloidal magnetic field from the toroidal one, the production of the toroidal magnetic field from the poloidal one and the relaxation process which converts the small scale magnetic field into the large scale field.

The classical mean field dynamo theory has been criticized in several ways (eg. Kulsrud and Anderson 1992; Vainshtein and Cattaneo 1992). An attractive solution of the whole bunch of problems concerning the bases of the galactic dynamo theory was presented by Parker (1992) who proposed to replace the concept of classical cascading turbulence by effects of buoyancy (Parker) instability and the fast magnetic reconnection. The Parker instability driven by cosmic rays in the rotating galactic disk contributes to an efficient generation of the poloidal magnetic field on scales of several hundreds of parsecs. On the other hand the Parker instability provides contact discontinuities in the perturbed magnetic field. The discontinuities lead to extremely high currents, plasma instabilities accompanied by an anomalous resistivity which drives magnetic reconnection. Such a chain of processes provides the dissipation and relaxation of the small scale magnetic fields, i.e. the necessary ingredients of the overall dynamo process.

## 2. Formulation of the problem:

We focus our attention on the 3D evolution of the Parker instability in the presence of the Coriolis force and fast magnetic reconnection, which is supposed to operate in the ISM with a rate of the order of a few percent of the Alfven speed (Parker 1992, Kahn and Brett 1993). Our main goal is to investigate the changes of magnetic field topology. We perform isothermal, resistive numerical simulations of the Parker instability in a 3D computational domain embedded in a rigidly rotating galactic disk stratified by uniform vertical gravity and pervaded by an azimuthal initial magnetic

TABLE 1. Table 1

| quantities in A, B and C | in X | in Y | in Z |
|---|---|---|---|
| number of grid points: | 80 | 180 | 80 |
| scale length of the box in pc: | 600 | 1800 | 600 |
| case | reconnection | uniform diffusion | |
| A | no | no | |
| B | yes | no | |
| C | no | yes | |

field. Cartesian coordinates are used, with X, Y and Z corresponding respectively to radial, azimuthal and vertical directions. We use the Zeus3D MHD code (Stone & Norman, 1992a, 1992b) with modifications aimed to account for the anomalous resistivity of the form:

$$\eta(j) = \eta_2(j^2 - j_{\rm crit}^2)(\Theta(j^2 - j_{\rm crit}^2)) \tag{1}$$

in the induction equation

$$\partial\vec{B}/\partial t = {\rm rot}(\vec{v} \times \vec{B}) - {\rm rot}(\eta(j)\ {\rm rot}\vec{B}) \tag{2}$$

where $\eta_2$ is a constant coefficient, $j_{\rm crit}$ is the critical current above which the resistivity switches on and $\Theta$ is the step function (see Ugai 1992 and references therein).

We apply perturbations of the vertical velocity in lower part of the box for z < H = 150 pc, where H is the vertical scale height of the disk (Hanasz et al. 2000). The perturbations evolve in a way typical for the Parker instability. Magnetic arcs form and rise toward the galactic halo and thermal gas flows along inclined magnetic lines toward magnetic valleys (Hanasz et al. 2000). The input quantities are summarized in Table 1. The assumed isothermal sound speed is 7 $\rm kms^{-1}$, a vertical gravitational acceleration is $-2\times10^{-9}$ $\rm cms^{-2}$ and the angular velocity of galactic rotation is 0.025 $\rm Myr^{-1}$ corresponding to that of the orbit of the Sun (see Hanasz et al. 2000).

## 3. Results

We perform three basic numerical experiments: the ideal case A (without reconnection and diffusion), the case B with reconnection (both presented in Hanasz et al. 2000) and the run C with the uniform diffusion in the whole domain.

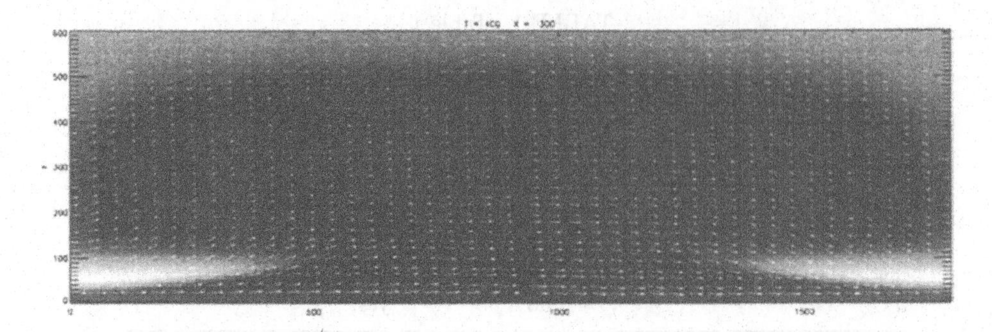

*Figure 1.* The vertical cut through the computational volume shows the deformation of the initially horizontal magnetic field (arrows) which is typical for the Parker instability. Gray scales represent the gas density perturbation $\Delta\rho/\rho_0$, light and dark grey shades correspond to large and small densities respectively. Note that the gas accumulates in magnetic valleys (see Hanasz et al. 2000 for more details).

Fig. 1 shows the magnetic field vectors for case A superimposed onto the gas density contrast map for T=400 Myr and x = 300 pc (Hanasz et al. 2000), where dark and light colors mark under-dense and over-dense regions, respectively. We can see that the magnetic lines form classical structure of the Parker instability. The rising motions of under-dense regions deform the magnetic field lines in a wave like fashion and the gas slides down along the magnetic field accumulating in the magnetic valleys making the rising regions lighter. Fig. 2 presents the set of the magnetic lines with the starting points at y = 0 pc and z = 50 pc at the same time step as Fig. 1. The lines are shaped in bows extending over the whole range of the Y (azimuthal) coordinate. Due to the Coriolis force the rising lines of magnetic field are twisted opposite to the large-scale rotation and lower parts are twisted in the sense of galactic rotation. In the ideal case A such a structure of magnetic field persists up to $T \sim 500$ Myr, however the effects of numerical resistivity are visible (more details see Hanasz et al. 2000).

A dramatically different magnetic field topology results from the experiment B in the presence of the anomalous resistivity (Hanasz et al. 2000). Figs 3 and 4 (left panels) present the magnetic field vectors shown in the XZ slice at y = 300 pc overlaid onto the current density grey plot at two subsequent time steps: 450 Myr and 650 Myr. Light and dark colors mark the maximal and minimal current density strength. Figs 3 and 4 (right panels) show the lines of force starting at y = 0 pc z = 5 pc at the same time stages. After 400 Myr the critical current density value is exceeded and reconnection starts to act. The previously thin sheets of gas and magnetic field start to form circular structures filled with gas blobs (Fig. 3, left panel). Magnetic reconnection working locally (in the brightest regions of the map) converts the smooth arcs from Fig. 2 into a family of helical

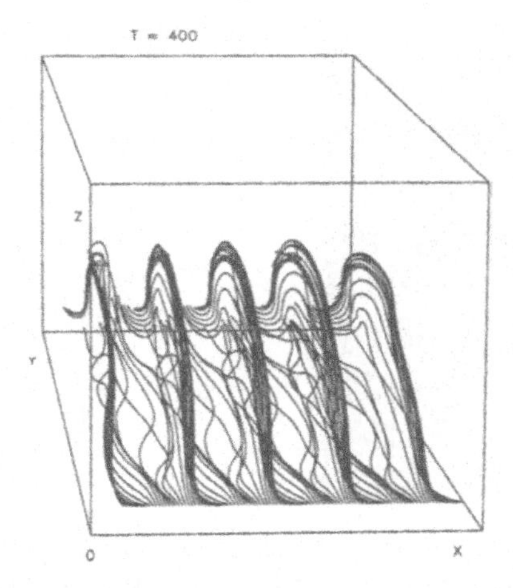


*Figure 2.* Topology of magnetic field lines at $T = 400$ Myr for the case A. The magnetic field lines represent a wavelike deformation typical for the Parker instability

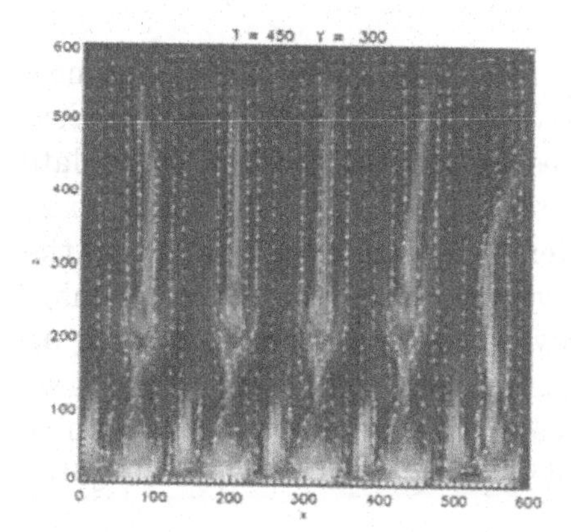


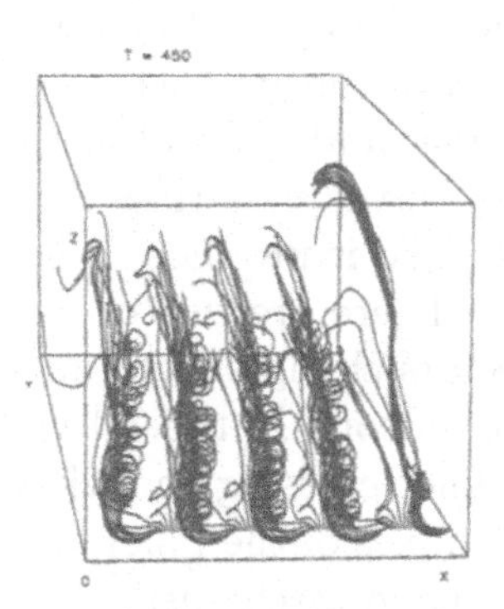


*Figure 3.* The magnetic field vectors for the case B overlaid onto the current density grey plot for T = 450 Myr. The light and dark regions denote high and low magnitude of current, respectively. The XZ cut is taken at y=300 pc (left panel). Topology of magnetic field lines at the same time stage (right panel).

flux tubes clearly visible after 450 Myr in Fig. 3 (right panel) in the 3D space. The apparent helical twist of field lines in the whole computational volume is due to the Coriolis force. The centers of tubes contain gas filaments collected in magnetic valleys before reconnection. The topological evolution of magnetic field lines proceeds in a way similar to that predicted by Parker (1992). The difference is however that the reconnection does not lead to the formation of closed magnetic loops but instead to the formation

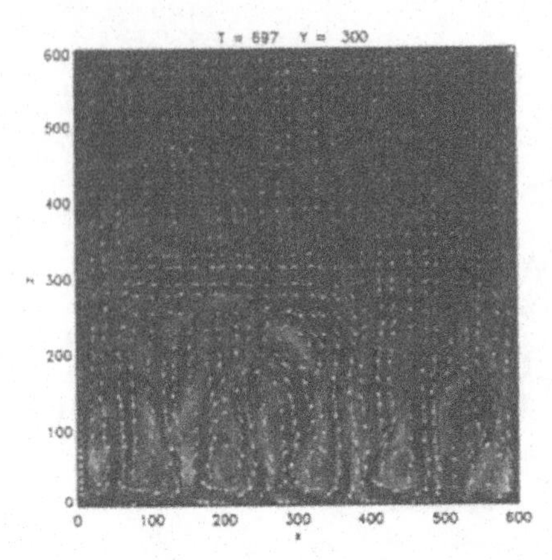
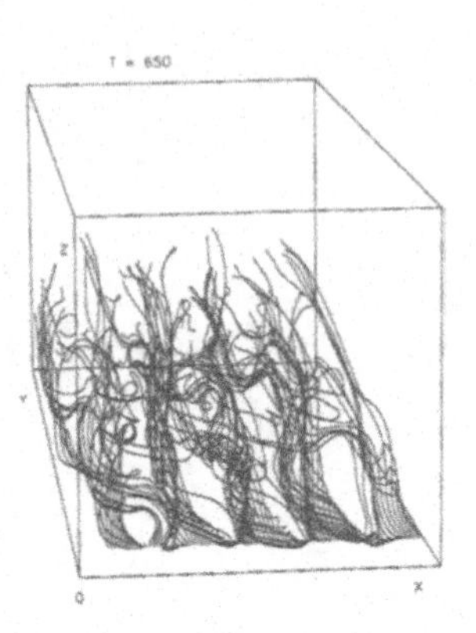

*Figure 4.* The magnetic field vectors for the case B overlaid onto the current density grey plot for T = 650 Myr. The light and dark regions denote high and low magnitude of current, respectively. The XZ cut is taken at y=300 pc (left panel). Topology of magnetic field lines at the same time stage (right panel).

of magnetic helices. After the next time interval of about 200 Myr a new generation of current sheets is formed between the adjacent closely packed helical tubes and due to the anomalous resistivity the flux tubes coagulate into larger structures (Fig. 4). The large-scale poloidal (along the X axis) magnetic field is produced possessing different directions depending on the Z height. There are two or even more levels at which the course of magnetic vectors changes from inward to outward the galactic center going up from the bottom. In the 3D view of magnetic lines (Fig. 4, right panel) we observe lines crossing the whole radial scale. The process of separation of opposite fluxes of the poloidal magnetic field component is required by the MHD dynamo mechanism.

The last experiment C is performed with the uniform diffusion assumed in the whole computational box (without the anomalous resistivity mechanism, see Table 1). The results show that the topology of magnetic field remains similar to the ideal case (A) to 500 Myr. Later on the evolution starts and due to the magnetic resistivity the arcs of magnetic field, resulting from the Parker instability, convert to partially helical structures. Fig. 5 (left panel) presents the XZ slice at y = 300 pc of the magnetic field vectors superimposed onto the current density grey plot at 650 Myr. We can see that there are at least two layers with a different mean orientation of vectors going along the X axis dependent on z, thus the large-scale poloidal component is produced as well. The lines of magnetic field (Fig. 5, right panel) have similar topology to the case B (Fig. 4, right panel), but they are apparently less helical. In this experiment the evolution goes much slower than in the case with reconnection (B) and the lines of force do not

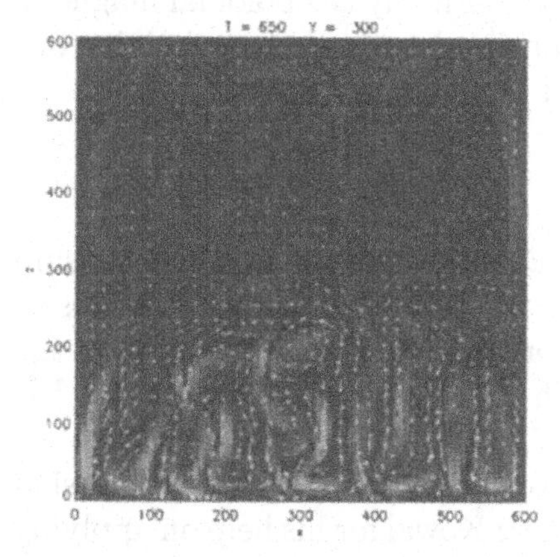


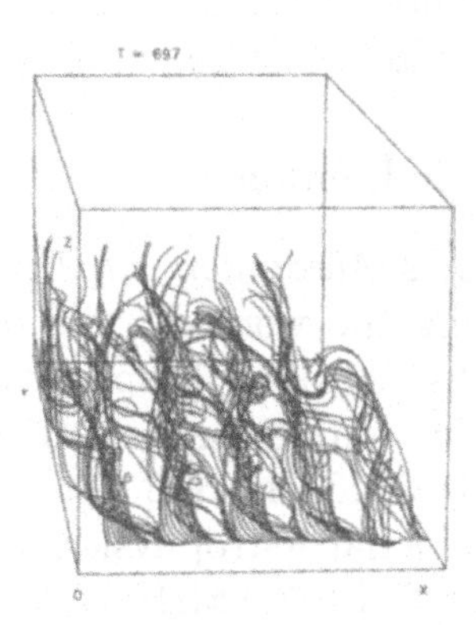


*Figure 5.* The magnetic field vectors for the case C overlaid onto the current density grey plot for T = 650 Myr. The light and dark regions denote high and low magnitude of current, respectively. The XZ cut is taken at y = 300 pc (left panel). Topology of magnetic field lines at the same time stage (right panel).

form helically twisted flux tubes visible so clearly in Fig. 3 (right panel).

## 4. Summary

We have performed three numerical experiments of the magnetic field evolution under the influence of Coriolis force and external disturbances: the ideal case (A) with no diffusion and reconnection, the experiment only with anomalous resistivity (B) and the case with uniform magnetic diffusion in the whole computational volume (C). In all simulations the arcs of magnetic field typical for the Parker instability are formed in the same rate (Fig. 1a and 2). The magnetic field topology in the ideal case A remains almost the same during the whole evolution, however signs of numerical reconnection became visible around $T \sim 500$ as an initial phase of the formation of helical structures. The calculation with the resistivity taken into account (run B) resulted in dramatical changes of the magnetic field structure in the whole simulation volume. The magnetic arcs twisted by the Coriolis force evolved toward a system of well defined helical flux tubes (Hanasz et al. 2000). This process produced gas filaments surrounded by helical magnetic field in centers of flux tubes. On a longer time scale the next generations of current sheets formed between adjacent flux tubes. The flux tube coalescence through the magnetic reconnection gave rise to the large-scale poloidal component of the magnetic field (in radial direction). Such behavior of magnetic field could be a part of the fast dynamo process in the galaxy. The simulations performed with the uniform diffusion coefficient in the whole computational box (run C) resulted in much slower topological evolution of magnetic field than in the case B. Finally the poloidal magnetic field has been formed as in the case of localized anomalous resistivity but almost without the intermediate phase of flux tube formation.

## Acknowledgements

MH thanks *Alexander von Humboldt Foundation* for the AvH Fellowship supporting the work on this paper and the München University Observatory, for the kind hospitality. This work was partly supported by the Polish Committee for Scientific Research (KBN) through the grants PB 457/P03/97/13 and PB 4264/P03/99/17. We thank the Laboratory for Computational Astrophysics, University of Illinois for the kind permission to use the ZEUS-3D MHD code and Grzegorz Kowal for his help on applying the software. The computations were partially performed on the Convex-SPP machines at the Academic Computer Centre 'Cyfronet' in Kraków (grants no. KBN/C3840/UJ/011/1996 and KBN/SPP/UJ/011/1996).

## References

1. Chyży K., Beck R., Kohle S., Klein U., Urbanik M.: 2000 *Astron. & Astrophys.*, 355, 128
2. Hanasz, M., & Lesch, H. 1998, *Astron. & Astrophys.*, 332, 77
3. Hanasz, M., & Lesch, H. 2000, *Astrophys. J.*, (Nov, 1st issue)
4. Hanasz, M., Otmianowska-Mazur, K., & Lesch, H. 2000, *Astrophys. J.*, (submitted)
5. Hartmann L., Geller M., Huchra J.: 1986, AJ 92, 1278
6. Klein U., Hummel E., Bomans D., Hopp U.: 1996, *Astron. & Astrophys.* 313, 396
7. Kahn, F.D., Brett, L. 1993, *Mon. Not. R. Astron. Soc.* , 263, 37
8. Kulsrud, R.M., and Anderson, S.W.: 1992, *Astrophys. J.* , **396**, 606-630.
9. Parker, E.N., 1992, *Astrophys. J.*, 401, 137
10. Stone, J.M., Norman, M.L, 1992a, *Astrophys. J.S.*, 80, 753
11. Stone, J.M., Norman, M.L, 1992b, *Astrophys. J.S.*, 80, 791
12. Ugai, M., 1992, Phys. Fluids B, 4, 2953
13. Vainshtein, S.I., and Cattaneo, F.: 1992 *Astrophys. J.* , **393**, 165-171.

# MAGNETIC HELICITY FLUX AND THE NONLINEAR GALACTIC DYNAMO

N. KLEEORIN
*Department of Mechanical Engineering*
*Ben-Gurion University of Negev*
*POB 653, 84105 Beer-Sheva, Israel*

D. MOSS
*Department of Mathematics, University of Manchester*
*Manchester M13 9PL, UK*

I. ROGACHEVSKII
*Department of Mechanical Engineering*
*Ben-Gurion University of Negev*
*POB 653, 84105 Beer-Sheva, Israel*

AND

D. SOKOLOFF
*Department of Physics, Moscow State University*
*Moscow 119899, Russia*

## Abstract

We demonstrate that by including a flux of helicity in the magnetic helicity balance in the nonlinear galactic dynamo equations, the magnetic field dynamics are changed radically. The large-scale magnetic field now saturates at approximately the equipartition level. This is in contrast to the situation without the flux of helicity, when the magnetic helicity is conserved locally, leading to substantially subequipartition values for the equilibrium large-scale magnetic field.

## 1. Introduction

It is widely believed that the large-scale magnetic fields of galaxies are generated by a galactic dynamo, resulting from the simultaneous action of the helicity of interstellar turbulence and differential rotation (see, e.g., Ruz-



*P. Chossat et al. (eds.), Dynamo and Dynamics, a Mathematical Challenge,* 247–252.

maikin et al. 1988). The evolution of a weak magnetic field with negligible effect on the turbulent flows (i.e. the kinematic stage of the dynamo) seems to be clear, but the nonlinear stage of dynamo evolution is a more contentious topic (for reviews, see Beck et al. 1996, Kulsrud 1999). The most hotly debated issue is the strength of the equilibrium magnetic field, at which dynamo action saturates.

Taking the naive viewpoint, that the saturation level for the *large-scale* magnetic field is given by the equipartition between kinetic energy and the energy of the large-scale magnetic field $\mathbf{B}$ (see, e.g., Zeldovich et al. 1983), leads to models of dynamo generated magnetic fields which are in basic agreement with the available observational information.

Vainshtein and Cattaneo (1992) formulated a more sophisticated argument, suggesting that the equilibrium magnetic field should be determined by a balance between the kinetic energy and the energy of the *total* magnetic field. The simplest models of dynamo generation then result in the estimate $b/B \sim Rm^{1/2}$, where $\mathbf{b}$ is the small-scale magnetic field, and the magnetic Reynolds number $Rm \approx 10^8$ for the interstellar turbulence. Thus the ideas of Vainshtein and Cattaneo lead to the conclusion that any dynamo generated large-scale galactic magnetic field must be negligible in comparison with that observed, and so the large scale observed field must be generated by some other mechanism. However, there are no immediate candidates to fill this role.

The conclusions of Vainshtein and Cattaneo (1992) do not seem inevitable. For example, a dynamo generated magnetic field can itself produce helicity, so the nonlinear effects can even amplify rather than suppress field generation at the initial stages of nonlinear evolution (Moss et al. 1999).

Here we demonstrate that by taking open boundaries, the approach of Vainshtein and Cattaneo (1992) yields basically the same estimate for the equilibrium magnetic field strength as is given by the naive viewpoint expressed above. The essence of our arguments is as follows. Vainshtein and Cattaneo argue that the suppression of dynamo action by the small-scale magnetic field (inevitably generated together with the large-scale) is connected with the magnetic helicity of the small-scale magnetic field. Because the total magnetic helicity is an inviscid invariant of motion, the magnetic helicity of the small-scale magnetic field can be connected with the magnetic helicity of the large-scale magnetic field. Kleeorin and Ruzmaikin (1982; see the discussion by Zeldovich et al., 1983) formulated the corresponding governing equation for the magnetic helicity, which was investigated by Kleeorin et al. (1995) for stellar dynamos, and self-consistently derived by Kleeorin and Rogachevskii (1999). During nonlinear stages of the dynamo, the $\alpha$-effect is thought to be determined by the hydrodynamic and magnetic helicities, so a closed system of equations can be obtained for the evolution of the magnetic field and the $\alpha$-coefficient (see below, Sect. 2). Kinematic galactic dynamo action can only occur if there is a turbulent flux of magnetic field through the surface of the galactic disk (see, e.g., Zeldovich et al. 1983, Ch. 11). It is quite natural to believe that this flux can transport magnetic helicity to the outside of the disc. Using the methods of Kleeorin and Rogachevskii (1999), we can introduce the corresponding term into the governing equations for the galactic dynamo. We demonstrate by numerical simulations, and to some extent analytically, that this term leads to a drastic change in the magnetic field evolution. Now the steady-state large-scale magnetic field strength is approximately in equipartition with the kinetic energy of the interstellar turbulence.

## 2. Dynamics of the magnetic helicity

Using arguments from the magnetic helicity conservation law, Kleeorin and Ruzmaikin (1982) parameterized the back-reaction of dynamo generated magnetic field in terms of a differential equation for the magnetic part of the $\alpha$ effect:

$$\partial\alpha^h/\partial t + \alpha^h/T = 4(h/l)^2(R_\alpha^{-1}\mathbf{B}\cdot\mathbf{curl}\,\mathbf{B} - \alpha B^2)\,, \tag{1}$$

where $l$ is the maximum scale of turbulent motions. We adopt here the standard dimensionless form of the galactic dynamo equation from Ruzmaikin et al. 1988; in particular, the length is measured in units of the disc thickness $h$, the time is measured in units of $h^2/\eta_T$ and $B$ is measured in units of the equipartition energy $B_{\rm eq} = \sqrt{4\pi\rho}\,u$, the magnetic part of the $\alpha$-effect, $\alpha^h$, and the total $\alpha$ (which includes hydrodynamic and magnetic parts) are measured in units of $\alpha_*$ (the maximum value of the hydrodynamic part of the $\alpha$ effect). Here $\rho$ is the density, $u$ is the characteristic turbulent velocity in the scale $l$, $\eta_T = lu/3$ is the turbulent magnetic diffusivity, $T = (1/3)(l/h)^2 Rm$, $Rm = lu/\eta$, $\eta$ is the magnetic diffusion due to the electrical conductivity of the fluid, and $R_\alpha = l\alpha_*/\eta_T$. It is important that, when applied to galaxies, Eq. (1) contains a large factor $4(h/l)^2 \sim 100$ typically. Also, then the term $\alpha^h/T$ is very small and can be dropped. When $\partial\alpha^h/\partial t = 0$ and $R_\alpha^{-1}\mathbf{B}\cdot\mathbf{curl}\,\mathbf{B} \ll \alpha B^2$, Eq. (1) yields $\alpha = \alpha^v/[1 + (4/3)RmB^2]$ (see, e.g., Vainshtein and Cattaneo , 1992). However, the latter estimate is not valid for galaxies because $\partial\alpha^h/\partial t \gg \alpha^h/T$. In addition, the condition $R_\alpha^{-1}\mathbf{B}\cdot\mathbf{curl}\,\mathbf{B} \ll \alpha B^2$ seems not to be valid for galaxies.

In Kleeorin and Rogachevskii (1999), the analysis was extended to include a flux of magnetic helicity. Based on Eq. (13) of that paper, the approximate relation

$$\partial\alpha^h/\partial t \quad = \quad 4(h/l)^2\{\mathbf{B}\cdot\mathbf{curl}\,\mathbf{B}R_\alpha^{-1} - \alpha(B)B^2$$

$$+ \quad (\partial/\partial z)[\alpha^v(z)hf_1(z)\phi(B)B^2]\} \tag{2}$$

can be formulated. In Eq. (2), $f_1(z)$ describes the inhomogeneity of the turbulent diffusivity, and we define $f(z) = \alpha^v(z)f_1(z)$. The profile $f(z)$ depends on details of the galactic structure. Also, $\alpha(B)$ is the total $\alpha$ effect and $\alpha = \alpha^v\phi(B) + \alpha^h\phi_1(B)$, where $B = |\mathbf{B}|$. Here $\alpha^v$ is the hydrodynamic part of the $\alpha$ effect, with $\alpha^v\phi(B)$ its modification due to nonlinear effects. Correspondingly, $\alpha^h$ is the magnetic part of the $\alpha$ effect, and $\alpha^h\phi_1(B)$ is the modification caused by nonlinear effects. The functions $\phi_1(B) = (3/8B^2)(1-\arctan(\sqrt{8}B)/\sqrt{8}B)$ and $\phi(B) = (4/7)\phi_1(B)+(3/7)\{1-16B^2+128B^4\ln[1+(8B^2)^{-1}]\}$ (see Rogachevskii and Kleeorin, 2000). The last term in Eq. (2) is related to the turbulent flux of magnetic helicity which causes a drastic change in the dynamics of the large-scale magnetic field. In Eq. (2) $z$ is dimensional. For simplicity we replace the flux divergence in the right hand side of Eq. (2) by a decay term, i.e. we replace $\partial/\partial z$ by $1/h$ (in principle, there is no problem in treating this point more carefully).

## 3. The equilibrium magnetic field configuration

We now present some asymptotic expansions for galactic dynamo models with the nonlinearity (2). As the parameter $4(h/l)^2$ in the right hand side of Eq. (2) is large, we can make the approximation

$$\alpha(B) = f(z)\phi(B) + R_\alpha^{-1}B^{-2}\mathbf{B}\cdot\mathbf{curl}\,\mathbf{B}, \tag{3}$$

where the function $\phi(B)$ has the following asymptotic behaviour: $\phi(B) = 1/(4B^2)$ for $B \gg 1/\sqrt{8}$ and $\phi = 1 - (48/5)B^2$ for $B \ll 1/\sqrt{8}$ (see, Rogachevskii and Kleeorin, 2000). Then insertion of the $\alpha$-coefficient given by Eq. (3) into the local disc dynamo problem yields the following equations:

$$\partial b_r/\partial t = -(\alpha(B)B_\phi)' + b_r'' , \tag{4}$$

$$\partial B_\phi/\partial t = Db_r + B_\phi'' , \tag{5}$$

where $B_r = R_\alpha b_r$ and $\partial F/\partial z \equiv F'$. In a steady-state Eqs. (3)-(5) yield $[(B')^2]' + 2Df(z)\phi(B)B^2 = 0$, where we consider only the fields of quadrupole symmetry in the $\alpha\Omega$ dynamo, so that $B \approx B_\phi$. This assumption is justified if $|D| >> R_\alpha$, i.e. $|R_\omega| >> 1$. For $B \gg 1/\sqrt{8}$ this equation has an exact solution for arbitrary profile $f(z)$. The steady solution of this equation with the boundary conditions $B_\phi(z=h) = 0$ and $B'_\phi(z=0) = 0$ and negative dynamo number is given by $B_\phi \approx B(z) = (|D|/2)^{1/2}\{\int_z^h[|\int_0^{z'} f(\tilde{z})\,d\tilde{z}|]^{1/2}\,dz'\}$. In particular, for the specific choice of helicity profile $f(z) = \sin\pi z$ we have

$$B_\phi = (2\sqrt{|D|}/\pi^{3/2})B_{\rm eq}\cos(\pi z/2) , \tag{6}$$

$$B_r = -(\sqrt{\pi R_\alpha/|R_\omega|}/2)B_{\rm eq}\cos(\pi z/2) , \tag{7}$$

(Kleeorin et al., 2000), where we have restored the dimensional factor $B_{\rm eq}$. This solution is remarkably close to the results from the naive *Ansatz* $\alpha = \alpha_0(1-(B/B_{\rm eq})^2)$ or $\alpha = \alpha_0/(1+(B/B_{\rm eq})^2)$, or from the model of Moss et al. (1999). For example, the pitch angle of the magnetic field lines is $p = -\arctan(\pi^2/4|R_\omega|) \approx 14°$ for $|D| = 10$ and $R_\alpha = 1$.

## 4. Numerical results

Numerical solution of the equations verified that an initially weak magnetic field approaches the equilibrium configuration (6) with accuracy 1% for $|D| > 1000$, and an accuracy of 50% for $|D| > 10$. As suggested in Sect. 3, near to conditions of marginal excitation, the behaviour of the solutions may be more complicated. The threshold value for the nonlinear solution of Eqs. (4) and (5) is $D \approx -3.14$, while the linear threshold value is $D \approx -8$. This is because the nonlinear solution arranges itself so that the term $\mathbf{B}\cdot\mathbf{curl}\,\mathbf{B}/B_\phi^2$ in $\alpha$ (see Eq. (3)) is of order 1. Thus, for the nonlinear solution with $D = -8$, the maximal value of $\alpha$ is about 1.25, whereas for $D = -5$, the maximal value is about 1.76. For $|D| \lesssim 10$ we obtain numerically $B_\phi(0) \approx 0.23|D - D_{\rm cr}|^{0.52}$, where $D_{\rm cr}$ is the nonlinear threshold value. As $|D - D_{\rm cr}|$ increases towards 10, the slope increases slightly, but the equation for $B_\phi(0)$ remains a reasonable estimate.

This result is robust under variations of the helicity profile. For $f(z) = z$ we find $D_{\rm cr} = -7.49$ in the nonlinear case, while the linear threshold value is $D_{\rm cr} = -12.5$ and $B_\phi(0) \approx 0.15|D - D_{\rm cr}|^{0.50}$ near $D = D_{\rm cr}$, i.e. again a square root dependence to within the errors of our procedure. Further, with $f(z) = z/|z|$ in $|z| > 0.2$, and a smooth interpolation to zero in $|z| \le 0.2$, we find $D_{\rm cr} \approx -2.41$ and $B_\phi(0) \approx 0.25|D - D_{\rm cr}|^{0.50}$, again closely the same dependence. (In this case the linear threshold value is $D_{\rm cr} = -6.53$.)

## 5. Discussion

If local helicity conservation is assumed (Eq. (1)), then the result is that the magnetic field decays, after a stage of kinematic growth. If the molecular diffusivity of the magnetic field is taken into account, this decay is followed by a stabilization at a very low magnetic field strength, corresponding to the estimate of Vainshtein and Cattaneo (1992). The overall scenario of magnetic field and helicity dynamics can then be described as follows. Large-scale dynamo action produces large-scale magnetic helicity. Due to the local conservation of helicity, suppression of field generation results. An equilibrium is possible if molecular diffusivity is present, so the equilibrium magnetic field strength is very low.

On the other hand, if transport of helicity out of the dynamo region is allowed (Eq. (2)), the local value of the helicity changes during magnetic field evolution. The scenario of magnetic field and helicity dynamics can be presented as follows. As usual, magnetic helicity of the large-scale magnetic field is produced but now, however, the total magnetic helicity is not conserved locally, rather the magnetic helicity of the small-scale magnetic field is redistributed by a helicity flux. The equilibrium state is given by a balance between helicity production and transport. The helicity conservation law now expresses the conservation of an integral of the helicity over the galactic disc. However this conservation law is trivial, because the integral vanishes identically as helicity is an odd function with respect to $z$. Our main result is that, if the arguments leading to Eq. (2) are accepted, then the equilibrium strength of the large-scale magnetic field is of order that of the equipartition field, as given by naive estimates of alpha-quenching; galactic mean field dynamos can produce large scale magnetic fields of the observed strengths.

*Acknowledgements* This work was partially supported by INTAS (Grant No. 99-348), NATO (Grant PST.CLG 974737), and RFBR (grant 00-05-64062a). DS was aided financially by Ben-Gurion University and the Royal Society.

## References

1. Beck, R., Brandenburg, A., Moss, D., Shukurov, A. and Sokoloff, D. (1996) Galactic Magnetism: Recent Developments and Perspectives, *Ann. Rev. Astron. Astrophys.* **34**, 155–206.
2. Kleeorin, N., Moss, D., Rogachevskii, I. and Sokoloff D., (2000) Helicity Balance and Steady-State Strength of Dynamo Generated Galactic Magnetic Field. *Astron. and Astroph.*, **361**, L5–L8.
3. Kleeorin, N. and Rogachevskii, I. (1999) Magnetic Helicity Tensor for an Anisotropic Turbulence, *Phys. Rev. E.* **59**, 6724–6729.
4. Kleeorin, N., Rogachevskii, I. and Ruzmaikin, A. (1995) Magnitude of Dynamo - Generated Magnetic Field in Solar - Type Convective Zones, *Astron. Astrophys.* **297**, 159–167.
5. Kleeorin, N.I. and Ruzmaikin, A.A. (1982) Dynamics of the Mean Turbulent Helicity in a Magnetic Field *Magnetohydrodyn.* **18**, 116–122.
6. Kulsrud R. (1999) A Critical Review of Galactic Dynamos, *Ann. Rev. Astron. Astrophys.* **37**, 37–64.
7. Moss, D., Shukurov, A. and Sokoloff, D. (1999) Galactic Dynamos Driven by Magnetic Buoyancy, *Astron. Astrophys.* **343**, 120–131.
8. Rogachevskii, I. and Kleeorin, N. (2000) Electromotive Force for an Anisotropic Turbulence: Intermediate Nonlinearity, *Phys. Rev. E.* **61**, 5202–5210.
9. Ruzmaikin, A., Shukurov, A. and Sokoloff, D. (1988) *Magnetic Fields of Galaxies*, Kluwer, Dordrecht.
10. Vainshtein, S.I. and Cattaneo, F. (1992) Nonlinear Restrictions on Dynamo Action, *Astrophys. J.* **393**, 165–171.
11. Zeldovich, Ya.B., Ruzmaikin, A.A. and Sokoloff, D.D. (1983) *Magnetic Fields in Astrophysics*, Gordon and Breach, New York.

# HELIOSEISMIC TESTS OF DYNAMO MODELS

A.G. KOSOVICHEV
*W.W.Hansen Experimental Physics Laboratory,*
*Stanford University, USA*

**Abstract**

Helioseismology provides important input and test data for dynamo theories of solar activity by measuring variations of the internal structure and dynamics of the Sun with the activity cycle. Recent results from the GONG network and MDI/SOHO space experiment obtained in 1996-2000 cover the period of transition from the 'old' solar cycle 22 to the 'new' cycle 23. These data have revealed correlated variations of zones of generation of the solar magnetic fields and zonal shear flows in the convection zone. An attempt is made to detect solar-cycle variations in the tachocline region at the base of the convection zone, which is believed to be the main cite of the solar dynamo. However, the current results are controversial. By comparing the internal rotation profile with the rotation rates of the 'old' and 'new' magnetic fluxes it has been suggested that both fluxes were generated in a low-latitude zone of the tachocline. Zonal flows and structures migrating towards the equator and probably associated with dynamo waves have been detected in the solar interior. They show a curious sudden displacement towards higher latitudes in the second-half of 1999. This is not explained by the current dynamo theories and indicates a complex dynamical behavior of the solar dynamo. Recently developed methods of local helioseismology have allowed us to investigate processes of formation of active regions and sunspots. In particular, converging vortex flows have been found beneath a sunspot in agreement with Parker's theory.

## 1. Introduction

Helioseismic diagnostics are based on observations of acoustic waves on the Sun. The acoustic waves are excited in a thin subsurface layer of turbulent convection and propagate through the inner layers. Because the sound speed



*P. Chossat et al. (eds.), Dynamo and Dynamics, a Mathematical Challenge,* 253–260.

increases with depth the waves are refracted and reappear on the surface. There are two basic techniques for inferring the information about internal structures and flows from observed properties of these waves. One of these techniques is based on representing the acoustic waves in terms of normal modes of solar oscillations and measuring the mode frequencies, and the other method is to represent the waves in terms of propagating wave packets and determine the travel times of these packets. The former approach is called sometimes 'global helioseismology', and the later is called 'local-area helioseismology', 'time-distance helioseismology', or 'heliotomography'. The method of normal modes allows us to determine the axisymmetrical (zonal) structures and differential rotation. It is limited to 2D inferences, but it is well developed and provides robust information from the solar surface down to the energy-generating core. The time-distance technique is capable to determine 3D structures and flows in the Sun, but much less developed than the normal mode approach and requires substantial computing resources. It has been mostly employed for studying MHD processes in the upper convection zone, such as convection, formation of active regions and dynamics of sunspots. Thus, these two main helioseismology tools complement each other providing the information about both 'global' dynamo related to the solar cycle and 'local' dynamo which may produce magnetic network and support sunspots. Some models and mathematical tools of helioseismology are described by Kosovichev (1999).

## 2. Differential rotation inside the sun and dynamics of solar magnetic flux

Rotation rate of the solar convection zone is one of the most important input parameters of the dynamo theory. In Figure 1 we show the rotation rate as a function of radius at three latitudes, $0^\circ$ (equator), $15^\circ$, and $30^\circ$, obtained from the Michelson Doppler Imager (MDI) data on SOHO spacecraft (Kosovichev *et al.*, 1997; Schou *et al.*, 1998). These results show that the latitudinal differential rotation, well-known from observations of the solar surface, extends through the convection zone, and that the radiative zone ($r < 0.7R$) rotates almost rigidly. The zone of transition between the differentially rotating convection zone and rigidly rotating radiative core is called 'tachocline'. Most of this zone is located in the convectively stable radiative core, and its characteristic width is 0.05–0.1 $R$ (Kosovichev, 1996; Charbonneau *et al.*, 1999). Frequency inversions for the radial structure of the Sun show a sharp peak in the sound-speed profile at this zone (Fig. 1b), which is likely to result from material mixing, probably, caused at least partially by convective overshoot. The overshoot layer is likely to be the place where the solar dynamo operates (Parker, 1993).

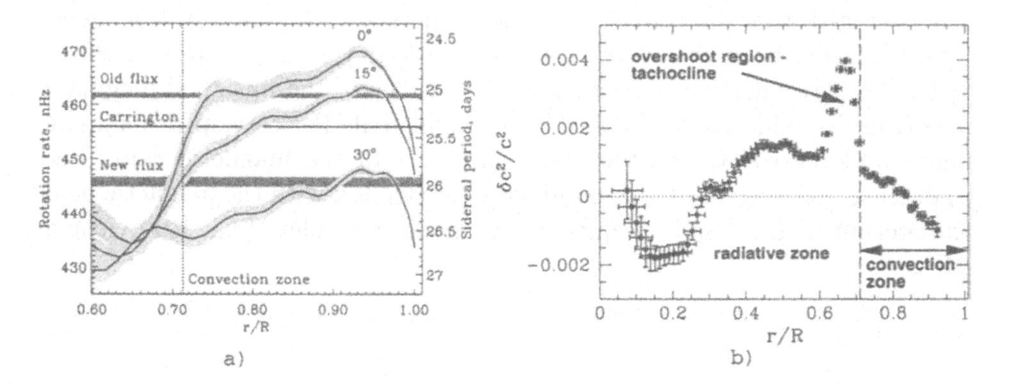


*Figure 1.* a) Rotation rate inside the Sun determined by helioseismology (Kosovichev *et al.*, 1997; Schou *et al.*, 1998) as a function of radius at three latitudes, 0°, 15°, 30° (solid curves with light gray areas indicating $1\sigma$-error estimates). The horizontal shaded areas show the rotation rates of the old magnetic flux in the latitude range $1-5°$, and the new magnetic flux in the range $28-32°$ (Benevolenskaya *et al.*, 1999). The horizontal solid line shows the Carrington rotation rate, 456.03 nHz (sidereal period 25.38 days). The vertical dotted line shows the lower boundary of the solar convection zone. $R$ is the solar radius. b) The relative differences between the squared sound speed in the Sun and a standard solar model as inferred from the SOHO/MDI data. The horizontal bars show the spatial resolution, and the vertical bars are error estimates.

In Fig. 1a we compare the the internal rotation rate with the surface rotation rate of the solar magnetic fluxes during the transition from the previous ('old') solar cycle 22 to the current ('new') cycle 23 in 1996-97. The 'old' flux appeared in a low-latitude zone, $\pm 5°$, and rotated faster than the 'new' flux which appeared at $\sim 30°$ latitude. Fig. 1a shows that if the 'new' flux was generated in the tachocline region and anchored there then it was probably generated in a low-latitude zone, $\pm 15°$, because this flux rotated faster than the plasma outside this latitudinal zone of the tachocline.

It has been also found that the 'new' toroidal magnetic flux (with reversed polarities according to Hale's law) appeared first at the same active longitudes where the 'old' flux was mostly concentrated (Benevolenskaya *et al.*, 1999). This means that the solar dynamo is highly non-uniform in both latitude and longitude. Studying the longitudinal structure of the tachocline is one the current goals of helioseismology.

## 3. Solar-cycle variations of the internal structure of the Sun

The latitudinal structure of the Sun determined by helioseismology reflects variations of temperature and magnetic field distributions associated with the dynamo processes. In Figure 2 we compare the latitudinal variations of the averaged over the convection zone aspherical component of the sound speed inferred from the MDI data during a four-year period from 1996 to 2000 (Fig.2a) with the corresponding variations of the unsigned magnetic flux (Fig. 2b) which is determined by averaging Kitt Peak synoptic magnetic maps and represents the toroidal component of the magnetic field. The aspherical component of the sound speed associated with the magnetic field is determined from split frequencies of mode multiplets (e.g. Kosovichev, 1999).

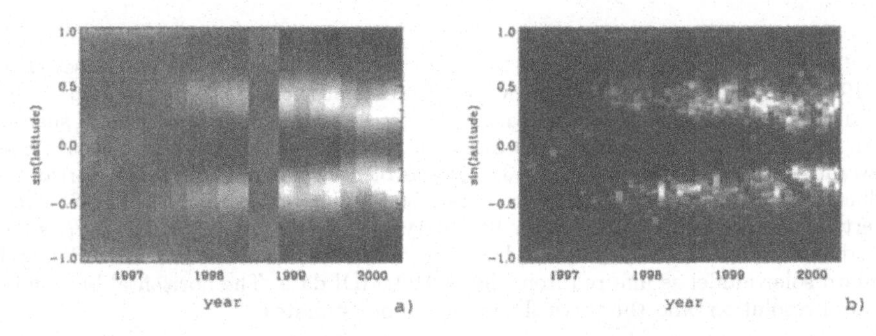


*Figure 2.* a) Variations of the aspherical component of the sound speed averaged for all depths through the convection zone with the solar cycle. The bright colors show positive variations, and the dark colors show negative variations. The result is obtained from the SOHO/MDI data. b) Temporal variations of the latitudinal distribution of the unsigned surface magnetic flux from Kitt Peak Observatory data.

It is clear that the the sound-speed signal is correlated with the magnetic field, and therefore, it is likely to be caused mostly by the magnetic field. The aspherical component of the sound speed is of the order of 0.01%, and it increases with the increased magnetic activity. These variations are likely to be associated with solar dynamo waves. An intriguing feature is a reverse migration of the activity belts to the poles which started in the late 1999. This variation might be related to the 2-year component of the solar cycle (e.g. Benevolenskaya, 1998), and generally reflects a complicated dynamic nature of solar dynamo. The shift of the sunspot formation zone towards the poles probably indicates that the current solar cycle may be longer than usual.

The depth structure of the sound-speed variations with the solar cycle has not been fully established. Our current results show that while most of the radial variations occur close to the surface, at $0.95-1R$, there might be some significant non-spherical variations deeper in the convection, including the tachocline region (Dziembowski *et al.*, 2000).

## 4. Torsional oscillations

So-called 'torsional oscillations' of the Sun represent bands of zonal mass flows moving faster or slower with respect to a smooth differential rotation profile (Howard & LaBonte, 1981). These bands originate in a mid-latitude zone at the beginning of a solar cycle and migrate towards the equator as the 11-year cycle progresses, following the sunspot zones. It is unclear if these flows with a typical speed of $\pm 5$ m/s are simply a result of the Lorentz force associated with bands of the toroidal field (Yoshimura, 1981; Kliorin & Ruzmaikin, 1984), or if they play a more fundamental role in solar dynamo. In any case, these zonal flows provide an important information about the dynamo because they are observed in deeper layers on the Sun by helioseismology (Kosovichev & Schou, 1997; Schou *et al.*, 1998).

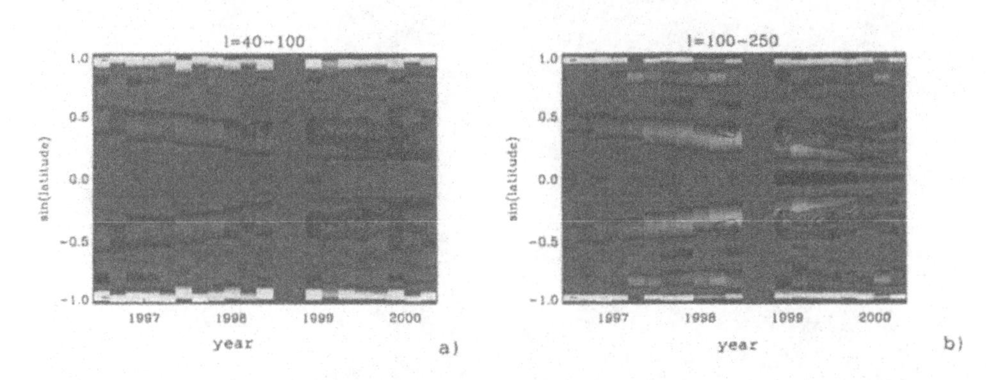


*Figure 3.* Variations of the zonal azimuthally averaged flows in the convection zone - so-called torsional oscillations. Near the surface these flows are typically about 5 m/s, slowing with depth. The red color shows higher velocities and blue - lower velocities. The zonal flows are correlated with the aspherical component of the solar structure and the surface magnetic flux. The contour lines shows the largest sound-speed variations which appear where near the maxima of the velocity gradients. Panel a) shows the inferences from the acoustic modes of angular degree $l = 40 - 100$, which sample the whole convection zone. Panel b) show the results for the modes of higher degree, $l = 100 - 250$, which propagate only close to the surface.

Fig. 3 shows the latitudinal distribution of the zonal flows inferred from two sets of normal modes. The left panel which represents an average through the convection zone including deep layers provides a hint of a second set of zonal flows at higher latitudes, which started in 1999. This might be related to a second wave of solar activity. The recent result have shown that in the upper 5% of the solar convection zone these flows have the properties similar to the surface flows. However, there is an indication that at the base of the convection zone the zonal flows are varying with a much shorter period of 1.3 years (Howe *et al.*, 2000). If confirmed this can indicate on non-linear dynamics of the solar dynamo (Covas *et al.*, 2000).

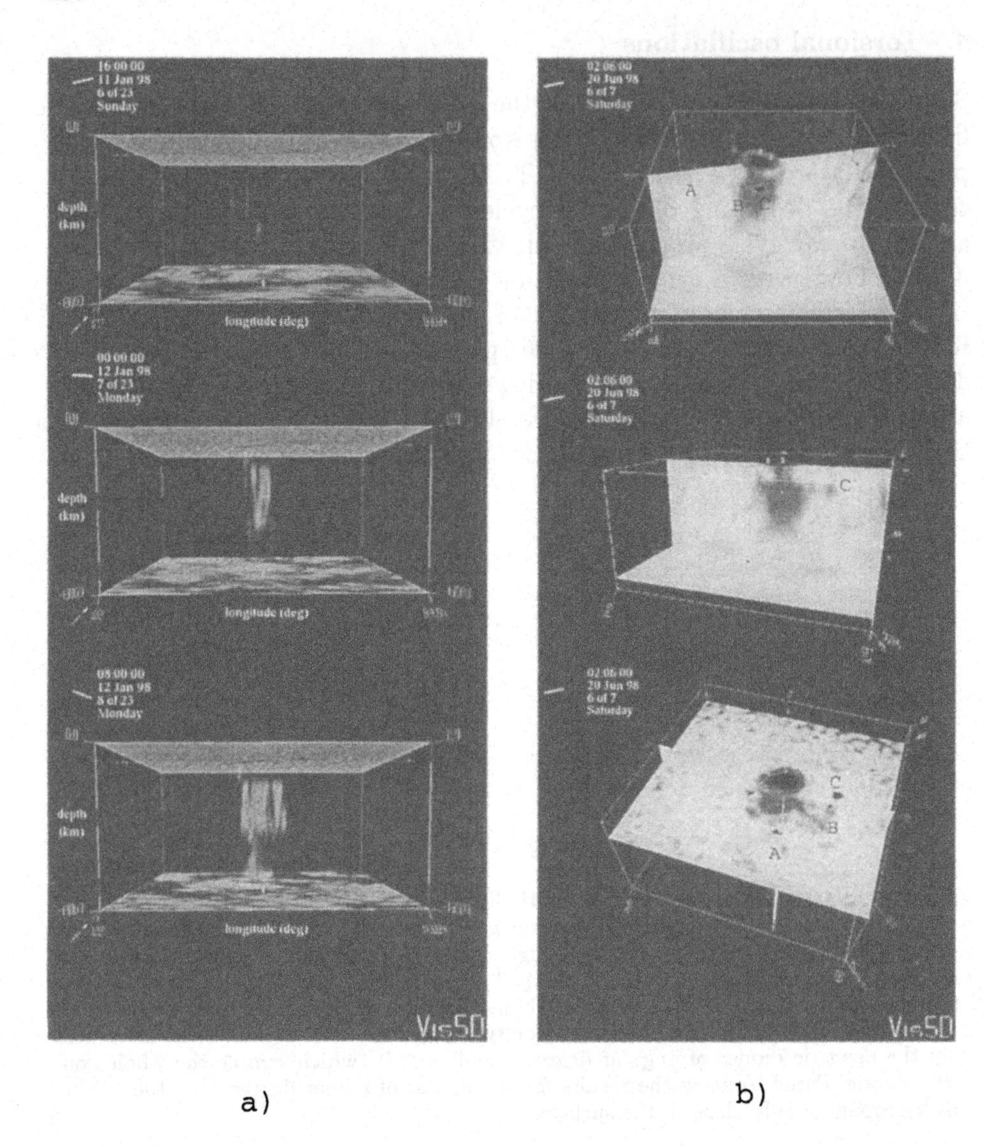


*Figure 4.* a) Images of sound- speed perturbations caused by an emerging active region in January 1998. The horizontal size of the box is approximately 38 degrees (460 Mm), the vertical size is 18 Mm. The panels on the top are MDI magnetogram showing the surface magnetic field of positive (red) and negative (blue) polarities. The perturbations of the sound speed are approximately in the range from -1 to +1 km/s. The positive variations are shown in red, and the negative ones in blue. b) An example of the internal structure (in three different projections) of a large sunspot observed on June 17, 1998. An image of the spot taken in the continuum is shown at the top. The wave-speed perturbations in the spot are much stronger than in the emerging flux. The typical perturbations range from 0.3 to 1 km/s. At a depth of 4 Mm, a 1 km/s wave-speed perturbation corresponds to a 10% temperature variation (about 2.800 K) or to a 18 kG magnetic field.

## 5. Diagnostics of emerging active regions and sunspot formation

Solar magnetic fields which are thought to be generated by dynamo processes in the tachocline propagate through the convection zone and emerge on the surface forming active regions and sunspots. We have attempted to image an emerging magnetic flux in the upper convection zone using the time-distance technique (Fig.4a). We have concluded that the emerging flux propagated through the characteristic depth of 10 Mm in approximately 2 hours. This gives an estimate of the speed of emergence $\sim$ 1.3 km/s in the upper convection zone (Kosovichev, Duvall and Scherrer, 2000). This is somewhat higher than predicted by theories. The observed development of the active region seems to suggest that the sunspots are formed as a result of the concentration of magnetic flux close to the surface.

It is also found that beneath the spot the perturbation is negative in the subsurface layers and becomes positive in the deeper interior (Fig. 4b). One can suggest that the negative perturbations of the wave speed beneath the spot are, probably, due to the lower temperature. These data also show 'fingers' - narrow long perturbations which are 4-5 Mm deep and connect to the spot two pores (marked B and C). These pores have the same magnetic polarity as the sunspot. Pore A which has the opposite magnetic polarity is not connected to the spot.

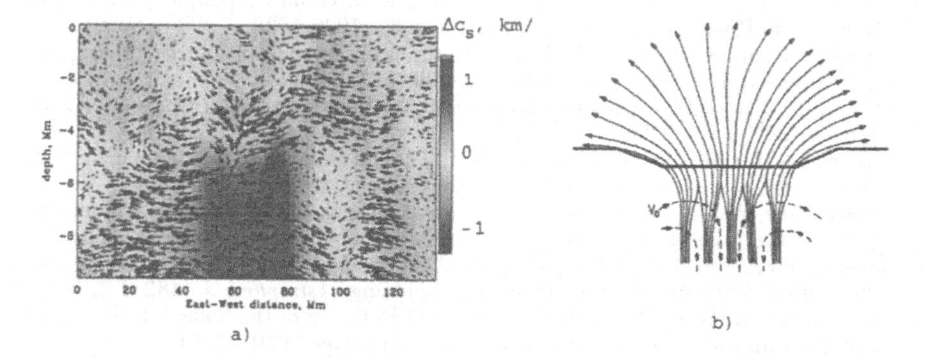


*Figure 5.* a) Flow pattern around the sunspot in a vertical cut through the upper convection zone. The length of the arrows is proportional to flow velocity. The maximum velocity is approximately 1 km/s. The background color image shows the sound-speed variations in the vertical plane. b) Parker's (1979) model of sunspots.

Figure 5a shows the sound speed distribution and the flow field in a vertical cut through the sunspot shown in Fig. 4b. This result provides evidence of converging flows beneath the spot. Parker (1979) suggested that the field of a sunspot divides into many separate tubes within the first 1000 km below the surface, and that a downdraft beneath the sunspot holds the separate tubes in a loose cluster (Fig. 5b).

## 6. Conclusion

Helioseismology has provided important information about the structure and dynamics of the tachocline - a narrow layer of strong rotational shear beneath the adiabatically stratified convection zone. There is evidence of additional mixing in this layer which is likely to be due to convective overshoot. This provides the necessary conditions for dynamo processes in the tachocline. However, direct evidence of the dynamo action in this layer has not been established.

Recently developed time-distance technique has provided 3-D images of emerging magnetic flux, sunspots and associated flows, and allows us to study dynamo processes in the upper convection zone.

## References


1. Benevolenskaya, E.E. (1998) A model of the double magnetic cycle of the sun, *Astrophys. J.*, **509**, L49–L52
2. Benevolenskaya, E. E., Hoeksema, J. T., Kosovichev, A. G. and Scherrer, P. H. (1999) The Interaction of New and Old Magnetic Fluxes at the Beginning of Solar Cycle 23, *Astrophys. J.*, **517**, L163–L166
3. Charbonneau, P. *et al.* (1999) Helioseismic Constraints on the Structure of the Solar Tachocline, *Astrophys. J.*, **527**, 445–460
4. Covas, E., Tavakol, R., Moss, D. and Tworkowski, A. (2000) Torsional oscillations in the solar convection zone, *Astronomy and Astrophysics*, **360**, L21–L24
5. Dziembowski, W. A., Goode, P. R., Kosovichev, A. G. and Schou, J. (2000) Signatures of the Rise of Cycle 23, *Astrophys. J.*, **537**, 1026–1038
6. Howard, R. and Labonte, B. J. (1981) The sun is observed to be a torsional oscillator with a period of 11 years, *Astrophys. J.*, **239**, L33–L36
7. Howe, R. *et al.* (2000) Dynamic Variations at the Base of the Solar Convection Zone, *Science*, **287**, 2456–2460
8. Kliorin, N. I. and Ruzmaikin, A. A. (1984) The nature of the 11-year solar torsional oscillations, *Sov. Astr. Lett.*, **10**, 390–392
9. Kosovichev, A. G. (1996) Helioseismic Constraints on the Gradient of Angular Velocity at the Base of the Solar Convection Zone, *Astrophys. J.*, **469**, L61–L64
10. Kosovichev, A. G. and Schou, J. (1997) Detection of Zonal Shear Flows beneath the Sun's Surface from f-Mode Frequency Splitting, *Astrophys. J.*, **482**, L207–L210
11. Kosovichev, A. G. *et al.* (1997) Structure and Rotation of the Solar Interior: Initial Results from the MDI Medium-l Program, *Solar Phys.*, **170**, 43–61
12. Kosovichev, A. G. (1999) Inversion Methods in Helioseismology and Solar Tomography, *J. Comp. Appl. Math*, **109**, 1–39
13. Kosovichev, A. G., Duvall, T.L., Jr and Scherrer, P.H. (2000) Time-Distance Inversion Methods and Results, *Solar Phys.*, **192**, 159–176
14. Parker, E. N. (1979) Sunspots and the physics of magnetic flux tubes. I - The general nature of the sunspot. *Astrophys. J.*, **230**, 905–923
15. Parker, E. N. (1993) A solar dynamo surface wave at the interface between convection and nonuniform rotation, *Astrophys. J.*, **408**, 707–719
16. Schou, J., *et al.* (1998) Helioseismic studies of differential rotation in the solar envelope by the solar oscillations investigation using the Michelson Doppler Imager , *Astrophys. J.*, **505**, 390–417
17. Yoshimura, H. (1981) Solar cycle Lorentz force waves and the torsional oscillations of the sun, *Astrophys. J.*, **247**, 1102-1112

# ASYMPTOTIC WKBJ-STUDIES OF SOLAR DYNAMO WAVES: 1D AND 2D CASES

*Solar Internal Rotation, Strong Generation. Dynamo Waves, WKBJ-solutions*

K.M. KUZANYAN
*IZMIRAN, Troitsk*
*Moscow Region, 142190 RUSSIA*

## 1. Introduction

Systematic observations of tracers of the solar magnetic activity such as sunspots, sunspot groups, active regions, polar faculae etc. indicate that this activity has a form of a travelling dynamo wave. It consists of two wings over a given hemisphere at every 11 yr sunspot cycle. They are low-latitudinal equatorward and high-latitudinal poleward waves whose propagation time approximately equal to the period of the cycle and which are shifted in time with respect to each other by approximately a half of the period. The waves represent toroidal and poliodal magnetic fields generated in the solar convective zone.

The simplest model of mechanism of this generation was suggested by Parker (1955). It assumes presence of the so-called (in mean field theory) $\alpha$-effect (mean helicity) and the differential rotation shear ($\Omega$-effect). The corresponding to these effects dimensionless Reynolds numbers are large and their product (field regeneration rate) can be used in asymptotic analysis as a large parameter. Often the solar dynamo can be represented by the so-called $\alpha\Omega$-model which is extensively studied. The complexity increases with approach to the $\alpha^2\Omega$ model (Griffiths et al., 2000) but then various regimes of the field generation are represented more faithfully.

The use of WKBJ enables us to resolve short waves in thin layers which arise in such fast dynamo generation mechanism. In the simplest 1D case suggested by Proctor and Spiegel (1991) the linear problem is asymptotically solved for an inhomogeneous medium by Kuzanyan and Sokoloff (1995). Even a kinematic approach yields a general structure of the solution and fixes the frequency of the cycle (Kuzanyan and Sokoloff, 1997). Provided knowledge on the generation sources ($\alpha$ and $\Omega$-effects) we are able to outline the shape of the dynamo wave. Nonlinear development of



*P. Chossat et al. (eds.), Dynamo and Dynamics, a Mathematical Challenge*, 261–269.

such solution essentially heritages some properties of the linear wave but includes finite wave amplitudes and refines the structure of the wave front (Meunier et al., 1997; Bassom et al., 1999). This also enables some relation between the period and amplitude of the dynamo waves. Asymptotic solution of the weakly-nonlinear problem yields the transition to the finite amplitude state. This gives information of the type of bifurcation.

Recent advances in helioseismology (project SOHO-MDI) provide reliable information on the internal differential rotation profile of the Sun. These data allow us to estimate gradients of angular velocity, i.e., the strength of the $\Omega$-effect. A careful account of the $r$-dependence provides development of a 2D asymptotic dynamo model which is better consistent with the conditions in the convective shell of the Sun and other stars. This model reveals spatial locations of the dynamo waves (Belvedere et al., 2000), which appeared near the bottom of the convection zone. The solution is shown to possess properties of traveling dynamo waves obeying Yoshimura's (1975) law (dynamo waves propagate along lines of constant angular velocity). In our approach two non-overlapping independent dynamo waves are obtained. The first wave propagates equatorwards over low latitudes, while the second one propagates polewards over high latitudes.

Further studies are focused on the interaction between these two waves in one given hemisphere, or the waves in both the hemispheres with each other. This yields splitting the frequency of the cycle for dipole and quadrupole modes, and so may cause global modulation of the solar cycle. Further nonlinear development of these studies should involve not only algebraic but dynamic nonlinear back-reaction of the magnetic field to rotation and convection (e.g., Malkus-Proctor effect). These produces complicated multiperiodic oscillatory behaviour of mean fields. Such theoretical considerations are subject to comparison with available observational signatures and tracers.

## 2. Basic equations

Here we consider a linear (kinematic) dynamo problem, i.e. the Mean Field Electrodynamics equations governing axially symmetric magnetic field dynamo generation (Krause & Rädler 1980) in a differentially rotating spherical shell. We denote $B$ is the azimuthal component of the large-scale magnetic field, $\hat{A}$ is proportional to the azimuthal component of the magnetic vector potential, $r$ and $\theta$ are radius and latitude of the spherical coordinate system with the solar (stellar) centre in the origin, and $\theta = 0$ corresponds to the equator. Let $\alpha$ denote the $\alpha$-effect (mean turbulent helicity) which provides regeneration of the poloidal field from the toroidal one, $\beta$ is the coefficient of turbulent diffusivity assumed here to be homogeneous, and the angular velocity $\Omega$ is a function of the coordinates as well. We introduce the dimensionless quantities (the dynamo number and other magnetic Reynolds numbers) as follows:

$$D = R_\alpha R_\omega, \qquad \text{where} \qquad R_\alpha = \frac{\alpha^0 R}{\beta}, \qquad R_\omega = \frac{R^3 G^0}{\beta}, \tag{1}$$

where $G^0 = \max |\nabla \Omega|$ and $\alpha^0 = \max \alpha$. Below, we use rescaled quantities for $\alpha \to \alpha/\alpha^0$ and for the radial $G \to G/G^0$ and latitudinal $F \to F/G^0$. components of the vector $|\nabla \Omega|$.

Then, in terms of dimensionless quantities, upon suitable rescaling and droping the $\alpha^2$-effect terms, we have the following equations:

$$\frac{\partial A}{\partial t} = \alpha B + \frac{1}{r}\frac{\partial^2 (rA)}{\partial r^2} + \frac{1}{r^2}\frac{\partial}{\partial \theta}\left[\frac{1}{\cos\theta}\frac{\partial (A\cos\theta)}{\partial \theta}\right], \tag{2}$$

$$\frac{\partial B}{\partial t} = -D\left[G\frac{\partial (rA\cos\theta)}{\partial \theta} + F\frac{\partial (rA\cos\theta)}{\partial r}\right] \tag{3}$$

$$+\frac{1}{r}\frac{\partial^2 (rB)}{\partial r^2} + \frac{1}{r^2}\frac{\partial}{\partial \theta}\left[\frac{1}{\cos\theta}\frac{\partial (B\cos\theta)}{\partial \theta}\right].$$

The main assumption in the use of asymptotic WKBJ methods is that the absolute value of the dynamo number $|D|$ is large, i.e. the spatial scale of the solution is small. The dynamo number can be positive as well as negative. Here, let $D$ be negative/positive provided that the product $\alpha G$ is negative/positive at low latitudes in the Northern/Southern hemisphere at the bottom of the convective zone. One can see that for simple dynamo models such choice results in a basically equatorward dynamo wave.

## 3. Asymptotic Solution for 1D case

For extremely large dynamo numbers the equations governing the magnetic field (2-3) can be reduced to a simple 1D system (Parker, 1955, see also Proctor & Spiegel, 1991):

$$\frac{\partial A}{\partial t} = \alpha(\theta) B + \frac{\partial^2 A}{\partial \theta^2}, \tag{4}$$

$$\frac{\partial B}{\partial t} = -\hat{D} G(\theta)\cos\theta \frac{\partial A}{\partial \theta} + \frac{\partial^2 B}{\partial \theta^2}.$$

We seek an exponentialy growing with time solution in the form $\sim e^{\gamma t}$, where $\mathrm{Re}\,\gamma$ is the growth rate of the magnetic field, $\mathrm{Im}\,\gamma$ is proportional to the frequency of the magnetic field oscillations. We abopt the simplest form of the helicity ($\alpha$-effect) function $\alpha(\theta) = \sin\theta$ and the uniform differential rotation $G = 1$. Now the asymptotic WKBJ-solution for equations (4) follows:

$$\begin{pmatrix} A \\ |D|^{-2/3} B \end{pmatrix} = \exp\left(i|D|^{1/3} S\right)(\mathbf{f}_0 + |D|^{-1/3}\mathbf{f}_1 + ...),$$

$$\gamma = |D|^{2/3}\Gamma_0 + |D|^{1/3}\Gamma_1 + ..., \tag{5}$$

where $S$ and the vectors $\mathbf{f}_0$, $_1, \ldots$ are complex functions of position. The condition of non-trivial solvability of the dynamo wave equations yields the dispersion equation for determination of function $S(\theta)$ and the constant $\Gamma_0$

$$(\Gamma_0 + k^2)^2 - i\,k\hat{\alpha}(\theta) = 0. \tag{6}$$

where $\hat{\alpha} = \alpha(\theta)\cos\theta$, and $S' = k$ Let $\hat{\alpha}_*$ be a maximum value of the function $\hat{\alpha}$, which is attained at $\theta = \theta_0$, and so we may assume that $\hat{\alpha}$ varies between 0 and $\hat{\alpha}_*$. Equation (6) is an algebraic fourth order equation with respect to $k$, where $\hat{\alpha}(\theta)$ is a parameter, therefore, for a given value of $\hat{\alpha}$ it has in general four different complex solutions. With variation of $\hat{\alpha}$ they draw four curves in the complex plane (Fig. 1-left).

As the boundary conditions we require the solution to vanish beyond the domain where the generation sources are localized. For construction of our solution we would like to choose a branch for which in the vicinity of the boundaries of the considered domain the solution decays asymptotically, i.e., the imaginary part of $S$ increases in absolute value and positive, namely:

$$\operatorname{Im} k|_{\theta\to\pi/2} > 0, \qquad \operatorname{Im} k|_{\theta\to 0} < 0. \tag{7}$$

It is shown by Kuzanyan and Sokolof (1995), that for the only values

$$\Gamma_0 = 3\cdot 2^{-11/3}(1 \pm i\sqrt{3})\,\hat{\alpha}_*^{2/3}. \tag{8}$$

it is possible to obtain function $S$ which is continuous at the turning point and for which conditions (7) are satisfied by choosing branch 3 for $\theta > \theta_0$ and branch 4 for $\theta < \theta_0$). (see Fig. 1-left). Thus, we can solve the equation for the leading order of asymptotics exactly for all values of $\theta$.

The obtained solution has two remarkable properties. Firstly, he maximum of the obtained solution is not situated at the turning point $\theta_0$, where the generation sources are maximum, but shifted towards the direction of the dynamo wave propagation to the point, where the value of $\operatorname{Im} k$ vanishes and the function $\operatorname{Im} S(\theta)$ has a minimum (Fig. 1-right-B). This point $\theta_1 < \theta_0$ is situated at branch 4, and its location is determined by the following condition:

$$\hat{\alpha}_1 = \frac{\hat{\alpha}(\theta_1)}{\hat{\alpha}_*} = \frac{9\sqrt{3}}{16\sqrt{2}\sqrt{\sqrt{3}-1}} \simeq 0.8052\,. \tag{9}$$

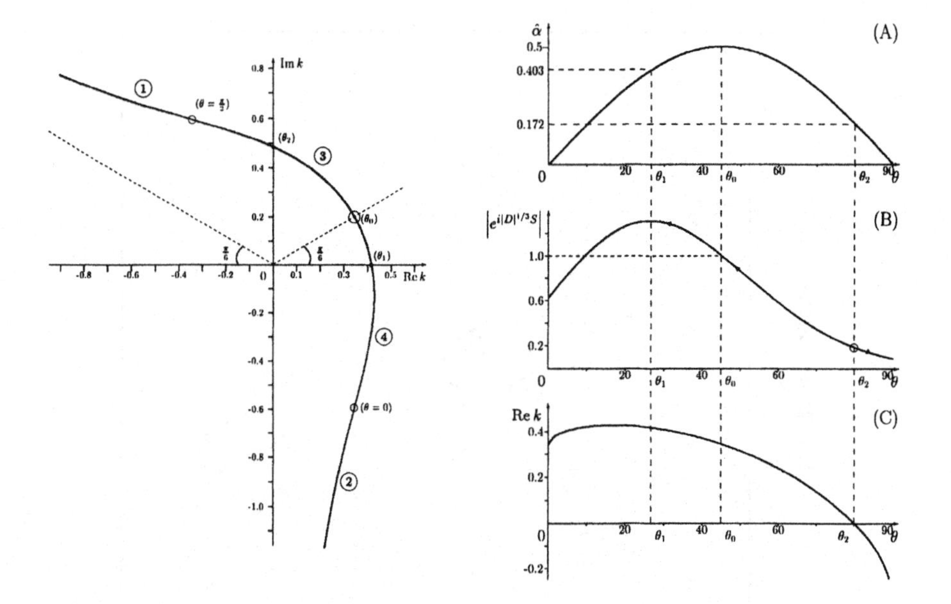

*Figure 1.* LEFT PANEL: Location of roots $k$ of dispersion equation in the complex plane as a function of the parameter $\theta$. The points of matching of different branches are circled, and corresponding values of the variable $\theta$ are given in brackets. The circled digits indicate the branch numbers. RIGHT PANEL: Asymptotic solution. The following dependences upon latitude $\theta$ are shown: (A) – function $\hat{\alpha}(\theta)$; (B) – amplitude of exponential term of solution of equations (2) for $D = -10^3$; (C) – real part of the wave number of the dynamo wave $\mathrm{Re}\, k$. The point of reversal of the dynamo wave propagation is circled (B). To the left from this point the dynamo wave propagates equatorwards, and to the right – polewards (for the northern Solar hemisphere). The direction of the dynamo wave propagation is shown by the arrows (B). [Figures of Kuzanyan & Sokoloff (1995) are used.]

For the simplest form of helicity $\hat{\alpha}(\theta) = \frac{1}{2}\sin 2\theta$ the value of $\theta_1$ is equal to $0.468 \simeq 26.8°$. This is a nontrivial result which is linked to the concept of advection at the group velocity of dynamo wave (Kuzanyan and Sokoloff, 1995, see also Soward & Jones , 1983).

Secondly, under our assumptions the dynamo wave propagates mostly equatorwards. However, even for the sign-constant $\Omega$-effect the direction of the dynamo wave propagation reverses at high latitudes at point $\theta_2 > \theta_0$ belonging to branch 3, which is determined by the following condition:

$$\hat{\alpha}_2 = \frac{\hat{\alpha}(\theta_2)}{\hat{\alpha}_*} = \frac{9\sqrt{6}}{64} \simeq 0.3445, \tag{10}$$

the $\mathrm{Re}\, S'(\theta)$ changes its sign, and for $\theta > \theta_2$ the dynamo wave reverses. For $\alpha(\theta) = \sin\theta$ we calculate $\theta_2 \simeq 1.39 \sim 80°$. This poleward dynamo wave

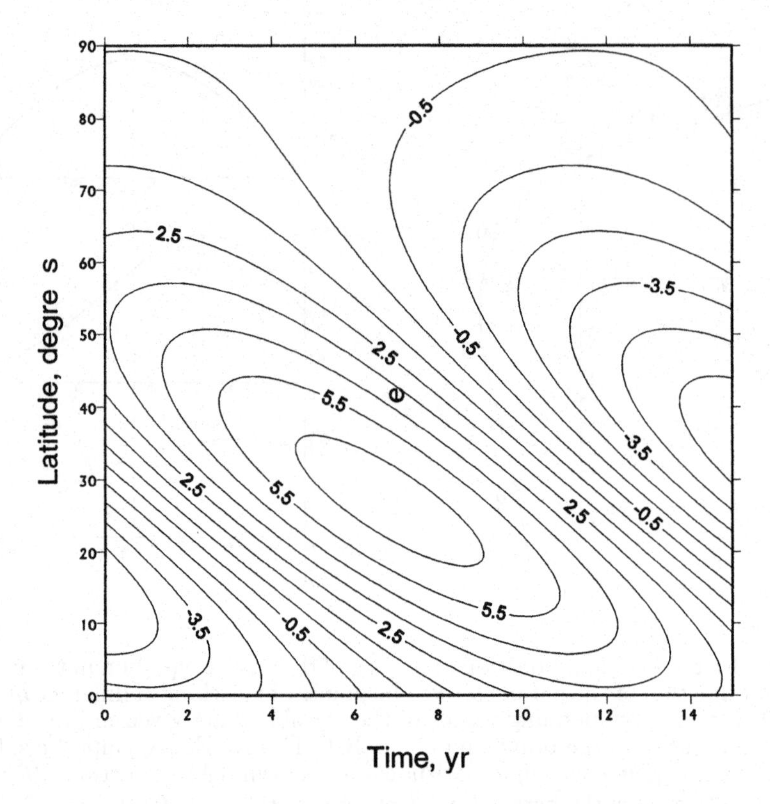


*Figure 2.* Maunder Butterfly diagram plotted for the 1D solution with the values used in Fig. 1. [Figure 2 of Kuzanyan & Sokoloff (1997) is used.]

has an amplitude much less than the equatorward one, and decays with propagation.

This property is consistent with solar observations, e.g., Makarov and Sivaraman (1983). The existence of these two dynamo waves has been traditionally connected with complicated structure of the solar convective shell, and the dynamo waves of different directions have been supposed to be generated within different depth intervals (Makarov et al., 1987; Ruzmaikin and Starchenko, 1987). Our results, however, show that the dynamo waves propagating in different directions can be obtained in principle within the framework of models with simple solar convective shell structure.

A qualitative picture of the dynamo wave propagation is shown in Fig. 2.

## 4. Results for 2D case

This one dimensional solution can be generalized for 2D case. The asymptotic envelope of solution of equations (2-3) is calculated in Belvedere et al. (2000), see Fig. 3. As the sources of generation we adopted the inter-

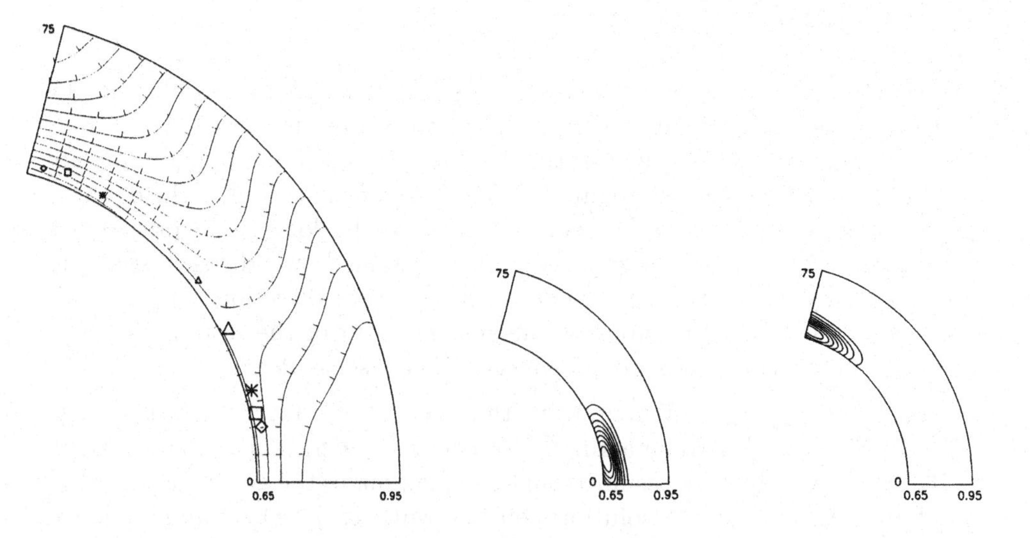


*Figure 3.* LEFT PANEL: Contour plot of the angular velocity $\Omega$ versus radius ($0.65 \leq r \leq 0.95$ in units of the solar radius) and latitude ($0 \leq \theta \leq 75°$) calculated in the model analytic fit after the data obtained by helioseismologists (Schou et al. 1998). The points of the dynamo wave maxima and reversal points in the one-dimensional model are shown as diamonds and triangles, respectively. Stars indicate the maxima of the sources [turning points], and square maxima of the two-dimensional asymptotic solution. The larger symbols indicate the solution at low latitudes, and the smaller ones at high latitudes. RIGHT PANEL: Contour plot of the envelope of the asymptotic solution calculated for the value of dynamo number $|D| = 10^3$ for the solution at low (middle) and high (right) latitudes. [Figures of Belvedere et al. (2000) are used.]

nal differential rotation, deduced from helioseismological data within the convection zone (above r=0.65), and for not too high latitudes (below 75°), which we represented by an analytic fitting function. The knowledge on the $\alpha$-effect is rather limited. For latitudinal dependence we adopted an estimate given by Krause (1967) presumes that it is proportional to $\sin\theta$. Such an idea is in agreement with observations of helicities of magnetic fields (Seehafer 1994, Pevtsov & Canfield 1994) which may be a signature of the $\alpha$-effect (Bao & Zhang 1998, Kuzanyan et al. 2000). For the radial dependence of $\alpha$-effect we simply assumed that it changes its sign near the bottom of the convection zone, due to the dominance of converging/diverging flows in the lower/upper part of the convection zone. This is in agreement with other studies and numerical simulations (e.g., Brummel et al., 1998).

## 5. Discussion

So we obtained an asymptotic solution of 2D solar dynamo problem in the form of a traveling wave, which is shown to possess properties consistent with Yoshimura's (1975) law stating that dynamo waves propagate mainly along the lines of constant angular velocity. Specific locations of the dynamo wave maxima and reversals were calculated to be located mainly at the bottom of the convection zone. We have obtained two distinct spatially separated dynamo waves propagating in opposite directions. This result is a consequence of the input of the specific form of the internal angular velocity in the convective zone deduced by helioseismology.

This approach is different from the one carried out by Ruzmaikin et al. (1988), and also Ruzmaikin & Starchenko (1988), and Makarov et al. (1987). They implicitly assumed that the maxima of the generation sources and the maxima of the solution coincide with each other, and obtained conditions to determine the constant $\Gamma$ and the functions $S$ in the form of power series expansions up to the second order terms. However, as is shown by Kuzanyan & Sokoloff (1995), this assumption is not appropriate for a linear $\alpha\Omega$-dynamo. Here we determine $\Gamma$ and the functions $S$ in the zero-th and first order from the stationary point conditions (advection at the group velocity of dynamo wave), and use the second order expansion only to determine the second order coefficients in the expansion of function $S$. Therefore, we do not assume the maxima of the sources and the maxima of the solutions to coincide, or to be close from each other. Again, we impute this fact due to the properties of our solution which are consequence of the conditions of advection at the group velocity of dynamo wave at the stationary point. Furthermore, we expect a finite shift between their locations, and so we have a primary intension to calculate these locations and the shift between them. Using this method for the one-dimensional case, one can reproduce the results of the asymptotic analysis given by Kuzanyan & Sokoloff (1995). Note, however, that this problem does not luckily arise for an $\alpha^2$-dynamo model, which WKBJ-asymptotics was calculated by Sokoloff et al. (1983) and Ruzmaikin et al. (1990).

**Acknowledgements:** The author would like to thank Alexander Kosovichev for providing the data on the solar internal rotation. His thanks go to the orgainzers of the NATO ARW in Cargese and CNRS for supporting his attendance the meeting. The work is supported by INTAS grant 99-00348, NATO Collaborative Linkage Grant PST.CLG.976557, RFBR grants 00-02-17854, 99-02-18346a, and the Young Researchers' grant of Russian Academy of Sciences. An essential part of this work has been conceived while the author was a guest of the Istituto di Astronomia dell'Universitá di Catania (Italy).

## References

1. Bao S.D., Zhang H.Q. (1998) Patterns of Current Helicity for Solar Cycle 22. *Astrophys. J.* **496**, L43–L46
2. Bassom A.P., Kuzanyan K.M., Soward A.M. (1999) A nonlinear dynamo wave riding on a spatially varying background, *Proc. Roy. Soc. A*, **455**, 1443–1481
3. Belvedere, G.M., Kuzanyan, K.M. and Sokoloff, D.D. (2000) A two-dimensional asymptotic solution for a dynamo wave in the light of the solar internal rotation, *Mon. Not. R. astr. Soc.* **315**, No. 4, 778–790
4. Brummell N.H., Hurlburt N.E., Toomre J. (1998) Turbulent Compressible Convection with Rotation. II.Mean Flows and Differential Rotation, *Astrophys. J.* **493**, 955–969
5. Griffiths, G., Bassom, A. P., Soward, A. M., and Kuzanyan, K. M. (2000) Nonlinear $\alpha^2\Omega$-dynamo waves in stellar shels: I. General structure, *Geophys. Astrophys. Fluid Dyn.*, in press,
6. Krause F. (1967) Habilitationsschrift, Univ. Jena [translated into English by Roberts, P. H., & Stix, M., 1971, *The Turbulent Dynamo*, NCAR Technical Note TN/IA-60 ]
7. Krause, F. and Rädler, K.–H. (1980) *Mean-Field Magnetohydrodynamics and Dynamo Theory*, Pergamon, Oxford
8. Kuzanyan K.M., Sokoloff D.D. (1995) A dynamo wave in an inhomogeneous medium *Geophys. Astrophys. Fluid Dynam.*, **81**, 113–129
9. Kuzanyan K.M., Sokoloff D.D. (1997) Half-width of a solar dynamo wave in Parker's migratory dynamo, *Solar Phys.*, **173**, 1–14
10. Kuzanyan K.M., Zhang H., Bao S. (2000) Probing Signatures of the Alpha-effect in the Solar Convection Zone, *Solar Phys.*,**191/2**, 231–246
11. Makarov, V.I. and Sivaraman, K.R. (1983) Poleward Migration of the Magnetic Neutral Line and Reversals of the Polar Fields on the Sun, *Solar Phys.* **85**, 215–226
12. Makarov, V.I., Ruzmaikin, A.A. and Starchenko, S.V. (1987) Magnetic Waves of Solar Activity, *Solar Phys.* **111**, 267–277
13. Meunier N., Proctor M.R.E., Sokoloff D., Soward A.M., Tobias S. (1997) Asymptotic Properties of a nonlinear $\alpha\omega$-dynamo wave: Period, Amplitude and Latitude Dependence, *Geophys. Astrophys. Fluid Dynam.*, **86**, 249–285
14. Parker, E.N. (1955) Hydromagnetic Dynamo Models, *Astrophys. J.***122**, 293–314
15. Pevtsov A.A, Canfield R.C. (1994) Patterns of helicity in solar active regions. *Astrophys. J.* **425**, L117–L119
16. Proctor, M.R.E. and Spiegel, E.A. (1991) Waves of Solar activity. *Proc. IAU Colloquium 130 'The Sun and Cool Stars: Activity, Magnetism, Dynamos' (I. Tuominen, ed.)* pp. 117–128. Springer Lecture Notes in Physics 380.
17. Ruzmaikin A.A., Starchenko S.V. (1988) Magnetic Manifestations of Solar Rotation, *Sov. Astron.*, **31**, 552 [translated from 1987, *Astron. Zh.*, **64**, 1057–1065]
18. Ruzmaikin A.A., Sokoloff D.D., Starchenko S.V. (1988) Excitation of Non-Axially Symetric Modes of the Sun's Mean Magnetic Field, *Solar Phys.*, **115**, 5–15
19. Schou J., Antia H.M., Basu S. et al., (1998) Helioseismic Studies of Differential Rotation in the Solar Envelope by the solar Oscillations Investigation using the Michelson Doppler Imager, *Astrophys. J.* **505**, 390–417
20. Seehafer, N. (1994) Alpha-effect in the Solar atmosphere, *Astron. Astrophys.* **284**, 593–598
21. Sokoloff D., Shukurov A., Ruzmaikin A. (1983) Asymptotic Solution of the $\alpha^2$-Dynamo Problem *Geophys. Astrophys. Fluid Dyn.*, **25**, 293–307
22. Soward A.M., Jones C.A. (1983) *Q. J. Mech. appl. Math.*, **36**, 19–42
23. Yoshimura H. (1975) Solar-cycle dynamo wave propagation, *Astrophys. J.* **201**, 740–748

# DYNAMO EFFECT WITH INERTIAL MODES IN A SPHERICAL SHELL ?

M. RIEUTORD
*Observatoire Midi-Pyrénées and Institut Universitaire de France*
*14 av. E. Belin, F-31400 Toulouse, France*

AND

L. VALDETTARO
*Dipartimento di Matematica, Politecnico di Milano*
*Piazza L.da Vinci, 32, 20133 Milano, Italy*

**Abstract.** We present preliminary results of integrations of the induction equation using the velocity field produced by inertial modes in a spherical shell. The results indicate a possible dynamo action at rather large Reynolds number ($R_e > 500$).

## 1. Introduction

Inertial waves are waves specific to rotating fluids as the Coriolis force is their restoring force. Many times in the past literature on the dynamo problem, they have been invocated as good kinematic dynamos: first from theoretical arguments Moffatt (1970) who noticed their helicity and second from their possible role in the geodynamo: indeed it has been advocated ([3, 2]) that such modes can be destabilized by the elliptic instability (through the tidal deformation of the Core-Mantle Boundary) and thus play a part (yet unknown) in the geodynamo (see also Rieutord 2000).

In this paper we therefore explore the possibility of making kinematic dynamos with inertial modes of a fluid contained in spherical shell. In section 2 we recall the rather peculiar properties of these modes and then present some preliminary results of numerical simulations integrating the induction equation.



*P. Chossat et al. (eds.), Dynamo and Dynamics, a Mathematical Challenge,* 271–277.

## 2. Inertial modes in a spherical shell

Let us consider an incompressible inviscid fluid in solid body rotation and contained in a spherical shell whose outer radius is $R$ and inner radius is $\eta R$ ($\eta < 1$). Small amplitude pertubations of frequency $\omega$ verify the momentum and mass conservation equations:

$$i\omega\vec{u} + \vec{e}_z \times \vec{u} = -\vec{\nabla}p, \qquad \vec{\nabla}\cdot\vec{u} = 0 \tag{1}$$

where $\vec{e}_z$ is the unit vector along the rotation axis, $p$ the reduced pressure and $\vec{u}$ the non-dimensional velocity field. We used $(2\Omega)^{-1}$ as the time scale where $\Omega$ is the angular velocity of the solid body rotation. From this equation we can eliminate the velocity field in favour of the pressure and obtain the Poincaré equation, namely

$$\Delta p - \frac{1}{\omega^2}\frac{\partial^2 p}{\partial z^2} = 0 \tag{2}$$

These equations are completed by the boundary conditions $\vec{u}\cdot\vec{e}_r = 0$ on both spheres ($\vec{e}_r$ is the unit radial vector). These boundary conditions may also be expressed with the pressure ([1]), but the important point here is that Poincaré's equation is hyperbolic (because $\omega < 1$, see Greenspan 1969) and therefore the problem is ill-posed mathematically.

This mathematical property of the inviscid problem implies that in general solutions are singular. The appearance of singular solutions depends on the shape of the container and while the full sphere has only smooth regular solutions, the spherical shell owns essentially singular solutions. The singularities are such that the velocity field is not square-integrable; they come from the fact that the characteristics associated with the hyperbolic equation generically converge towards an attractor made of a periodic orbit of characteristics. These attractors play an important part in the shape of inertial modes in a spherical shell as shown in figures 1 and 2, even when viscosity is taken into account. We refer the reader to our papers Rieutord et al. 2000ab, where a detailed analysis of inertial modes is presented.

## 3. Magnetic field evolution

We thus solve numerically, with pseudo-spectral methods based on spherical harmonics (for the angular part) and Chebyshev polynomial (radial part) the induction and flux conservation equation, namely

$$\begin{cases} \partial_t\vec{B} = \vec{\nabla}\times(\vec{u}\times\vec{B}) + R_m^{-1}\Delta\vec{B} \\ \vec{\nabla}\cdot\vec{B} = 0 \end{cases} \tag{3}$$

We decompose the magnetic field into radial $B^{(r)}_{lm}$, poloidal $B^{(h)}_{lm}$ and toroidal $B^{(T)}_{lm}$ components:

$$\vec{B}(r,\theta,\phi) = \sum_{l=0}^{L}\sum_{m=-l}^{l} B^{(r)}_{lm}(r)\vec{R}_{lm} + B^{(h)}_{lm}(r) r\vec{S}_{lm} + B^{(T)}_{lm}(r) r\vec{T}_{lm}$$

with:

$$\vec{R}_{lm} \equiv Y_{lm}\vec{e}_r, \quad \vec{S}_{lm} \equiv \nabla Y_{lm}, \quad \vec{T}_{lm} \equiv \nabla \times (\vec{e}_r Y_{lm})$$

The fields $B^{(r)}_{lm}(r)$, $B^{(h)}_{lm}(r)$ and $B^{(T)}_{lm}(r)$ are further developed in series of Chebyshev polynomials for the rescaled radial variable $(2r-1-\eta)/(1-\eta)$.

The boundary conditions are such that the inner boundary is a perfect conductor and the outer boundary is a perfect insulator, namely

$$B^{(r)}_{lm} = 0, \quad r\frac{dB^{(h)}_{lm}}{dr} + B^{(h)}_{lm} = 0, \quad r\frac{dB^{(T)}_{lm}}{dr} + B^{(T)}_{lm} = 0 \quad \text{at} \quad r = \eta$$

$$r\frac{dB^{(r)}_{lm}}{dr} + (2+l)B^{(r)}_{lm} = 0, \quad r\frac{dB^{(h)}_{lm}}{dr} + (2+l)B^{(h)}_{lm} = 0, \quad B^{(T)}_{lm} = 0 \quad \text{at} \quad r = 1$$

The velocity fied is taken to be a given superposition of inertial modes. We have considered two inertial modes having an azimuthal wavenumber $m = 2$ (the kinetic helicity in a meridional plane is shown in figures 1 and 2). The Ekman number ($E = \frac{\nu}{2\Omega R^2}$) was $E = 3 \times 10^{-6}$. The temporal evolution of the inertial modes was assumed to be oscillatory with no damping. The frequency of oscillation of each mode was taken proportional to the actual frequency of the mode, and it was fixed to 0.5 for the first one. We have found that the frequency is an important parameter for the dynamo action: frequencies larger than the one chosen give faster decays of magnetic field. Lower frequencies are difficult to be simulate because the computing time required to cover some oscillations becomes prohibitive.

In the first run the magnetic Reynolds number is $R_m = 500$. We have found that it was necessary to take an angular resolution much larger than the radial one. A resolution of $L_{\max} = 200$, $N_{\max} = 50$ was chosen.

It is useful to define the Chebyshev and Legendre spectra of the field $u$ with spectral components $u(l,m,n)$ and radial dependence of the spherical harmonics coefficients $u(l,m,r)$ in the following way:

$$C_k = \sqrt{\sum_{l,m} |u(l,m,n)|^2} \qquad L_k(r) = \sqrt{\sum_{l,m,k-1/2 \leq \sqrt{l^2+m^2} \leq k+1/2} |u(l,m,r)|^2}$$

From figures 5 and 6 we see that the magnetic energy spectrum is much steeper for the Chebyshev expansion than for the angular one. A reason for this is that the kinetic helicity of the inertial modes chosen tends to concentrate near the poles (figures 1 and 2). As a consequence the magnetic field tends to be excited mostly at the poles (figure 4), and a large number of spherical harmonics are needed to resolve it.

In the second run the magnetic Reynolds number was the same as before ($R_m = 500$) but hyperviscosity was used to damp the largest harmonics of the angular decomposition. The resolution could be decreased to $L_{\max} = 150$. As a consequence we were able to run the simulation to a much longer time.

From figure 3 we see that in both cases the magnetic energy decreases with time. The two curves are not dissimilar and this gives confidence on the results of the run with hyperviscosity. The magnetic field is excited mostly in the region where the kinetic helicity is maximum, that is at the poles, as we can clearly see from figure 4.

Other runs using different inertial modes have given qualitatively the same result. Therefore we conclude that at $R_m = 500$ no dynamo effect is excited. However the tendency is that the decay rate decreases as the Reynolds number increases. The decay obtained at $R_m = 500$ (shown in figure 3) is quite small and this indicates that we are not far from the critical magnetic Reynolds number. We plan to obtain and investigate the supercritical regime in a forthcoming work.

## References

1. Greenspan, H. P. (1969). *The theory of rotating fluids.* Cambridge University Press.
2. Kerswell, R. (1994). Tidal excitation of hydromagnetic waves and their damping in the Earth. *J. Fluid Mech.*, 274:219–241.
3. Malkus, W. (1989). An experimental study of global instabilities due to the tidal (elliptical) distortion of a rotating elastic cylinder. *Geophys. Astrophys. Fluid Dyn.*, 48:123–134.
4. Moffatt, K. (1970). Dynamo action associated with random inertial waves in a rotating conducting fluid. *J. Fluid Mech.*, 44:705–719.
5. Rieutord, M. (2000a). A note on inertial modes in the core of the Earth. *Phys. Earth Plan. Int.*, 117:63–70.
6. Rieutord, M., Georgeot, B., and Valdettaro, L. (2000a). Inertial waves in a rotating spherical shell: attractors and asymptotic spectrum. *to appear in J. Fluid Mech.; physics/0007007.*
7. Rieutord, M., Georgeot, B., and Valdettaro, L. (2000b). Waves attractors in rotating fluids: a paradigm for ill-posed cauchy problems. to appear in *Phys. Rev. Let.*

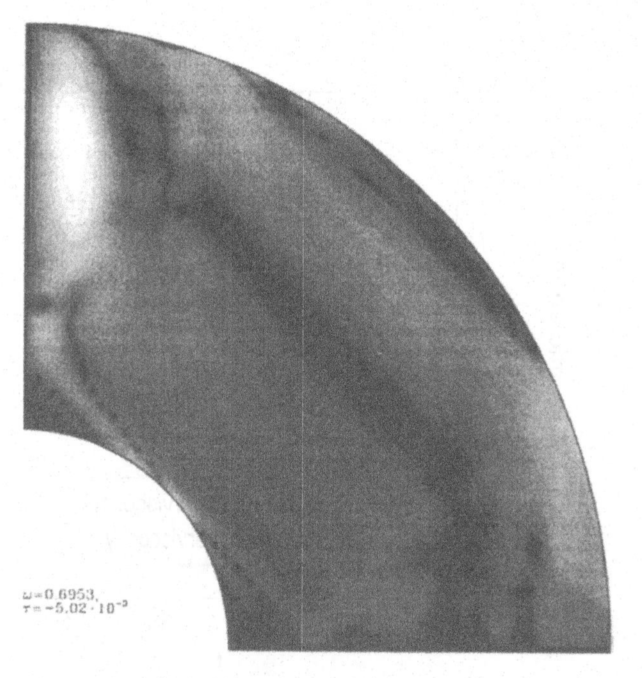


*Figure 1.* Contours of kinetic helicity in a meridional plane for inertial mode N.1. Largest values are in white color, smallest values in black.

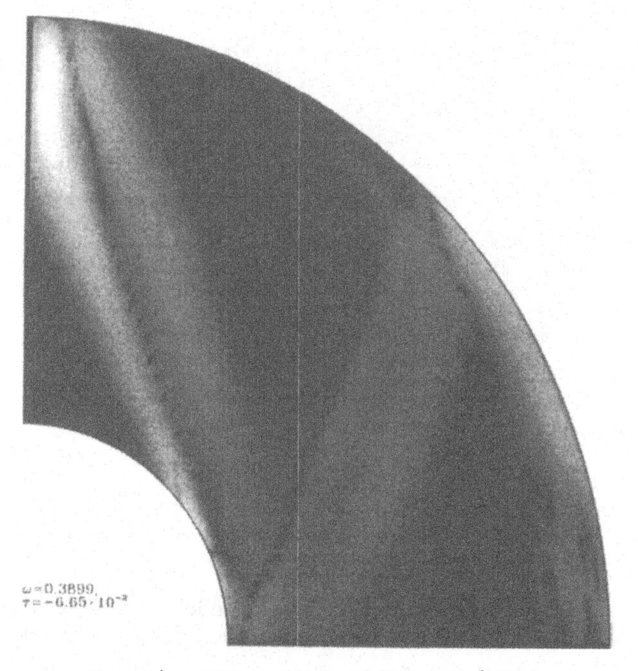


*Figure 2.* Contours of kinetic helicity for inertial mode N.2. Note the concentration of

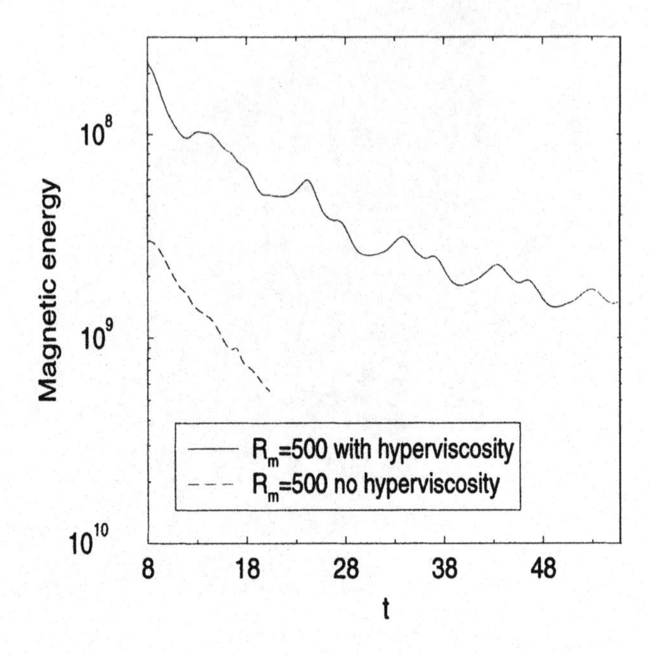


*Figure 3.* Magnetic energy versus time for both runs. Amplitude is arbitrary.

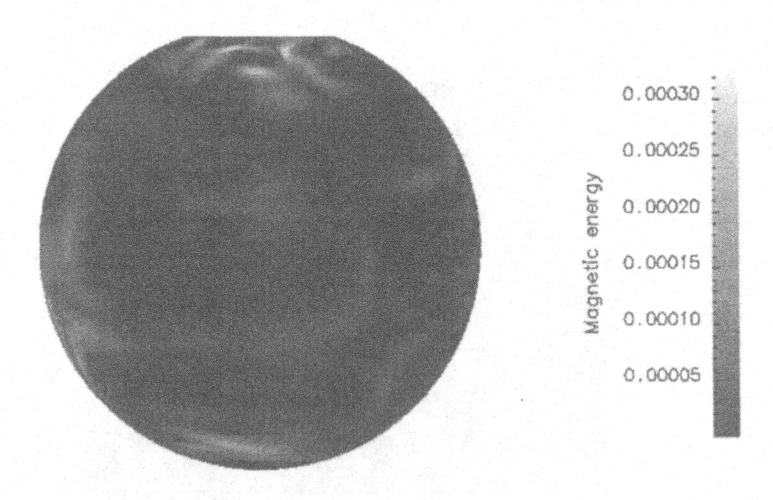


*Figure 4.* Isocontours of magnetic energy at middle radius, at time $t = 36$ for run 2. Note the large intensity near the poles, where the helicity of the velocity field is largest.

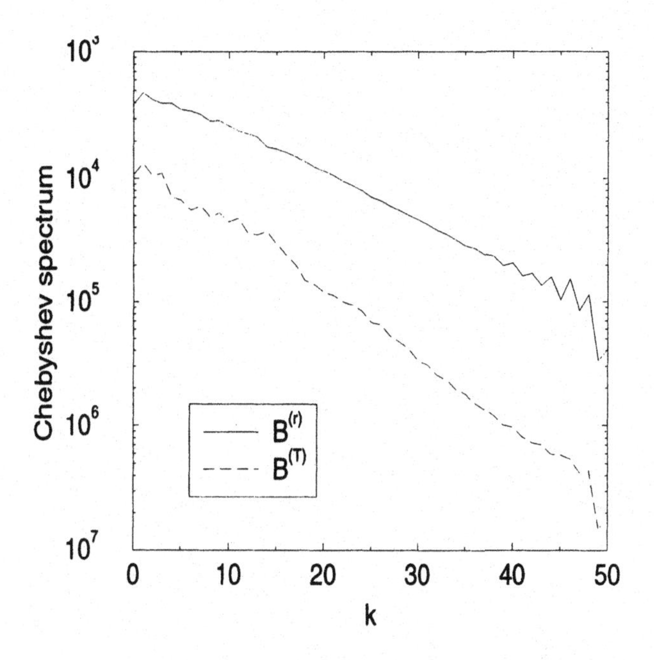


*Figure 5.* Chebyshev spectrum of magnetic energy versus time for run 1

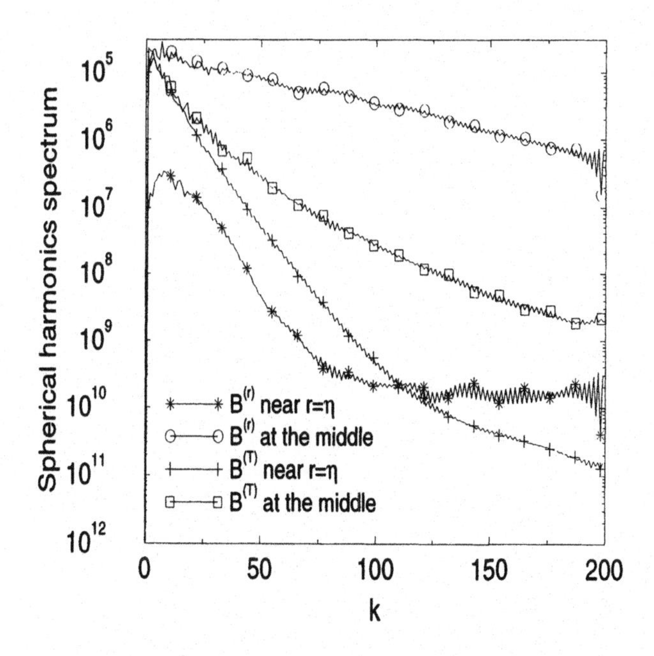


*Figure 6.* Angular spectrum of magnetic energy for run 1

# TWO TYPES OF NONLINEARITIES IN MAGNETIC DYNAMO

I. ROGACHEVSKII AND N. KLEEORIN
*Department of Mechanical Engineering*
*Ben-Gurion University of Negev*
*POB 653, 84105 Beer-Sheva, Israel*

## Abstract

Two types of nonlinearities (algebraic and dynamic) are discussed. The algebraic nonlinearity implies a nonlinear dependence of the mean electromotive force on the mean magnetic field. The dynamic nonlinearity is determined by a differential equation for the magnetic part of the $\alpha$-effect. It is shown that the algebraic nonlinearity alone (which includes the nonlinear $\alpha$-effect, the nonlinear turbulent diffusion, etc) cannot saturate the dynamo generated mean magnetic field while both, the algebraic and dynamic nonlinearities limit the mean magnetic field growth.

The nonlinear mean electromotive force is calculated for an anisotropic background turbulence (i.e., the turbulence with zero mean magnetic field) with one preferential direction. It is shown that the toroidal and poloidal magnetic fields have different nonlinear turbulent diffusion coefficients. It is demonstrated that even for homogeneous turbulence there is an effective nonlinear velocity which exhibits diamagnetic or paramagnetic properties depending on anisotropy of turbulence and level of magnetic fluctuations in the background turbulence. The diamagnetic velocity results in the field is pushed out from the regions with stronger mean magnetic field, while the paramagnetic velocity causes the magnetic field tends to be concentrated in the regions with stronger field. Analysis shows that anisotropy of turbulence strongly affects the nonlinear turbulent diffusion coefficients and the nonlinear effective velocity.

## 1. Introduction

Turbulent motions of a conducting fluid can generate large-scale (mean) magnetic field and small-scale magnetic fluctuations. Many dynamo mod-



*P. Chossat et al. (eds.), Dynamo and Dynamics, a Mathematical Challenge,* 279–287.

els (see, e.g., Krause and Rädler, 1980) are kinematic (i.e., they predict a magnetic field that grows without limit). In order to find, e.g., the equilibrium magnetic field configuration, the nonlinear effects which limit the field growth must be taken into account. The nonlinearities in turbulent mean-field dynamo imply an effect of a mean magnetic field on the $\alpha$ effect, turbulent magnetic diffusion, turbulent diamagnetic velocity, etc. The mean magnetic field $\mathbf{B}$ is determined by an induction equation

$$\partial \mathbf{B}/\partial t = \nabla \times [\mathbf{V} \times \mathbf{B} + \mathcal{E}] + \eta \Delta \mathbf{B} \tag{1}$$

where $\mathbf{V}$ is a mean velocity (e.g., the differential rotation), $\eta$ is the magnetic diffusion due to the electrical conductivity of fluid, $\mathcal{E} = \langle \mathbf{u} \times \mathbf{b} \rangle$ is the turbulent electromotive force, $\mathbf{u}$ and $\mathbf{b}$ are fluctuations of the velocity and magnetic field, respectively, angular brackets denote averaging over an ensemble of turbulent fluctuations. The turbulent electromotive force in kinematic dynamo for an isotropic turbulence is given by

$$\mathcal{E} = \alpha_0^{(v)} \mathbf{B} + \mathbf{U}_0 \times \mathbf{B} - \eta_T \nabla \times \mathbf{B} \tag{2}$$

(see, e.g., Krause and Rädler, 1980), where $\alpha_0^{(v)} = -(1/3)\langle \tau \mathbf{u} \cdot (\nabla \times \mathbf{u}) \rangle$ is the hydrodynamic part of the $\alpha$ effect, $\mathbf{U}_0 = -(1/2)\nabla \langle \tau \mathbf{u}^2 \rangle$ is the turbulent diamagnetic velocity, $\eta_T = (1/3)\langle \tau \mathbf{u}^2 \rangle$ is the turbulent magnetic diffusion, $\tau$ is the correlation time of the turbulent velocity field.

In the nonlinear stage of evolution of the mean magnetic field the $\alpha$ effect, turbulent diffusion and turbulent diamagnetic velocity depend on the mean magnetic field $\mathbf{B}$. The total $\alpha$ effect in nonlinear dynamo is splitted into hydrodynamic $\alpha^{(v)}$ and magnetic $\alpha^{(h)}$ parts, where $\alpha^{(h)} = (1/3\mu_0 \rho)\langle \tau \mathbf{b} \cdot (\nabla \times \mathbf{b}) \rangle$ and $\mu_0$ is the magnetic permeability of the fluid with the density $\rho$ (see Pouquet et al., 1976). Such splitting of the $\alpha$ effect is introduced in nonlinear dynamo because the growing magnetic field reacts differently on the hydrodynamic and the magnetic parts of the $\alpha$ effect (see Kleeorin and Ruzmaikin, 1982; Kleeorin et al., 1995; Kleeorin and Rogachevskii, 1999). The back reaction of the mean magnetic field on the hydrodynamic part of the $\alpha$ effect is almost instantaneous (of the order of a characteristic correlation time of the turbulence, $\tau_0 = l_0/u_0$, where $u_0$ is the characteristic turbulent velocity in the maximum scale of turbulent motions $l_0$). However, the characteristic time $T$ of the back action of the mean magnetic field on the magnetic part of the $\alpha$ effect is much larger than $\tau_0$ for large magnetic Reynolds numbers. Recent study (see Rogachevskii and Kleeorin, 2000) for an anisotropic turbulence demonstrated that the hydrodynamic $\alpha^{(v)}$ and magnetic $\alpha^{(h)}$ parts of the $\alpha$ effect are nonlinearized in the form of a quenching, i.e. $\alpha^{(v)} = \alpha_0^{(v)} \Phi^{(v)}(B)$ and $\alpha^{(h)} = \alpha_0^{(h)} \Phi^{(h)}(B)$, where $\Phi^{(v)}(B)$ and $\Phi^{(h)}(B)$ are the decreasing functions of the mean magnetic field (see below).

Since $T \gg \tau_0$, the back reaction of the magnetic field on the magnetic part of the $\alpha$ effect cannot, in general, be only reduced to a simple quenching but must be described by a differential equation (see Kleeorin and Ruzmaikin, 1982; Kleeorin et al., 1995; Kleeorin and Rogachevskii, 1999).

Thus there are two main types of nonlinearities for the $\alpha$ effect: a quenching of the total $\alpha$ effect in the form of the decreasing algebraic functions (Rogachevskii and Kleeorin, 2000) and a nonlinear evolution of the magnetic part of the $\alpha$ effect which is determined by a differential equation (see Kleeorin and Ruzmaikin, 1982; Kleeorin et al., 1995; Kleeorin and Rogachevskii, 1999). In spite of the fact the nonlinear $\alpha$ effect is a subject of active study, an influence of the mean magnetic field on the turbulent diffusion, turbulent diamagnetic velocity and on other coefficients defining the mean electromotive force are still poorly understood.

In the present paper a nonlinear electromotive force for an anisotropic turbulence with one preferential direction in the case of intermediate nonlinearity is calculated. The intermediate nonlinearity implies that the mean magnetic field is not enough strong in order to affect the correlation time of turbulent velocity field. It is found that the toroidal and poloidal magnetic fields have different nonlinear turbulent diffusion coefficients. It is shown that anisotropy of turbulence strongly affects the nonlinear turbulent diffusion coefficients and the nonlinear effective velocity. It is demonstrated that even for homogeneous turbulence there is a nonlinear effective velocity which exhibits diamagnetic or paramagnetic properties depending on the anisotropy of turbulence and the level of magnetic fluctuations in the background turbulence (i.e., the turbulence with zero mean magnetic field).

## 2. The procedure of the derivation of equation for the nonlinear mean electromotive force

(a). In order to derive equations for the mean electromotive force we use a mean field approach. By means of the momentum equation and the induction equation for the turbulent fields $\mathbf{u}$ and $\mathbf{b}$ we derive equations for the second moments:

$$
\begin{aligned}
f_{ij}(\mathbf{k},\mathbf{R}) &= \int \langle u_i(\mathbf{k}+\mathbf{K}/2) u_j(-\mathbf{k}+\mathbf{K}/2)\rangle \exp{(i\mathbf{K}\mathbf{R})}\, d^3K \,, \\
h_{ij}(\mathbf{k},\mathbf{R}) &= \int \langle b_i(\mathbf{k}+\mathbf{K}/2) b_j(-\mathbf{k}+\mathbf{K}/2)\rangle \exp{(i\mathbf{K}\mathbf{R})}\, d^3K/\mu_0\rho \,, \\
g_{ij}(\mathbf{k},\mathbf{R}) &= \int \langle b_i(\mathbf{k}+\mathbf{K}/2) u_j(-\mathbf{k}+\mathbf{K}/2)\rangle \exp{(i\mathbf{K}\mathbf{R})}\, d^3K \,.
\end{aligned}
$$

where the coordinate $\mathbf{R}$ and the wave vector $\mathbf{K}$ correspond to the large scales, and $\mathbf{r}$ and $\mathbf{k}$ to the small scales, i.e., $\mathbf{R} = (\mathbf{x} + \mathbf{y})/2$, $\mathbf{r} = \mathbf{x} - \mathbf{y}$, $\mathbf{K} = \mathbf{k}_1 + \mathbf{k}_2$, $\mathbf{k} = (\mathbf{k}_1 - \mathbf{k}_2)/2$. We assumed that $\boldsymbol{\nabla} \cdot \mathbf{u} = 0$. We consider the case of large hydrodynamic ($\mathrm{Re} = l_0 u_0/\nu \gg 1$) and magnetic ($\mathrm{Rm} = l_0 u_0/\eta \gg 1$) Reynolds numbers, where $\nu$ is the kinematic viscosity.

(b). We split all correlation functions (i.e., $f_{nm}, h_{nm}, g_{nm}$) into two parts, e.g., $h_{nm} = h_{nm}^{(N)} + h_{nm}^{(S)}$, where $h_{nm}^{(N)} = [h_{nm}(\mathbf{k}, \mathbf{R}) + h_{nm}(-\mathbf{k}, \mathbf{R})]/2$ and $h_{nm}^{(S)} = [h_{nm}(\mathbf{k}, \mathbf{R}) - h_{nm}(-\mathbf{k}, \mathbf{R})]/2$. The tensor $h_{nm}^{(S)}$ describes the helical part of the tensor, whereas $h_{nm}^{(N)}$ describes the nonhelical part of the tensor. Such splitting is caused, e.g., by different times of evolution of the helical and nonhelical parts of the magnetic tensor. In particular, the characteristic time of evolution of the tensor $h_{nm}^{(N)}$ is of the order $\tau_0 = l_0/u_0$ while the relaxation time of the component $h_{nm}^{(S)}$ is of the order of $\tau_0 \mathrm{Rm}$.

(c). Equations for the second moments contain higher moments and a problem of closing the equations for the higher moments arises. The simplest closure procedure is the $\tau$ approximation which allows us to express the third moments in terms of the second moments: $M_{nm} - M_{nm}^{(0)} = -(f_{nm} - f_{nm}^{(0)})/\tau(k)$, $R_{nm}^{(N)} - R_{nm}^{(0)} = -(h_{nm}^{(N)} - h_{nm}^{(0)})/\tau(k)$, $C_{nm} - C_{nm}^{(0)} = -(g_{nm} - g_{nm}^{(0)})/\tau(k)$, where $M_{nm}$, $R_{nm}$, $C_{nm}$ are the third moments in equations for $f_{nm}, h_{nm}$ and $g_{nm}$, respectively. The superscript (0) corresponds to the background turbulence (it is a turbulence with zero mean magnetic field) and $\tau(k)$ is the characteristic relaxation time of the statistical moments.

(d). We assume that the characteristic time of variation of the mean magnetic field $\mathbf{B}$ is substantially longer than the correlation time $\tau(k)$ for all turbulence scales. This allows us to get a stationary solution of the equations for $f_{nm}, h_{nm}$ and $g_{nm}$. Using these equations we calculate the electromotive force $\mathcal{E}_i(\mathbf{r} = 0) = \int \mathcal{E}_i(\mathbf{k})\, d\mathbf{k}$, where $\mathcal{E}_i(\mathbf{k}) = (1/2)\varepsilon_{imn}(g_{nm}^{(N)}(\mathbf{k}, \mathbf{R}) - g_{mn}^{(N)}(\mathbf{k}, \mathbf{R}))$.

(e). For the integration in $\mathbf{k}$-space we have to specify a model of the background turbulence. The second moments for turbulent velocity and magnetic fields of the background turbulence are given by

$$\begin{aligned} \tau c_{ij}(\mathbf{k}) &= (5/4)\{P_{ij}(k)[(2/5)\tilde{\eta}_T^{(a)}(\mathbf{k}) - \mu_{mn}^{(a)}(\mathbf{k}) k_{nm}] + 2[\mu_{ij}^{(a)}(\mathbf{k}) \\ &+ \delta_{ij}\mu_{mn}^{(a)}(\mathbf{k}) k_{nm} - \mu_{im}^{(a)}(\mathbf{k}) k_{mj} - k_{im}\mu_{mj}^{(a)}(\mathbf{k})]\} \end{aligned} \tag{3}$$

(see, Rogachevskii and Kleeorin, 1997; Rogachevskii and Kleeorin, 1999; Rogachevskii and Kleeorin, 2000), where $c_{ij} = f_{ij}^{(0N)}$ when $a = v$, and $c_{ij} = h_{ij}^{(0N)}$ when $a = h$, the anisotropic part of this tensor $\mu_{mn}^{(a)}(\mathbf{k})$ has the properties: $\mu_{mn}^{(a)}(\mathbf{k}) = \mu_{nm}^{(a)}(\mathbf{k})$ and $\mu_{pp}^{(a)}(\mathbf{k}) = 0$. Inhomogeneity of the background turbulence is assumed to be weak. Here $\tilde{\eta}_T^{(v)}(\mathbf{k}) = \tau f_{pp}^{(0N)}(\mathbf{k})$, $\tilde{\eta}_T^{(h)}(\mathbf{k}) = \tau h_{pp}^{(0N)}(\mathbf{k})$, $P_{ij}(k) = \delta_{ij} - k_{ij}$, $\delta_{mn}$ is the Kronecker tensor and $k_{ij} = k_i k_j / k^2$. To integrate over $k$ we use, for example, the Kolmogorov spectrum of the background turbulence, i.e., $\tau f_{pp}^{(0N)}(\mathbf{k}) = \eta_T^{(v)} \varphi(k)$, $\tau h_{pp}^{(0N)}(\mathbf{k}) = \eta_T^{(h)} \varphi(k)$ and $\mu_{mn}^{(a)}(\mathbf{k}) = \mu_{mn}^{(a)}(\mathbf{R}) \varphi(k)/3$, where $\varphi(k) = (\pi k^2 k_0)^{-1} (k/k_0)^{-7/3}$, $\tau(k) = 2\tau_0 (k/k_0)^{-2/3}$ and $k_0 = l_0^{-1}$.

(f). Next, we consider an anisotropic background turbulence with one preferential direction, say along unit vector $\mathbf{e}$. Thus the tensor $\eta_{ij}^{(v)}(\mathbf{B} = 0) = \langle \tau v_i v_j \rangle$ is given by

$$\eta_{ij}^{(v)}(\mathbf{B} = 0, \mathbf{R}) = \eta_T^{(v)} \delta_{ij} + \mu_{ij}^{(v)} = \delta_{ij} [\eta_T^{(v)} - (1/3) \varepsilon_\mu^{(v)}] + e_{ij} \varepsilon_\mu^{(v)} , \qquad (4)$$

where $\varepsilon_\mu^{(v)} / \eta_T^{(v)}$ is a degree of an anisotropy of the turbulence and $e_{ij} = e_i e_j$. The anisotropic part $\mu_{ij}^{(v)}$ is given by $\mu_{ij}^{(v)} = \varepsilon_\mu^{(v)} [e_{ij} - (1/3) \delta_{ij}]$. Note that the components of the tensor $\eta_{ij}^{(v)}(\mathbf{B} = 0)$ should be positive and it yields the condition $-3/2 \leq \varepsilon_\mu^{(v)} / \eta_T^{(v)} \leq 3$. The equations for the corresponding magnetic tensors are obtained after the change $v \to h$ in the above formulas. For the magnetic fluctuations we also have $-3/2 \leq \varepsilon_\mu^{(h)} / \eta_T^{(h)} \leq 3$. For galaxies, e.g., the preferential direction $\mathbf{e}$ is along rotation (which is parallel to the effective gravitational field). For axisymmetric $\alpha\Omega$–dynamo for large magnetic Reynolds numbers the toroidal magnetic field is much larger than the poloidal field. Therefore, the value $\mathbf{e} \cdot \mathbf{B}$ is very small and it can be neglected because $\mathbf{B}$ is approximately directed along the toroidal magnetic field.

## 3. The nonlinear mean electromotive force in anisotropic background turbulence with one preferential direction

The mean electromotive force $\boldsymbol{\mathcal{E}}$ is given by

$$\boldsymbol{\mathcal{E}} = \boldsymbol{\alpha} \mathbf{B} + \mathbf{V}^{(N)} \times \mathbf{B} - \boldsymbol{\eta} (\boldsymbol{\nabla} \times \mathbf{B}) - \boldsymbol{\kappa} \boldsymbol{\partial} \hat{\boldsymbol{B}} , \qquad (5)$$

where $(\partial \hat{B})_{jk} = (B_{jk} + B_{kj})/2$, $B_{ij} = \nabla_j B_i$, and the general structure of the tensors and vectors defining the mean electromotive force for an anisotropic background turbulence with one preferential direction is $\eta_{ij}(\mathbf{B}) = N_\eta \delta_{ij} + N_e e_{ij} + N_\beta \beta_{ij}$, $\mathbf{V}^{(N)}(\mathbf{B}) = (1/B^2)[N_V^{(1)} \boldsymbol{\nabla} B^2 + N_V^{(2)} \mathbf{e} (\mathbf{e} \cdot \boldsymbol{\nabla}) B^2]$, $\boldsymbol{\kappa}(\mathbf{B}) \boldsymbol{\partial} \hat{\boldsymbol{B}} = N_\kappa \mathbf{e} \times (\mathbf{e} \cdot \boldsymbol{\nabla}) \mathbf{B}$, where $\mathbf{e} \cdot \mathbf{B} = 0$ and $\beta_{ij} = B_i B_j / B^2$. The $\alpha$-tensor is given by $\alpha_{ij}(\mathbf{B}) = \alpha \delta_{ij}$, where $\alpha = \alpha_0^{(v)} \Phi^{(v)}(\beta) + \alpha_0^{(h)}(\mathbf{B}) \Phi^{(h)}(\beta)$, $\beta = 4B / (u_0 \sqrt{2 \mu_0 \rho})$, $\Phi^{(v)}(\beta) = (1/7)[4 \Phi^{(h)}(\beta) + 3 L(\beta)]$, $\Phi^{(h)}(\beta) = (3/\beta^2)(1 - \arctan(\beta)/\beta)$ and $L(\beta) = 1 - 2\beta^2 + 2\beta^4 \ln(1 + \beta^{-2})$ (Rogachevskii and

Kleeorin, 2000). The magnetic part $\alpha_0^{(h)}(\mathbf{B})$ of the $\alpha$-effect is determined by the differential equation

$$\frac{\partial \alpha_0^{(h)}}{\partial t} + \frac{\alpha_0^{(h)}}{T} + \nabla \cdot (\mathbf{V}^{\mathrm{eff}} \alpha_0^{(h)} + \mathbf{F}_{\mathrm{flux}}) = -\frac{4}{9 \eta_T \mu_0 \rho} \boldsymbol{\mathcal{E}}(\mathbf{B}) \cdot \mathbf{B} \tag{6}$$

(see Kleeorin and Rogachevskii, 1999; Kleeorin et al., 2000), where $\mathbf{V}^{\mathrm{eff}}$ is the effective velocity which is proportional to the mean fluid velocity $\mathbf{V}$ (e.g., for isotropic turbulence $\mathbf{V}^{\mathrm{eff}} = \mathbf{V}$), and $\mathbf{F}_{\mathrm{flux}}$ is the flux of the magnetic helicity which is not related with the mean fluid velocity $\mathbf{V}$. Equations (6) determines the dynamic nonlinearity.

## 4. Mean-field equations for the thin-disk axisymmetric $\alpha\Omega$–dynamo

Using the equation for the nonlinear mean electromotive force we derive the mean-field equations for the thin-disk axisymmetric $\alpha\Omega$–dynamo:

$$\partial B / \partial t = (\eta_B B')' + G A' , \tag{7}$$

$$\partial A / \partial t = \eta_A A'' - V_A A' + \alpha B , \tag{8}$$

where $\mathbf{B} = B \mathbf{e}_\varphi + \nabla \times (A \mathbf{e}_\varphi)$, $B' = \partial B / \partial z$, and $\eta_A = N_\eta + N_\kappa + N_\beta$, $\eta_B = N_\eta + N_\kappa - 2 N_V$, $V_A = (\eta_A - \eta_B)(\ln |B|)'$, and $N_V = N_V^{(1)} + N_V^{(2)}$. The parameter $G = -R(\partial \Omega / \partial R)$ describes the differential rotation. It is seen from the equations for $\eta_A$, $\eta_B$ and $V_A$ that the contributions to the turbulent diffusion coefficients $\eta_A$ and $\eta_B$ are from the tensor of turbulent diffusion $\eta_{ij}(B)$, the tensor $\kappa_{ijk}(B)$ and the velocity $\mathbf{V}^{(N)}(\mathbf{B})$. On the other hand, contributions to the effective velocity $V_A$ are from the tensor of turbulent diffusion $\eta_{ij}(B)$ and the nonlinear velocity $\mathbf{V}^{(N)}(\mathbf{B})$. The functions $\eta_A(B)$, $\eta_B(B)$ and $V_A(B)$ are given by

$$\eta_A(B) = \tilde{\eta}(B) + [(10/9)(2C_1 - A_1)\varepsilon_\mu - A_1 \eta_T]^{(*)} , \tag{9}$$

$$\begin{aligned} \eta_B(B) &= \tilde{\eta}(B) + [(5/18)(8C_1 + C_2 + 10C_3 - 4A_1 - 4A_2)\varepsilon_\mu \\ &\quad -(A_1 + A_2)\eta_T]^{(*)} + (\sqrt{2}\beta/24)\Psi(\sqrt{2}\beta) , \end{aligned} \tag{10}$$

$$\begin{aligned} V_A(B) &= \{[(5/18)(4A_2 - C_2 - 10C_3)\varepsilon_\mu + A_2 \eta_T]^{(*)} \\ &\quad -(\sqrt{2}\beta/24)\Psi(\sqrt{2}\beta)\}(\ln |B|)' , \end{aligned} \tag{11}$$

where $\Psi(x) = 12[A_1'(x) + (1/2)A_2'(x)]\eta_T^{(-)} + \Psi_0'(x)\varepsilon_\mu^{(-)}$, $\tilde{\eta}(B) = (A_1 + (1/2)A_2)\eta_T^{(+)} + (\varepsilon_\mu^{(+)}/12)\Psi_0$, $\Psi_0 = (5/3)(4A_1 + 3A_2 + 4C_1 - C_3)$ and $S^{(*)}(\beta) = S^{(v)}(\beta) - S^{(-)}(\sqrt{2}\beta)$, $S^{(\pm)}(\beta) = S^{(v)}(\beta) \pm S^{(h)}(\beta)$. Here the functions $A_k(\beta)$ and $C_k(\beta)$ are given by

$$A_1(\beta) = \frac{6}{5}\left[\frac{\arctan \beta}{\beta}\left(1 + \frac{5}{7\beta^2}\right) + \frac{1}{14}L(\beta) - \frac{5}{7\beta^2}\right] ,$$

$$A_2(\beta) = -\frac{6}{5}\left[\frac{\arctan\beta}{\beta}\left(1+\frac{15}{7\beta^2}\right) - \frac{2}{7}L(\beta) - \frac{15}{7\beta^2}\right],$$
$$C_1(\beta) = \frac{3}{10}\left[\frac{\arctan\beta}{\beta}\left(1+\frac{10}{7\beta^2}+\frac{5}{9\beta^4}\right) + \frac{2}{63}L(\beta) - \frac{235}{189\beta^2} - \frac{5}{9\beta^4}\right],$$
$$C_2(\beta) = \frac{3}{2}\left[\frac{\arctan\beta}{\beta}\left(\frac{3}{5}+\frac{30}{7\beta^2}+\frac{35}{9\beta^4}\right) + \frac{16}{315}L(\beta) - \frac{565}{189\beta^2} - \frac{35}{9\beta^4}\right],$$
$$C_3(\beta) = -\frac{3}{2}\left[\frac{\arctan\beta}{\beta}\left(\frac{1}{5}+\frac{6}{7\beta^2}+\frac{5}{9\beta^4}\right) - \frac{8}{315}L(\beta) - \frac{127}{189\beta^2} - \frac{5}{9\beta^4}\right].$$

For $\beta \ll 1$ the functions $\eta_A \approx \eta_* - O(\beta^2)$, $\eta_B \approx \eta_* - O(\beta^2)$ and $V_A(B) \sim O(\beta^2)$, where $\eta_* = \eta_T^{(v)} + (2/3)\varepsilon_\mu^{(v)}$. For $\beta \gg 1$ the functions $\eta_A \propto 1/\beta$, $\eta_B \propto 1/\beta$, $V_A(B) \propto 1/\beta$ and $\alpha \propto 1/\beta^2$. The nonlinear dependencies of – (A) the turbulent diffusion coefficients $\eta_A(B)/\eta_T^{(v)}$ and $\eta_B(B)/\eta_T^{(v)}$; (B) the effective velocity $V_A(B)/(B^2)'$; and (C) the nonlinear dynamo number $D(B)/D_*$ are presented in Figures 1-2. Here $D_* = \alpha_* G h^3/\eta_*^2$, $D(B) = \alpha^{(v)}(B) G h^3/[\eta_A(B)\eta_B(B)]$, $\alpha_*$ is the maximum value of the hydrodynamic part of the $\alpha$ effect, $h$ is the disc thickness and $\alpha^{(v)}(B) = \alpha_0^{(v)}\Phi^{(v)}(B)$.

In order to separate the study of the algebraic and dynamic nonlinearities we defined the nonlinear dynamo number using only the hydrodynamic part of the $\alpha$ effect. We considered two cases: an anisotropic background turbulence ($\varepsilon_\mu^{(v)} = -1.35\eta_T^{(v)}$; $\varepsilon_\mu^{(h)} = 0$) without magnetic fluctuations (Fig. 1) and an isotropic ($\varepsilon_\mu^{(v)} = \varepsilon_\mu^{(h)} = 0$) background turbulence with equipartition of hydrodynamic and magnetic fluctuations (Fig. 2). The negative degree of anisotropy $\varepsilon_\mu^{(v)}$ implies that the vertical (along axis $z$) size of turbulent elements is less than the horizontal size. Figures 1-2 and the equations for $\eta_A(B)$, $\eta_B(B)$ and $V_A(B)$ show that the toroidal and poloidal magnetic fields have different nonlinear turbulent diffusion coefficients. In isotropic background turbulence (Fig. 2) the nonlinear effective velocity $V_A(B)$ is negative. The latter implies that it is diamagnetic velocity, which results in the field is pushed out from the regions with stronger mean magnetic field. In the anisotropic background turbulence (Fig. 1) the nonlinear effective velocity is positive, (i.e., paramagnetic velocity which causes the magnetic field tends to be concentrated in the regions with stronger field). The dependencies of the nonlinear dynamo number $D(B)/D_*$ on the mean magnetic field $B/B_{\rm eq}$ demonstrate that the algebraic nonlinearity alone (i.e., quenching of both, the nonlinear $\alpha$ effect and the nonlinear turbulent diffusion coefficients) cannot saturate the growth of the mean magnetic field (where $B_{\rm eq} = \sqrt{\mu_0\rho}\, u_0$). Indeed, for anisotropic background turbulence without magnetic fluctuations (Fig. 1) the nonlinear dynamo number $D(B)/D_*$ is a nonzero constant for $\beta \gg 1$, i.e., it is independent on $\beta$. This is because for $\beta \gg 1$ the functions $\eta_A \propto 1/\beta$, $\eta_B \propto 1/\beta$ and $\alpha \propto 1/\beta^2$

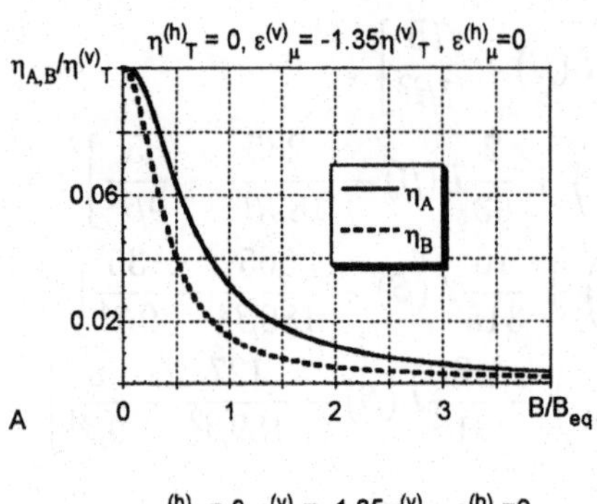


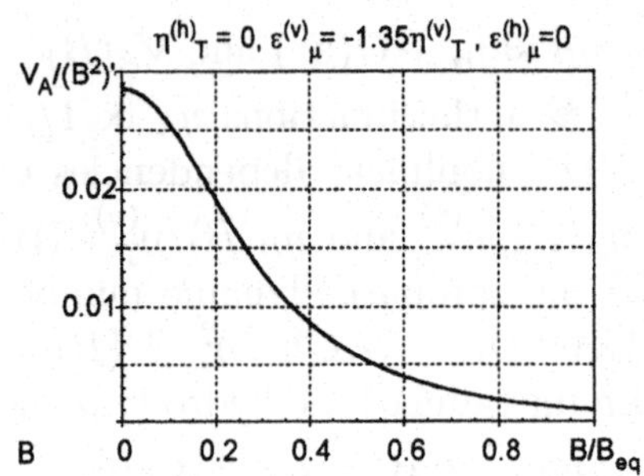


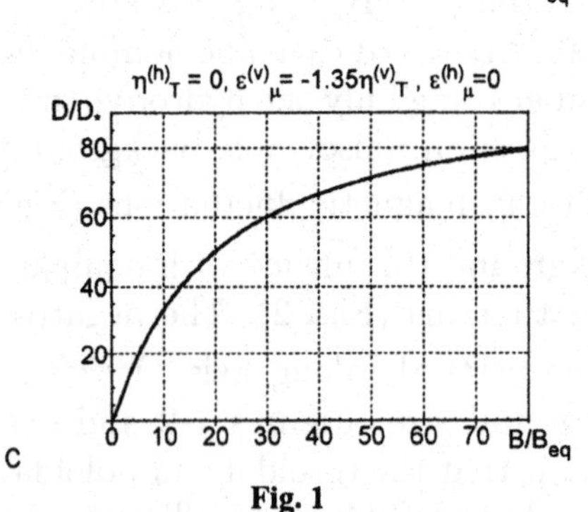


**Fig. 1**

(A) The nonlinear turbulent diffusion coefficients; (B) The nonlinear effective velocity; (C) The nonlinear dynamo number for $\eta_T^{(h)} = 0;\ \varepsilon_T^{(v)} = -1.35\,\eta_T^{(v)};\ \varepsilon_T^{(h)} = 0$

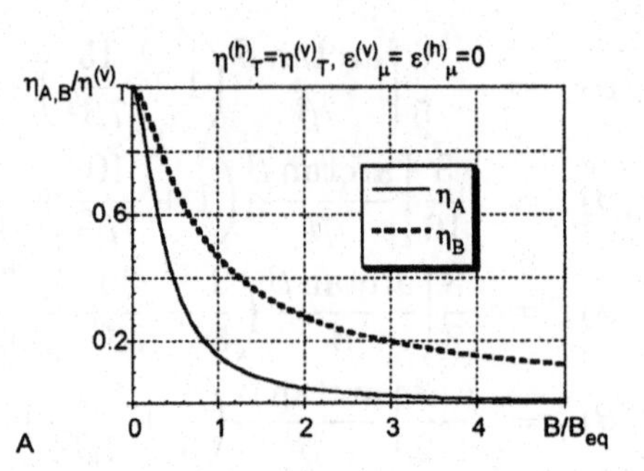


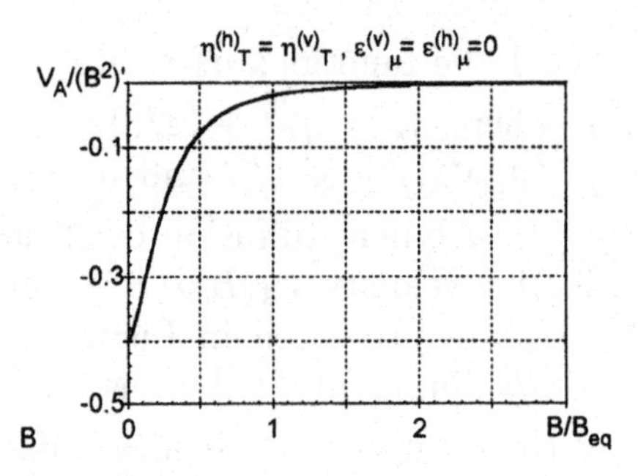


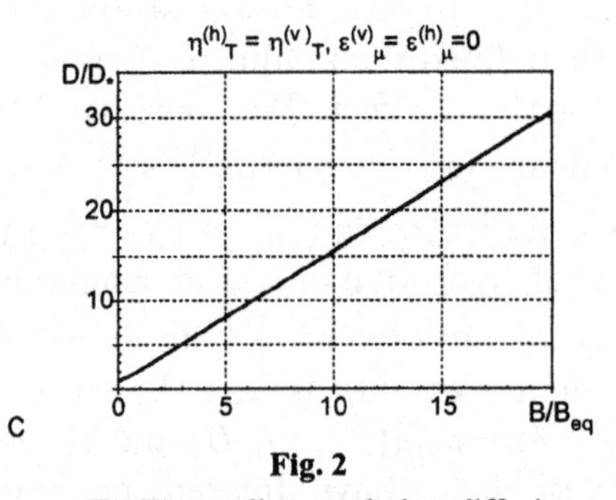


**Fig. 2**

(B) The nonlinear turbulent diffusion coefficients; (B) The nonlinear effective velocity; (C) The nonlinear dynamo number for $\eta_T^{(h)} = \eta_T^{(v)};\ \varepsilon_T^{(v)} = \varepsilon_T^{(h)} = 0$

In the case of isotropic background turbulence with equipartition of hydrodynamic and magnetic fluctuations (Fig. 2) the nonlinear dynamo number $D(B)/D_* \propto \beta$ for $\beta \gg 1$ because the functions $\eta_A \propto 1/\beta^2$, $\eta_B \propto 1/\beta$ and $\alpha \propto 1/\beta^2$. Note that the saturation of the growth of the mean magnetic field can be achieved when the nonlinear dynamo number $D(B)/D_* \rightarrow 0$ for $\beta \geq 1$. We expect that both, the algebraic and dynamic nonlinearities can limit the growth of the mean magnetic field.

*Acknowledgements* This work was partially supported by INTAS (Grant No.99-348).

## References

1. Kleeorin, N., Moss, D., Rogachevskii, I. and Sokoloff D. (2000) Helicity Balance and Steady-State Strength of Dynamo Generated Galactic Magnetic Field. *Astron. and Astroph.*, **361**, L5–L8.
2. Kleeorin, N. and Rogachevskii, I. (1999) Magnetic Helicity Tensor for an Anisotropic Turbulence, *Phys. Rev. E.* **59**, 6724–6729.
3. Kleeorin, N., Rogachevskii, I. and Ruzmaikin, A. (1995) Magnitude of Dynamo - Generated Magnetic Field in Solar - Type Convective Zones, *Astron. Astrophys.* **297**, 159–167.
4. Kleeorin, N.I. and Ruzmaikin, A.A. (1982) Dynamics of the Mean Turbulent Helicity in a Magnetic Field *Magnetohydrodyn.* **18**, 116–122.
5. Krause, F. and Rädler, K. H. (1980) *Mean-Field Magnetohydrodynamics and Dynamo Theory*, Pergamon, Oxford.
6. Pouquet, A., Frisch, U. and Leorat, J. (1976) Strong MHD Turbulence and the Nonlinear Dynamo Effect, *J. Fluid Mech.* **77**, 321–354.
7. Rogachevskii, I. and Kleeorin, N. (1997) Intermittency and Anomalous Scaling for Magnetic Fluctuations, *Phys. Rev. E* **56**, 417–426.
8. Rogachevskii, I. and Kleeorin, N. (1999) Anomalous Scaling and Dynamics of Magnetic Helicity, *Phys. Rev. E* **59**, 3008–3011.
9. Rogachevskii, I. and Kleeorin, N. (2000) Electromotive Force for an Anisotropic Turbulence: Intermediate Nonlinearity, *Phys. Rev. E.* **61**, 5202–5210.

# THE SOLAR DYNAMO: AXIAL SYMMETRY AND HOMEGENEITY BROKEN

A. RUZMAIKIN
*Jet Propulsion Laboratory, California Institute of Technology*
*4800 Oak Grove Drive, Pasadena, CA 91109, USA*

## 1. Introduction

The Sun is a natural site for a dynamo. In fact, the dynamo concept was introduced by Larmor in his 1919 report to the British Association for Advanced Science entitled "How could a rotating body as the Sun become magnetic?" Cowling's famous anti-dynamo theorem appeared in his paper "Magnetic fields of sunspots" (MNRAS. **94**, 39 ,1934). Yet the origin of the Sun's magnetic field is not well understood. Some scientists still challenge the dynamo as the source of solar magnetic field [7].

Because solar activity manifests itself on a global scale – in systematic migration of sunspots and changing polarity of the global magnetic field– consideration of the largescale, mean magnetic field was a wise idea [16, 14]. Mean-field dynamo models are mostly kinematic, the regular velocity, such as rotation, enters explicitly, the turbulent velocity is parameterized by smooth functions. This parameterization includes turbulent diffusion and mean kinetic helicity. The solar dynamo works in the convection zone that occupies an outer third of the Sun's volume. With the parameters typical for the convection zone, the mean-field dynamo solutions are oscillating which fits well to the observed cyclic nature of solar activity. In the commonly used axisymmetric approximation, the mean field is represented by toroidal and poloidal components. The former is associated with the sunspot magnetic field, the latter with the global field reversing its sign in the course of the solar cycle.

The observed solar magnetic fields, however, strongly lacks axial symmetry, in particular at maxima of solar activity. The activity itself tends to cluster, i.e. distributed very inhomogeneously. These two problems are discussed here in relation to the mean-field dynamo and beyond.



*P. Chossat et al. (eds.), Dynamo and Dynamics, a Mathematical Challenge,* 289–296.

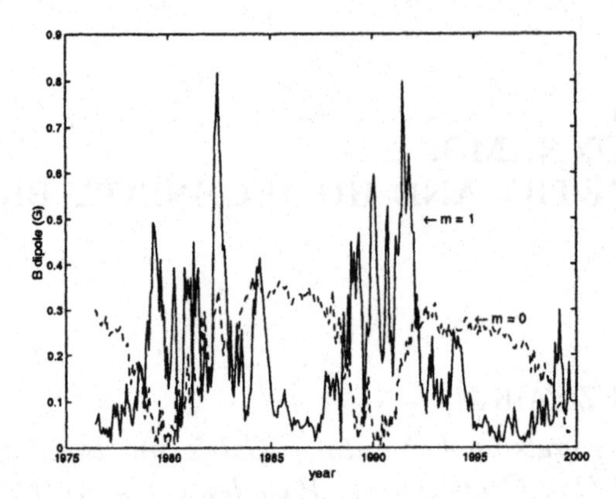


*Figure 1.* Amplitudes of the first two modes of the solar field for Carrington Rotations 1642 to 1953 corresponding to 1976-1999. The $m = 1$ non-axisymmetric mode is shown by solid curve, the axisymmetric $m = 0$ mode is shown by the dash curve.

## 2. Broken Axial Symmetry: Preferred Longitudes

The pressure, temperature and density levels that determined the basic equilibrium of the Sun are nearly spherically symmetric. This is well confirmed by helioseismic observations [13]. The spherical symmetry is "slightly" broken by the relatively weak solar rotation. The rotation makes the North and the South distinguishable but retains the axial symmetry. The sources of the mean field (rotation, helicity and meridional circulation) are also axisymmetric. One expects then that the leading dynamo solutions be axisymmetric. Observations of the magnetic fields on the solar surface tell us, however, a different story. Figure 1 shows amplitudes of the first axisymmetric and non-axisymmetric mode of the field (from data obtained at the Wilcox Solar Observatory for the last 24 years). We see that the non-axisymmetric mode not only is always present but also strongly dominates at maxima of the 11-year cycles.

There is even more interesting observation. The tip of the main dipole mode of the non-axisymmetric field selects a certain longitude. Surprisingly enough, this preferred longitude can persist for a long time apparently exceeding the duration of a solar cycle. This was first noticed on the grounds that the sunspot number is rotationally modulated in a way consistent with preferred longitudes [22], and in longitudinal clustering of solar energetic proton events and x-ray flares [9]. There are indications of preferred longitudes in the distribution of the photospheric magnetic field [10]. The solar magnetic field extends into the solar wind. For example, the two-sector structure of the interplanetary magnetic field [20] is associated with the tilted dipole field at the Sun, which is represented as a sum of a polar and an equatorial dipole, i.e. of $m = 0$ and $m = 1$ modes. Recently, Neugebauer et al. (2000) using the 36 year long spacecraft record of 6-hour average interplanetary field has found a persistent magnetic structure rotating with (synodic) period of 27.03 days (Figure 2).

The term "preferred" is to be understood as a coherent phenomenon, i.e. the field maximum tends to persist at some longitude for a time well exceeding the rotation period of the Sun. Both solar and interplanetary studies indicate persistence of magnetic structures with preferred longitudes over several solar cycles. The selection of a particular longitude could be accidental, more important is the persistence (duration) of the coherency. The preferred longitudes can be found from the spherical harmonic expansion of the magnetic field observed on the photosphere [1]. The observations give the distribution of radial magnetic field on the solar surface called "magnetograms". Thus, the radial field corresponding to the first non-axisymmetric mode (an equatorial dipole) is represented by

$$B_r(t,\theta,\varphi) = [g_1(t) sin\varphi + h_1(t) cos\varphi] sin\theta.$$

The angle at which the pole tipped is then $\varphi_0(t) = \text{artan}(h_1/g_1)$. The magnetograms are obtained in a coordinate system rotating with the so-called Carrington period $P_0 = 27.275$ days, which is close to the period of equatorial rotation of the Sun. However, the period of rotation of a magnetic structure does not necessarily coincide with the Carrington period. In search for the actual rate of rotation, we write $\Omega = \Omega_0 + \delta\Omega$ where $\Omega_0 = 2\pi/P_0$ is the Carrington rate. First, we calculate $\varphi_0$ from the data, as the inverse tangent of the ratio $h_1/g_1$. This function is confined to the interval $[0, 2\pi]$; the specific quadrant is determined by the signs of $h_1$ and $g_1$. Then for a given $P$ we calculate $\varphi_P$ shifted from $\varphi_0$ according to this rotation rate and plot histograms of distribution of $\varphi_P$. If the histogram shows a peak at certain longitude stable to small changes in the value of $P$, we identify this peak as indication of a magnetic structure rotating with the period $P$. A positive result of the search is shown in Figure 2.

The non-axisymmetric fields are consistent with the mean-field theory. Although the sources of the field generation are axisymmetric, the magnetic field can have both axisymmetric and non- axisymmetric components in accord with general principles [4] and earlier dynamo simulations [21, 12]. The differential rotation tends toward the axial symmetry but mean helicity ($\alpha$) favors non-axial symmetry. Preferred generation of non-axisymmetric modes is then possible if maxima of $\nabla\Omega$ and $\alpha$ occur close to each other and the ratio of their amplitudes is small [21].

Earlier mean-field dynamo models assumed that the differential rotation was the strongest source and therefore the leading modes were found to be axisymmetric. Another frequent assumption was a description of convective motions without taking into account the transition layer between the con-

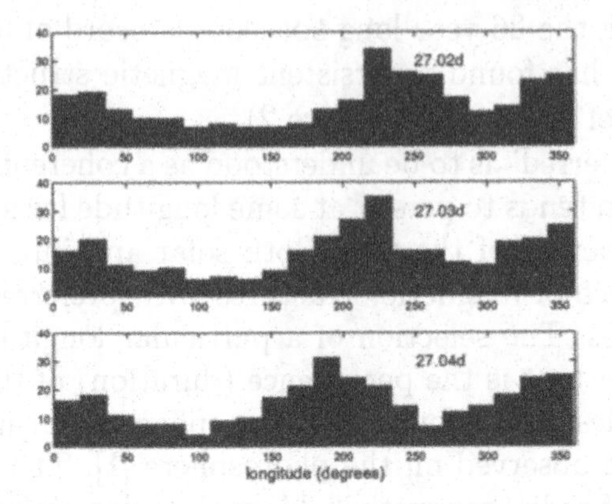


*Figure 2.* The distribution of pole of the m =1 mode at the period of rotation 27.03 days (Ruzmaikin, 2000). The same period has been found in the independent study of the interplanetary magnetic field (Neugebauer et al., 2000).

vection zone and the radiative core called " tachocline". We now know that gradients of rotation in the convection zone are only modest. According to analysis of the SOHO helioseismic observations, the maximum of the radial gradient of rotation is located near the bottom of convection zone, at about $0.3R_{\odot}$ below the solar surface [13]. The distribution of the kinetic helicity ($\alpha$) is less certain and at present can only be evaluated on model grounds. If the convection could be approximated as well-mixed turbulence $\alpha$ should be located close to the solar surface [24]. If, on the other hand, the kinetic helicity were generated by the processes in the tachocline, the maximum of $\alpha$ should be near the base of the convection zone [3]. The dominance of the non-axisymmetric mode reported above favors this location of $\alpha$.

A characteristic, which has to be calculated and compared with the observations is the drift rate of the non-axisymmetric modes, as has been done earlier by Stix (1974). There can be modes with a positive drift rate (faster than the Carrington rotation) and modes with a negative drift rate (slower than the Carrington rotation), in full analogy with the westward and eastward drifts of geomagnetic anomalies. The magnitudes of these rates depend on distributions of rotation and helicity in the convection zone. Our result (27.03 days) indicates the positive drift. The mode with negative drift is definitely weaker, at least we were not able to identify it.

Analyses of helioseismic observations (c.f. Howe et al., 2000) show that the solar rotation as a function of depth is different at different latitudes but converges to a unique solid rotation at the bottom of the convection zone. The synodic period of this rotation is 28.7 days. The period at the surface near the equator is the Carrington period. The period 27.03 days found is shorter than either of these periods. An intriguing question, along with explanation of the persistence of preferred longitudes, is why we observed the advanced mode if the field is generated near the slowly rotating tachocline?

## 3. Broken Homogeneity: Clustering of Emerging Flux

Magnetic fields generated in the convection zone emerge at the solar surface. In spite of the homogeneous distribution of the convection, the emerging fields are not even statistically homogeneous. New magnetic flux tends to emerge where there is old flux, leading to clustering. Statistical studies have demonstrated that the clustering is not random [8]. Gaizauskas et al. (1983) described a pattern which lasted for 6 solar rotations and contained 29 major active regions. An example of a cluster is shown in Figure 4. The clustering can lead to hazards at the Earth. Thus, the March 1989 cluster induced 25 coronal mass ejections leading to energetic solar particles and geomagnetic storms (Feynman, 1997).

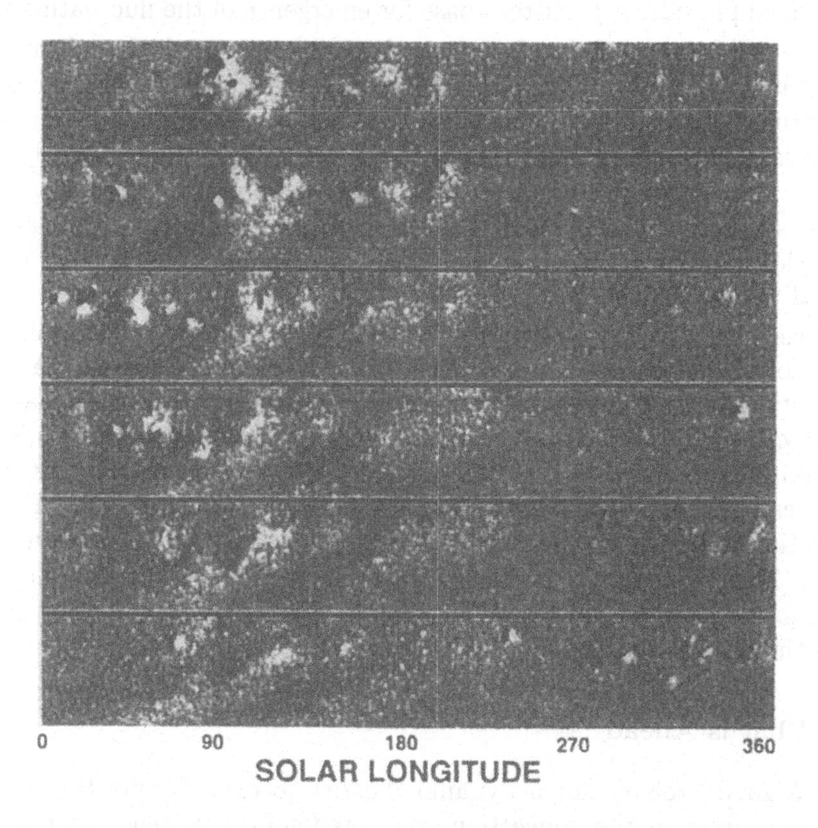


*Figure 3.* A magnetic cluster identified by Feynman (1997). Stacked magnetograms for 4 solar rotations starting from December 14, 1993 are shown. The earliest rotation is at the top. Latitudes range between $0 - 40°$ along the ordinate axis.

A percolation model for clustering of emerging magnetic field was suggested by Wentzel and Sieden (1992). The model starts by randomly designating 1% of cells on a 2-D region as active. An active cell then can induce activity in its neighbor cell - like a tree permits a fire to hop to another tree causing a forest fire – with a prescribed stimulated probability. The crucial assumption of the hopping was however not justified.

In an alternative approach (Ruzmaikin, 1998) no assumption is made about the influence of an active cell on its neighbor. "Active cells" are considered as independent, i.e. the new flux "does not know anything" about the old flux. The persistence of the flux emergence and hence its clustering arises due to large-scale subsurface non-axisymmetric field and sufficiently strong coherent fluctuating fields. The mean-field, assumed to be generated by the mean-field dynamo, does not have enough strength to emerge on the solar surface. It was suggested that the magnetic field emerges at the solar surface when the mean field plus a fluctuation (total magnetic field) exceeds the threshold for buoyancy. A non-axisymmetric mean field provides a persistent base for emergence of the fluctuating field.

The importance of the deviations from the mean field is indicated by observations. For example, no single large flux tube emerges when sunspots are formed. Instead, the sunspot magnetic field is assembled through the progressive gathering of many flux tubes (Zwaan, 1978). Hence, the dynamo process that produces the mean field must also produce fluctuating fields which is the case in the solar convection zone [16, 14, 24]. For high magnetic Reynolds number in the convection zone, mean-squared deviations greatly exceed the mean field.

Over the past few years, it has become possible to model collective behavior of turbulent magnetic fields. An example is the reorganization of magnetic fields into large-scale structures by the inverse cascade of magnetic helicity. Recent simulations by Brandenburg (2000), Figure 5, demonstrate that initially chaotic, small-scale magnetic fields can self-organize into large-scale features extending over the size of the computational domain. Pattern formation in a strongly stratified convection layer is studied by Rucklidge et al., (2000). The nature of transitions from steady convection to a chaotic behavior is described in this paper as symmetry breaking - in good accord with the theme of this Workshop.

## 4. What is Ahead

It is a good problem for a dynamo theorist to explain why the axisymmetric sources in the convection zone predominantly generate the non-axisymmetric magnetic fields at solar maxima. The standard $\alpha\Omega$ dynamo does not lead to the observed relationship between the axisymmetric and

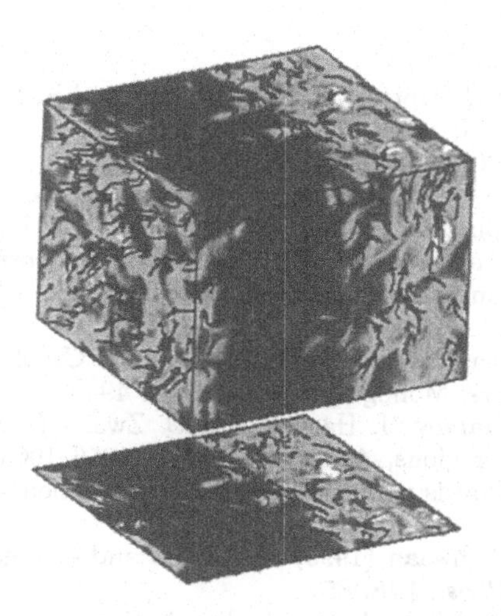

*Figure 4.* Example of collective behavior in 3-D MHD turbulence with helicity (Brandenburg, 2000). The flow is driven at small scales (k=5), however a large-scale pattern emerges after long times. Shown is the vector magnetic field superimposed on a gray scale representation of one component of this field.

non-axisymmetric components. Perhaps, a correct relationship can be found in the $\alpha^2\Omega$ dynamos. The future model should give the magnitude of advanced and retarded drifts to compare with the observed drift(s) of magnetic patterns on the solar surface and in interplanetary space.

But it is a great challenge to explain why the tip of the main non-axisymmetric mode (an equatorial dipole) seems to remember its longitude for several solar cycles, i.e. the persistence of preferred longitudes. The idea of a weak relict magnetic field frozen in the radiative core comes to mind. However, it is unclear how the non-axisymmetric component of this field could survive the destructive action of the differential rotation in the transition region between the radiative core and the convection zone. A more promising idea is to explore the enormous difference between the diffusion times in the convection zone and the radiative core.

Another challenge is the origin of clustering of the magnetic flux emerging at the solar surface. It is certainly a coherent non-linear phenomenon that can not be described by the mean field only. Fluctuating fields should play a key role in the clustering. 3-D MHD modeling will be the best way to attack this problem.

## Acknowledgements

This research was conducted in part at the Jet Propulsion Laboratory, California Institute of Technology, under contract with NASA.

## References


1. Altschuler, M. D., D. E. Trotter, G. Newkirk, Jr., and R. Howard (1974) The large-scale solar magnetic field, *Solar Phys.*, **39**, 3–17.
2. Brandenburg, A. (2000), *Astrophys. J* (submitted)
3. Charbonneau, P., and K. B. MacGregor (1997) Solar interface dynamos. II. Linear, kinematic models in spherical geometry, *Astrophys. J.*, *486*, 502-514.
4. Chossat, P., and R. Lauterbach (2000) *Methods in equivariant bifurcations and dynamical systems*, Advanced Series in Nonlinear Dynamics - Vol. 15, World Scientific Publs.
5. Feynman, J. (1997) in *Coronal Mass Ejections*, N. Crooker, J. A. Joselyn and J. Feynman, eds., Geophys. Monograph 99, AGU, p. 49.
6. Gaizauskas, V., K. Harvey, J. Harvey, and C. Zwaan (1983) Large-scale patterns formed by solar active regions, *Astrophys. J*, **265**, 1056–1065.
7. Gough, D. (2000) Towards understanding solar convection and activity, *Solar Phys.*, **192**, 3-26.
8. Harvey, K. L. and C. Zwaan (1993) Properties and emergence patterns of bipolar active regions, *Solar Phys.*, **148**, 85.
9. Haurwitz, M. W. (1968) Solar longitude distributions of proton flares, meter bursts, and sunspots,*Astrophys. J.*, **151**, 351–364.
10. Hoeksema, J. T. and P. H. Scherrer (1987) Rotation of the coronal magnetic field, *Astrophys. J.*, **318**, 428–436.
11. Howe, R., J. Christensen-Dalsgard, F. Hill, R. W. Komm, R. M. Larsen, J. Schou, M. J. Thompson, and J. Toomre (2000) Dynamic variations at the base of the solar convection zone, *Science*, **287**, 2456-2460.
12. Ivanova, T.S., and A. Ruzmaikin (1985) Three-dimensional model for generation of the mean solar magnetic field. *Astron. Nachr.*, **306**, 177-186.
13. Kosovichev, A. G. and 33 co-authors.: 1997, Structure and rotation of the solar interior: initial results from the MDI medium program,*Solar Phys.*, **170**, 43–61.
14. Krause, F., and K.-H. Rädler, Mean-Field Electrodynamics and Dynamo, Springer Verlag, 1981.
15. Neugebauer, M., E. J. Smith, J. Feynman, A. Ruzmaikin and A. Vaughan (2000) The solar magnetic field and the solar wind: Existence of preferred longitudes, *J. Geophys. Res.*, *105*, 2315-2324.
16. Parker, E. N. (1979) *Solar Magnetic Fields*, Oxford Univ. Press, Oxford.
17. Rucklidge, A. M., N. O. Weiss, D. P. Brownjohn, P. C. Matthews and M. R. E. Proctor, (2000) Compressible magnetoconvection in three dimensions: pattern formation in a strongly stratified layer, *J. Fluid Mech.*, **419**, 283-323.
18. Ruzmaikin, A. (1998) Clustering of emerging magnetic flux, *Solar Phys.*, 181, 1-12.
19. Ruzmaikin, A., J. Feynman, Neugebauer, M., and E. J. Smith (2000) Preferred solar longitudes with signatures in the solar wind, *J. Geophys. Res.*, submitted
20. Svalgaard, L., and J. M. Wilcox, Long term evolution of solar sector structure, *Solar Phys.*, *41*, 461–475, 1975.
21. Stix, M. (1974) Comments on the solar dynamo, *Astron. & Astrophys.*, **37**,121–133.
22. Vitinsky, Yu. (1960), *Izvest. Atsron. Obs. Poulkovo*, **21**, N163, p.96.
23. Wentzel, D. G., and P. E. Sieden (1992) Solar active regions as a percolation phenomenon, *Astrophys. J.*, **390**, 280–289.
24. Zeldovich, Ya. B., A. Ruzmaikin and D. D. Sokoloff (1983) *Magnetic Fields in Astrophysics*, Gordon and Breach, London–Paris–New York.
25. Zwaan, C. (1978) *Solar Phys.*, **169**, 265.

# ALPHA-QUENCHED $\alpha^2\Omega$-DYNAMO WAVES IN STELLAR SHELLS

ANDREW SOWARD AND ANDREW BASSOM
*School of Mathematical Sciences, University of Exeter, North Park Road, Exeter, Devon EX4 4QE, UK*

AND

YANNICK PONTY
*Observatoire de la Côte d'Azur, Laboratory Cassini - CNRS UMR 6529, B. P. 4229, 06304 Nice Cedex 4, France*

## 1. Introduction

The generally accepted explanation of the sunspot cycle is in terms of a kinematic $\alpha\Omega$-dynamo wave propagating with fixed period from the pole to the equator (see Parker [1]); for a recent review see Rüdiger and Arlt [2]. Since such oscillatory behaviour is a robust feature generic to all $\alpha\Omega$-dynamo models, the simplicity of the idea is compelling. Both solar and stellar dynamos generally operate in convective spherical shells. There are two limiting cases, namely thick or thin shells as characterised by the ratio $\varepsilon$ of the shell thickness to shell radius. In the thick shell limit, it is necessary to consider the full partial differential equations involving the radial and latitudinal dependence. Conversely in the thin shell limit $\varepsilon \ll 1$, it is possible to average the dynamo equations radially leaving a one-dimensional system dependent on the latitude $\theta$ alone.

Numerical integrations (Moss *et al.* [3]) of the full partial differential equations governing axisymmetric $\alpha\Omega$-dynamos in the thin shell limit $\varepsilon \ll 1$, upon which we will focus, indicate that there is a short latitudinal length scale comparable to the shell depth. Advantage of this feature was taken by Kuzanyan and Sokoloff [4], who employed WKBJ methods to solve the one-dimensional kinematic $\alpha\Omega$-dynamo system with latitudinally dependent $\alpha$-effect and differential rotation. This non-uniform background is characterised by local magnetic Reynolds numbers with functional forms $-\varepsilon^{-1}R_\alpha f(\theta)$ and $\varepsilon^{-2}R_\Omega g(\theta)$ respectively; the product of the



*P. Chossat et al. (eds.), Dynamo and Dynamics, a Mathematical Challenge,* 297–304.

dimensionless parameters $-\varepsilon^{-1}R_\alpha$ and $\varepsilon^{-2}R_\Omega$ defines the dynamo number $-\varepsilon^{-3}D := -\varepsilon^{-3}R_\alpha R_\Omega$. The quasi-kinematic extension, in which the $\alpha$-effect is quenched, was investigated both analytically and numerically by Meunier *et al.* [5]. Strictly the $\alpha\Omega$-dynamo case corresponds to $R_\alpha \to 0$. When $R_\alpha$ is finite, the dynamo is of $\alpha^2\Omega$-type, for which Griffiths *et al.* [6] undertook the corresponding analytic development (but see also Meunier *et al.* [7]). In this paper we summarise some of their key results and outline further recent developments, which include new supporting numerical evidence.

Our $\alpha^2\Omega$-dynamo waves are governed by the model equations

$$\left.\begin{aligned} \frac{\partial A}{\partial t} &= \alpha B + \varepsilon^2\frac{\partial^2 A}{\partial\theta^2} - A, \\ \frac{\partial B}{\partial t} &= \varepsilon D g(\theta) - R_\alpha^2\left[\varepsilon^2\frac{\partial}{\partial\theta}\left(\alpha\frac{\partial A}{\partial\theta}\right) - \alpha A\right] + \varepsilon^2\frac{\partial^2 B}{\partial\theta^2} - B, \end{aligned}\right\} \qquad (1)$$

where, in suitable dimensionless units, $t$ is time, $B$ and $-\varepsilon R_\alpha \partial A/\partial\theta$ are the azimuthal and radial magnetic fields respectively ($|\mathbf{B}|$ is the total field), and the scaled dynamo number

$$D := R_\alpha R_\Omega \qquad (2)$$

has the opposite sign to that often employed. We restrict our discussion to the particular case $f(\theta) := \sin\theta$ and $g(\theta) := 2\cos\theta$ and so introduce the scaled $\alpha$-effect Reynolds and Dynamo numbers

$$\mathcal{R}(\theta) = R_\alpha \sin\theta \qquad \text{and} \qquad \mathcal{D}(\theta) = D\sin(2\theta), \qquad (3)$$

respectively.

The reduced problem is characterised by the two parameters $R_\alpha$ and $D$. To understand the nature of the solutions that our numerical integrations of (1) reveal, it is helpful to note the simple steady state solutions with long azimuthal length scales. For them, all $\theta$ derivatives in (1) are negligible except in thin boundary layers which have no significant consequences. So restricting attention to the Northern hemisphere, the zero amplitude state bifurcates to $R_\alpha A = B = B_S(\theta)$, where

$$B_S(\theta) := \begin{cases} \pm\sqrt{(\mathcal{R}(\theta)-1)/2} & \text{on a polar cap} \quad \theta_S < \theta < \pi/2, \\ 0 & \text{elsewhere on} \quad 0 \le \theta < \theta_S, \end{cases} \qquad (4)$$

and $\theta_S = \sin^{-1}(1/R_\alpha)$. There is a polar boundary layer of latitudinal length scale $\mathrm{O}(\varepsilon)$ across which adjustments of $A$ and $B$ are made to meet the polar boundary conditions. There is a boundary layer also at $\theta_S$ leading to exponentially small values of the magnetic field at the equator $\theta = 0$. Consequently the nature of our asymptotics means that we cannot comment on symmetry as we are unable to distinguish between dipole and quadrupole parity. This is a general feature of the thin shell limit which is not restricted to these particular $\alpha^2$-dynamo modes. Only for thick shells can these important parity issues be addressed (see e.g. Jennings [8] and Tobias [9]).

Our main concern is with the bifurcation to short length scale travelling waves either as a primary bifurcation from the zero amplitude state or as a secondary bifurcation from the steady finite amplitude state (4). For fixed $R_\alpha$ that bifurcation occurs at some critical value $D_c(R_\alpha)$ of the dynamo number $D$ with some critical frequency $\omega_c$. The latitudinal length scale of these modes is $O(\varepsilon)$. We also determine the nature of the fully developed finite amplitude travelling wave states. These are localised at mid-latitudes and can be identified simply by demanding that the magnetic field associated with them decays to zero as both the pole and equator are approached. They also have the generic feature that these (Parker [1]) waves evaporate smoothly at some latitude $\theta_P$ at the equatorial end but are terminated abruptly across a front of width $O(\varepsilon)$ (the wave length scale) at some latitude $\theta_F$ at the polar end. The presence of fronts in this class of dynamo problems was first identified by Worledge *et al.* [10] in the case of a uniform background state. Much of the underlying analytical theory for our non-uniform background is reviewed by Soward [11].

## 2. Linear Theory

Since the waves of interest have short $O(\varepsilon)$ length scale, we consider a WKBJ representation of small perturbations locally proportional to

$$\exp(\mathrm{i}\xi)\,, \qquad \text{where} \qquad \xi := \omega t + \varepsilon^{-1}\textstyle\int k \, \mathrm{d}\theta\,. \tag{5}$$

In the case of perturbations to the non-magnetic basic state, the complex frequency $\omega$ is related to the complex wave number $k(\theta)$ by the dispersion relation

$$(\mathrm{i}\omega + 1 + k^2)^2 - \mathcal{R}^2(1 + k^2) - \mathrm{i}\mathcal{D}k = 0\,. \tag{6}$$

For perturbations to the steady $\alpha^2$-dynamo (4), we write

$$[R_\alpha A, B] = B_S(\theta)[1,1] + [R_\alpha a(\theta), b(\theta)]\exp(\mathrm{i}\xi) \tag{7}$$

and the most important consequence of $B_S \neq 0$ is that the $\alpha$-effect is quenched. As a result the coefficient of the exponential of the linearised perturbation to

$$R_\alpha \alpha \begin{bmatrix} R_\alpha A \\ B \end{bmatrix} \quad \text{is} \quad \frac{1}{\mathcal{R}} \begin{bmatrix} 1 & -(\mathcal{R}-1) \\ -(\mathcal{R}-1) & 1 \end{bmatrix} \begin{bmatrix} R_\alpha a \\ b \end{bmatrix}. \tag{8}$$

Thus linearising (1) leads, for $\mathcal{R} \geq 1$, to the local dispersion relation

$$[\mathrm{i}\omega + 1 + k^2 + (1 - \mathcal{R}^{-1})]^2 - \mathcal{R}^{-1}[1 + k^2 - (1 - \mathcal{R}^{-1})] - \mathrm{i}\mathcal{R}^{-2}\mathcal{D}k = 0 \quad (9)$$

which replaces (6) at locations $\theta$, where $B_S(\theta) \neq 0$. Of course, both relations are identical at $\theta_S$, where $\mathcal{R}(\theta_S) = 1$.

In order that the WKBJ solution is a uniformly valid approximation everywhere between the equator and the pole it is necessary that both the complex group velocity and complex phase mixing vanish,

$$\omega_{,k} = 0\,, \qquad \omega_{,\theta} = 0\,, \qquad \mathrm{Im}\,\{\omega\} = 0\,, \qquad (10)$$

at some $\theta_c$ and $k_c$ possibly complex (see Huerre and Monkewitz [12]). These conditions have been applied to the dispersion relation (6). When $R_\alpha = 0$, $\theta_c$ is located at $\pi/4$, the local dynamo number maximum, but $k_c$ is complex (see Kuzanyan and Sokoloff [4]). This means that the maximum of the generated magnetic field is localised elsewhere at a lower latitude $\theta_M$ (say). As $R_\alpha$ is increased, $\theta_c(R_\alpha)$ moves off the real axis and becomes fully complex like $k_c(R_\alpha)$ (see Griffiths *et al.* [6]). Otherwise, the characteristics of the solution are similar to the localised $R_\alpha = 0$ $\alpha\Omega$-dynamo wave.

## 3. Nonlinear Analytic Theory

Weakly nonlinear theory is difficult to implement in the small $\varepsilon$-limit and leads to surprisingly complicated results (see Griffiths *et al.* [6], Bassom *et al.* [13]). Its range of applicability is limited to a very small region of parameter space and the results that can be obtained are not that helpful (see Le Dizès *et al.* [14]).

A more fruitful approach is to study fully developed finite amplitude states. Unlike the marginal small amplitude solutions which are localised in the neighbourhood of some latitude $\theta_M$, the finite amplitude waves exist over the extended range $\theta_P < \theta < \theta_F$. These travelling wave solutions depend locally on the single variable $\xi$ (see (5)) and have amplitude and wavelength $2\pi/k$ (real) which vary with latitude $\theta$. They merge smoothly with the linear WKBJ solutions at $\theta_P$. On the other hand, the simple wave structure just described is lost at $\theta_F$, where the frontal structure depends on both $(\theta - \theta_F)/\varepsilon$ and $t$ explicitly.

In fact the role of the front may be thought of in terms of a wave transition problem. The arrival of the finite amplitude wave at the front leads to linear transmitted and reflected waves. The linear transmitted waves provide the key to the existence of the front itself. According to both (6) and (9), there are four roots for $k$ leading to four distinct WKBJ type solutions. Only those which decay towards the pole are acceptable and the others that grow have to be rejected. This is only achieved when two of the roots coincide, which is equivalent to the vanishing of the complex group velocity

$$\omega_{,k} = 0 . \tag{11}$$

The importance of this frontal condition was identified by Dee and Langer [15] and is now known as the Dee-Langer condition. The double root feature means that the two corresponding WKBJ solutions are disentangled in a thicker transition layer ahead of the front (see Meunier *et al.* [5], Bassom *et al.* [13], Bassom and Soward [16] and also Pier *et al.* [17]).

To determine the finite amplitude solution, the Dee-Langer condition is applied to the linear dispersion relation. For given $R_\alpha$ and $D$, this leads to explicit values of real $\omega$ and $\theta_F$ but complex $k(\theta_F)$, when they exist. In one sense the value of the frequency $\omega$ is the key to the solution as well as being the focal point of physical applications (see e.g. Rüdiger and Arlt [18]). Once it is determined, the remaining characteristics follow.

For given $R_\alpha$, the minimum value $D_{\text{min}}$ at which frontal solutions exist is of interest. For $R_\alpha = 0$, it coincides with the critical value; $D_{\text{min}} = D_c$ but for $R_\alpha > 0$, the frontal solutions are subcritical; $D_{\text{min}} < D_c$. From a mathematical point of view this subcriticality is traced to the complex value of $\theta_c$. The fact that $\theta_c$ moves further from the real axis signals the stabilising influence of phase mixing on the linear solution; the nonlinear frontal solution relaxes the phase mixing constraint because the Dee-Langer condition is not concerned with $\omega_{,\theta}$.

## 4. Nonlinear Numerical Results

We integrated the governing equations (1) as an initial problem from an arbitrary seed field. The numerical solutions were computed until the transients had died away leaving a periodic solution. The numerical method used a pseudo–spectral tau–collocation method with the Chebyshev polynomials. The time stepping is performed with a Crank–Nicholson method for the diffusion terms, and the Adams–Bashford method for the remaining terms. All the selected results reported here are for the case

$$\varepsilon = \pi/600 , \tag{12}$$

which was employed in the original $R_\alpha = 0$ calculations of Meunier *et al.* [5]. This value is certainly far smaller than is appropriate for solar applications for which $\varepsilon$ is about 1/3. We adopt the small value $\pi/600$ to emphasise the asymmetry of the solutions and the frontal structure.

For the case $R_\alpha = 0$, the asymptotics predicts the values $D_c \approx 8.71$, $\omega_c \approx 1.73$, and $D_{\text{min}} = D_c$, $\omega_{\text{min}} = \omega_c$. In figure 1 we display the numerically computed amplitude of $B$ at some fixed time for $D = 9.0$. The $\varepsilon \to 0$ analytic theory, predicts that $\omega = \omega_c$ and $2\theta_F/\pi = 0.58$. Our numerical

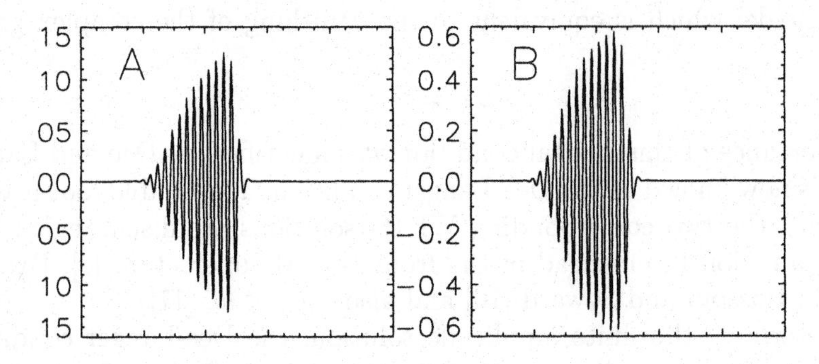


*Figure 1.* The case $R_\alpha = 0$, $D = 9.0$.

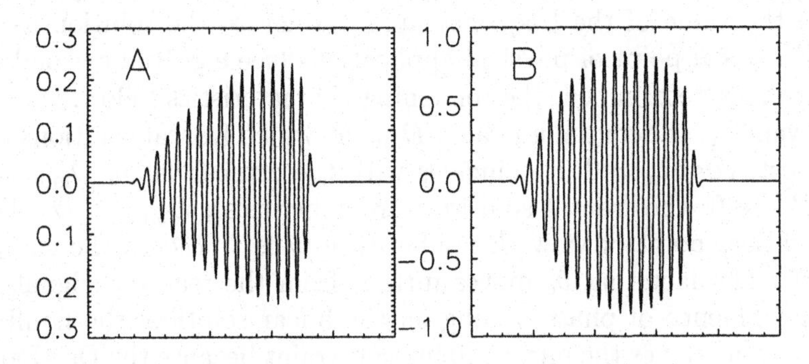


*Figure 2.* The case $R_\alpha = 1.0$, $D = 9.0$.

results, which are comparable to those reported by Meunier *et al.* [5], determine $\omega = 1.747$, while the analytic value of $\theta_F$ is consistent with figure 1. The front width is seen to be about two wavelengths as illustrated also by Tobias *et al.* [19]. This is a robust feature which is not particularly sensitive to the value of $D$. On the other hand, the latitudinal range broadens and the amplitude of the magnetic field increases with $D$.

For the case $R_\alpha = 1$, the asymptotics predicts the values $D_c \approx 7.26$, $\omega_c \approx 1.44$, and $D_{\min} \approx 6.98$, $\omega_{\min} \approx 1.35$. In figure 2, we again plot the numerically integrated $B$, and it illustrates essentially the same features as figure 1. The realised frequency $\omega = 1.25$ is close to the analytic prediction $\omega = 1.148$, while the front location is visably close to the analytical prediction $2\theta_F/\pi = 0.777$.

The case $R_\alpha = 1.4$ illustrated in figure 3 is particularly interesting because the basic state has bifurcated to the steady $\alpha^2$-state. The defect visible at the pole is due to the application of zero boundary conditions there. It may be removed by application of more physically realistic boundary conditions but does not affect the results elsewhere. The front is clearly visible riding on the steady state solution. Applying the Dee-Langer con-

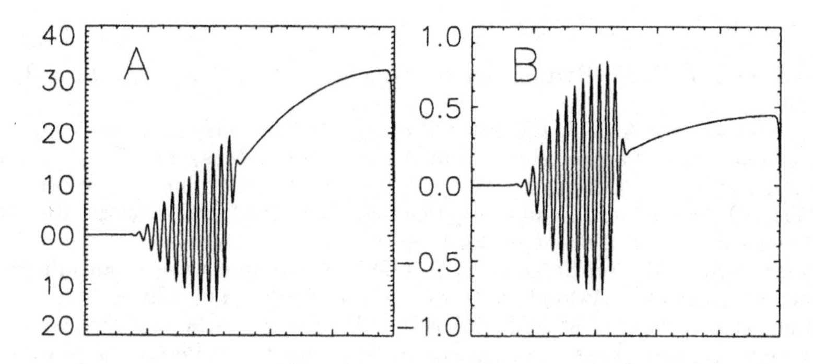


*Figure 3.* The case $R_\alpha = 1.4$, $D = 9.0$.

dition to formula (9) appropriate to that finite amplitude state determines $\omega = 1.381$, $2\theta_F/\pi = 0.5778$. The numerical results yield $\omega = 1.446$, which together the visibly estimated value of $\theta_F$ provide a healthy agreement between the theory and numerics.

Preliminary numerical calculations at higher values of $D$ for various values of $R_\alpha$ suggest that the solution bifurcates and introduces second frequencies; further bifurcations lead on to chaos. An interesting feature is that for given $R_\alpha$ and $D$, the different frequencies might dominate at different latitudes. Nevertheless, the dominant frequency at the front was always found to be that predicted by our use of the Dee-Langer criterion.

## Acknowledgements

Y.P. gratefully acknowledges the support of a Leverhulme Fellowship, grant no. F/144/AH, at the University of Exeter between August, 1997 and August, 2000. The numerical calculations were performed using the computing facilities of the laboratory Cassini, Observatoire de Nice (France), provided by the program "(SIVAM)" and the computing facilities of IDRIS (Palaiseau, France).

## References

1. Parker, E.N. (1955) Hydromagnetic dynamo models, *Astrophys. J.*, **122**, pp. 293–314
2. Rüdiger, G. and Arlt, R. (2000) Physics of the solar cycle, in: *Advances in Nonlinear Dynamos*, (Ed. Manuel Núñez and Antonio Ferriz Mas) The Fluid Mechanics of Astrophysics and Geophysics. Vol. 00, pp. 000–000, Gordon and Breach
3. Moss, D., Tuominen, I. and Brandenburg, A. (1990) Buoyancy-limited thin shell dynamos, *Astron. Astrophys.*, **240**, pp. 142–149
4. Kuzanyan, K.M. and Sokoloff, D.D. (1995) A dynamo wave in an inhomogeneous medium, *Geophys. Astrophys. Fluid Dynam.*, **81**, pp. 113–129
5. Meunier, N., Proctor, M.R.E., Sokoloff, D.D., Soward, A.M. and Tobias, S.M. (1997) Asymptotic properties of a nonlinear $\alpha\omega$-dynamo wave: Period, amplitude and latitude dependence, *Geophys. Astrophys. Fluid Dynam.*, **86**, pp. 249–285
6. Griffiths, G.L., Bassom, A.P., Soward, A.M. and Kuzanyan, K.M. (2000) Nonlinear $\alpha^2\Omega$-dynamo waves in stellar shells: I. General structure, *Geophys. Astrophys. Fluid Dynam.*, **93**, pp. 000–000
7. Meunier, N., Nesme-Ribes, E. and Sokoloff, D.D. (1996) Dynamo wave in a $\alpha^2\omega$ dynamo, *Astron. Rep.*, **40**, pp. 415–423
8. Jennings, R.L. (1991) Symmetry breaking in a nonlinear $\alpha\omega-$dynamo, *Geophys. Astrophys. Fluid Dynam.*, **57**, pp. 147–189
9. Tobias, S.M. (1997) Properties of nonlinear dynamo waves, *Geophys. Astrophys. Fluid Dynam.*, **86**, pp. 287–343
10. Worledge, D., Knobloch, E., Tobias, S. and Proctor, M.R.E. (1997) Dynamo waves in semi-infinite and finite domains, *Proc. R. Soc. Lond.* A, **453**, pp. 119–143
11. Soward, A.M. (2000) Thin aspect ratio $\alpha\omega$-dynamos in galactic discs and stellar shells, in: *Advances in Nonlinear Dynamos*, (Ed. Manuel Núñez and Antonio Ferriz Mas) The Fluid Mechanics of Astrophysics and Geophysics. Vol. 00, pp. 000–000, Gordon and Breach
12. Huerre, P. and Monkewitz, P.A. (1990) Local and global instabilities in spatially developing flows, *Annu. Rev. Fluid Mech.*, **22**, pp. 473–537
13. Bassom, A.P., Kuzanyan, K.M. and Soward, A.M. (1999) A nonlinear dynamo wave riding on a spatially varying background, *Proc. R. Soc. Lond.* A, **455**, pp. 1443–1481
14. Le Dizès, S., Huerre, P., Chomaz, J.-M. and Monkewitz, P.A. (1993) Nonlinear stability analysis of slowly-diverging flows: Limitations of the weakly nonlinear approach, in: *Bluff-Body Wakes, Dynamics and Instabilities*, (Ed. H. Eclelmann, J.M.R. Graham, P. Huerre and P.A. Monkewitz) Proceedings of IUTAM Symposium, pp. 147–152, Springer, Berlin
15. Dee, G. and Langer, J.S. (1983) Propagating pattern selection, *Phys. Rev. Lett.*, **50**, pp. 383–386
16. Bassom, A.P. and Soward, A.M. (2000) A nonlinear dynamo wave riding on a spatially varying background: II. Transition from weakly nonlinear to fully nonlinear states, *Proc. R. Soc. Lond.* A, submitted
17. Pier, B., Huerre, P., Chomaz, J-M. and Couairon, A. (1998) Steep nonlinear global modes in spatially developing media, *Phys. Fluids*, **10**, pp. 2433–2435
18. Rüdiger, G. and Arlt, R. (1996) Cycle times and magnetic amplitudes in nonlinear 1D $\alpha^2\Omega$-dynamos, *Astron. Astrophys.*, **316**, pp. L17–L20
19. Tobias, S.M., Proctor, M.R.E. and Knobloch, E. (1997) The rôle of absolute instability in the solar dynamo, *Astron. Astrophys.*, **318**, pp. L55–L58

# TWO-DIMENSIONAL DISK DYNAMOS WITH VERTICAL OUTFLOWS INTO A HALO

B. VON REKOWSKI, W. DOBLER AND A. SHUKUROV
*Department of Mathematics, University of Newcastle*
*Merz Court, Newcastle upon Tyne NE1 7RU, UK*

AND

A. BRANDENBURG
*Nordita, Blegdamsvej 17, DK-2100 Copenhagen Ø, Denmark*

**Abstract.** We study the effects of vertical outflows on mean-field dynamos in disks. These outflows could be due to thermal winds or magnetic buoyancy. We analyse numerical solutions of the nonlinear mean-field dynamo equations using a two-dimensional finite-difference model. Contrary to expectations, a modest vertical velocity can enhance dynamo action. This can lead to super-exponential growth of the magnetic field and to higher magnetic energies at saturation in the nonlinear regime.

## 1. Introduction

Large scale magnetic fields are often considered to be an important factor in accretion disks (including generation of turbulence), crucial in launching winds or jets. The origin of such large scale magnetic fields in accretion disks is still unclear: they may be advected from the surrounding medium or be generated by a dynamo inside the disk. Advection appears unlikely as turbulence leads to enhanced viscosity and magnetic diffusivity, so that the two are of the same order, i.e. the magnetic Prandtl number is of order unity [1]. In this case the turbulent magnetic diffusivity can compensate the dragging of the field by viscously induced accretion flow [2, 3, 4]. Turbulent dynamo action is a plausible mechanism for producing large scale magnetic fields in accretion disks [5, 6]. Dynamo magnetic fields can launch winds from accretion disks [7, 8, 9]. However, the wind can also affect the dynamo. In particular, a dynamo enhancement by winds was suggested earlier for



*P. Chossat et al. (eds.), Dynamo and Dynamics, a Mathematical Challenge,* 305–312.

galactic dynamos [10]. In the same context the effects of shear in the vertical velocity on the dynamo was considered [11].

We assume here that the magnetic fields are generated by a dynamo acting in a relatively thin accretion disk. We show how vertical flow can enhance the dynamo, allowing for a larger growth rate and leading to super-exponential growth of the magnetic field and to enhanced saturation levels of magnetic energy.

The vertical velocities can have several origins. For example, they can be due to a thermally driven wind emanating from the disk or magnetic buoyancy in the disk. Note that here we invoke magnetic buoyancy as a driver of vertical outflows; it can itself contribute to the dynamo effect [12], but such effects are not taken into account here.

## 2. The model

The equation which we solve is the mean-field induction equation which we evolve in terms of the vector potential $\mathbf{A}$, where $\mathbf{B} = \nabla \times \mathbf{A}$,

$$\frac{\partial \mathbf{A}}{\partial t} = \mathbf{V} \times (\nabla \times \mathbf{A}) + \alpha \nabla \times \mathbf{A} - \eta \mu_0 \mathbf{j}, \tag{1}$$

where $\mathbf{j} = \nabla \times \mathbf{B}/\mu_0$ is the current density, $\mu_0$ the magnetic permeability, $\eta$ the turbulent magnetic diffusivity, $\alpha$ the $\alpha$-effect and $\mathbf{V}$ the mean velocity. We neglect the radial component of the mean velocity. Assuming the $\alpha$- and $\eta$-tensors to be isotropic, we can consider scalar quantities. We do not make the thin-disk approximation but solve the general equations. We adopt cylindrical coordinates $(r, \varphi, z)$ and restrict ourselves to axisymmetric solutions. Equation (1) is solved using a sixth order finite-difference scheme in space and a third order Runge-Kutta time advance scheme.

Our computational domain contains a disk embedded in its surrounding halo. We take a disk aspect ratio of $h_{\text{disk}}/R_{\text{disk}} = 0.1$ and a halo with $h_{\text{halo}}/h_{\text{disk}} \approx 6$.

As an initial condition for the magnetic field we choose a purely poloidal field in the disk of either dipolar or quadrupolar symmetry.

On the boundaries of the computational domain we impose pseudo-vacuum conditions. However, since the boundaries are in the halo far away from the disk, the choice of boundary conditions is not crucial.

The $\alpha$-coefficient $\alpha(r, z)$ is antisymmetric about the disk midplane and vanishes outside the disk. We adopt

$$\alpha(r, z) = \begin{cases} \alpha_0(r) \sin\left(\pi \frac{z}{h}\right) \xi_\alpha(r) & \text{for } |z| \le h, \\ 0 & \text{for } |z| > h. \end{cases} \tag{2}$$

The $\xi_\alpha$-profile cuts off the $\alpha$-effect at the outer radius of the disk as well as at an inner radius, where the rotational shear is very strong.

As appropriate for accretion disks, we adopt a softened Keplerian angular velocity profile in $r$ in the disk as well as the halo,

$$\Omega(r) = \sqrt{\frac{GM}{r^3}} \left[1 + \left(\frac{r_0}{r}\right)^n\right]^{-\frac{n+1}{2n}}, \tag{3}$$

where $G$ is Newton's gravitational constant, $M$ is the mass of the central object, $r_0 = 0.05$ is the softening radius, and $n = 5$. At $r = 0$, $\Omega$ vanishes.

The turbulent magnetic diffusivity is given by

$$\eta(r, z) = \eta_{\text{halo}} + (\eta_{\text{disk}} - \eta_{\text{halo}})\xi(r, z). \tag{4}$$

The profile $\xi(r, z)$ defines the disk: $\xi$ is equal to unity inside the disk and vanishes in the halo but with a smooth transition between. We carried out computations for two cases: homogeneous conductivity, $\eta_{\text{halo}}/\eta_{\text{disk}} = 1$, and low conductivity in the halo, $\eta_{\text{halo}}/\eta_{\text{disk}} = 20$.

The profile for $V_z$ is linear in $z$, $V_z(z) = V_{z0}z/h$, where $V_{z0}$ is a characteristic vertical velocity.

Our dynamo problem is controlled by the three magnetic Reynolds numbers related to the $\alpha$-effect, the differential rotation and the vertical velocity,

$$R_\alpha = h\alpha_0/\eta_{\text{disk}}, \quad R_\omega = h^2 S/\eta_{\text{disk}}, \quad R_v = hV_{z0}/\eta_{\text{disk}}, \tag{5}$$

where $S(r) \equiv r d\Omega/dr \approx -3/2\ \Omega(r)$ is the rotational shear. The dynamo number is defined as $\mathcal{D} \equiv R_\alpha R_\omega$. Since the dynamo number is approximately constant, $|\mathcal{D}| \simeq \alpha_{\text{SS}}^{-2} \simeq 10^2 - 10^4$ in thin accretion disks [5], we assume $\mathcal{D}$ to be constant by setting $\alpha_0(r) = \mathcal{D}\eta_{\text{disk}}^2/S(r)h^3$.

Because of the strong differential rotation we assume that the magnetic field in the disk is generated by a standard $\alpha\omega$-dynamo. The two control parameters are then the dynamo number $\mathcal{D}$ and the vertical magnetic Reynolds number $R_v$. The value of $R_v$ obviously depends on the origin of the vertical outflow. With magnetic buoyancy, a rough estimate can be done assuming that the vertical velocities are comparable to the Alfvén speed. Estimating the Alfvén speed from magnetic equipartition, one gets a vertical magnetic Reynolds number $R_v$ of order unity.

## 3. Linear results for $R_v = 0$

For $\eta_{\text{halo}} = \eta_{\text{disk}}$, the growth rates of dipolar and quadrupolar modes should be exchanged when $\mathcal{D}$ reverses sign, i.e. the graph of $Re(\gamma)$ as a function of $\mathcal{D}$ should be symmetric with respect to the vertical axis $\mathcal{D} = 0$ with quadrupolar and dipolar modes exchanging their rôles [13]. Our numerical simulation shows this symmetry with a good precision (Fig. 1a). This is because the ratio between $h_{\text{halo}}$ and $h_{\text{disk}}$ is large enough, about 6.

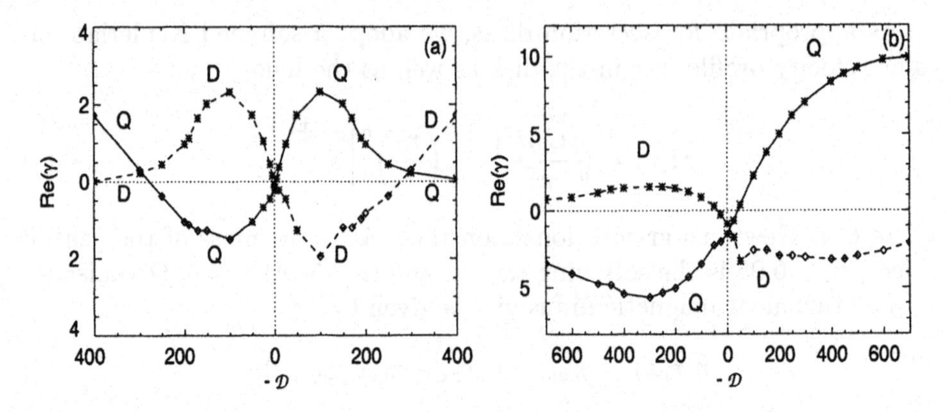


*Figure 1.* Real part of the growth rate of the magnetic field as a function of $-\mathcal{D}$ with $R_v = 0$. Asterisks denote non-oscillatory, diamonds oscillatory solutions. Solid lines denote quadrupolar (Q), dashed lines dipolar (D) modes. (a) is with homogeneous conductivity, (b) for low-conducting halo. Note the symmetry of (a) with respect to $\mathcal{D} = 0$.

For $\mathcal{D} < 0$ ($R_\alpha > 0$) the first leading growing mode is quadrupolar whereas for $\mathcal{D} > 0$ ($R_\alpha < 0$) it is dipolar; both are non-oscillatory. We thus study the effect of vertical velocities for $\mathcal{D} < 0$ in quadrupolar and for $\mathcal{D} > 0$ in dipolar symmetry at dynamo numbers up to $\pm 300$.

For $\eta =$ const., the critical dynamo number is $|\mathcal{D}_{\rm crit}| \simeq 5$. The dominant mode changes its symmetry and becomes oscillatory at $|\mathcal{D}| \simeq 300$.

The qualitative behaviour in the case of low conductivity in the halo is quite similar to that of homogeneous conductivity. Also the quadrupolar non-oscillatory modes first become dominant for $R_\alpha > 0$ and the dipolar non-oscillatory modes for $R_\alpha < 0$ (compare Figs. 1a and b).

But there are quantitative differences. The value of $|\mathcal{D}|$, where the symmetry of the leading mode changes, is larger than 700. Also, the diagram in Fig. 1b is not symmetric. The critical dynamo number is $\mathcal{D} \simeq -50$ (50) for quadrupolar (dipolar) modes.

## 4. Linear behaviour with vertical velocities

As discussed in section 3, we only consider quadrupolar symmetry for $\mathcal{D} < 0$. As shown in Fig. 2, the growth rate of the magnetic field $\mathrm{Re}(\gamma)$ is a non-monotonous function of $R_v$, for $\eta_{\rm halo}/\eta_{\rm disk}$ equal to both 1 and 20. A maximum in $\mathrm{Re}(\gamma)$ occurs, however, only for $|\mathcal{D}|$ large enough, and the smaller the magnetic diffusivity of the halo, the smaller is the required $|\mathcal{D}|$. The larger $|\mathcal{D}|$, the more pronounced is the maximum. The maximum growth rate occurs for $R_v$ of the order of 1 to 10.

In Fig. 3 we show the effect of the dynamo number and the vertical velocity on the magnetic field configuration for $\eta_{\rm halo} = \eta_{\rm disk}$. Increasing $|\mathcal{D}|$

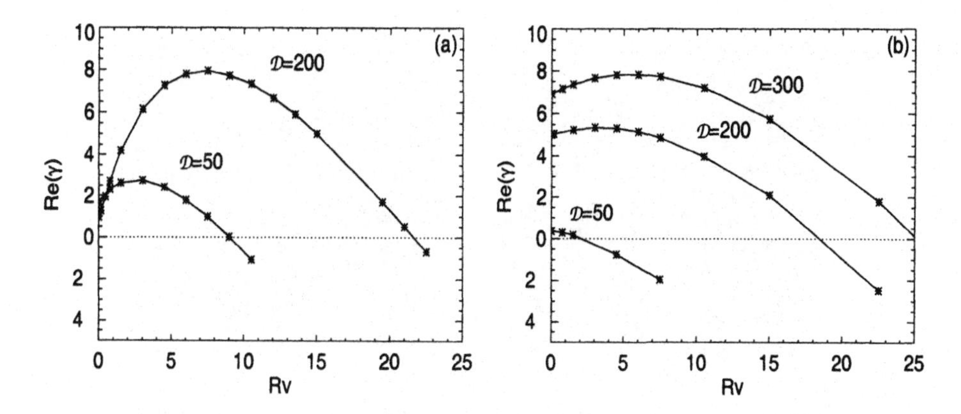


*Figure 2.* Real part of the growth rate of the magnetic field as a function of $R_v$ for non-oscillatory quadrupolar modes. (a) $\eta_{\rm halo}/\eta_{\rm disk} = 1$, (b) $\eta_{\rm halo}/\eta_{\rm disk} = 20$.

from 50 to 200 in the absence of any vertical velocity results in the magnetic field becoming concentrated at larger radii in the disk (Fig. 3a and b).

As can be seen in Fig. 3b, c and d, a vertical velocity with $R_v < 7.5$ leads to a wider vertical distribution of magnetic field which reduces magnetic diffusion and enhances the dynamo action. At still larger $R_v$, the wind aligns the poloidal field lines with the lines of constant rotation, and the $\Omega$-effect is slowly switched off leading to a decrease in the growth rate, which eventually becomes negative. The mode structures at a maximum of Re($\gamma$), reached at $R_v = 7.5$ for $\mathcal{D} = -200$) and $R_v = 3$ for $\mathcal{D} = -50$, are very similar to each other.

The growth rate for dipolar non-oscillatory modes ($\mathcal{D} > 0$) has no maximum as $R_v$ increases; at $R_v$ of order 1 or even less the dynamo is switched off. The magnetic field is advected outwards and aligned with the $\Omega$-contour lines in vertical direction very quickly.

## 5. Nonlinear behaviour with vertical velocities

We consider nonlinear, saturated solutions in a model with homogeneous conductivity, $\eta_{\rm halo} = \eta_{\rm disk}$, negative dynamo number and quadrupolar symmetry (the dominant symmetry for moderate $|\mathcal{D}|$ with $R_v = 0$). All saturated magnetic fields are non-oscillatory, as the corresponding linear modes.

### 5.1. MAGNETIC BUOYANCY

We parameterize the effect of magnetic buoyancy by assuming the vertical velocity to be proportional to the maximum magnetic field strength. Thus,

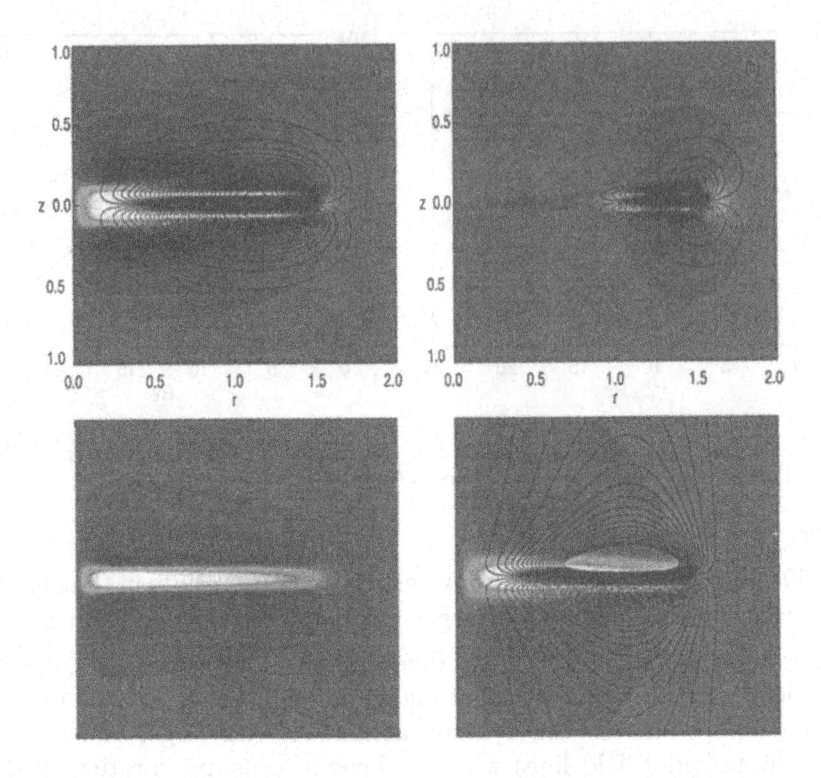


*Figure 3.* The effect of dynamo number and vertical velocity on the magnetic field configuration, for the linear model with $\eta_{\text{halo}} = \eta_{\text{disk}}$. Shown are the poloidal field lines (solid is clockwise, dotted is counter-clockwise) and the toroidal field (grey scales: bright is in positive, dark in negative azimuthal direction). (a): $\mathcal{D} = -50$, $R_v = 0$; (b): $\mathcal{D} = -200$, $R_v = 0$; (c): $\mathcal{D} = -200$, $R_v = 0.75$; (d): $\mathcal{D} = -200$, $R_v = 7.5$ (where $\text{Re}(\gamma)$ is maximum). Comparing (a) and (b) shows the effect of increasing $\mathcal{D}$ while $R_v = 0$. Panels (b) to (d) follow the upper curve in Fig. 2a. All modes are quadrupolar and non-oscillatory.

the vertical magnetic Reynolds number becomes time-dependent,

$$R_v = R_{v0} \max_{\mathbf{x}}(B_r, B_\varphi)/|B_0|, \quad R_{v0} \equiv V_{z0}h/\eta_{\text{disk}}, \tag{6}$$

where $B_0$ is a characteristic field strength.

As time increases, $|\mathbf{B}|$ grows and therefore $R_v$ is increasing and the growth rate follows, e.g., the upper curve in Fig. 2a for $\mathcal{D} = -200$. Thus, as long as $R_v$ is less than the position of the $\gamma$-maximum, $\gamma$ increases with time, which results in a super-exponential growth. After the maximum, at $R_v = 7.5$, $\gamma$ decreases and eventually the magnetic field approaches its saturation level. This is reached when $R_v = R_{v*}$, where $R_{v*}$ is the zero of $\gamma(R_v)$. According to Eq. (6), the saturation value of magnetic energy will thus be $E_{\text{mag}} \propto 1/R_{v0}^2$.

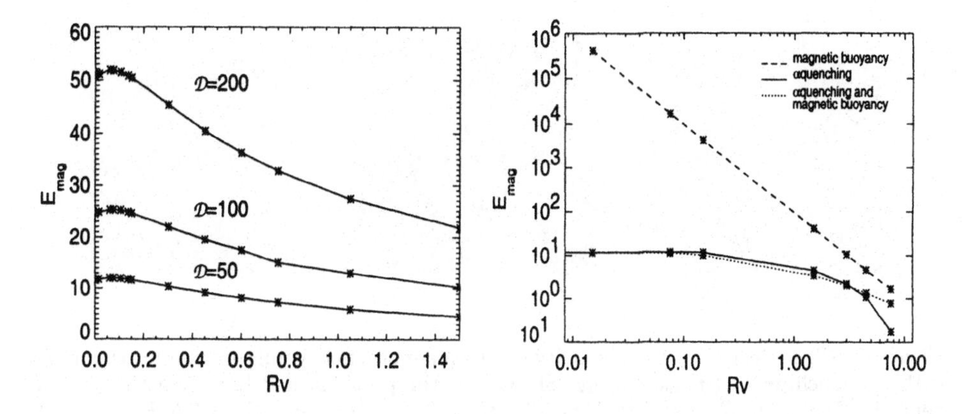


*Figure 4.* Magnetic energy as a function of $R_v$ for the model with $\alpha$-quenching (left) and for the models with $\alpha$-quenching and/or magnetic buoyancy (right) for $\eta_{\text{halo}} = \eta_{\text{disk}}$. All modes are quadrupolar and non-oscillatory. The right panel shows the case $\mathcal{D} = -50$.

### 5.2. $\alpha$-QUENCHING

We consider the back-reaction of the magnetic field on the $\alpha$-effect, introducing the nonlinearity $1/(1 + B^2/B_{\text{eq}}^2)$ as a factor in the $\alpha$-effect; $B_{\text{eq}}$ is the equipartition field.

The magnetic energy has a maximum at a certain value of $R_v$ (Fig. 4, left), but this maximum occurs at $R_v \approx 0.1$, a value smaller than where a maximum of $\text{Re}(\gamma)$ occurs in the linear regime, and it is also less pronounced. The value of $R_v$ where the maximum occurs, is independent of $\mathcal{D}$. The magnetic energy scales roughly with the dynamo number, $E_{\text{mag}} \propto \mathcal{D}$.

The effect of vertical velocity on the magnetic field for the model with $\alpha$-quenching is shown in Fig. 5; the structure changes only weakly with the dynamo number. With increasing $R_v$ the poloidal field lines become vertical, i.e. aligned with the lines of constant rotation. Hence, shear has no effect, and the $\Omega$-effect is switched off, as in the linear model.

### 5.3. $\alpha$-QUENCHING TOGETHER WITH MAGNETIC BUOYANCY

When $\alpha$-quenching and magnetic buoyancy are combined, the vertical magnetic Reynolds number depends on position and time and takes the form $R_v = R_{v0}\sqrt{B_r^2 + B_\varphi^2}/|B_{\text{eq}}|$. Without $\alpha$-quenching, since $R_v$ is now nonlocal, the temporal behaviour of the growth rates at short times does not have to be super-exponential. Surprisingly, again we find $E_{\text{mag}} \propto 1/R_{v0}^2$ (see Fig. 4, right, dashed line).

In the second panel of Fig. 4 magnetic energy is plotted against $R_v$ for $\mathcal{D} = -50$ for the three nonlinear models. $\alpha$-quenching appears to be the dominant nonlinearity in the model considered.

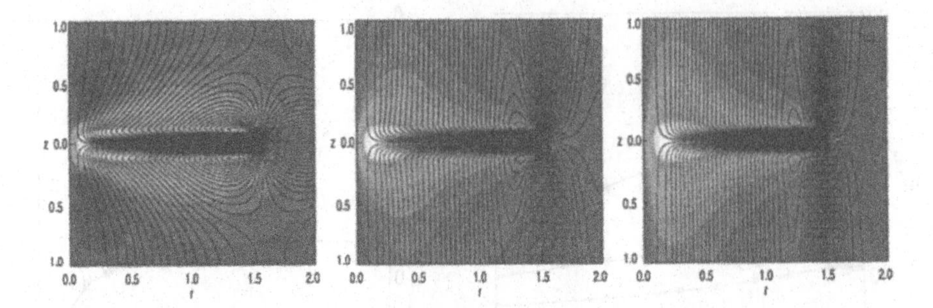

*Figure 5.* The effect of vertical velocity on the magnetic field configuration for the model with $\alpha$-quenching and $\eta_{\text{halo}} = \eta_{\text{disk}}$. Shown are the poloidal field lines and the toroidal field. (a): $\mathcal{D} = -50$, $R_v = 0$; (b): $\mathcal{D} = -50$, $R_v = 3$; (c): $\mathcal{D} = -50$, $R_v = 7.5$. Panels (a) to (c) follow the lower curve in Fig. 4, left. All modes are quadrupolar and non-oscillatory.

## Acknowledgements

We acknowledge financial support from PPARC (Grant PPA/G/S/1997/00284) and the Leverhulme Trust (Grant F/125/AL). The use of the PPARC supported GRAND parallel computer is acknowledged.

## References

1. Pouquet, A., Frisch, U. and Léorat, J. (1976) Strong MHD helical turbulence and the nonlinear dynamo effect, *J. Fluid. Mech.*, **77**, 321–354
2. van Ballegooijen, A.A. (1989) Magnetic fields in the accretion disks of cataclysmic variables, *in G. Belvedere (ed.), Accretion disks and magnetic fields in astrophysics, Kluwer Academic Publishers*, 99–106
3. Lubow, S.H., Papaloizou, J.C.B. and Pringle, J.E. (1994) Magnetic field dragging in accretion discs, *MNRAS*, **267**, 235–240
4. Heyvaerts, J., Priest, E.R. and Bardou, A. (1996) Magnetic field diffusion in self-consistently turbulent accretion disks, *ApJ*, **473**, 403–421
5. Pudritz, R.E. (1981) Dynamo action in turbulent accretion discs around black holes – II. The mean magnetic field, *MNRAS*, **195**, 897–914
6. Stepinski, T.F. and Levy, E.H. (1988) Generation of dynamo magnetic fields in protoplanetary and other astrophysical accretion disks, *ApJ*, **331**, 416–434
7. Campbell, C.G., Papaloizou, J.C.B. and Agapitou, V. (1998) Magnetic field bending in accretion discs with dynamos, *MNRAS*, **300**, 315–320
8. Campbell, C.G. (1999) Launching of accretion disc winds along dynamo-generated magnetic fields, *MNRAS*, **310**, 1175–1184
9. Brandenburg, A., Dobler, W., Shukurov, A. and von Rekowski, B. (2000) Pressure-driven outflow from a dynamo active disc, *A&A*, submitted
10. Brandenburg, A., Donner, K.J., Moss, D., Shukurov, A., Sokoloff, D.D. and Tuominen, I. (1993) Vertical magnetic fields above the discs of spiral galaxies, *A&A*, **271**, 36–50
11. Elstner, D., Golla, G., Rüdiger, G. and Wielebinski, R. (1995) Galactic halo magnetic fields due to a 'spiky' wind, *A&A*, **297**, 77–82
12. Moss, D., Shukurov, A. and Sokoloff, D. (1999) Galactic dynamos driven by magnetic buoyancy, *A&A*, **343**, 120–131
13. Proctor, M.R.E. (1977) On the eigenvalues of kinematic alpha-effect dynamos, *Astron. Nachr.*, **298**, 19–25

# STRUCTURALLY STABLE HETEROCLINIC CYCLES AND THE DYNAMO DYNAMICS

DIETER ARMBRUSTER
*Department of Mathematics Arizona State University*
*Tempe AZ 85287-1804, USA*

PASCAL CHOSSAT
*I.N.L.N., C.N.R.S. and Université de Nice-Sophia Antipolis*
*1361 route des lucioles, 06560 Valbonne, France*

AND

IULIANA OPREA
*Department of Mathematics Colorado State University*
*Ft. Collins, Co, USA, and*
*Faculty of Mathematics, University of Bucarest, Romania*

## 1. Introduction

### 1.1. STRUCTURALLY STABLE HETEROCLINIC CYCLES

Heteroclinic cycles, i.e. trajectories that connect a finite number of saddle points of a dynamical system until they eventually come back to the same saddle point, are structurally unstable. They occur as bifurcation phenomena. However it has been shown that additional structure in the dynamical systems may lead to structurally stable behavior of these cycles. This is typically the case for Hamiltonian systems where it has been well known for a long time. In addition, symmetry in the equations will also force heteroclinic cycles to be structurally stable. This fundamentally is accomplished by the fact that symmetric systems will have invariant subspaces. Hence a connection between two saddles will become structurally stable if the restriction of one of the saddles to an invariant subspace leads to a sink in that subspace and hence the restriction of the flow to the invariant subspace may generate a saddle - sink connection. For a bibliography on the subject see [8] in the present volume and for a comprehensive introduction, see [6]. The prototypical example has been studied by Busse et al [4] and analysed



*P. Chossat et al. (eds.), Dynamo and Dynamics, a Mathematical Challenge,* 313–322.

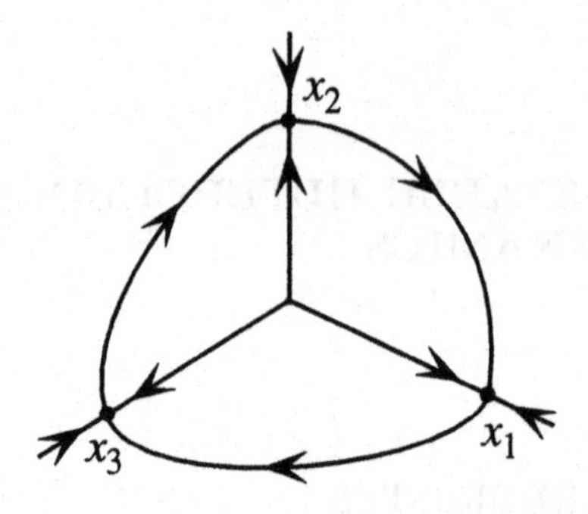


*Figure 1.* The prototypical structurally stable heteroclinic cycle

as a robust heteroclinic cycle by Guckenheimer and Holmes [7]. Figure 1 illustrates the case: We consider a 3-d system where all coordinate planes and all coordinate axes are invariant subspaces. This can for instance be obtained for a system that has the symmetry group generated by reflections through the planes of coordinates and by cyclic permutation of these axes of coordinates. Now, assume that there exists a saddle on a coordinate axis with a 2-d stable manifold and a 1-d unstable manifold. By the permutation symmetry we will have saddles on each of the axes and if there is a heteroclinic orbit connecting 2 saddles in one invariant subspace there will be a whole cycle of these orbits connecting a saddle back to itself (see figure 1). In some parameter regimes the cycle is attracting. A time series for any of the three variables will show the variable to level off at a particular value until it transits to another value in a very short time where it will stay again for a long time etc. Upon addition of noise a "stochastic limit cycle" is created, whereby the transition times between saddles is exponentially distributed and unlike an attracting heteroclinic cycle without noise, there exists a finite mean period [11]. This dynamical behavior of relatively long quiescent behavior randomly followed by a quick transition to another long quiescent behavior makes this an attractive model for the behavior of magnetic reversals. The following sections will discuss our program to flesh out this model with more and more concrete physical details.

## 1.2. CONVECTION WITH INTERMITTENT REVERSALS OF AXISYMMETRIC FLUID FLOW

Here we summarize some important facts about the occurence of robust heteroclinic cycles in spherical convection. A more detailed description is given by [8] in the present volume. We consider the problem of convection in a spherical shell when the ratio $\eta$ of the two shell radii is allowed to vary. There exist values $\eta_0$ at which two critical modes $\ell_0$ and $\ell_0+1$ coexist which lead to the onset of convection. This *mode interaction* has been studied in great detail in the case $\ell_0 = 1$ in [1]. In this case a center manifold reduction to a system of 8 differential equations in $x_j$, $j = -1, 0, 1$ (amplitudes in the $\ell = 1$ modes) and $y_k$, $k = -2, \ldots, 2$ (amplitudes in the $\ell = 2$ modes) can be performed. Since the system is equivariant with respect to $O(3)$, the symmetry group of the sphere, we can use symmetric bifurcation theory to analyze this system. The most important symmetry operations are

- $R_\varphi$ the rotation around the $z$-axis of angle $\varphi$;
- $\chi$ the rotation by $\pi$ around the $x$-axis;
- $\sigma$ the reflection through the origin;
- $\kappa$ the reflection in $R^3$ through the vertical plane $Oyz$;
- $\nu$ the reflection through the horizontal plane.

The following subspaces are fixed point subspaces of these operations and hence are flow invariant:

- $S = Fix(\kappa) = \{(x_0, x_{1r}, y_0, y_{1r}, y_{2r})\}$;
- $S' = Fix(\nu) = \{(x_{\pm 1}, y_0, y_{\pm 2})\}$ (symmetric from $S$);
- $D = Fix(R_\pi, \kappa) = \{(x_0, y_0, y_{2r})\}$;
- $P_0 = Fix(R_\varphi,\ \varphi \in S^1, \kappa) = \{(x_0, y_0)\}$;
- $P_2 = Fix(\kappa, \nu, \sigma) = \{(y_0, y_{2r})\}$;
- $L_0 = P_0 \cap P_2$.

Moreover, rotations by $\pi/2$ around the horizontal coordinate axes $Ox$, $Oy$ in physical space transform $P_2$ into itself, the axis $L_0$ being transformed into two axes in $P_2$ which we call $L_0'$ and $L_0''$, while the axis $Ox_0$ is transformed respectively into the axes $Ox_{1r}$ and $Ox_{1i}$. It follows that $P_0$ is transformed into flow-invariant planes $P_0'$ and $P_0''$ which intersect $P_2$ along $L_0'$ and $L_0''$ and include the coordinates $x_{1r}$ and $x_{1i}$ respectively.

In [1] we show that two types of equilibria $\alpha$ and $\beta$ bifurcate from zero within $L_0$ (take $\alpha$ be such that $y_0 < 0$ by convention). This implies that we have additional equilibria in $P_2$ called $\alpha', \beta'$ and $\alpha'', \beta''$, on the rotated axes $L_0'$ and $L_0''$ respectively. These equilibria are axisymmetric steady-states for the convective flow. By acting with $O(3)$ on each of them, one obtains a *group orbit* of equilibria which is a compact 2-manifold (homeomorphic to the 2-dimensional real projective space).

In addition there exists a saddle-sink connection from $\alpha$ to $\beta$ in $P_0$ and saddle-sink connections $\beta \to \alpha', \alpha''$ in $P_2$. In fact in $S$, the unstable manifolds of $\beta, \beta', \beta''$ lie in the stable manifolds of $\alpha, \alpha', \alpha''$. Note that, while $\alpha$, $\beta$ and $P_0$ are axisymmetric, the connections in $P_2$ are not.
These connections close up to form a structurally stable heteroclinic cycle, as sketched in Figure 2. By acting on this cycle with all transformations in $O(3)$, this results in a structurally stable heteroclinic cycle connecting the group orbit of $\alpha$ to the group orbit of $\beta$. A further study by Chossat

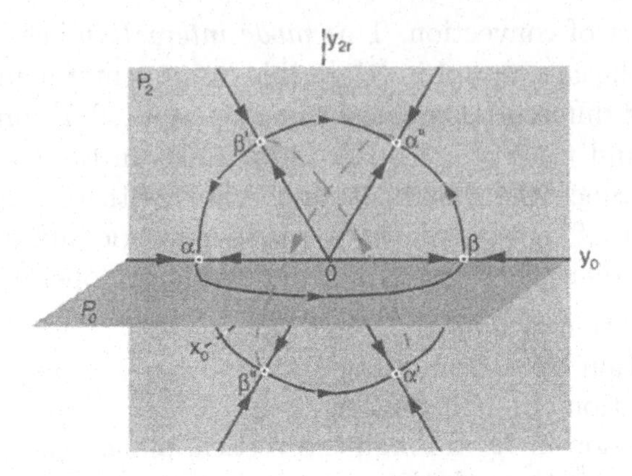


*Figure 2.* The heteroclinic cycles in the $\ell = 1, \ell = 2$ O(3) symmetric mode interaction. Phase portrait of $P_0$ and $P_2$. Dotted trajectories indicate connections from $\alpha'$ to $\beta'$ and from $\alpha''$ to $\beta''$, which lie in the planes $P_0'$ and $P_0''$ respectively.

et al [5] has shown that this cycle is in fact part of a higher dimensional "generalized" heteroclinic cycle which, moreover, is asymptoticaly stable in an open range of parameter values.

The paper by Chossat et al [5] proves that for weak rotation a lot of this structure will still persist: The subspaces $P_0, Q_2 = \{(y_0, y_{\pm 2})\}$ and $S'$ are still flow invariant and as a consequence there exist structurally stable heteroclinic cycles involving connections in these subspaces between $\alpha$, $\beta$ and the rotating wave generated by $\alpha_1$ in $Q_2$. A picture of this object is provided in citeLau00.

## 1.3. DYNAMO BIFURCATION FROM AN AXISYMMETRIC STEADY-STATE

As an attempt to relate the dynamics of the convection problem discussed in the last section to the generation of magnetic fields we study the induction equation in which the imposed velocity field is that of the axisymmetric steady-states $\alpha$ or $\beta$.

Bosch-Vivancos et al [3] determined that the first flow to generate an instability is the one with upwelling at the poles (i.e. $\alpha$). In that case the unstable magnetic modes have the form $B(r, \theta)e^{i\varphi}$ and complex conjugate, i.e. the bifurcated mode has an $m = 1$ azimutal structure. Note that the magnetic problem has the additional symmetry $s : \mathbf{B} \to -\mathbf{B}$. The structure of the unstable modes imply that the symmetries $\sigma$ and $s$ are broken, while $\chi$, $\nu = \sigma \cdot R_\pi$ and $\sigma \cdot s$ are not (and therefore lie in the symmetry group, often called the *isotropy subgroup*, of the magnetic field). Hence this bifurcation reduces to the standard bifurcation with $O(2)$ symmetry from the analytical point-of-view. However there are some additional symmetries, broken or not broken, that play a role in the choice of "typical" magnetic modes in the next section.

## 2. A low-dimensional model for the dynamo produced by bifurcation from a heteroclinic cycle

In order to build up a simple dynamical system that accounts for the heteroclinic cycles of the convection problem and for the unstable magnetic modes of the induction problem near the bifurcation from the purely convective states, we augment the 8-dimensional center manifold of the convection problem with a finite combination of spherical magnetic modes. The constraints are that (i) the resulting system must be $O(3)$ symmetric (i.e. invariant under the action of this group), (ii) a bifurcation from the purely convective to the magnetic modes must occur, which respects the symmetry properties of the magnetic instability described above. Then we consider the simplest possible configuration which is consistent with these constraints. The resulting dynamical system is then analyzed for the dynamics of the magnetic field and in particular for polar reversals.
We make some specific assumptions that are consistent with all previous work to reduce the free parameter space: For $B = 0$ the system has the purely convective invariant subspace which in our case is represented by the 8-d center manifold of the $\ell = 1, \ell = 2$ interaction. We assume a supercritical dynamo bifurcation and consider the simplest possible spherical magnetic modes inducing the same symmetry-breaking as in Section 1.3. These are standard spherical modes with $\ell = 1$, denoted by $z_0, z_{\pm 1}$. We will couple these modes to the equations for $x_j$ and $y_k$ in a way, compatible with the spherical symmetry of the problem and with the symmetry $s$.

The resulting system does not attempt to be quantitatively correct as a genuine Galerkin projection would be (with a sufficiently large number of modes). However, we chose the modeled modes in an attempt to capture the qualitative behavior of the interaction based on the eigenmode analysis of the kinematic dynamo problem in [3] and their interaction with the convection problem forced by the symmetries of the problem. We believe that this is a new and fruitful approach.

The resulting equations up to order 3 are as follows:

$$\begin{aligned}
\dot{x}_0 &= f_0(x,y) + \alpha x_0 \|z\|^2 + \beta z_0 (x_0 z_0 + x_1 \bar{z}_1 + \bar{x}_1 z_1) \\
\dot{x}_1 &= f_1(x,y) + \alpha x_1 \|z\|^2 + \beta z_1 (x_0 z_0 + x_1 \bar{z}_1 + \bar{x}_1 z_1)
\end{aligned}$$

$$
\begin{aligned}
\dot{y}_0 &= g_0(x,y) + \gamma(z_0^2 - z_1\bar{z}_{-1}) + \delta y_0\|z\|^2 \\
\dot{y}_1 &= g_1(x,y) + \sqrt{3}\gamma z_0 z_1 + \delta y_1\|z\|^2 \\
\dot{y}_2 &= g_2(x,y) + \sqrt{3/2}\gamma z_1^2 + \delta y_2\|z\|^2 \\
\dot{z}_0 &= z_0[\lambda + b\|x\|^2 + b'\|y\|^2 + b''\|z\|^2] + c(z_0 y_0 + \sqrt{3}Re(y_1\bar{z}_1)) + \\
&\quad 2c'[z_0(3/2y_0^2 + 2y_1\bar{y}_1 - y_2\bar{y}_2) + \sqrt{3}/2(z_1 y_0\bar{y}_1 + \bar{z}_1 y_0 y_1)] \\
\dot{z}_1 &= z_1[\lambda + b\|x\|^2 + b'\|y\|^2 + b''\|z\|^2] + c/2(-y_0 z_1 + \sqrt{3}y_1 z_0 + \sqrt{6}y_2\bar{z}_1) + \\
&\quad c'(z_1 y_1\bar{y}_1 + \sqrt{3}z_0 y_0 y_1 + 3\sqrt{2}z_0\bar{y}_1 y_2 + 3\bar{z}_1 y_1^2 - \sqrt{6}/2\bar{z}_1 y_0 y_2 + z_1 y_2\bar{y}_2)
\end{aligned}
\tag{1}
$$

where $f_j$ and $g_k$ denote the vector field for the purely convective amplitude equations ([5]).

In principle the parameters $\alpha, \beta, etc$ are order 1 parameters that could be calculated from the detailed interaction of the numerically known critical eigenfunctions of the convection problem and the critical eigenfunctions of the magnetic bifurcation problem. However, since we do not have a simultaneous bifurcation of the convection and the magnetic problem (which would be codimension 3) we would have to take into account the nonlinear evolution of the convection problem. Since we cannot do that the model does not have a rigorous justification. However, several coefficients have unique relationships to the physical problem and can therefore be restricted. Notice that all the interaction terms $b, b', b''$ as well as $\alpha, \beta$ and $c$ are all proportional to the magnetic Reynolds number. Hence since we are assuming a supercritical dynamo bifurcation from the $\alpha$ or $\beta$ convection cell (or both) we need $b'' < 0$ and $b' > 0$.

Choosing parameters for the convection problem that lead to structurally stable heteroclinic cycles and varying the coupling parameters we find different dynamical regimes for equations 1. We begin by simulating a weakly rotating system that has a heteroclinic cycle connecting the $\alpha$ convection state to the $\beta$ convection state and to the rotating waves of $\alpha$ structure back to the $\alpha$ convection state ($\alpha \to \beta \to RW_\alpha \to \alpha$). Figure 3a shows the resulting system when the interaction with the magnetic modes is included: The $\alpha$ convection becomes unstable to a magnetic bifurcation but the $\beta$ convection is still stable. As a result we seem to get a relatively stable magnetic fixed point $B_\alpha$ in the $z_0$ direction axisymmetric with the $\alpha$ state and a magnetic excitation $B_{RW_\alpha}$ in the $z_1$ direction related to the instability of the rotating wave $RW_\alpha$. The resulting heteroclinic cycle has the form $\alpha \to B_\alpha \to \beta \to RW_\alpha \to BRW_\alpha \to \alpha$. Note the change of sign of $z_0$ after one heteroclinic cycle which would indicate a reversal of the magnetic dipole. This situation is somewhat unsatisfactory since at the level of the resolution of the model we have a bifurcation to a magnetic field that has the same symmetry axis as the convection problem. How-

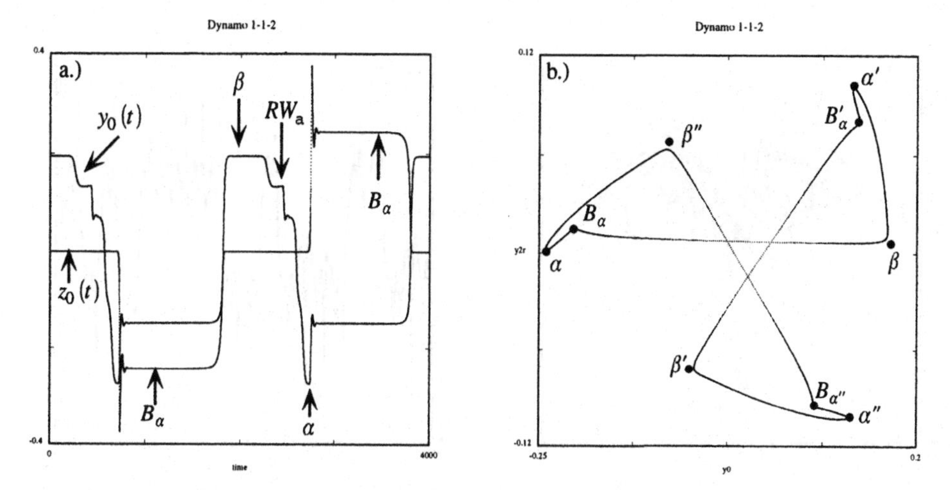


*Figure 3.* Simulation of Eq. 1 with weak rotation and a supercritial bifurcation of the $\alpha$ fixed point into the magnetic direction. a) the axisymmetric components $y_0(t)$ and $z_0(t)$ of the velocity and the magnetic field, respectively. b) Phase space projection into the plane $P_2$ of a simulation of Eq. 1 without rotation and with a supercritial bifurcation of the $\alpha$ fixed point into the magnetic direction.

ever, we should keep in mind that higher modes might lead to small tilts of the dipolar axis. Figure 3b generates a heteroclinic cycles of the type $\alpha \rightarrow B_\alpha \rightarrow \beta \rightarrow \alpha' \rightarrow B_{\alpha'} \rightarrow$ etc. In this case the bifurcation is to a magnetic dipole that is orthogonal to the rotation axis ($B_\alpha$ is in the $z_1$ direction). Note however, that the system is not rotating in that case.

Finally, in Figure 4 we show a situation that is very reminiscent of Lorenz chaos. The system irregularly switches back and forth between two opposite dipole states. In that case we have included a relatively strong rotation for which we cannot guarantee the theoretical existence of a heteroclinic cycle in the convection problem. As a result the previous $\alpha$ and $\beta$ convection states are not recognizable in the trajectory of the solution.

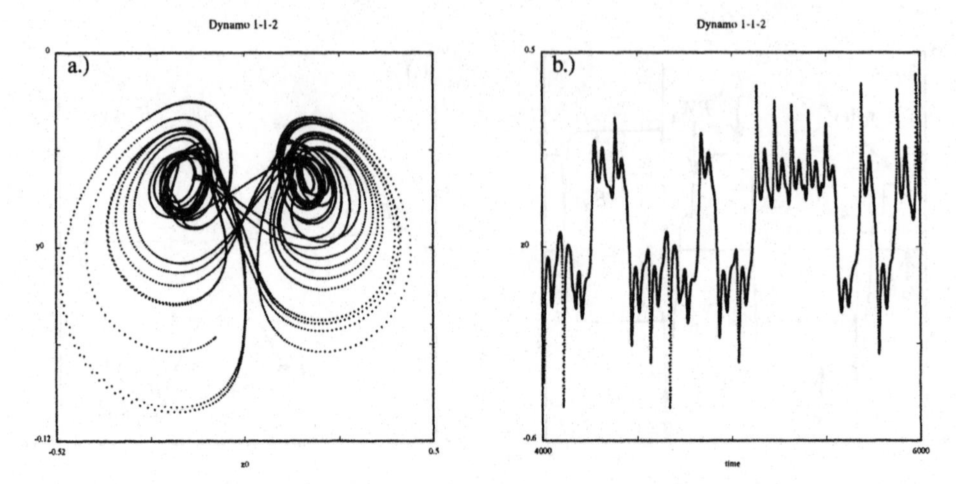


*Figure 4.* Simulation of Eq. 1 with stronger rotation. a) a phase space projection into the plane spanned by $(y_0, z_0)$, b) a time series for the magnetic component $z_0(t)$.

Nevertheless, it is interesting that the interaction of these modes creates a time series that was suggested about 25 years ago as a model for the reversals of the geodynamo [10]. Obviously, the parameter space is too rich for a full exploration and as a result we find different types of dynamical behavior many of which probably do not have physical significance. However, with this type of study we have moved the paradigm of magnetic reversals as heteroclinic dynamics from the case of the non-rotating, driven kinematic dynamo [9] via the rotating, driven kinematic dynamo [2] to the rotating, weakly interacting intermediate dynamo model and hence closer to physical reality.

## 3. An intermediate model

Previously we focused on the dynamics of the magnetic field in the kinematic dynamo driven by heteroclinic cycles in the convection problem [9], [2]. The previous section discussed an attempt to go beyond that and model the interaction of the convection problem with the magnetic problem at the level of an extended center manifold. An alternative approach tries on solve the induction equation

$$\frac{\partial \mathbf{B}}{\partial t} = \nabla^2 \mathbf{B} + R_m \nabla \times (\mathbf{v} \times \mathbf{B}) \tag{2}$$

$$\nabla \cdot \mathbf{B} = 0, \tag{3}$$

exactly (numerically) and to model the interaction by a projection onto the center manifold of the convection problem. Specifically, we consider the Lorentz force $\mathbf{J} \times \mathbf{B}$, where $\mathbf{J} = \nabla \times \mathbf{B}$ and project it onto the 8-dimensional center manifold of the convection problem. We then study the joint evolution of the dynamics on the center manifold and the induction equation

The projection can be written as

$$\mathbf{J} \times \mathbf{B} = \sum_{l=1}^{2} \sum_{m=-l}^{l} < \mathbf{J} \times \mathbf{B}, \xi_m^l > \xi_m^l, \tag{4}$$

where the dot product $< \cdot, \cdot >$ is given by

$$< \mathbf{f}, \mathbf{g} >= \frac{1}{8\pi^2} \int_0^{2\pi} \int_0^{\pi} \int_{r_i}^{r_0} \mathbf{f}(r, \theta, \phi) \cdot \overline{\mathbf{g}(r, \theta, \phi)} \sin \theta d\phi d\theta dr. \tag{5}$$

with $\xi_m^l, l = 1, 2, m = -l..l$ the 8 eigenfunctions spanning the center manifold of the convective problem.

To simulate equation 2 we use an expansion of $\mathbf{B}$ and $\mathbf{v}$ into *generalized spherical functions* described in [9]. The ODE part of the model is solved numerically with an adaptive Runge Kutta method. We study the dynamics of this intermediate system for different values of the parameters and compare it with the evolution of the driven kinematic dynamo. We are specifically interested to see if, and for what parameter values, the kinematic solutions can be found in the intermediate model.

The analysis of this extended system has only just begun. We have considered parameters values for the convection problem that generate the chaotic velocity and the heteroclinic cycles, respectively [9]. In the case in which the chaotic velocity generates an intermittent magnetic field in the kinematic model, preliminary numerical calculations show a periodic solution $(\mathbf{v}, \mathbf{B})$ for the extended model. In the second case, in which the heteroclinic velocity field generates a similar heteroclinic kinematic behaviour for the magnetic field, the extended system has a transient long period orbit that settles down in a steady state $(\mathbf{v}, \mathbf{B})$, at values for the critical $R_m$ close to the kinematic case. In general the structure of the solutions set appears to be very rich. When the main parameters $Ra$ and $R_m$ are varied, many different states can be obtained numerically. Indications are that intermittent as well as heteroclinic behaviour like the "dynamo-1-1-2" cycle from Fig. 3 is possible.

## 4. Acknowledgment

This work was supported by a NATO travel grant. D.A. acknowledges NSF support under DMS-0075041.

## References

1. Armbruster, D., Chossat, P. (1991) Heteroclinic orbits in a spherically invariant system, *Physica D* **50**, pp. 155-176.
2. Armbruster, D., Oprea, I. (2000) Dynamical systems and the kinematic dynamo *in Nonlinear Instability, Chaos and Turbulence Vol2*, WIT Press, L. Debnath, D.N. Riahi, (eds), pp. 162- 192.
3. Bosch-Vivancos, I., Chossat, P., Laure, P. (1999) Symmetry- breaking convective dynamos, *J. of Nonlinear Science* **9(2)**, pp. 169-196.
4. Busse, F.H., Heikes, K.E. (1980) Convection in a rotating layer: a simple case of turbulence, *Science* **208**, pp. 173-175.
5. Chossat, P., Guyard, F., Lauterbach, R. (1999) Generalized heteroclinic cycles in spherically invariant systems and their perturbations, *J. Nonlinear Sci.* **9**, pp. 479-524.
6. Chossat P., Lauterbach R. (2000) Methods in equivariant bifurcations and dynamical systems, *Adv. Series in Nonlin. Dyn.* **15**, World Scientific, Singapur.
7. Guckenheimer, J., Holmes, P. (1988) Structurally stable heteroclinic cycles, *Math. Proc. Cambridge Philos. Soc.* **103** , pp 189-192.
8. Lauterbach, R. (2001) Heteroclinic cycles and fluid motions in rotating spheres, *this volume.*
9. Oprea, I., Chossat, P., Armbruster, D. (1997) Simulating the kinematic dynamo forced by heteroclinic convective velocity fields, *Theoret. Comput. Fluid Dynamics* **9**(3/4), pp. 293 - 310.
10. Robbins, K.A. (1977) A new approach to subcritcal instability and turbulent transitions in a simple dynamo, *Math. Proc. Camb. Phil. Soc.* **82**, pp. 309- 325.
11. Stone, E., Holmes, P. (1990) Random perturbations of heteroclinic attractors, *SIAM J. Appl. Math.* **50** (3), pp. 726 - 743.

# TWO-COMPONENT DYNAMICAL MODEL OF THE SOLAR CYCLE

E. E. BENEVOLENSKAYA
*Pulkovo Astronomical Observatory, St. Petersburg, Russia;*
*W. W. Hansen Experimental Physics Laboratory,*
*Stanford University, USA*

**Abstract.** The solar magnetic cycle shows a complicated multi-mode behaviour. At least, two components are found in the distribution of the magnetic field on the solar photosphere. The existence of these two components can be explained by a non-linear model based on the Parker's dynamo theory with two sources. The properties of this dynamical system are investigated numerically, and it shown that this model can qualitatively reproduce the observed behaviour of the two dynamo components.

## 1. Double Magnetic Cycle

The 22-yr cycle of solar activity is a magnetic cycle which consists of two 11-years sunspot cycles. The poloidal magnetic field, or background field ( $B_r$ -component), also shows a 22-yr periodicity ([8]). Other studies ([15]) have suggested that the low-latitude and polar fields similarly show a 22-yr periodicity . Furthermore, these studies have hinted at a shorter term periodicity of about 2 yr (high-frequency component).

Similar periodicities were found in variations of radio flux on 10.7cm , flares and sunspot areas ([1]). Bao and Zhang (1998) have found that the magnetic (current) helicity observed during the solar cycle 22 (1988-1997) varied with time. The first maximum in the current helicity was in 1991, the second was in 1989 June and the third — in 1990. These variations could be interpreted as a quasi-biennial periodicity. Therefore, the observations indicate two main periodicities of helicity variations on the Sun: 11 yrs and 1.5–2.5 yrs.

It has been noted that during periods of three-fold polar magnetic field reversals which occur during some sunspot cycle maxima, the temporal



*P. Chossat et al. (eds.), Dynamo and Dynamics, a Mathematical Challenge,* 323–330.

separation of zones of alternated polarity of the magnetic field, determined from $H_\alpha$ charts is approximately equal to 1.5-2.5 years ([17]; [10]). In order to explain the three-fold polar magnetic field reversals it has been suggested that the solar magnetic cycle consists of two main periodic components: a low-frequency component (Hale's 22-year cycle) and a high-frequency component (quasi-biennial cycle)([3]). The existence of the double magnetic cycle on the Sun is confirmed using Stanford, Mount Wilson and Kitt Peak magnetograph data from 1976 to 1996 (solar cycles 21 and 22) (Benevolenskaya, 1995, 1996). In the current paper I present some results obtained from the Kitt Peak magnetograph data.

Each of the Kitt Peak magnetic synoptic maps is represented by the values of $B_{\|}$ as a function sine latitude and Carrington longitude. The observed line-of-sight components ( $B_{\|}$ ) averaging over all longitudes for each Carrington Rotation is represented in Figure 1b. The time interval in the series of $B_{\|}$ corresponds to Carrington rotation period ( $P = 27.2753$ days). The relative sunspot number is shown in Figure 1a. Component $B_r$ (Figure 1c), was calculated from observational data assuming that the true average field direction is radial and $B_\theta \cong 0$. The radial component can be easily found at all latitudes besides near the poles:

$$B_r = B_{\|}/\sin\theta, \tag{1}$$

where $\theta$ is colatitude.

To separate the high-frequency component in the data, we apply a difference filter by computing $\Delta B_r/\Delta t$ for different time intervals $\Delta t$. This is a reasonable procedure if the poloidal magnetic field consists of two components and is represented as two dynamo waves:

$$B_r(t,\theta) = b_r(\theta)\sin\omega_c t + \frac{b_r(\theta)}{A}\sin(\omega_\delta t + \varphi). \tag{2}$$

where $b_r(\theta)$ is the amplitude of the low-frequency radial component of the magnetic field; $A$ is ratio between amplitudes of low-frequency and high-frequency components; $\omega_c$ is the frequency of the Hale's solar cycle; $\omega_\delta$ is the frequency of biennual cycle; $\varphi$ is a phase.

The expression for the $\Delta B_r/\Delta t$ can be written as

$$\frac{\Delta B_r(t,\theta)}{\Delta t} = b_r(\theta)\omega_c\cos\omega_c t + \frac{b_r(\theta)}{A}\omega_\delta\cos(\omega_\delta t + \varphi). \tag{3}$$

When $\omega_\delta \cong 10\omega_c$, $A \cong 2$ as it took place in cycle 20, then the low-frequency term dominates in the expression for $B_r$ (equation 2) and the high-frequency term prevails in the expression for $\Delta B_r/\Delta t$ (equation 3). Thus, presuming the actual existence of the double magnetic cycle, one finds that the low-frequency term prevails in the $B_r$-component, while the

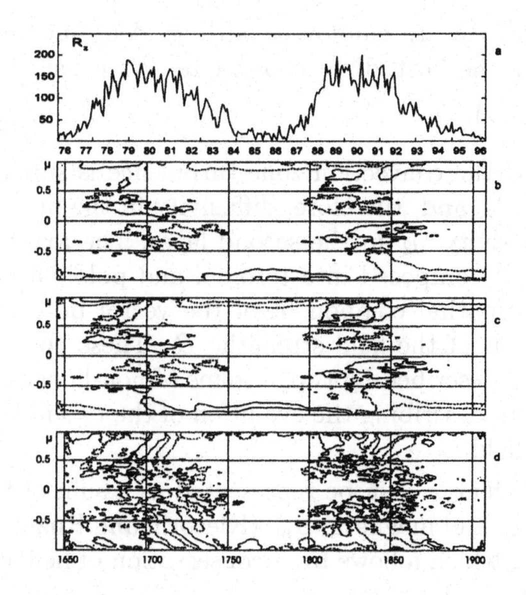


*Figure 1.* (a) The values of the sunspot number for Solar Cycles 21 and 22 as function of time (in years). (b) The line-of-sight component of the solar magnetic field as function of Carrington rotation number. (c) The radial component of the solar magnetic field (blue lines correspond to isolines for +3, +10 $G$, red lines, for -3, -10 $G$). (d) The values of the $\Delta B_r/\Delta t$ for $\Delta t = 6$ solar rotations.

high - frequency term dominates in $\Delta B_r/\Delta t$. Contours plots (diagrams) of $\Delta B_r/\Delta t$ as a function $\mu$ (or $\cos\theta$) and time, $t$, measured in Carrington rotation are represented in Figure 1d, for $\Delta t = 6P$. During Solar Cycles 21 and 22 the zones of increasing and decreasing strength of the surface magnetic field appear in both the $N$ and $S$ hemispheres (see Figures 1d). The width of these zones is approximately equal to 2 years.

## 2. Model

The possible explanation of the double magnetic cycle is that the magnetic fields are generated by Parker's dynamo acting in convective zone ([12]). The low-frequency component is generated at the base of the convective zone due to large scale radial shear $\left(\frac{\partial\Omega}{\partial r}\right)$, $\Omega$ is the angular velocity. The high-frequency component may be generated in subsurface regions due to latitudinal shear $\left(\frac{\partial\Omega}{\partial\theta}\right)$ or due to radial shear $\left(\frac{\partial\Omega}{\partial r}\right)$. The recent investigations of solar interior rotations show the significant radial gradient of angular velocity exists in subsurface of the convective zone together with latitudinal gradient of the angular velocity ([14]). For simplicity we used only latitudinal shear for generation of the high-frequency component.

Cartesian coordinates are employed, with $x$ denoting the radial, $y$ the azimuthal and $z$ the latitudinal coordinate. We consider axisymmetrical solutions $\left(\frac{\partial}{\partial\varphi} = 0\right)$.

At the base of the convection zone turbulence is suppressed by strong magnetic field ([13]) and, therefore, diffusivity in the first layer, $\eta_1$, could be less then diffusivity $\eta_2$ in the second layer. The axisymmetrical mean magnetic field is decomposed into toroidal and poloidal parts, and represented by the azimuthal component of the vector-potential, $A$, and the toroidal component of the field strength, $B$. These characteristics of the magnetic field are described by two systems of non-linear differential equations, each of them describing the evolution of the two independent sources of the magnetic field.

Helicity coefficient $\alpha$ can be represented as a sum of two components: kinetic helicity $\alpha_h$ and magnetic $\alpha_m$ (Kleeorin and Ruzmaikin, 1982). The equation for $\alpha_m$, which follows from conservation of helicity, generally can be written in as:

$$\frac{\partial \alpha_m}{\partial t} = -\nu\alpha_m + pAB - q(\alpha_h + \alpha_m)B^2, \tag{4}$$

where $\nu$ is a parameter characterizing dissipation of the magnetic helicity; and $p$ and $q$ are parameters characterizing the non-linear influence of the magnetic field on the magnetic helicity. One of the first non-linear models using this equation was investigated by Malinetskii et al. (1986).

In our two-component model it is suggested that $p = -1$ and $q = 0$. This is reasonable because coefficient $\alpha_m$ is proportional to an ensemble average of the fluctuations of magnetic helicity, $\langle \mathbf{a}\cdot\mathbf{b}\rangle$, where $\mathbf{a} = \mathbf{A} - \langle\mathbf{A}\rangle$, and $\mathbf{b} = \mathbf{B} - \langle\mathbf{B}\rangle$. If the total helicity is conserved then $\langle\mathbf{A}\cdot\mathbf{B}\rangle = \langle\mathbf{A}\rangle\cdot\langle\mathbf{B}\rangle + \langle\mathbf{a}\cdot\mathbf{b}\rangle = \text{const}$ (e.g. [16]). That means that time variations of $\alpha_m$ are proportional to the variations of $\langle\mathbf{A}\rangle\cdot\langle\mathbf{B}\rangle$. This explains the second term in the RHS of equation (4).

Then, the dynamo equations describing our two-zone model are:

$$\begin{cases} \dfrac{\partial A_1}{\partial t} &= (\alpha_{h1} + \alpha_{m1})B_1 + \sigma\Delta A_1 \\[2ex] \dfrac{\partial B_1}{\partial t} &= -\dfrac{\partial A_1}{\partial z}G_x + \sigma\Delta B_1 \\[2ex] \dfrac{\partial \alpha_{m1}}{\partial t} &= -\nu\alpha_{m1} + pA_1B_1 \end{cases} \tag{5}$$

$$\begin{cases} \dfrac{\partial A_2}{\partial t} & = & (\alpha_{h2} + \alpha_{m2})B_2 + \Delta A_2 \\ \dfrac{\partial B_2}{\partial t} & = & \dfrac{\partial A_2}{\partial x} G_z + \Delta B_2 \\ \dfrac{\partial \alpha_{m2}}{\partial t} & = & -\nu \alpha_{m2} + pA_2 B_2 \end{cases} \tag{6}$$

where $\alpha_{h1}$ and $\alpha_{h2}$ are the coefficients proportional to the kinetic helicity in the two zones, and $\sigma = \eta_1/\eta_2$ is the ratio of diffusivity in these zones.

If the two sources of magnetic field were independent then in the frame of this model it is difficult to explain the observed variations of the high-frequency components of the magnetic field and helicity. However, the erupted low-frequency magnetic field can influence the physical conditions in the region where the high-frequency component operates, through modifying helicity in this region. This couples the dynamo zones and allows the high-frequency component vary with time. In this case, the equation for the variable part of coefficient $\alpha_{m2}$ (which is proportional to the magnetic helicity) in the region of generation of the high-frequency component becomes

$$\frac{\partial \alpha_{m2}}{\partial t} = -\nu \alpha_{m2} + p(A_2 + aA_1)(B_2 + aB_1), \tag{7}$$

where parameter $a$ represents the influence of the low-frequency component of magnetic field, $B_1$, on the helicity in region 2 where the high-frequency component is generated.

Adopting equation (7) and reducing the systems of the partial differential equations (5) and (6) to ordinary differential equations following Weiss et al. (1984), we obtain two coupled systems of differential equations, which are solved numerically in non-dimensional units (Benevolenskaya, 1998). Since there is an underlying periodicity in all calculations, we use this period as a relative unit of time, and employ the following transformations: $t \longrightarrow t' = \eta_2 k^2 t \alpha_o$, $\alpha_h \longrightarrow \alpha_h' = \alpha_h/\eta_2 k^2 \alpha_o$, $\sigma = \eta 1/\eta 2$, $\alpha_m \longrightarrow \alpha_m' = \alpha_m/\eta_2 k^2 \alpha_o, G_z \longrightarrow G_z' = G_z/\eta_2 k \alpha_o$ and $G_x \longrightarrow G_x' = G_x/\eta_2 k \alpha_o$, where $\alpha_o$ is a characteristic mean value of $\alpha$, and $k$ is the radial wavenumber. With these transformations, $G_x$ and $G_z$ represent the dynamo numbers in our model. The period of the main cycle, 22 years, is chosen as the unit of time.

If $a = 0$ then these two systems are independent, and for the low-frequency component described by system (5) we obtain 11-yr variations for coefficient $\alpha_{m1}$ as described in Introduction. For the high-frequency component in this case we find coefficient $\alpha_{m2} \sim \sin 2\omega_2 t$ that gives 1-yr

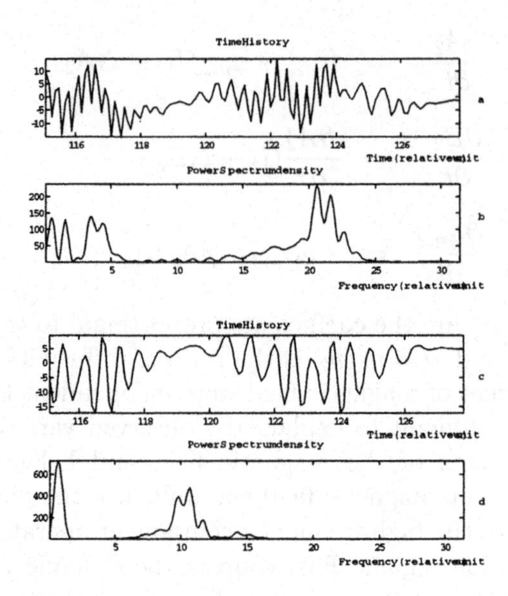


*Figure 2.* Solution of the decoupled systems (2) and (3) ( $a = 0$ ): (a) coefficient $\alpha_{m2}$ (which is proportional to magnetic helicity in the upper layer) as a function of time in relative units; (b) Power spectrum of $\alpha_{m2}$ ; (c) Real part of the total azimuthal vector-potential, $A = A_1 + A_2$ , as a function of time; (d) Power spectrum of $A$ in relative units.

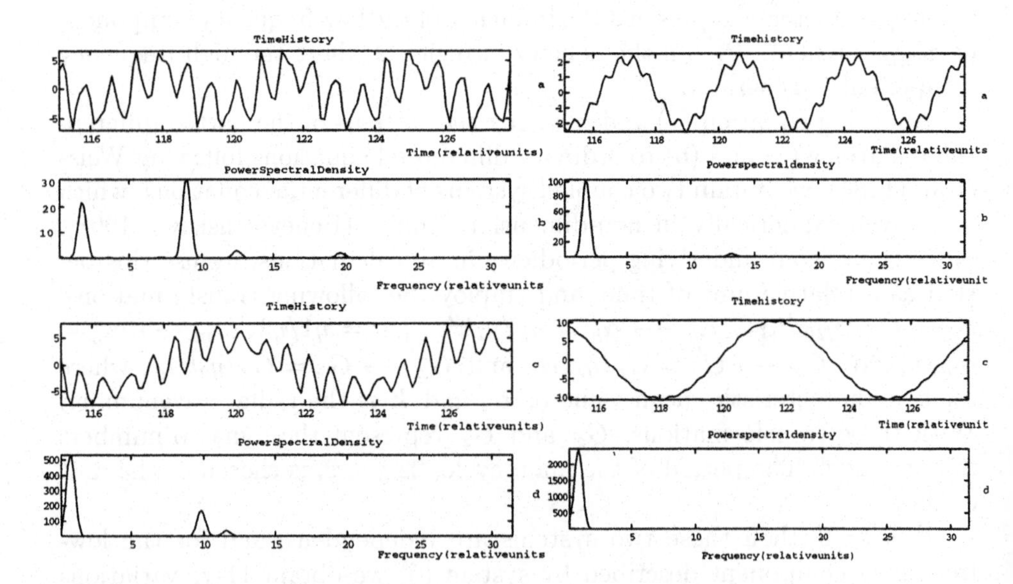


*Figure 3.* The same as in Fig.2, but for the coupling coefficient $a = 0.17$ (left four panels) and for $a = 0.2$ (right four panels).

variations if the magnetic field varies with a 2-yr period (frequency $\omega_2$), because $B_2 \sim \sin\omega_2 t$ and $A_2 \sim \cos\omega_2 t$ and $\omega_2 = 2\pi/2$ yr. However, if these systems are coupled ($a \neq 0$) due to the influence of the low-frequency magnetic field on helicity in the high-frequency zone then the 2-yr periodicity in helicity can be obtained because of the non-linear cross-terms, $aA_1B_2$ and $aA_2B_1$. We have investigated solutions of the two non-linear systems for the coupling parameter, $a$, in the interval $0 \leq a \leq 2$, and $\alpha_{h1} = \alpha_{h2} = 1$, $\nu = 0.5$, $p = -1$, $\sigma = 0.1$, $G_x = 0.3$, $G_z = 20$, which are in the reasonable range for the Sun ([6]).

In the case $a = 0$ (no coupling) coefficient $\alpha_{m2}$ shows the dominant variations with the frequency $f \approx 20$ (Figures 2a and 2b). Since our frequency unit is $1/22$ yr$^{-1}$ this frequency corresponds to a 1-year period. Multiple low-frequency peaks in Fig. 2b at low-frequencies represent some non-linear noise. The total vector-potential, $A$, (Figures 2c and 2d) has two main frequencies, $f \approx 1$ and $f \approx 11$, corresponding to periods 22 yr and 2 yr. Therefore, the solution with no coupling between the two dynamo zones shows the quasi-periodic variations of magnetic field, corresponding to the observations, but gives the incorrect period of the helicity variations in the upper layer (1 year instead of the observed 2 years).

With the coupling, when the low-frequency component of the magnetic field generated at the bottom of the convection zone influences the helicity in the upper layer, the helicity variations become more regular and clearly show two components with the frequencies $f \approx 2$ and $f \approx 10$, which correspond to 11-yr and 2-yr periods in accordance with the observations (Fig. 3, left panels a and b) (Benevolenskaya, 2000). The magnetic field variations in this case also have correct periods of $T \approx 22$ and $T \approx 2$ years, with the second component being much weaker. When the influence of the low-frequency mode becomes stronger the high-frequency component in the helicity variations decreases. This situation is illustrated in Fig. 3 (right panels) which shows the results for $a = 0.2$. This may explain why the high-frequency component varies considerably with the magnitude of the main sunspot cycle.

## 3. Conclusion

Using a simple non-linear two-zone dynamo model we have demonstrated that the observed properties of the 11-year and 2-year components of the solar cycle can be qualitatively explained by non-linear interaction between the two zones, caused by helicity variations in the high-frequency zone by the erupted magnetic field generated in the low-frequency zone. This model explains the observed periodicities for both magnetic field and helicity as well as the amplitude relations between these components.

This work was supported by JURRISS Program NASA NRA 98-OSS-08 and by the Russian Federal Programme 'Astronomy', Grant 1.5.3.4.

## References


1. Akioka M., Kubota J., Suzuki M. et al., (1987) The 17-month periodicity of sunspot activity, *Solar Phys.*, **112**, 313–316
2. Bao, S. and Zhang, H., (1998) Patterns of current helicity for solar cycle 22, *Astrophys.J.*, **496**, L43–L46
3. Benevolenskaya, E.E., (1991) A topological model of the solar magnetic field reversals, in *The Sun and Cool Stars: Activity, Magnetism, Dynamos*, Tuominen, I., Moss, D. and Rüdiger, G. (eds.), Springer-Verlag, 234–236
4. Benevolenskaya, E.E., (1995) Double magnetic cycle of the solar activity, *Solar Phys.*, **161**, 1–8
5. Benevolenskaya, E.E., (1996) Origin of the polar magnetic field reversals , *Solar Phys.*, **167**, 47–55
6. Benevolenskaya, E.E., (1996) A model of the double magnetic cycle of the sun, *Astrophys. J.*, **509**, L49–L52
7. Benevolenskaya, E.E., (2000) A mechanism of helicity variations on the sun, *Solar Phys.*, **191**, 247–255
8. Howard, R. and LaBonte, B.J., (1981) Surface magnetic fields during the solar activity cycle, *Solar Phys.*, **74**, 131–145
9. Kleeorin, N.I. and Ruzmaikin, A.A., (1982) Dynamics of the mean turbulent helicity in the magnetic field, *Magnitnaya Gidrodinamika*, No 2, 17–24
10. Makarov, V.I. and Sivaraman K.R., (1989) Evolution of latitude zonal structure of the large-scale magnetic field in solar cycles, *Solar Phys.*, **119**, 35–44
11. Malinetskii, G.G., Ruzmaikin, A.A., Samarskii, A.A., (1986) Model of the longterm variations of solar activity, *Preprint Institute of Applied Mathematics*, **170** 1–28
12. Parker E.N., (1955) Hydrodynamic dynamo models, *Astrophys. J.*, **122**, 293–314
13. Parker, E. N., (1993) A solar dynamo surface wave at the interface between convection and nonuniform rotation, *Astrophys. J.*, **408**, 707–719
14. Schou, J., et al., (1998) Helioseismic studies of differential rotation in the solar envelope by the solar oscillations investigation using the Michelson Doppler Imager , *Astropys. J.*, **505**, 390–417
15. Stenflo, J.O., (1988) Global wave pattern in the Sun's magnetic fields, in *Astrophys. Space Sci.*, **144**, 321–336
16. Vainshtein, S.I., (1983) Magnetic field in space, ed. Stepanov, B.E., Nauka, Moskva, 237
17. Waldmeier, M., (1973) A secondary polar zone of solar prominences, *Solar Phys.*, **28**, 389–398
18. Weiss, N.O., Cattaneo, F., Jones, C.A., (1984) Periodic and aperiodic dynamo waves, *Geophys. Astrophys. Fluid. Dyn.*, **30**, 305–341

# SYMMETRIES OF THE SOLAR DYNAMO: COMPARING THEORY WITH OBSERVATION

JOHN M. BROOKE
*MRCCS, Manchester Computing, University of Manchester, Oxford Road, Manchester, M13 9PL, United Kingdom*
JAAN PELT
*Tartu Observatory, Tõravere, 61602, Estonia*
PENTTI PULKKINEN
*Academy of Finland, P.O. Box 99, 00501 Helsinki, Finland*
AND
ILKKA TUOMINEN
*Astronomy Division, University of Oulu, P.O. Box 333, 90571 Oulu, Finland*

## 1. Introduction

Since the discovery by Schwabe in 1843 of the cyclic nature of sunspot activity, there has been a great deal of attention paid to the nature of these cycles. In 1849 Wolf established a measure of sunspot activity, the Wolf number, defined as $W = k(10g + f)$ where $g$ is the number of sunspot groups, $f$ is the total number of spots and $k$ is a factor allowing comparison between different observation series. Based on this measure we have the Zurich series of Wolf numbers which extends back to 1749. Since the counts for $W$ are taken over the whole visible surface of the sun, all information about the spatial distribution of the sunspot activity is lost. Realising the importance of such spatial information, Carrington in 1853 began the recording of the latitudes and longitudes of spots and this eventually resulted in the Greenwich series of photographic recordings of the solar surface which lasted from 1874 to 1976. Maunder introduced the famous Maunder or butterfly diagram (so called because the pattern of the cycles resembles the wings of a butterfly) which shows spot activity as a travelling wave beginning in the higher latitudes and proceeding to the equator.



*P. Chossat et al. (eds.), Dynamo and Dynamics, a Mathematical Challenge,* 331–338.

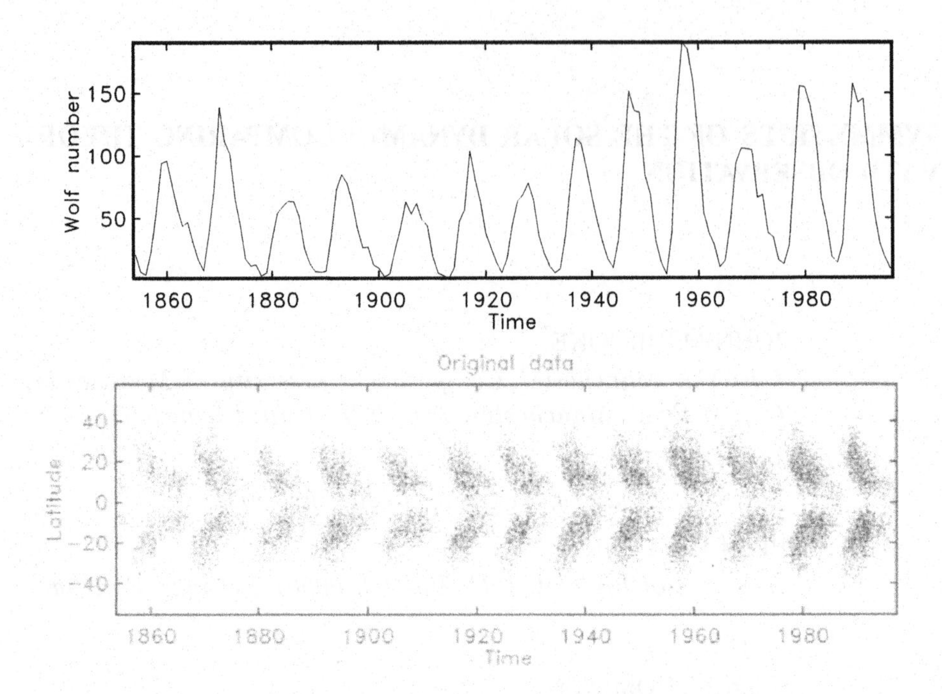


*Figure 1.* Upper panel: Time series showing Wolf numbers vs time (1853-1996). The minima are often some distance from zero due to overlapping between old and new cycles. Lower panel: A butterly diagram showing latitude $\theta$ vs time (1853-1996). The cycles are separated by clearly defined gaps.

A comparison between the one-dimensional and two-dimensional representations of the solar cycle can be seen in Figure 1. A central point of our analysis is that these representations of the Schwabe cycle are not equivalent; the Wolf number representation loses all information about the spatial location of the spots. One consequence can be seen immediately from visual inspection of Figure 1, in the Wolf number series the cycle minima are not precisely defined and are a perceptible distance from zero. In the Maunder diagram it is clear that there is a void region between each cycle, however the time intervals for this zero occurence of sunspots vary progressively with latitude. It is now clear why the Wolf number series does not show precise minima very close to zero; spots from adjacent cycles are counted together, the new spots at high latitudes and the older spots near the equator.

The inclusion of spatial information also gives information about the spatial and spatiotemporal symmetries of the solar magnetic field. It is well known that the symmetry group of a system has important consequences for its dynamical behaviour (e.g. [5]). The symmetry group of the axisymmetric mean-field dynamo equations and its effects on the bifurcation sequence as the dynamo number increases has been the subject of recent study [6, 7, 12, 3, 13]. There is strong evidence that there was a major departure from equatorial symmetry in the distribution of sunspots at the end of the Maunder minimum [14]. However, the equatorial symmetry of the sunspot cycle over the last 200 years has been the subject of some debate (e.g. [15, 4]). It is clear from the records that there are large departures from equatorial symmetry over timescales of a few years, and evidence that whole cycles have a dominant hemisphere which can change between cycles. It is not agreed, however, to what extent this is a statistical effect of a noisy component of the sunspot spatiotemporal distribution, or whether there are cyclic changes in the symmetry of the solar field over a time scale of several solar cycles, as predicted by several recent dynamo models. We can use our spatiotemporal definition of the solar cycle to address these issues.

There are severe problems in making deductions about the dynamics of the solar cycle from such a limited time series. Regular observations of sunspot position are not available before 1853, apart from the French observations during the Maunder minimum (1660-1719). Proxy records, such as $^{10}Be$ and $^{14}C$ records, can be extended back for much longer but do not contain information as to the equatorial symmetry of the solar field.

## 2. A new definition of the solar cycle

We use the sunspot data over the whole range of modern observations, starting from Carrington's (1853-1861) and Spörer's (1861-1894) data. To extend this record to the present day (i.e. 1874-1996), the data from the Greenwich Photoheliographic Results and the Solar Optical Observing Network have been combined[1]. These data sets cover completely sunspot cycles 10 to 22 of which cycles 10 and 11 are taken from Carrington/Spörer data and cycles 12 to 22 from the Greenwich data. For more details of this data see [10].

Our method commences with treating the wings of the butterfly diagram as the fundamental definition of a cycle and trying to identify the distribution in space and time of the regions void of sunspots which lie in between. Thus we define a solar cycle as a spatiotemporal rather than a purely temporal entity and we seek to define regions on the $\theta$-$t$ diagram (where $\theta$ is the solar latitude) that separate the cycles uniquely (Figure 1). We call these regions "sunspot voids" or "voids" for short.

As a first step we try to investigate whether an averaged period for the cycles can be usefully defined. The method of doing this is via a phase-process diagram as shown in Figure 2. We determine the phase of each

[1]The 1874-1996 data is obtainable at the internet site http://wwwssl.msfc.nasa.gov/ssl/pad/solar/greenwch.htm

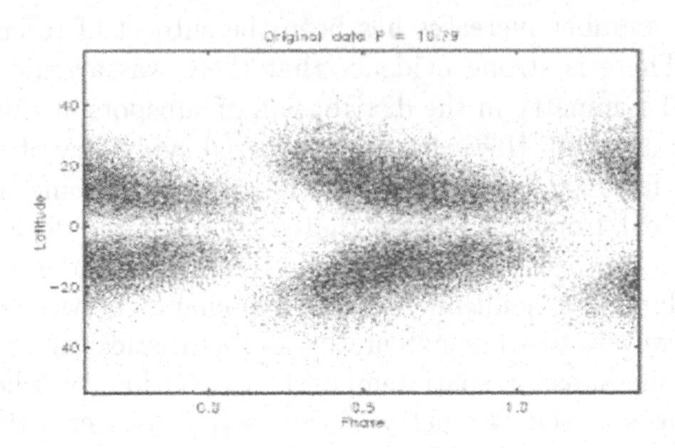


*Figure 2.* Phase process diagram for the data shown in Figure 1 for the optimum period 10.8yr. The clearly defined shape of the envelope of all the cycles is distinctly seen, thus indicating that the concept of an averaged period has real relevance.

point in the time series for a trial period $P$ defined by

$$\phi = t/P - [t/P], \tag{1}$$

and plot the latitude $\theta$ of each spot against its phase $\phi$.

The utility of the phase-process diagram is based on the observation that periodic components in a timeseries show up as phase-process diagrams with reduced dispersion (in the case of ordinary time series) or as a well concentrated pattern (in the case of spatiotemporal distribution). By this we mean that for a selected trial period, if it happens to be near the real period, we will get a well defined smooth phase curve or an easily discernable pattern, but for an arbitrary trial period the diagram will be randomly scattered. This is what defines the phase dispersion in the PPD analysis. We used the **ISDA** software package to determine the optimal period (Pelt [8]). Figure 2 is thus a superposition of all the butterfly cycles sampled at the optimum period as determined by ISDA. It already shows an important result, even when we superimpose in this way we still have a void region separating the ends of all of the cycles from the beginnings of all the cycles. This justifies our initial assumption that the butterfly cycles are sufficiently regular to allow the concept of an average period to be meaningful.

The void region defining the separation between the cycles appears to curve around the wing of the butterfly. This represents the sunspot minimum as a spatiotemporal curve rather than as an epoch. We seek next a nonlinear transformation of the $\phi$ axis in the $\theta$-$\phi$ plane that will transform this curve into an epoch or uniquely defined phase in the transformed $\theta$-$\phi$ plane. To speak more precisely, we actually seek a rectangular strip in the new $\theta$-$\phi$ plane with a given extent on the transformed phase axis and the

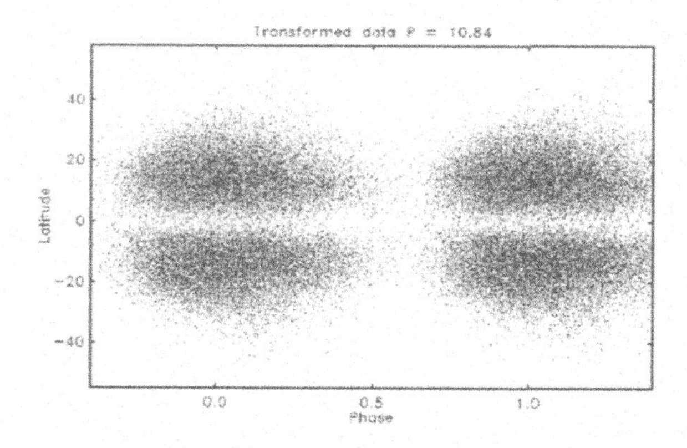


*Figure 3.* Superposition of the cycles using the optimisation (2). A clearly defined activity minimum can be seen at phase 0.66. It can be seen that the onset of activity is now represented by a line of constant phase.

nonlinear transformation is optimised to minimise the number of spots in this strip. Essentially we consider the original void region as an area to be transformed to a rectangular strip. This will, of course, change the shape of the cycle on the PPD but our initial intention here is to clarify the distinction between the cycles, rather than the form of the cycles themselves.

We show in Figure 3 the effects of such a nonlinear transformation by choosing the width of the strip, $\Delta = 0.05$, the linear slope parameter $\kappa = 0.0187$ and optimal period $P = 10.8$. In each trial case in the optimization procedure the form of the nonlinear transformation was the same,

$$\phi^* = \phi + \kappa P|\theta|cos(\pi\theta/180) \tag{2}$$

where $\kappa$ is constant for a given choice of $\Delta$. The $P$ is included in the correction term to make the slope parameter $\kappa$ independent of the trial periods used in the optimization. For an optimal set of transform parameters the phase for centre of strip occurred at $\phi = 0.659$ and there were only 84 spots inside the optimal strip (from 32410 spots in total). The points on the strip can be plausibly considered as the precursors of the starting phase or the leftovers from the ending phase of the wings. The width parameter $\Delta$ of the stripes is somewhat arbitrary, however by slightly changing it we find only modest changes in the set of optimal parameters.

We can apply the nonlinear transform of the phase (2) to the actual time $t$

$$t^* = t + \kappa P|\theta|cos(\pi\theta/180) \tag{3}$$

This transformation applies the phase correction (2), derived from the optimised averaged period, to the time series as a whole. In Figure 4 we see that this gives a unique epoch defining the onset of each solar cycle using

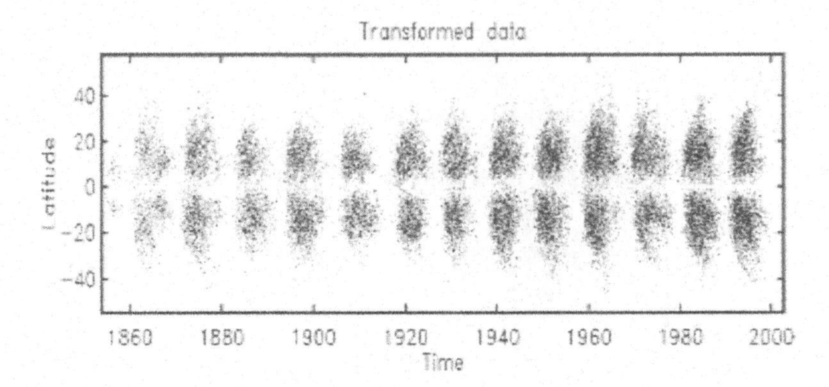


*Figure 4.* The butterfly diagram using the redefined time $t^*$. The onset of cycle activity is now represented by a straight line very nearly parallel to the $\theta$ axis.

this revised "time" axis. This has a clear physical interpretation as a travelling wave of onset sweeping from higher latitudes to the equator (see [9] for a detailed analysis). We also have a unique epoch for the sunspot minima defined by the void regions in between the cycles. We draw particular attention to the remarkable result that a cycle which is often described as chaotic, nonetheless demonstrates this coherence of phase and period over 12 consecutive cycles.

## 3. Oscillations of the solar magnetic equator

It is known that the distribution of sunspots between the northern and southern hemispheres is asymmetric in the short time scale and may undergo longer and more systematic variation (as discussed above). The usual index of sunspot asymmetry is $AS = (N - S)/(N + S)$ where $N$ and $S$ are the total activity counts in the northern and southern hemispheres. Such a measure loses all information about the latitudinal variation of the sunspots and from Figure 1 it can be seen that the latitudinal distribution is an important component of the Schwabe cycle.

As the mean latitude of sunspots, or the whole sunspot belt, is varying between cycles and hemispheres, their relative distance from the equator is changing too. This can be seen, when we calculate the latitude of the "magnetic equator" defined by sunspot latitudes. This is defined as the sum of the sunspot latitudes. The latitude is a signed quantity ($\langle\theta(S)\rangle$, the latitude south of the equator, is negative) and we can form an average mean latitude of the spots in either hemisphere at a given time. We then define the magnetic equator as $M = \langle\theta(N)\rangle + \langle\theta(S)\rangle$.

An interesting pattern is seen in Figure 5 where this sum is plotted as

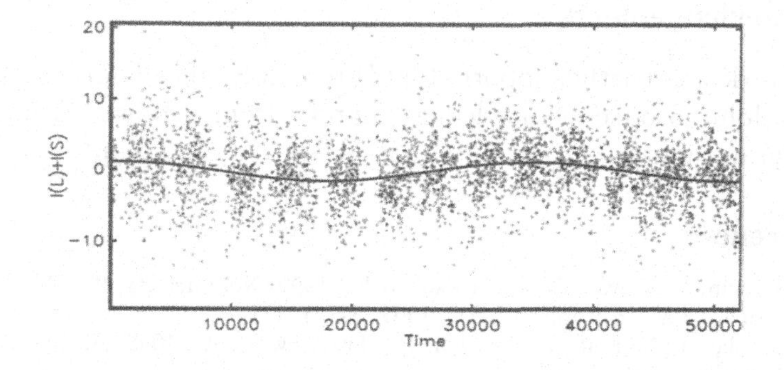


*Figure 5.* Variation of the magnetic equator of the sunspots

10-day averages with an 11 point smoothing filter. In symmetric case this sum should be zero, but this band, although wide, is clearly moving up and down rather systematically. If this is fitted to a sinusoidal profile (solid line in Figure 2), we get a variation in the magnetic equator $M$

$$M = 1.29(\pm 0.10)\cos(2\pi t/P) + 0.20(\pm 0.21)\sin(2\pi t/P), \qquad (4)$$

where time $t$ is measured in days, $t = 0$ being Nov 8, 1853, the first measurement by Carrington. The period of this variation $P = 33900 \pm 950$ days, and the amplitude $1.31 \pm 0.13$ degrees. For further detailed analysis see [11] where it is demonstrated that both the period and the form of the oscillation is robust with respect to the averaging interval. Thus we consider it to be a genuine effect.

## 4. Challenges to dynamo theory

More detailed analysis indicates that the symmetry and cycle length changes are synchronized [9]. This represents a dynamical evolution very closely approximated by a $T^2$ torus with a basic period of just under 11 years modulated in length and symmetry over a longer period of approximately 90 years which represents approximately 8.5 times the basic period. There currently exist two possible classes of mechanism that could potentially explain this behaviour, the parity modulation associated with the Type I modulation of Knobloch and Landsberg [7] and the newly identified form of intermittency described as "spiralling" or "in-out" intermittency [2, 1]. It is of theoretical interest that the amplitudes of the cycles appear to be much more irregular (see paper by Kurths in this volume). A challenge for dynamo theory is to link this picture of a dynamical system derived from observations with the physical mechanisms behind the solar dynamo.

## Acknowledgements

The work was partly supported by Estonian Science Foundation grant No. 2628. John Brooke acknowledges support from a Royal Society exchange visit grant and the hospitality of the University of Oulu.

## References

1. Ashwin, P., Covas, E.O., Tavakol, R.K., 1999, Nonlinearity, **9**, 563
2. Brooke, J.M., Europhys. Lett., 1997, **37**(3), 171
3. Brooke, J.M., Pelt, J., Tavakol, R., Tworkowski, A., 1998, Astron. & Astrophys., **332**, 339
4. Carbonell, M., Oliver, R., Ballester, J.L., 1993, Astron. & Astrophys. **274**, 497
5. Golubitsky, M., Stewart, I., Schaeffer, M., *Singularities and groups in bifurcation theory*, Springer Verlag, London, 1988
6. Jennings R.L., Weiss N.O., 1991, Mon. Not. R. Ast. Soc. **252**, 249
7. Knobloch E., Landsberg A.S., 1996, Mon. Not. R. Ast. Soc., **278**, 294
8. Pelt, J., *Irregularly Spaced Data Analysis*, 1992, Helsinki University Press.
9. Pelt, J., Brooke, J.M., Pulkinnen, P., Tuominen, I., to appear in Astron. Astrophys.
10. Pulkkinen, P., Finnish Meteorological Report No 23, Helsinki, 1998 (also Ph.D. thesis for University of Helsinki)
11. Pulkinnen, P., Brooke, J.M., Pelt, J., Tuominen, I., 1999, Astron. & Astrophys., **341**, L43
12. Tobias, S.M., 1997, Astron. & Astrophys., **322**, 1007
13. Knobloch, E., Tobias, S.M., Weiss, N.O., 1998, Mon. Not. R. Ast. Soc., **297**, 1123
14. Ribes, J.C., Nesme-Ribes, E., 1993, Astron. & Astrophys., **276**, 549
15. Verma, V.K., 1993, Astrophys. J., **403**, 797

# SUICIDAL AND PARTHENOGENETIC DYNAMOS

H. FUCHS, K.-H. RÄDLER AND M. RHEINHARDT
*Astrophysikalisches Institut Potsdam*
*D-14489 Potsdam, Germany*

**Abstract.** Numerical studies of a laminar dynamo model have revealed two remarkable phenomena. We consider a spherical body of an electrically conducting incompressible fluid which is surrounded by free space. The fluid shows an inner motion due to a given force and satisfies the no-slip condition at the boundary. For some investigations in addition to the forcing a rotation of the body is also considered. The full interaction of magnetic field and motion is taken into account.

Starting from a fluid motion capable of dynamo action and a weak magnetic field it was observed that the growing magnetic field can destroy the dynamo property of the motion so that it decays, and that the system ends up in a state with another motion incapable of dynamo action. However, for sufficient high magnetic Reynolds numbers a dynamo with a different symmetry may switch on after the 'suicide' of the original one.

In another case with a motion unable to prevent small magnetic fields from decay it proved to be possible that strong magnetic fields deform it so that a dynamo starts to work which enables the system to approach a steady state with a finite magnetic field. This 'parthenogenetic' dynamo is a genuine magneto-hydromagnetic state and has no kinematic counterpart.

## 1. Introduction

In many cases kinematic dynamo models developed in view of cosmic bodies have been extended by including the back-reaction of the magnetic field on the fluid motion. The simplest effect expected from introducing this kind of nonlinearity is the limitation of the growth of the magnetic field. There are, however, quite a few other effects such as changes in the dominating symmetry properties of magnetic field and motion or transitions to oscillatory, intermittent or chaotic behaviours. Remarkably enough, as we want



P. Chossat et al. (eds.), Dynamo and Dynamics, a Mathematical Challenge, 339–346.

to exemplify in this paper, the back–reaction of the magnetic field can also change the fluid motion resulting from a given distribution of forces such that it looses its dynamo capability with respect to the original magnetic field. ([6], see also [3, 10, 4]). Depending on the magnetic Reynolds number the magnetic field either goes to zero or may evolve to a new dynamo with different symmetry properties.

Recently several investigations have been carried out on the possibility that a fluid motion which is stable with respect to non–magnetic perturbations and incapable of dynamo action can be destabilized by a finite magnetic field, and that then motion and magnetic field evolve toward a dynamo state. In particular the so–called Balbus–Hawley instability [1] has been considered in this sense [2, 5]. In the following we will describe another example in which a fluid motion maintained by given forces and lacking dynamo capability gains this property in an evolution initialized by a sufficiently strong magnetic field which then ends up with a steady dynamo.

## 2. The model

We consider a spherical body of an electrically conducting incompressible fluid in non–conducting surroundings. It is assumed that the magnetic flux density, $\boldsymbol{B}$, and the fluid velocity, $\boldsymbol{u}$, inside the body are governed by

$$\frac{\partial \boldsymbol{B}}{\partial t} = \eta \Delta \boldsymbol{B} + \nabla \times (\boldsymbol{u} \times \boldsymbol{B}), \quad \nabla \cdot \boldsymbol{B} = 0, \tag{1}$$

$$\frac{\partial \boldsymbol{u}}{\partial t} + (\boldsymbol{u} \cdot \nabla)\boldsymbol{u} = -\nabla P + \nu \Delta \boldsymbol{u} - 2\,\boldsymbol{\Omega} \times \boldsymbol{u} + \frac{1}{\mu\rho}(\nabla \times \boldsymbol{B}) \times \boldsymbol{B} + \boldsymbol{\mathcal{F}}, \quad \nabla \cdot \boldsymbol{u} = 0. \tag{2}$$

$P$ is a modified pressure, and $\boldsymbol{\mathcal{F}}$ stands for an external body force. We refer to a rotating frame with $\boldsymbol{\Omega}$ being the angular velocity that defines the Coriolis force. As usual, $\eta$ is the magnetic diffusivity, $\nu$ the kinematic viscosity, $\rho$ the mass density and $\mu$ the magnetic permeability, all considered as constant. According to the assumption on non–conducting surroundings $\boldsymbol{B}$ has to continue as a potential field in outer space. Furthermore, the no–slip condition $\boldsymbol{u} = \boldsymbol{0}$ is assumed at the boundary. We use a spherical coordinate system $(r, \vartheta, \varphi)$ in the following such that its origin $r = 0$ coincides with the center of the fluid body and the axis $\vartheta = 0$ with the rotation axis. $R$ denotes the radius of the body.

We specify the force $\boldsymbol{\mathcal{F}}$ so that equations (2) for vanishing rotation and magnetic field, that is $\boldsymbol{\Omega} = \boldsymbol{B} = \boldsymbol{0}$, allow a steady solution $\boldsymbol{u}$ with a given flow pattern. Supposing symmetry of $\boldsymbol{u}$ about both the axis $\vartheta = 0$ and the equatorial plane $\vartheta = \pi/2$ we define it by

$$
\begin{aligned}
\boldsymbol{u} &= U_0 \boldsymbol{u}_0 \,, \quad \boldsymbol{u}_0 = \frac{1}{\sqrt{2}}(\boldsymbol{u}_0^{\mathrm{P}} + \boldsymbol{u}_0^{\mathrm{T}}) \,, \\
\boldsymbol{u}_0^{\mathrm{P}} &= -N^{\mathrm{P}} \frac{1}{\xi} \Big( f^{\mathrm{P}}(\xi)(3\cos^2\vartheta - 1)\, \boldsymbol{e}_r + \frac{d}{d\xi}(\xi f^{\mathrm{P}}(\xi)) \cos\vartheta \sin\vartheta \, \boldsymbol{e}_\vartheta \Big) \,, \\
\boldsymbol{u}_0^{\mathrm{T}} &= -N^{\mathrm{T}} f^{\mathrm{T}}(\xi) \sin\vartheta \, \boldsymbol{e}_\varphi \,, \\
f^{\mathrm{P}}(\xi) &= \mathrm{j}_2(\mu_{33}\,\xi) - \mathrm{j}_2(\mu_{33})\,\xi^2 \,, \\
f^{\mathrm{T}}(\xi) &= \mathrm{j}_1(\mu_{13}\,\xi) \,. 
\end{aligned} \tag{3}
$$

$U_0$ is a constant with the dimension of a velocity and $\boldsymbol{u}_0^{\mathrm{P}}$ and $\boldsymbol{u}_0^{\mathrm{T}}$ denote poloidal and toroidal fields, respectively. $N^{\mathrm{P}}$ , $N^{\mathrm{T}}$ are positive constants such that the r.m.s. values of $\boldsymbol{u}_0^{\mathrm{P}}$ and $\boldsymbol{u}_0^{\mathrm{T}}$ and, consequently, of $\boldsymbol{u}_0$ are equal to unity. Furthermore, $\xi$ is the fractional radius, $\xi = r/R$, the $\mathrm{j}_n$ are spherical Bessel functions of the first kind and the $\mu_{nl}$ their positive zeros, $\mu_{13} = 10.904122$ and $\mu_{33} = 13.698023$. We denote this solution of (2) by (o).

We may express the intensity of the forcing $\boldsymbol{\mathcal{F}}$ by the magnitude $U_0$ of the velocity which it is able to maintain, or by the corresponding Reynolds number $U_0 R/\nu$. We will use the latter as dimensionless forcing parameter $F$.

Models similar to that described here have been considered in earlier papers [7, 8]. For the numerical investigation of our equations (1), (2) we used the code explained in [9].

## 3. Fluid motions

The solutions of (2) for $\boldsymbol{B} = \boldsymbol{0}$ and $\mathrm{Ta} = 0$ as given by (3) proved to be stable only for $F < F_{\mathrm{crit}} = 15.4$ . At $F = F_{\mathrm{crit}}$ a bifurcation occurs. For $F > F_{\mathrm{crit}}$ there are stable solutions deviating from (3). For all following considerations we choose a forcing with $F = 17.27$. Two stable steady solutions $\boldsymbol{u}$ have been found for $\mathrm{Ta} = 0$, which differ only slightly in their total kinetic energies but more in the distribution of the energy onto the poloidal and toroidal motions. We denote them by (ia) and (ib). For $\mathrm{Ta} = 1000$ one stable solution was found only. It is denoted by (ii). In all cases $\boldsymbol{u}$ is again symmetric about both the axis of rotation and the equatorial plane.

In Table 1 some properties of these solutions are listed. For comparison the unstable solution (o) is included too. For all cases the Reynolds number $\mathrm{Re} = UR/\nu$ is given with $U$ being the r.m.s. value of $\boldsymbol{u}$. The flow patterns of the solutions (o), (ia) and (ib) are shown in Fig. 1.

| Solution | Ta | Re | $E_K^P/E_K^T$ | $\|h\|$ | Rm$_{crit}$ S1 | Rm$_{crit}$ S3 |
|---|---|---|---|---|---|---|
| (o) | 0 | 17.27 | 1.000 | 8.18 | 51.23 | 612.8 |
| (ia) | 0 | 17.29 | 0.991 | 7.95 | 48.93 | 616.2 |
| (ib) | 0 | 17.46 | 1.385 | 6.13 | 179.16 | 210.0 |
| (ii) | 1000 | 17.19 | 0.985 | 7.52 | 77.54 | 746.5 |

TABLE 1. Characteristics of the solutions of (2) with the forcing $F = 17.27$. $E_K^P$ and $E_K^T$ are the kinetic energies of the poloidal and toroidal parts of the inner motion, and $h$ is a specific helicity, $h = \langle \boldsymbol{u} \cdot \nabla\times\, \boldsymbol{u}\rangle R/\langle \boldsymbol{u}^2\rangle$, with $\langle\cdot\rangle$ denoting the average over one hemisphere. For the definition of Rm$_{crit}$ see Section 4.

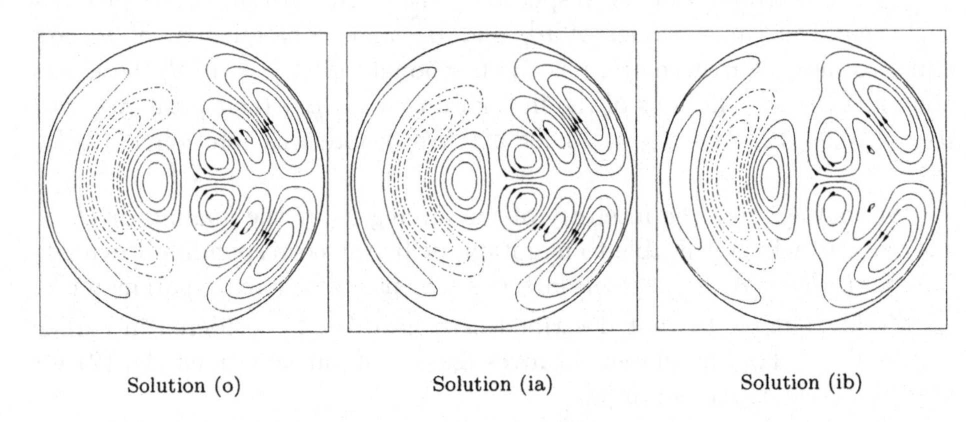

Solution (o) Solution (ia) Solution (ib)

*Figure 1.* The flow patterns of the solutions of (2). Solution (o) is given by (3), for (ia) and (ib), see above. For each picture: right half — streamlines of the poloidal (i.e. meridional) part, left half — isolines of the toroidal (i.e.azimuthal) part with solid lines indicating flow out of, broken lines flow into the paper plane.

## 4. Dynamos

Let us now admit magnetic fields, too. The influence of the fluid motion on them is characterized by the magnetic Reynolds number $\mathrm{Rm} = UR/\eta$ with $U$ as explained above. Clearly, $\mathrm{Rm} = \mathrm{RePm}$, with $\mathrm{Pm} = \nu/\eta$ being the magnetic Prandtl number.

We first restrict our attention on cases with weak magnetic fields only the influence of which on the fluid motion is negligible. Let us introduce the Alfvén number $A = B/\sqrt{\mu\rho}U$ with $B$ being the r.m.s. value of $\boldsymbol{B}$ inside the fluid body. In that sense we consider first the limit $A \to 0$.

Within this framework the question of stability or instability of the

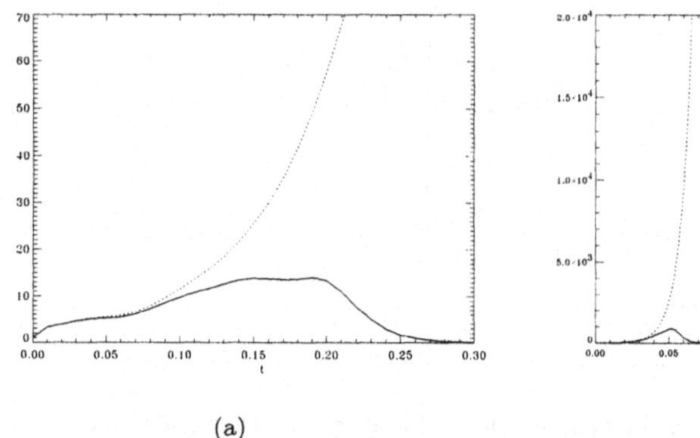

(a)

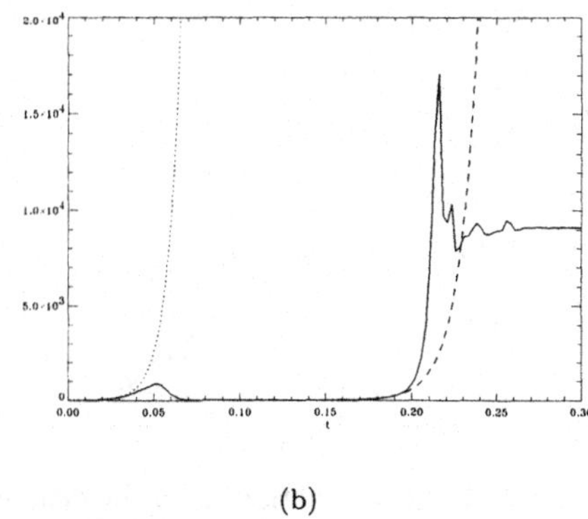

(b)

*Figure 2.* The magnetic energy for the evolution starting with a fluid flow of type (ia) in units of $\rho\eta^2 R$ (solid lines). The time is measured in units of the magnetic diffusion time $R^2/\eta$. Dotted/dashed lines describe the evolution of S1 type dynamos with the back–reaction of the magnetic field on the motions switched off. (a): Pm = 4, (b): Pm = 14.4. The dashed curve in (b) starts at $t = 0.068$.

non–magnetic state of our system described by (1) and (2) is just the central question asked in the kinematic dynamo theory. In all cases listed in Table 1 the fluid motion is capable of dynamo action, that is, allows small magnetic fields to grow if only the electric conductivity of the fluid, or Rm, is sufficiently high. More precisely, this condition reads $\mathrm{Rm} > \mathrm{Rm_{crit}}$ with the value of $\mathrm{Rm_{crit}}$ depending on the symmetry type of the magnetic field. Here we restricted ourselves on S1 and S3 type fields, that is, fields symmetric with respect to the equatorial plane and containing only the first or the third harmonic with respect to the azimuth $\varphi$, respectively. For both symmetries the values of $\mathrm{Rm_{crit}}$ given in Table 1 differ significantly for the two solutions (ia) and (ib). Generally, the $\mathrm{Rm_{crit}}$ for the S3 type fields exceed significantly those for the S1 type ones. Only for solution (ib) they are of the same order of magnitude.

In the following we will present two remarkable phenomena intrinsically connected with the interaction between magnetic fields and fluid motions.

To describe the first one we assume Ta = 0 and Pm = 4. Then a dynamo instability appears in the cases (0) and (ia) only but not in the case (ib). We now start from a non–magnetic state of our system with the fluid flow of type (ia) and disturb it by a small magnetic field of S1 type. As indicated by the solid line in Fig. 2(a) initially the magnetic field grows. After reaching a sufficient strength it destroys the dynamo capability of the fluid flow and begins to decay. The evolution ends up with a zero magnetic field and a fluid flow of type (ib). That is, *the dynamo kills itself.* Some details of this *suicidal* behaviour are given in Fig. 3.

In another run with the same forcing but a magnetic Prandtl number

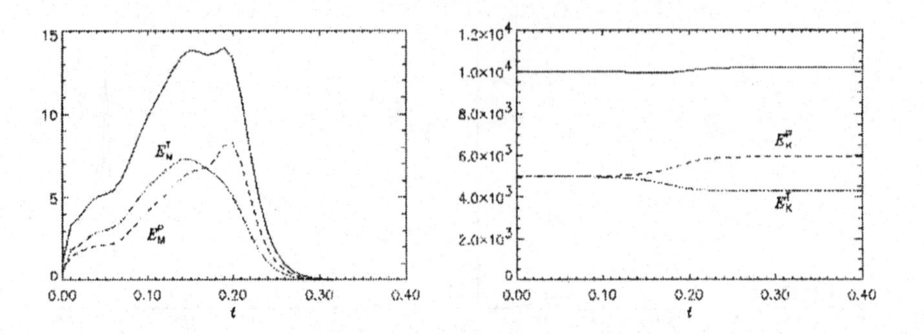


*Figure 3.* **The total energy $E_M$ of the magnetic field (for all space), the total kinetic energy $E_K$ and the energies $E_M^P, E_M^T, E_K^P, E_K^T$ of the poloidal and toroidal parts of magnetic field and motion during the evolution starting from a nearly non–magnetic state with a fluid flow of type (ia). Units as in Fig. 2, Pm = 4.**

Pm = 14.4 the magnetic field shows first a quasi–suicidal behaviour (see Fig. 2 (b)). An intermediate nearly non-magnetic state with flow pattern of solution (ib) is reached. However, the magnetic Reynolds number is now Rm $\approx$ 250. It exceeds the critical magnetic Reynolds numbers for both S1 and S3 type fields. The magnetic field grows again and the system reaches a stationary state, in which the magnetic field is dominated by its S3–symmetric parts.

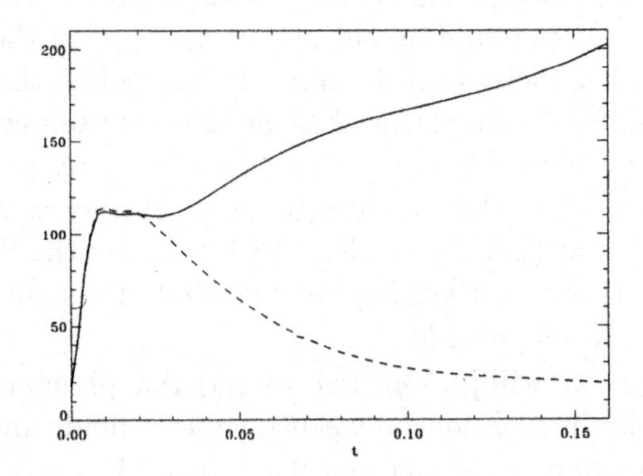


*Figure 4.* **The magnetic energy for the evolution starting with a fluid flow of type (ii).The dashed line describes the evolution without back–reaction of the magnetic fields on the motions. Units as in Fig. 2.**

To describe the second phenomenon we consider first a non–magnetic state of our system with a fluid flow of type (ii) with Pm = 4, Ta = 1000. Then small magnetic perturbations are bound to decay. With stronger perturbations, however, the flow can be deformed so that it gets capable of dynamo action. Then an evolution of the magnetic field as indicated by the solid line in Fig. 4 takes place which finally leads to a steady state with non–zero magnetic field. That is, *the magnetic field creates a dynamo*. The conditions under which this *parthenogenetic* behaviour occurs are determined by the initial Alfvén number $A$ and the initial geometrical structure of the magnetic field. Considering initial magnetic fields of S1 type only we found that the critical initial values of $A$ are lower for purely poloidal than for purely toroidal field structures as depicted in Fig. 5. Note that for purely poloidal initial fields an upper critical Alfvén number occurs beyond which no parthenogenetic dynamo exists.

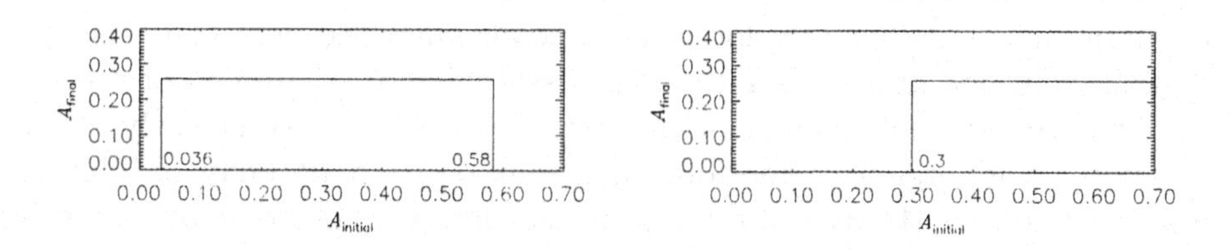


*Figure 5.* Dependence of the final state of the magnetic field on its initial geometry and strength for the flow (ii). Left: purely poloidal, right: purely toroidal initial field.

Fig. 6 presents some details for two examples with different initial toroidal fields. In the example in which initially $A = 0.306$ a steady state is reached with $A = 0.260$ whereas for an initial value $A = 0.298$ no dynamo was created. In the former case motion and magnetic field are finally still symmetric with respect to the equatorial plane. In addition to the dominating axisymmetric part in the motion the even harmonics with respect to $\varphi$ are present, and the magnetic field is enhanced by the higher odd harmonics.

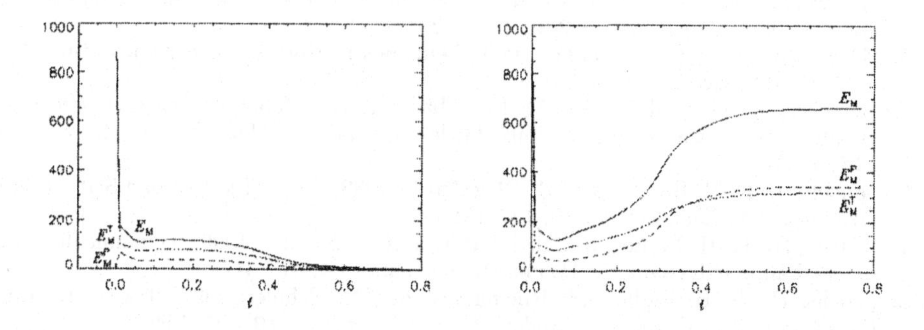


*Figure 6.* The energies $E_M$, $E_M^P$ and $E_M^T$ for evolutions starting from states with a fluid flow of type (ii) and toroidal fields with Alfvén numbers $A = 0.298$ (left) and $A = 0.306$ (right). For explanations see Fig. 3.

## 5. Conclusions

The first phenomenon described above provides us with an extreme example demonstrating that the behaviour of a system with genuine interaction of magnetic field and motion can drastically differ from the behaviour concluded from a kinematic model. Our finding implies a warning concerning simple parameterizations of the back–reaction of the magnetic field on the motion as used, e.g., in mean–field dynamo theory. The suicide of a dynamo as observed here is connected with the existence of more than one hydrodynamically stable states of the system, with and without stability with respect to magnetic perturbations. If the watershed between their basins of attraction is sufficiently low even a weak magnetic field can push the system from one basin to another, thus switching a dynamo on or off.

The second phenomenon is an example in which a sufficiently strong magnetic field organizes motions in favour of its maintenance. This requires, of course, the availability of kinetic energy, in our case delivered by forcing. Note that dynamos of that kind work only in a regime with the full interaction of magnetic field and motion and have no counterparts on the kinematic level. A related phenomenon has been discussed in the context of the Balbus–Hawley instability. It seems that such phenomena are not restricted to a very special situation but should be considered when discussing possibilities of the maintenance of magnetic fields.

## References

1. Balbus, S. A. and J. F. Hawley: 1998, 'Instability, Turbulence, and Enhanced Transport in Accretion Disks'. *Rev. Mod. Phys.* **70**, 1–53.
2. Brandenburg, A., A. Nordlund, R. F. Stein, and U. Torkelsson: 1995, 'Dynamo-generated Turbulence and Large-Scale Magnetic Fields in a Keplerian Shear Flow'. *Astrophys. J.* **446**, 741–754.
3. Brummell, N. H., F. Cattaneo, and S. M. Tobias: 1998, 'Linear and Nonlinear Dynamo Action'. *Phys. Lett. A* **249**, 437–442.
4. Demircan, A. and N. Seehafer: 2000, 'Dynamos in Rotating and Nonrotating Convection'. In: *These Proceedings*.
5. Drecker, A., G. Rüdiger, and R. Hollerbach: 1999, 'Global structure of selfexcited magnetic fields arising from Balbus-Hawley instability'. *Mon. Not. R. Astron. Soc.* **317**, 45–54.
6. Fuchs, H., K.-H. Rädler, and M. Rheinhardt: 1999, 'On Self-killing and Self-creating Dynamos'. *Astron. Nachr.* **320**, 127–131.
7. Fuchs, H., K.-H. Rädler, M. Rheinhardt, and M. Schüler: 1996, 'Dynamically Consistent Laminar Dynamos'. *Magnetohydrodynamics* **4**, 413–417.
8. Fuchs, H., K.-H. Rädler, M. Rheinhardt, and M. Schüler: 1997, 'Forced Laminar Dynamos'. *Acta Astron. et Geophys. Univ. Comenianae* **19**, 145–154.
9. Fuchs, H., K.-H. Rädler, and M. Schüler: 1993, 'A Numerical Approach to Dynamically Consistent Spherical Dynamo Models'. In: F. Krause, K.-H. Rädler, and G. Rüdiger (eds.): *The Cosmic Dynamo*. pp. 129 – 133.
10. Matthews, P.: 1999, 'Dynamo action in simple convective flow'. *Proc. R. Soc. London, Ser. A* **455**, 1829–1840.

# HETEROCLINIC CYCLES AND FLUID MOTIONS IN ROTATING SPHERES.

REINER LAUTERBACH
*Fachbereich Mathematik*
*Universität Hamburg*
*Bundesstrasse 55*
*20146 Hamburg*

## 1. Introduction

In their pioneering work BUSSE & HEIKES [2] looked at the classical Bénard problem in a rotating frame. They observed rolls with a certain axial direction which seem to be stable, ie they remain unchanged for long periods of time, but suddenly the behavior changes: new rolls appear which are rotated with respect to the original rolls by approximately 60 degrees. GUCKENHEIMER & HOLMES [15] looked at the Busse-Heikes problem from a theoretical point of view. They derived a three dimensional ODE exhibiting heteroclinic cycles. Thereafter many papers have dealt with various aspects of cycles: existence, stability, bifurcations and structural stability under certain settings: [1, 3, 4, 5, 7, 8, 16, 17, 18, 20, 21, 22, 25, 23, 24, 26, 27, 28, 29]. Due to the fact that solutions which pass near steady states remain there for a long time heteroclinic cycles serve as a model for metastable behavior. We see such a metastable behavior if we look at the polarity reversals of the magnetic field of the Earth, see for example GHIL & CHILDRESS [13]. Since the origin of the magnetic field and the mechanism of its reversals are unknown we try to look at it from a dynamical systems point of view and ask ourselves whether there are heteroclinic cycles in this problem. Here, we only look at the fluid mechanical part and not at full MHD-equations. In this paper we emphasize the rôle of rotation and the question what happens to heteroclinic cycles in the presence of rotations. Of course this approach dictates to look at the non rotating case first and treat the rotating case as a perturbation of the non rotating one. It is not clear whether such a approach is reasonable for studying the Earth' field but on the other hand numerical computations indicate that the region of validity of the results



*P. Chossat et al. (eds.), Dynamo and Dynamics, a Mathematical Challenge,* 347–354.

which are presented here exceeds the marginal speeds of rotation allowed by the usual perturbation methods.

## 2. Spherical Geometry

Here, we would like to present the basics of dynamical system theory in spherical geometry. Much more profound exposition of this material can be found in [10] or the recent monograph [9].

### 2.1. CENTER MANIFOLD REDUCTION

Relevant for the dynamo problem are MHD-equations for a velocity field $u$, pressure $p$ a temperature field $T$ and a magnetic field $B$ which read as

$$\left.\begin{aligned} \frac{\partial u}{\partial t} &= \nu\Delta u - \frac{1}{Pr}\langle u,\nabla\rangle u - \nabla p - \beta g T + \frac{1}{\mu}(\nabla\times B)\times B \\ \mathbf{div}\, u &= 0 \\ \frac{\partial T}{\partial t} &= \kappa\Delta T - \langle u,\nabla\rangle T - \langle u,\nabla T\rangle \\ \frac{\partial B}{\partial t} &= \mathbf{curl}\,(u\times B + \eta\Delta B), \end{aligned}\right\} \quad (1)$$

where $Pr$ denotes the Prandtl number, $g$ the gravitational force and these equations hold on a spherical shell $B(r,R) = \left\{x \in R^3 \,\middle|\, r < \|x\| < R\right\}$. Boundary conditions are such that the radial component of $u$ vanishes at the boundaries and $T$ assumes given values. One question one might ask is, whether the fluid motion bifurcating from the heat conducting solution supports a sustainable magnetic field. In order to explain field reversals one needs a mechanism for transient behavior, one such mechanism is the occurrence of heteroclinic cycles. We want to investigate this question however under the assumption $B = 0$. It is easily seen that $B = 0$ describes a subspace which is invariant for the dynamics of Equation (1). Of course, in order to find heteroclinic cycles one wants to use the machinery form nonlinear dynamics and therefore we want that Equation (1) determines a **dynamical system**. Of course the divergence equation and the fact that there is no equation for the dynamics of $p$ sheds some doubt onto whether this equation generates a dynamical system. However it is standard to reformulate this equation in the space of divergence free velocity fields. Doing so we get rid of the pressure term and we get a Partial Differential Equation generating a (semi-) dynamical system on the appropriate function space. This program is developed completely in the monograph by CHOSSAT & IOOSS [6]. It leads to a problem of the following type

$$\frac{\partial U}{\partial t} = L_\mu U + N(\mu, U), \quad (2)$$

where $L_\mu$ is a parameter dependent linear operator on an appropriate function space and $N$ contains the nonlinear terms. The center manifold reduction for this problem is explained in [6]. It is a constructive method to compute the lowest order terms for the vectorfield on the center manifold, a manifold which is tangential to the critical eigenspace of the linearization at the critical parameter value. Of course, here we have no space to expound this method. Its application to a similar problem is presented in [6].

## 2.2. CRITICAL MODES

Very closely related to the center manifold reduction are the critical modes. Consider the linearization $L_\mu$ of our problem. For some values of $\mu$ it will have parts of its spectrum on the imaginary axis. We call such parameter values *critical.* If $\mu$ is not just a single real parameter we might find surfaces of critical parameter values. In this case additional conditions are required to determine the physical meaningful critical parameters. We will briefly touch these issues. They are closely related to the effects of symmetry, so let us briefly discuss the symmetry in our problem.

The domain of the underlying equations is a spherical shell. Therefore the orthogonal group on $R^3$ maps the domain into itself. We denote this group by $\mathbf{O}(3)$. Solutions of our equations of course need not be invariant under the natural action of $\mathbf{O}(3)$ on the function space, however it is clear that solutions should be mapped onto solutions and therefore we require an equivariance condition, ie $L_\mu$ and $N(\mu, \cdot)$ commute with this action. This of course has consequences for the critical modes, ie those eigenfunctions corresponding to critical parameter values. Applications of the group elements map such eigenfunctions on other eigenfunctions corresponding to the same eigenvalue. So the eigenfunctions form linear spaces and they are representation spaces for the underlying group, in our case $\mathbf{O}(3)$. The representations of this group are ordered by integer $\ell$ and the $\ell$-th representation has dimension $2\ell+1$. The eigenmodes correspond to the *spherical harmonics.* Generically there is just one $\ell$ such that the corresponding modes are critical, compare [9]. In this case the presence of the group action implies a gradient like behavior implying simple dynamics, ie the existence of a certain number of equilibria and connecting orbits. The global attractor is just a stable equilibrium.

## 2.3. MODE INTERACTION

We have a discrete set of critical modes, however a continuum of parameters and therefore there are parameter regions where one or more modes are critical at the same time. In the spherical Bénard problem, the temperature difference across the fluid is a distinguished parameter. Variation of just one parameter leads generically to simple modes and hence simple dynamics. A second parameter in the problem is the aspect ratio

$$\rho = \frac{R_i}{R_o},$$

the quotient of the inner and the outer radius. This parameter decides on which mode is the first one to be critical (under variations of the temperature). So we have $\ell(rho)$, a step function on the unit interval. Of course there are values $\rho^*$ such that

$$\lim_{\rho \nearrow \rho^*} \ell(\rho) \neq \lim_{\rho \searrow \rho^*} \ell(\rho).$$

This case we call a *mode interaction.* The statement of gradient-like behavior does not apply and we have a chance of seeing complex dynamics.

## 3. Dynamics - the non rotating case

In this section we want to look a bit closer to the actual dynamics in the spherical Bénard problem. We know that the only cases allowing complex dynamics are the ones where we have a mode interaction. A special effect in dynamical systems are heteroclinic cycles. It is easily seen that heteroclinic cycles are not generic in systems without symmetry, but they are in systems with symmetry, see Field [11], Krupa [20] or [23].

### 3.1. HETEROCLINIC CYCLES IN MODE INTERACTIONS

Systems with mode interaction lack the gradient-like structure. Besides that the geometry of fixed point subspaces and the equivariant structure in the sense of the number of maps commuting with the group action is much richer than in then single mode case. Due to this fact one can hope the one finds heteroclinic cycles in such bifurcation problems. Systematic investigations are due to Chossat and his coworkers [1, 3, 4, 16]. Especially [4, 16] give lists of mode interactions of the type $\ell - (\ell + 1)$ and the subgroups of fixed point spaces such that one can find heteroclinic cycles. Among others on e can find a cycle in the $1 - 2$-mode interaction, which gives rise to an eight dimensional center manifold.

### 3.2. DYNAMICS IN THE 1-2-MODE INTERACTION

While the above list gives the pairs of modes where one can find heteroclinic cycles, we do not know yet, whether cycle exists for a large or a small set in parameter space, nor do we know whether these cycles are stable. One of the first cases that a heteroclinic cycle appeared in the context of the

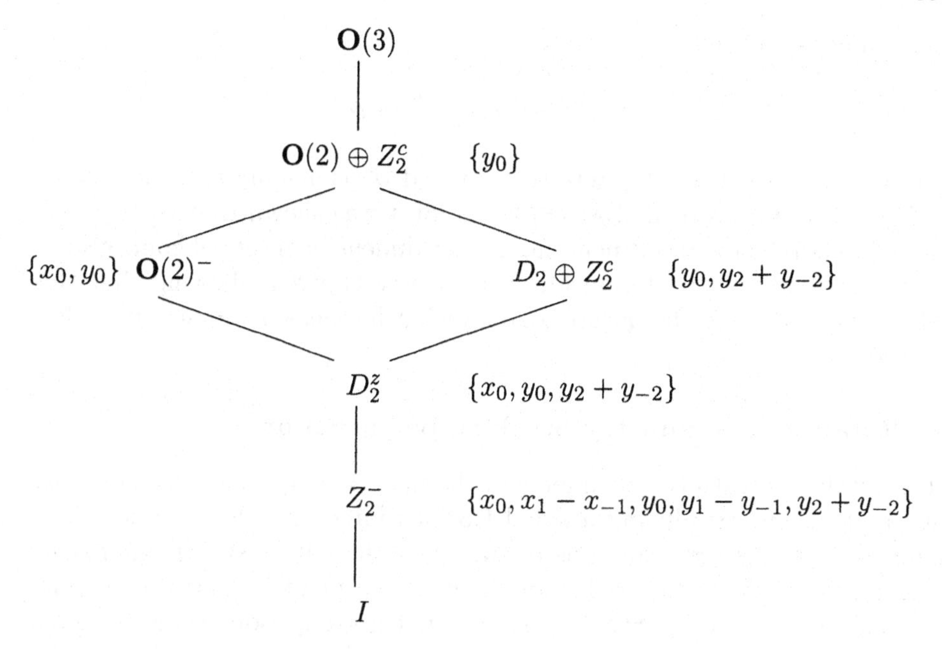


*Figure 1.* The partial ordered set of isotropy subgroups for the 1-2 mode interaction and the coordinates of the corresponding fixed point spaces. Here we write $x = (x_{-1}, x_0, x_1)$ for the coordinates on the $\ell = 1$ space and $x = (y_{-2}, y_{-1}, y_0, y_1, y_2)$ for the coordinates on the $\ell = 2$ space.

spherical Bénard problem was the paper by Friedrich and Haken [12]. This heteroclinic cycle was found by numerical simulations. In their paper they find that the dominant modes for this cycle are the $\ell = 1, \ell = 2$ modes. Armbruster and Chossat [1, 3] looked at precisely this mode interaction and made a theoretical study as well as a couple of simulations to find a cycle connecting the $\alpha$ and $\beta$-solution. They correspond to axisymmetric flows with inflow- outflow behavior at the equator and the poles. The analysis consists of the determining the equivariant maps in this representation and moreover the geometry of the fixed point spaces. Let us look at the second ingredient.

In Figure 1 we find the one dimensional fixed point space $\mathrm{Fix}(\mathbf{O}(2) \oplus Z_2^c)$ giving rise to the $\alpha$, $\beta$ solutions according to the Equivariant Branching Lemma (compare [14, 9, 19, 30]. Note that the Appendix B in [9] gives a translation table between the notation for the various groups used in standard bifurcation theory due to Ihrig and Golubitsky [19] and the notation commonly used in the Physics literature. ARMBRUSTER & CHOSSAT [1] proved that there exists an open region in parameter space, such that there exists a cycle of the form

$$\alpha \to \beta \to \alpha' \to \beta'' \to \alpha$$

where $\alpha', \alpha''$ are symmetry related to $\alpha$ and $\beta''$ is symmetry related to $\beta$. In CHOSSAT, GUYARD & LAUTERBACH [5] it was shown that this cycle is part of a more general set including higher dimensional sets of connections. Moreover a stability criterion for such a set was given, showing that this set can be stable. So the question arises what happens if we introduce slow rotation.

## 4. Rotation as symmetry breaking perturbation

Slow rotation breaks the symmetry of the underlyiong equations. Formally this is done by introducing a small Coriolis force into the equation. The symmetry of this term contains of two parts, first it is $\mathbf{SO}(2)$-symmetric, secondly it allwos certain reflection. The crucial point is, that the symmetry is an abelian group $\mathbf{SO}(2) \times Z_2^c$ and not a group isomorphic to $\mathbf{O}(2)$. This term introduces certain terms on the center manifold which can be computed.

### 4.1. PRINCIPAL EFFECTS

The main effects of such a perturbation on the dynamics is the effect on the $\alpha$, $\beta$ orbits of equilibria. On such a orbit, there is one point which is fixed under the new symmetry group. This point gives rise to an equilibrium which survives the perturbation. Another special feature on the group orbits are rotating waves, which are generatued by this perturbation. The dynamics on the perturbed group orbits of $\alpha$ or $\beta$-solutions can be studied following the lines of [25].

The main features of the dynamics on such a manifold are captured in the following figure.

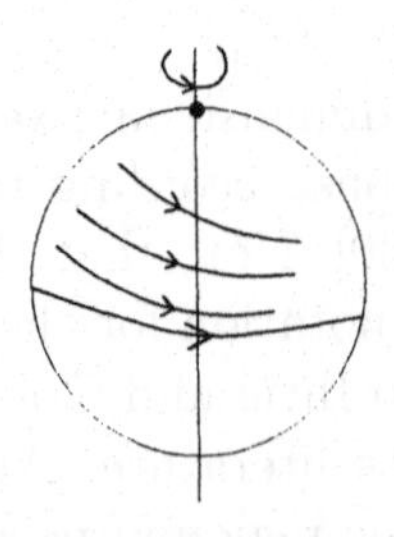

*Figure 2.* A sketch of a group orbit with the flow after perturbation. The equator corresponds to the rotating wave, the line indicates the axis of rotation, note, that the rotation takes place in $R^3$, while the group orbit sits in an eight-dmensional space.

### 4.2. HETEROCLINIC CYCLES

According to the previous discussion the equilibria on the group orbits of equilibria corresponding to the $\alpha, \beta$-solutions will be replaced by more complicated dynamics. The question is, whether we still find a heteroclinic cycle in the rotating system. We will write $\alpha$ for the single equilibrium surviving the rotation, and $RW_\alpha, RW_\beta$ for the unique rotating waves introduced on the group orbits by rotation. Then, one can show that for an open region in parameter space one finds heteroclinic cycles involving rotating waves, in particular we have a cycle of the type

$$\alpha \rightarrow \beta \rightarrow RW_\alpha \rightarrow \alpha.$$

In Figure 3 we see a plot of a simulation indicating this behavior.

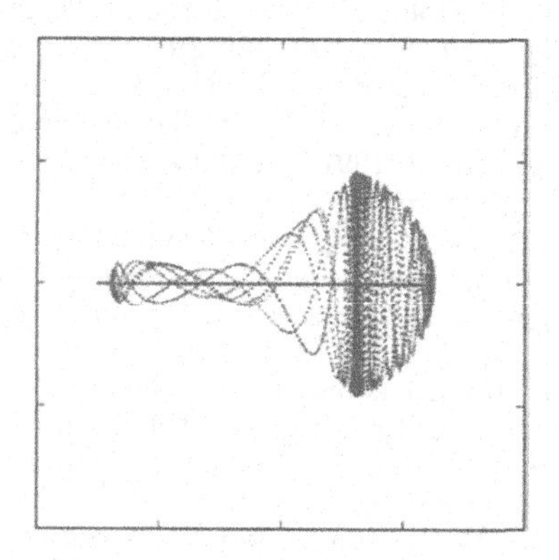

*Figure 3.*

## References


1. Armbruster, D. and P. Chossat: 1991, 'Heteroclinic cycles in a spherically invariant system'. *Physica D* **50**, 155–176.
2. Busse, F. H. and K. Heikes: 1980, 'Convection in a rotating layer: a simple case of turbulence'. *Science* **208**, 173.
3. Chossat, P. and D. Armbruster: 1991, 'Structurally Stable Heteroclinic Cycles in a System with **O**(3)-Symmetry'. In: M. Roberts and I. Stewart (eds.): *Singularity Theory and Its Applications, Warwick 1989, Part II.* pp. 38–62. Lecture Notes in Mathematics 1463.
4. Chossat, P. and F. Guyard: 1996, 'Heteroclinic cycles in bifurcation problems with **O**(3) symmetry and the Spherical Bénard Problem'. *J. Nonl. Sc.* **6**, 201–238.
5. Chossat, P., F. Guyard, and R. Lauterbach: 1999a, 'Heteroclinic sets in spherically invariant systems and their perturbations'. *J. Nonl. Sc.* **9**, 479–524.
6. Chossat, P. and G. Iooss: 1994, *The Couette-Taylor Problem*, No. 102 in Studies in Applied Mathematics. Springer Verlag.
7. Chossat, P., M. Krupa, I. Melbourne, and A. Scheel: 1997, 'Transverse bifurcations of homoclinic cycles'. *Physica D* **100**, 85–100.

8. Chossat, P., M. Krupa, I. Melbourne, and A. Scheel: 1999b, 'Magnetic dynamos in rotating convection- a dynamical systems approach'. *Dyn. of Cont., Discr. and Imp. Syst.* **5**, 327–340.
9. Chossat, P. and R. Lauterbach: 2000, *Methods in Equivariant Bifurcations and Dynamical Systems*, No. 15 in Advanced Series in Nonlinear Dynamics. World Scientific.
10. Chossat, P., R. Lauterbach, and I. Melbourne: 1991, 'Steady-State Bifurcation with **O**(3)-Symmetry'. *Arch. Rat. Mech. Anal.* **113**(4), 313–376.
11. Field, M.: 1977, 'Transversality in *G*-manifolds'. *Trans. Am. Math. Soc.* **231**(4), 429–450.
12. Friedrich, R. and H. Haken: 1986, 'Static, wavelike, and chaotic thermal convection in spherical geometries'. *Physical Rev. A* **34**, 2100–2120.
13. Ghil, M. and S. Childress: 1986, *Topics in Geophysical Fluid Dynamics: Athmospheric Dynamics, Dynamo Theory, and Climate Dynamics*, No. 39 in Applied Math. Sciences. New York: Springer Verlag.
14. Golubitsky, M., I. Stewart, and D. G. Schaeffer: 1988, *Singularities and Groups in Bifurcation Theory*, Vol. II. Springer Verlag.
15. Guckenheimer, J. and P. Holmes: 1988, 'Structurally stable heteroclinic cycles'. *Math. Proc. Cambridge Phil. Soc.* **103**, 189–192.
16. Guyard, F.: 1994, 'Interactions de mode dans les problèmes de bifurcation avec symétry sphérique'. Ph.D. thesis, Université de Nice Sophia – Antipolis.
17. Guyard, F. and R. Lauterbach: 1997, 'Forced symmetry breaking perturbations for periodic solutions'. *Nonlinearity* **10**, 291–310.
18. Guyard, F. and R. Lauterbach: 1999, 'Forced symmetry breaking: theory and applications'. In: *Pattern Formation in Continuous and Coupled Systems*. pp. 121–135.
19. Ihrig, E. and M. Golubitsky: 1984, 'Pattern Selection with **O**(3)-Symmetry'. *Physica 13D* pp. 1–33.
20. Krupa, M.: to appear, 'Robust heteroclinic cycles'. *J. Nonl. Sc.*
21. Krupa, M. and I. Melbourne: 1995, 'Asymptotic Stability of Heteroclinic Cycles in Systems with Symmetry'. *Ergod. Th. Dynam. Sys.* **15**(1), 121–147.
22. Krupa, M. and I. Melbourne: 1996, 'Nonasymptotically stable attractors in **O**(2) mode interactions'. In: W. Langdord and W. Nagata (eds.): *Normal Forms and Homoclinic Chaos*.
23. Lauterbach, R.: 1996, 'Symmetry Breaking in Dynamical Systems'. In: H. W. Broer, S. van Gils, I. Hoveijn, and F. Takens (eds.): *Nonlinear Dynamical Systems and Chaos*.
24. Lauterbach, R., S. Maier, and E. Reißner: 1996, 'A systematic study of heteroclinic cycles in dynamical system with broken symmetries'. *Proc. Roy. Soc. Edinburgh* **126A**, 885–909.
25. Lauterbach, R. and M. Roberts: 1992, 'Heteroclinic Cycles in Dynamical Systems with Broken Spherical Symmetry'. *J. Diff. Equat.* **100**, 428–448.
26. Maier-Paape, S. and R. Lauterbach: 1998, 'Heteroclinic cycles for reaction diffusion systems by forced symmetry breaking'. *Trans. Am. Math. Soc.* **(to appear)**.
27. Melbourne, I.: 1989, 'Intermittency as a codimension three phenomenon'. *J. Dyn. Diff. Equat.* **1**(4), 347–367.
28. Melbourne, I.: 1991, 'An example of a non-asymptotically stable attractor'. *Nonlinearity* **4**, 835–844.
29. Melbourne, I., P. Chossat, and M. Golubitsky: 1989, 'Heteroclinic cycles involving periodic solutions in mode interactions with **O**(2)-symmetry'. *Proc. Roy. Soc. Edinburgh* **113**(5), 315–345.
30. Vanderbauwhede, A.: 1982, *Local Bifurcation and Symmetry*, Vol. 75 of *Research Notes in Mathematics*. Pitman.

# CONVECTION-DRIVEN DYNAMOS, AMPLITUDE EQUATIONS AND LARGE-SCALE FIELDS

P.C. MATTHEWS
*School of Mathematical Sciences, University of Nottingham, University Park, Nottingham NG7 2RD, UK*

**Abstract.** This paper describes the weakly nonlinear behaviour of a dynamo driven by rotating convection in the form of two-dimensional rolls. The linear problem is separable and the onset of dynamo action occurs at small wavenumbers $m$ with a growth rate proportional to $m^2$. In the weakly nonlinear regime a band of wavenumbers is unstable, and an amplitude equation is obtained describing the nonlinear interactions between modes of different wavenumber. The behaviour of the amplitude equation shows an inverse cascade, with the mode of fastest growth rate giving way to solutions with longer and longer wavelength, over long timescales.

## 1. Introduction

There are two distinct approaches to the investigation of dynamo action driven by convection. The first, exemplified by the contributions by Busse, Cattaneo and Roberts in this volume, is to carry out fully nonlinear simulations of the governing equations at high numerical resolution, to model the magnetic field of the Earth or Sun as accurately as possible, within computational constraints. Although such simulations can give interesting results that are consistent with observations, it can sometimes be difficult to isolate the essential mechanisms at work. The alternative approach, taken here and in the contribution by Seehafer, is to study dynamo action in very simple convective systems. This allows a more thorough investigation of parameter space and facilitates the interpretation of the results.

Section 2 below extends the results of Matthews (1999) on dynamo action in two-dimensional convection rolls in a rotating layer. The nonlinear amplitude equation governing the weakly nonlinear behaviour of the dynamo is derived in section 3 and its behaviour is described in section 4.



*P. Chossat et al. (eds.), Dynamo and Dynamics, a Mathematical Challenge,* 355–362.

## 2. Dynamo action in rotating convection

The existence of convection-driven dynamos was established by Childress and Soward (1972) and Soward (1974). Their work was based on the simplifications that arise in the rapidly rotating limit, and showed that a single convection roll fails to generate a magnetic field, but a combination of two or more rolls can give dynamo action. This result is disappointing, firstly because experiments and theory indicate that rolls are commonly observed, and secondly because a two-dimensional roll flow simplifies the dynamo problem by allowing solutions that are separable in the third direction.

The aim of Matthews (1999) was to determine the simplest possible convection-driven dynamo. Clearly, two-dimensional convection rolls in a non-rotating layer must fail, since the flow is planar. In a rotating layer, convection can be steady or oscillatory, depending on the values of the Taylor number $T$ and the Prandtl number $\sigma$, but for $\sigma > 0.677$ convection is always steady at onset (Chandrasekhar, 1961). In a coordinate system with $z$ pointing vertically upwards, linear two-dimensional rolls with their axes aligned in the $y$-direction have the form $\boldsymbol{u} = (u, v, w)$ where

$$u = \frac{\pi}{k} \sin kx \cos \pi z, \quad v = -\frac{\tau\pi \sin kx \cos \pi z}{k(\pi^2 + k^2)}, \quad w = -\cos kx \sin \pi z. \tag{1}$$

Here, $\tau = \sqrt{T}$ and $k$ is the wavenumber of the rolls. The onset of convection occurs at the minimum of the marginal curve

$$R(k) = \frac{(k^2 + \pi^2)^3 + T\pi^2}{k^2}, \tag{2}$$

from which it follows that the critical value of $k$ obeys

$$T\pi^2 = (\pi^2 + k^2)^2(2k^2 - \pi^2), \tag{3}$$

and the critical Rayleigh number is $R_c = 3(\pi^2 + k^2)^2$. The flow (1) has three components, so it appears at first glance to be a possible candidate for dynamo action. In fact, however, the flow is confined to the planes $y = -\tau x/(\pi^2 + k^2) + \text{constant}$ (see fig. 23 of Chandrasekhar 1961) and therefore cannot function as a dynamo.

This negative result only applies to the linear eigenfunction for marginal convection. At finite amplitude, when $R = R_c + \epsilon^2 R_{c2}$, the amplitude of convection is of order $\epsilon$ and there are corrections to the flow of order $\epsilon^2$. In fact the second order correction only appears in the component $v$ along the axes of the rolls. The resulting flow, when scaled so that the maximum value of $w$ is one, has $u$ and $w$ given by (1) and

$$v = -\frac{\tau\pi}{k(\pi^2 + k^2)}(\sin kx \cos \pi z - \delta \sin 2kx) \tag{4}$$

where $\delta = \epsilon\pi/8k^2\sigma$.

This weakly nonlinear flow is two-dimensional (depending only on $x$ and $z$) but non-planar. Unlike most other known dynamos of this type, for example those of Roberts (1972), the streamlines are closed. The flow has the 'Boussinesq' up-down symmetry,

$$x \to x + \pi/k, \quad z \to 1 - z, \tag{5}$$

and a symmetry of rotation through $\pi$ about the $z$ axis,

$$x \to -x, \quad y \to -y. \tag{6}$$

Dynamo action in this flow was studied using a truncated Fourier series method (Matthews, 1999). Since the flow is two-dimensional, separable solutions can be sought in which the magnetic field $\boldsymbol{B}$ is proportional to $\exp imy$. The kinematic dynamo equation for the dynamo growth rate $\lambda$,

$$\lambda \boldsymbol{B} = \boldsymbol{\nabla} \times (\boldsymbol{u} \times \boldsymbol{B}) + \frac{1}{R_m} \nabla^2 \boldsymbol{B}, \tag{7}$$

then reduces to a two-dimensional eigenvalue problem. The boundary conditions were chosen so that $\boldsymbol{B}$ is horizontal at $z = 0$ and $z = 1$, with zero vertical derivative, ensuring that there is no flux of magnetic field through the boundaries. Note that these boundary conditions permit a large-scale horizontal magnetic field. Figure 1 shows the dynamo growth rate as a function of the wavenumber $m$ for four different values of the magnetic Reynolds number $R_m$. The Taylor number is 1000, so (3) gives $k = 3.71$. The parameter $\delta$, which depends on $\sigma$ and the amount of nonlinearity, was chosen to be 0.5. Only the eigenvalue with largest real part is plotted, and in all cases this eigenvalue is real.

The rotation symmetry (6) leads to a symmetry $m \to -m$, so only positive values of $m$ need to be considered. The Boussinesq symmetry (5) means that eigenfunctions occur in two types, according to whether (5) acts as $+1$ or $-1$ on $\boldsymbol{B}$. The kinks in the graphs of Figure 1 indicate a transition between these two different dynamo modes, where the eigenvalues cross. The mode that dominates at small $m$ is the '+1', case, which allows a large-scale mean horizontal field. The '−1' mode, in which a mean field is not allowed by the symmetry, dominates at a larger value of $m$.

As $R_m$ is increased, the first dynamo mode to appear is the $+1$ mode, at a critical value of $R_m$ equal to 19.74. There is in fact a small region of positive growth rates for small $m$ in the $R_m = 20$ curve of Figure 1.

It is no coincidence that the growth rate passes through zero at $m = 0$ for this dynamo mode. This result follows simply from conservation of

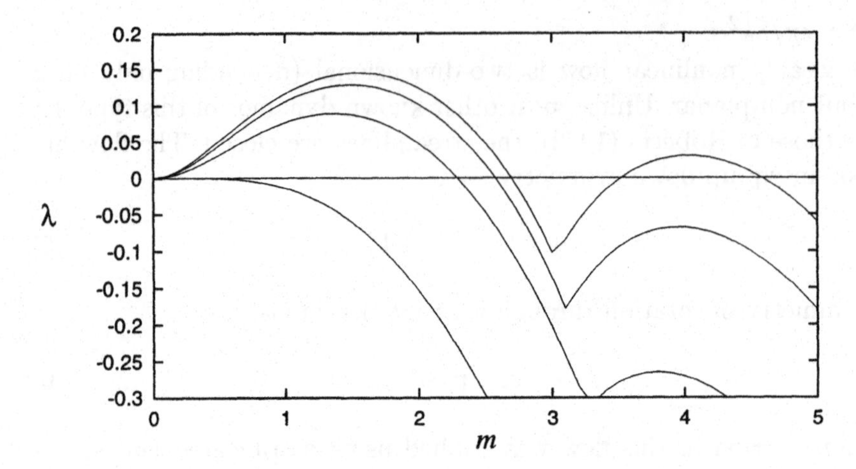


*Figure 1.* Dynamo growth rate $\lambda$ versus wavenumber $m$ for weakly nonlinear convection rolls with $T = 1000$ and $\delta = 0.5$. From top to bottom, the curves are for $R_m = 80$, 60, 40, 20.

magnetic flux. Consider integrating the induction equation (7) over a plane $x =$ constant. Applying Stokes' theorem and the boundary conditions gives

$$\lambda \iint B_x \, dy \, dz = 0. \tag{8}$$

In general this is satisfied because $\boldsymbol{B}$ is proportional to $\exp imy$. But in the limit $m \to 0$, we can deduce that either the growth rate or the net magnetic flux must vanish. For the $-1$ mode, the magnetic field is oppositely directed at the top and bottom of the layer and so the net flux is zero. But for the $+1$ mode, where in the limit $m \to 0$ there is a mean magnetic field, the net flux is non-zero and therefore the growth rate must be zero. Expressed more simply, this is just the statement that a uniform large-scale magnetic field does not grow or decay.

## 3. Derivation of the nonlinear amplitude equation

In this section the amplitude equation for the weakly nonlinear behaviour of the dynamo described above is derived. It is assumed that the system is unbounded in the $y$ direction. Suppose that the onset of dynamo action occurs at a critical magnetic Reynolds number, $R_m = R_{mc}$. Then, just beyond onset, at $R_m = R_{mc} + \epsilon^2 R_{m2}$, a band of wavenumbers is unstable. If the $-1$ mode is the first to become unstable, then the onset of dynamo action occurs at a finite wavenumber $m = m_c$, and the appropriate ansatz is to write $\boldsymbol{B} = \epsilon \boldsymbol{B}_0(x,z) \exp(im_c y) A(Y,T) + \text{c.c.}$, where $Y = \epsilon y$, $T = \epsilon^2 t$ and $A$ satisfies the Ginzburg–Landau equation $A_T = A + A_{YY} - |A|^2 A$.

However, it is clear from Figure 1 that the first modes to have positive growth rate are near to $m = 0$, and therefore the Ginzburg–Landau equation does not apply. In systems possessing conservation laws, there is a neutral large-scale mode and there are two possible cases for the onset of instability. If the critical wavenumber is non-zero, then the Ginzburg–Landau equation must be coupled to another equation for the large-scale mode. The resulting amplitude equations were derived and analysed by Matthews and Cox (2000), who showed that this situation applies to magnetoconvection and rotating convection, leading to new amplitude-driven instabilities in these systems. In the second case, which applies to the dynamo problem considered here, wavenumbers near $m = 0$ are unstable.

Consider the linear growth rate $\lambda$ as a function of the wavenumber $m$. Because of the symmetry $m \to -m$, only even powers of $m$ can appear. The leading term is therefore quadratic, and dynamo action occurs when the sign of the $m^2$ term becomes positive. Including the $m^4$ term to stabilise larger wavenumbers, the growth rate near onset can be written

$$\lambda = a(R_m - R_{mc})m^2 - bm^4 + O(m^6), \tag{9}$$

where $a$ and $b$ are positive constants. The appropriate scaling when $R_m = R_{mc} + \epsilon^2 R_{m2}$ is therefore $m = \epsilon m_1$, $\lambda = \epsilon^4 \lambda_4$. After this rescaling we can then write $\lambda_4 = aR_{m2}m_1^2 - bm_1^4$. Similar scalings were obtained by Chapman and Proctor (1980) for convection with poorly conducting boundaries. The ansatz for the magnetic field is now $\boldsymbol{B} = \epsilon \boldsymbol{B}_0(x,z)A(Y,T)$, where $Y = \epsilon y$ and $T = \epsilon^4 t$. Note that the amplitude $A$ is real, unlike the usual Ginzburg–Landau case. A nonlinear evolution equation for $A$ could be obtained by a step-by-step asymptotic expansion in powers of $\epsilon$, in which the amplitude equation would appear from a solvability condition at order $\epsilon^5$. However, this would not be straightforward, since even the leading-order problem, determining $B_0(x,z)$, cannot be done analytically.

Fortunately, it is possible to write down the form of the amplitude equation using just the symmetries and conservation law of the problem. The linear terms can be obtained from the above arguments for the linear growth rates, with each power of $m$ corresponding to a derivative with respect to $Y$. The possible nonlinear terms must involve only odd powers of $A$, since the governing equations are invariant under a sign change of $\boldsymbol{B}$, so the leading nonlinear term is cubic. Because of the conservation of magnetic flux (the time-dependent version of (8)), the nonlinear term must be an exact derivative. Finally, since the symmetry (6) reverses $Y$, the nonlinear term must include an even number of $Y$-derivatives. Combining these facts, we can deduce that the governing amplitude equation is

$$A_T = -A_{YY} - A_{YYYY} + (A^3)_{YY}$$

$$= -(A + A_{YY} - A^3)_{YY}. \qquad (10)$$

Note that all coefficients in (10) have been scaled to 1 by a suitable rescaling of $T$, $Y$ and $A$. It has been assumed that the bifurcation is supercritical in the choice of the sign of the cubic term (this is justified by the full numerical simulations of the convection-driven dynamo of Matthews 1999). The terms in (10) are all of consistent asymptotic order; all terms arise at order $\epsilon^5$ – recall that the first linear term is multiplied by $(R_m - R_{mc})$ in the unscaled equation, as in the usual Ginzburg–Landau equation.

The amplitude equation (10) is generic for systems with Euclidean symmetry combined with a conservation law and the symmetry $A \to -A$. It is therefore to be expected that this equation should arise in other contexts. For convection between boundaries at which the heat flux is fixed, the total heat content of the layer is conserved, and (10) was derived by Chapman and Proctor (1980). The equation (10) also arises in modelling phase changes in binary alloys, where it is often referred to as the Cahn–Hilliard equation (Novick-Cohen and Segel 1984). However, this is the first time that (10) has appeared in the context of dynamo theory.

## 4. Properties of the amplitude equation

In this section the behaviour of the amplitude equation (10) is described. Most of the following results were obtained by Chapman and Proctor (1980) or Novick-Cohen and Segel (1984).

The solutions $A = 1$ and $A = -1$ are stable. The solution $A = 0$ is unstable, with modes $A = A_0 \exp(\lambda t + imY)$ having a growth rate $\lambda = m^2 - m^4$. Thus all modes with $0 < m < 1$ are unstable, and the maximum growth rate is $\lambda = 1/4$ at $m = 1/\sqrt{2}$.

In the nonlinear regime, an important difference between (10) and the Ginzburg–Landau equation is that here, Fourier modes do not decouple, so nonlinear solutions should not be thought of in terms of Fourier modes.

Nonlinear, stationary solutions obey $A + A_{YY} - A^3 = cY + d$, where $c$ and $d$ are constants. In the dynamo context, it makes sense to consider periodic solutions in which the mean of $A$ is zero, with $c = d = 0$. One solution to this equation is $A = \tanh(Y/\sqrt{2})$, and periodic solutions can be written down in terms of the elliptic function sn. Periodic solutions exist with any period greater than $2\pi$.

In a large, periodic domain, there are therefore several possible nonlinear solutions, with different period. It was shown by Chapman and Proctor (1980) that any periodic solution is unstable to a solution of greater period. This leads to some interesting nonlinear dynamics, illustrated by the numerical simulation shown in figure 2. A spectral method was used, with 64 grid points and periodic boundary conditions with period 35. Because

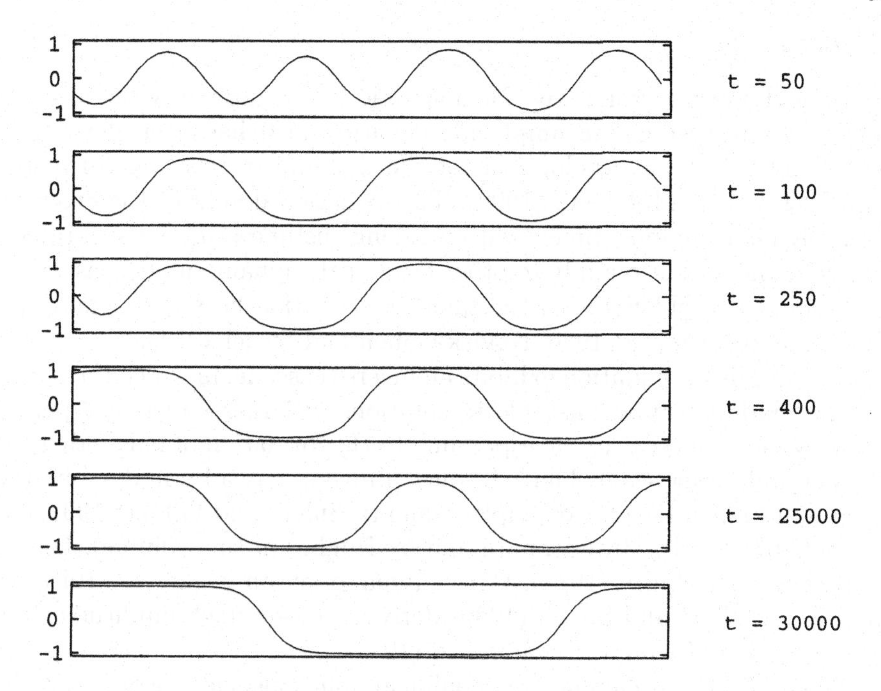


*Figure 2.* A numerical simulation of (10) in a periodic box of length 35, at six unequal time intervals.

of the fourth derivative in (10) and the very long time required to reach the final state, standard time-stepping methods are extremely inefficient. The method used solves the linear terms exactly and is second-order for the nonlinear terms (Cox and Matthews 2001), resulting in an allowable timestep approximately 50 times that of standard explicit methods.

The initial condition used was a small-amplitude random perturbation from the zero state, with zero mean. Because of the conservation law, the mean of $A$ must remain zero throughout the computation. The fastest-growing modes have $m = 1/\sqrt{2}$ and therefore a wavelength of 8.89, so in the initial stages ($t = 50$ in Figure 2) four waves are seen in the periodic box of length 35. The periodic solution with four waves in the box is unstable and gives way to a solution with three waves ($t = 100$) which in turn becomes unstable to a two-wave solution ($t = 400$). This state appears to be stable, but at very long times ($t = 25000$) evolves into the final state ($t = 30000$) in which the two locally stable solutions $A = \pm 1$ are connected through two tanh solutions. This very long timescale arises because the periodic solutions decay exponentially towards the stable solutions $A = \pm 1$, so the interaction between two pulses separated by a distance $L$ is proportional to $\exp(-cL)$ for some constant $c$, and so the timescale for the instability is of order $\exp(cL)$.

## 5. Discussion

By using the conservation law for magnetic flux combined with the symmetries of the system, an amplitude equation (10) has been derived for the weakly nonlinear development of the dynamo in rotating convection rolls described by Matthews (1999). This equation describes the evolution of the dynamo at long times and over long distances in the separable $y$ direction, and will also apply to other large-scale dynamo models as well as other physical problems with the same conservation laws and symmetries (Chapman and Proctor 1980, Novick-Cohen and Siegel 1984).

This amplitude equation exhibits an inverse cascade, in which the fastest-growing mode becomes unstable to solutions of larger and larger period, over very long timescales, corresponding to the formation of magnetic fields of larger and larger scale. The existence of inverse cascades in dynamo theory is well known (see for example, Galanti, Sulem and Gilbert 1991, and the contributions by Brandenburg and by Hughes in this volume). For the $\alpha^2$ dynamo of Roberts (1972), where dynamo action occurs at arbitrarily small $R_m$, Gilbert and Sulem (1990) derived two coupled amplitude equations that exhibit similar dynamics to (10).

Physically, the preference for nonlinear solutions such as those of Figure 2 in which large areas of uniform large-scale magnetic field are linked by fronts, is clear. Since uniform fields do not decay, no dynamo mechanism is needed to maintain such a field once it has been established.

## References


1. Chandrasekhar, S. (1961) *Hydrodynamic and hydromagnetic stability.* Oxford.
2. Chapman, C.J. and Proctor, M.R.E. (1980) Nonlinear Rayleigh–Benard convection between poorly conducting boundaries, *J. Fluid. Mech.*, **101**, 759–782.
3. Childress, S. and Soward, A.M. (1972) Convection-driven hydrodynamic dynamo, *Phys. Rev. Lett.*, **29**, 837–839.
4. Cox, S.M. and Matthews, P.C. (2000) Exponential time differencing for stiff systems, *J. Comp. Phys.*, submitted.
5. Galanti, B., Sulem, P.L. and Gilbert, A.D. (1991) Inverse cascades and time-dependent dynamos in MHD flows, *Physica* D, **47**, 416–426.
6. Gilbert, A.D. and Sulem, P.L. (1990) On inverse cascades in alpha effect dynamos, *Geophys. Astrophys. Fluid Dynamics*, **51**, 243–261.
7. Matthews, P.C. (1999) Dynamo action in simple convective flows, *Proc. R. Soc. Lond.* A, **455**, 1829–1840.
8. Matthews, P.C. and Cox, S.M. (2000) Pattern formation with a conservation law, *Nonlinearity*, **13**, 1293–1320.
9. Novick-Cohen, A. and Segel, L.A. (1984) Nonlinear aspects of the Cahn-Hilliard equation, *Physica* D, **10**, 277–298.
10. Roberts, G.O. (1972) Dynamo action of fluid motions with two-dimensional periodicity, *Phil. Trans. R. Soc. Lond.* A, **271**, 411–454.
11. Soward, A.M. (1974) A convection-driven dynamo, *Phil. Trans. R. Soc. Lond.* A, **275**, 611–630.

# A HETEROCLINIC MODEL OF GEODYNAMO REVERSALS AND EXCURSIONS

I. MELBOURNE
*Department of Mathematics*
*University of Houston, Houston, Texas 77204-3476 USA*

AND

M.R.E. PROCTOR AND A.M. RUCKLIDGE
*Department of Applied Mathematics and Theoretical Physics*
*University of Cambridge, Cambridge CB3 9EW, England*

**Abstract.** The Earth's magnetic field is by and large a steady dipole, but its history has been punctuated by intermittent excursions and reversals. This is at least superficially similar to the behaviour of differential equations containing structurally stable heteroclinic cycles. We present a model of the geodynamo that is based on the symmetries of velocity fields in a rotating spherical shell, and that contains such a cycle. Patterns of excursions and reversals that resemble the geomagnetic record can be obtained by introducing small symmetry-breaking terms.

## 1. Introduction

It has long been known that the Earth's magnetic field is due to dynamo action in its liquid core. An important part of the observational evidence for a self-excited field is the presence in palaeomagnetic data of reversals of the main axial dipolar part of the field. The timescale of these reversals ($10^5y$–$10^6y$) is very long compared with the natural diffusion time scale of the core, which is about $15\,000y$. There is an increasing amount of palaeomagnetic data on reversals, and it is clear that the distribution of reversal intervals is highly non-uniform – suggesting some sort of intermittent process [1]. More recently it has been discovered that between each reversal epoch there are many oscillations of the dipole direction and orientation which do not result in full reversals [2]; these have been termed *excursions*.



*P. Chossat et al. (eds.), Dynamo and Dynamics, a Mathematical Challenge,* 363–370.

It is only recently that detailed numerical simulations of the geodynamo have been able to show evidence of both these phenomena (see [3, 4, 5] and references therein). The time dependence of the fields in these experiments is spatially complex, and the difficulty of the calculation means that only a few reversals have been observed. It seems natural then to sacrifice spatial structure in favour of a model that can be integrated over many reversals. This has been done many times in the past, starting with the work of Bullard [6] and Rikitake [7], and continuing with many others [8, 9, 10, 11, 12]. A common feature is that the dynamo is modelled by low-order sets of ODEs that show oscillations and reversals of the field, but do not exhibit separation of timescales between quasi-steady polarity configurations and rapid reversal events. The equations are derived as highly truncated low-order models, or as descriptions of mechanical circuits very different from the Earth's core. Recent efforts [13, 14] to capture the separation of timescales rely on heteroclinic cycles in the underlying velocity field, which occur for convection in slowly rotating spheres. However, the Earth's rotation period is much less than the timescale for convection in the core.

Here we consider the more realistic situation of convection in a rotating sphere without restricting to slow rotation rates. We suppose that buoyancy forces drive a fully nonlinear flow, and focus on secondary dynamo instabilities. We use the symmetry of the velocity to write down a set of normal form equations as a model of reversals. Our model has structurally stable heteroclinic cycles, and when the symmetries are weakly broken, we can identify the residence time near these cycles with the excursion time. The reversal period in the model is many multiples of the excursion time, and appears to depend sensitively on details of the system.

## 2. Symmetries of the geodynamo

Our supposition is that the velocity field that leads to dynamo action has symmetries that can be used to distinguish different magnetic field patterns. Analytical [15] and numerical [5] convection studies yield 'cartridge belt' roll configurations with pairs of columnar cells aligned with the rotation axis, having definite rotation and reflection symmetries. There is some observational evidence (see [16]) that the actual velocity field $\mathbf{v}$ has this structure, though the number of rolls is not well-defined by the data. We look at the case of three pairs of rolls, though any number of pairs could be considered and would yield similar results. Then $\mathbf{v}$ is invariant under equatorial reflections ($\kappa$) and 120° rotations ($\rho$) about the Earth's axis. The growth of the magnetic field is described by the induction equation

$$\frac{\partial \mathbf{B}}{\partial t} = \nabla \times (\mathbf{v} \times \mathbf{B}) + \eta \nabla^2 \mathbf{B} \,, \tag{1}$$

which clearly has the symmetry $\mathcal{B}: \mathbf{B} \leftrightarrow -\mathbf{B}$. Because of the symmetries of $\mathbf{v}$, solutions to (1) can be of the following symmetry types:

| | | $\rho$ | $\kappa$ |
|---|---|---|---|
| Axial Dipole ($x_3$) | $D_a$ | $+1$ | $-1$ |
| Axial Quadrupole ($x_2$) | $Q_a$ | $+1$ | $+1$ |
| Equatorial Dipole ($z_1$) | $D_e$ | $e^{\frac{2i\pi}{3}}$ | $+1$ |
| Equatorial Quadrupole | $Q_e$ | $e^{\frac{2i\pi}{3}}$ | $-1$ |

The axial modes retain the 120° rotation symmetry of the underlying velocity field, while the equatorial modes break this symmetry.

One can imagine taking a flow with the symmetries above and changing one physical parameter (e.g. the magnetic diffusivity $\eta$) to investigate (codimension 1) bifurcations to dynamo action. Because of the symmetry $\mathcal{B}$ such bifurcations are either pitchfork or Hopf bifurcations (in fact the latter is generic for the equatorial modes with an odd number of pairs of rolls). It turns out, however, that for some dynamo models (based on Kumar–Roberts flows [17]), there can be a near degeneracy between the critical parameter for three modes, namely $D_a$, $D_e$ and $Q_a$ [18]. While there can be no rigorous proof of this (though see e.g. [19]), we shall make use of it as a motivation for investigating the effects of a codimension 3 bifurcation involving modes with these three symmetries. Three modes with different symmetries are required for a heteroclinic cycle.

We represent the magnetic field by the ansatz

$$\mathbf{B}(\mathbf{r},t) = z_1(t)\mathbf{D}_e(\mathbf{r}) + x_2(t)\mathbf{Q}_a(\mathbf{r}) + x_3(t)\mathbf{D}_a(\mathbf{r})\,. \tag{2}$$

Note that the $D_e$ mode is oscillatory. The action of the symmetries $\rho$ and $\kappa$ on $z_1$, $x_2$ and $x_3$ are in the table above. Supposing that the bifurcations are all supercritical, we have the following truncated normal form:

$$\dot{z}_1 = (\mu_1 + i\omega_1)z_1 + z_1(-|z_1|^2 + A_{12}x_2^2 + A_{13}x_3^2)\,, \tag{3}$$
$$\dot{x}_2 = \mu_2 x_2 + x_2(A_{21}x_1^2 - x_2^2 + A_{23}x_3^2)\,, \tag{4}$$
$$\dot{x}_3 = \mu_3 x_3 + x_3(A_{31}x_1^2 + A_{32}x_2^2 - x_3^2)\,. \tag{5}$$

These equations are nonlinear, in spite of the linearity of (1), because of the dynamical effects of the magnetic field on the flow. We note that at this stage the phase of $z_1$ decouples from the other variables (normal form symmetry), and so the dynamics may be described in terms of $x_1 = |z_1|$. It is easy to find conditions for a cycle of the standard type between three axial equilibria. For example, to achieve a connection between the equilibria $D_e = (\sqrt{\mu_1}, 0, 0) \rightarrow Q_a = (0, \sqrt{\mu_2}, 0)$, we need $\mu_2 + A_{21}\mu_1 > 0$, $\mu_3 + A_{31}\mu_1 < 0$. Similar connections can be found between the other pairs of equilibria under similar conditions. We can ensure that the cycle is attracting if the product of the moduli of the contracting eigenvalues at the fixed points is greater than the product of the corresponding expanding eigenvalues [20]. There is little difficulty in meeting all these conditions.

We have shown that with proper choices of the coefficients $A_{ij}$ the dynamics exhibits a structurally stable heteroclinic cycle, which takes the form of long period fluctuations in the amplitudes; these can perhaps be identified with excursions. Nonetheless, the model is unsatisfactory, not only because there are no reversals, but because an attracting cycle is characterised by ever increasing intervals between excursion-type events. Thus our definitive model consists of a refinement of the simple system (3–5).

### 3. A model for reversals

In order to obtain a model with the required properties we make the following changes: (a) we make the $z_1$ periodic orbit non-circular. This corresponds to breaking the normal form symmetry of the oscillatory $D_e$ mode – this is entirely natural since there are higher order harmonics generated naturally in the nonlinear regime. The effect of this is that the secular increase in transit times produced by the simpler system is replaced by chaotic/intermittent cycling behaviour [21]; (b) we weakly break the $\rho$ and $\kappa$ symmetries; this is also natural in the context of the Earth as there are lateral inhomogeneities in mantle heat fluxes, topography etc. This has the effect of allowing reversals of the main $D_a$ mode, on a timescale that differs from that for the excursions. The symmetry breaking terms in the model below are proportional to $\epsilon_1$ (normal form symmetry), $\epsilon_2$ (normal form and $\rho$), and $\epsilon_3$ ($\kappa$). The $\mathbf{B} \to -\mathbf{B}$ symmetry is of course retained. After some redefinition to bring out the role of the expanding and contracting eigenvalues near the fixed points, we obtain our model system

$$\begin{aligned} \dot{z}_1 &= (\mu_1 + \mathrm{i}\omega)z_1 - |z_1|^2 z_1 - \frac{c_2+\mu_1}{\mu_2}x_2^2 z_1 + \frac{e_3-\mu_1}{\mu_3}x_3^2 z_1 \\ &\quad + \epsilon_1 \bar{z}_1^5 + \epsilon_2 |z_1|^4 x_2 + \epsilon_3 z_1 x_2 x_3^3 , \end{aligned} \tag{6}$$

$$\begin{aligned} \dot{x}_2 &= \mu_2 x_2 - x_2^3 - \frac{c_3+\mu_2}{\mu_3}x_3^2 x_2 + \frac{e_1-\mu_2}{\mu_1}|z_1|^2 x_2 \\ &\quad + \epsilon_1 \,\mathrm{Re}(z_1^3)\, x_2^2 + \epsilon_2 x_1^5 + \epsilon_3 x_3^5 , \end{aligned} \tag{7}$$

$$\begin{aligned} \dot{x}_3 &= \mu_3 x_3 - x_3^3 - \frac{c_1+\mu_3}{\mu_1}|z_1|^2 x_3 + \frac{e_2-\mu_3}{\mu_2}x_2^2 x_3 \\ &\quad + \epsilon_1 \,\mathrm{Re}(z_1^3)\, x_2 x_3 + \epsilon_2 x_1^3 x_2 x_3 + \epsilon_3 x_2^5 , \end{aligned} \tag{8}$$

where $z_1 = x_1 + \mathrm{i}y_1$, $\mu_1$, $\mu_2$, $\mu_3$ and $\omega$ are growth rate parameters and a frequency, $c_1$, $c_2$ and $c_3$ are the contracting eigenvalues at the three fixed points of the cycle and $e_1$, $e_2$ and $e_3$ are the expanding eigenvalues.

In this preliminary report we focus on results for just one set of parameter values, selected to produce reversal-like behaviour. These are:

$$\begin{array}{lll} \lambda_1 = 0.3\ (\omega_1 = 1) & e_1 = 0.7 & c_1 = 1.1 \\ \lambda_2 = 0.2 & e_2 = 1.7 & c_2 = 1.2 \\ \lambda_3 = 0.4 & e_3 = 0.02 & c_3 = 1.3 \\ \epsilon_1 = 0.12 & \epsilon_2 = 0.1 & \epsilon_3 = 0.001 \end{array}$$

Figure 1(a) shows a typical long time series (details shown in figure 1b–d). The $x_3$ variable, measuring the dipole strength, vacillates many times without changing sign, punctuated by much rarer reversal events. The phase portrait in figure 1(e) shows that the dynamics is close to a heteroclinic cycle, while figure 1(f) shows the oscillatory nature of the $x_1$ variable. The probability distribution of reversals and excursions is in figure 2. While the excursion data show an unrepresentative peak at about 300 (arbitrary) time units, the reversal data is smoother and the cumulative picture is similar (though without 'superchron' outliers) to the real dataset (figure 3).

## 4. Discussion

In this paper we have shown some results from a normal form model based on the (approximate) symmetries of the full geodynamo problem. This model captures certain intermittent phenomena associated with the geodynamo in the form of excursions and field-reversals. Our model requires magnetic fields of three different symmetry types in order to have a heteroclinic cycle, and it differs from earlier low order models of reversals in that the symmetries are paramount in obtaining the form of the model equations. Of course, as for other low-order calculations, the parameters cannot be related to Earth-like quantities at this stage. That said, transitions between magnetic fields with different types of symmetry are a prominent feature of large-scale geodynamo calculations [3, 4]. In addition, our model suggests that weakly broken symmetries may play an important role in determining the ratio between excursion and reversal timescales. More detailed investigations of the model will be reported elsewhere [22].

**Acknowledgements.** We thank David Gubbins for helpful discussions. AMR is grateful for support from the EPSRC. The research of IM is supported in part by NSF Grant DMS-0071735. All three authors are grateful to NATO for support in attending the workshop in Cargèse.

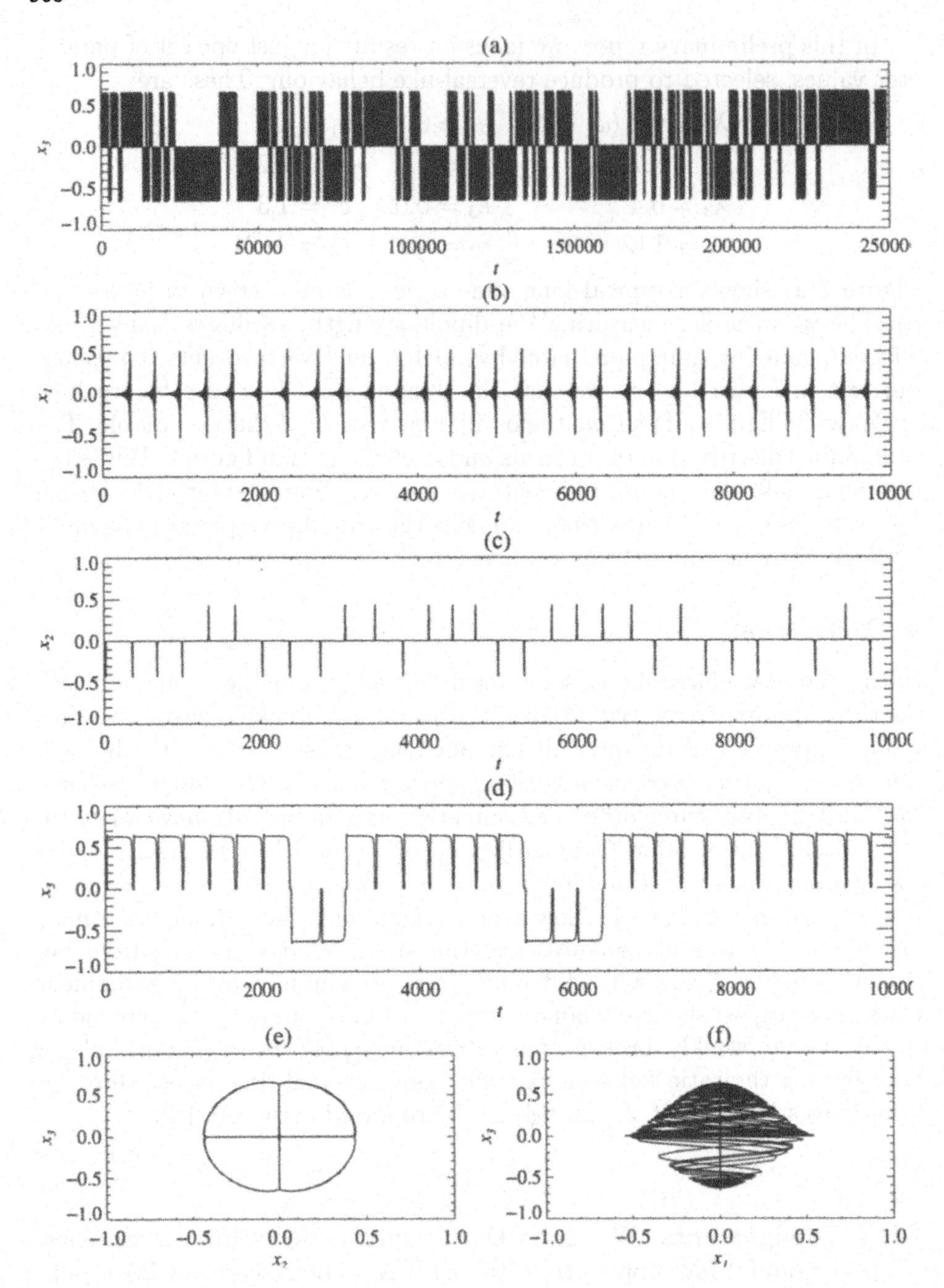


*Figure 1.* Behaviour of (6–8) for the parameters given in the text. (a) shows a long time series for the dipole mode $x_3$, and (b–d) show the $x_1$, $x_2$ and $x_3$ evolution in the first part of the time series. (e–f) show phase portraits: (e) $x_3$ vs. $x_2$ (f) $x_3$ vs. $x_1$.

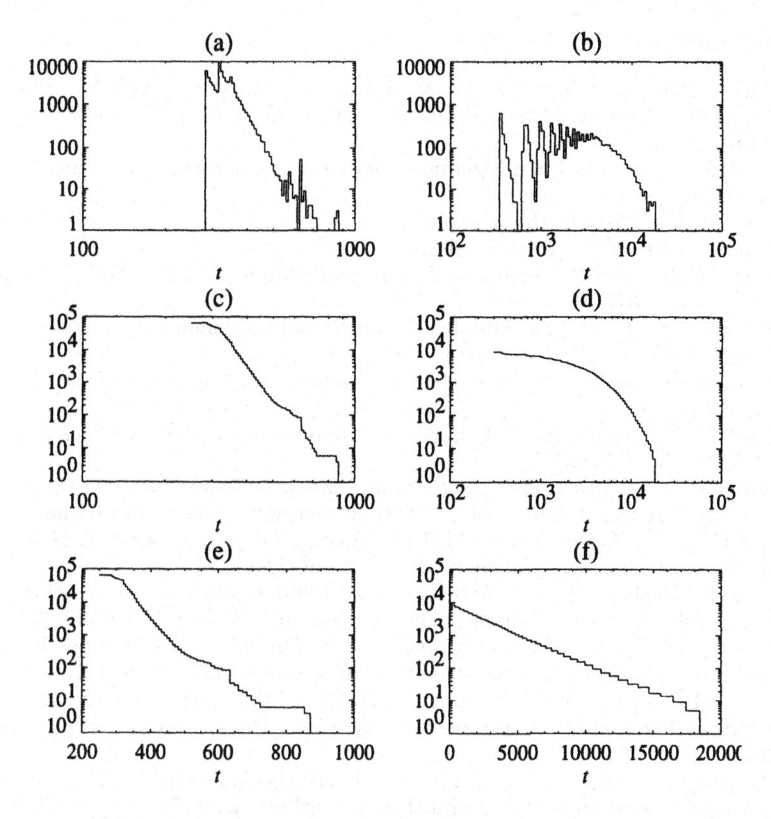


*Figure 2.* Distributions of durations of (a,c,e) excursions and (b,d,f) reversals. (a,b) histograms. (c,d) cumulative plot (log–log). (e,f) cumulative plot (log–linear).

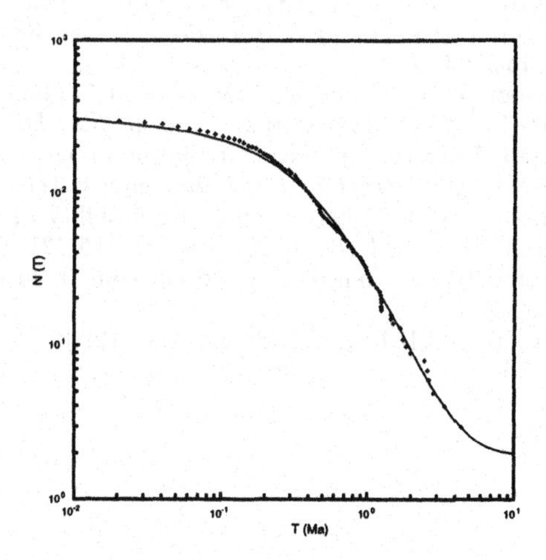


*Figure 3.* Distribution of reversal durations: cumulative plot (log–log) for the geomagnetic field (reproduced from [1]). Compare with figure 2(d).

## References


1. Merrill, R.L., McElhinny, M.W. & McFadden, P.L. (1996) *The Magnetic Field of the Earth: Paleomagnetism, the Core, and the Deep Mantle.* Academic Press: San Diego.
2. Gubbins, D. (1999) The distinction between excursions and reversals, *Geophys. J. Int.* **137** F1–F3
3. Busse, F.H. (2001) In this volume
4. Roberts, P.H. (2001) In this volume
5. Jones, C.A. (2000) Convection-driven geodynamo models, *Phil. Trans. R. Soc. Lond. A* **358** 873–897
6. Bullard, E.C. (1955) The stability of a homopolar dynamo, *Proc. Camb. Phil. Soc.* **51** 744–760
7. Rikitake, T. (1958) Oscillations in a system of disk dynamos, *Proc. Camb. Phil. Soc.* **54** 89–105
8. Allan, D.W. (1962) On the behaviour of systems of coupled dynamos, *Proc. Camb. Phil. Soc.* **58** 671–693
9. Robbins, K.A. (1975) *Disk Dynamos and Magnetic Reversal* (PhD thesis, M.I.T.)
10. Chui, A.Y.K. & Moffatt, H.K. (1993) A thermally driven disc dynamo. In *Solar and Planetary Dynamos* (ed. M.R.E. Proctor, P.C. Matthews & A.M. Rucklidge), pp. 51–58. Cambridge University Press: Cambridge
11. Hide, R., Skeldon, A.C. & Acheson, D.J. (1996) A study of two novel self-exciting single-disk homopolar dynamos: theory, *Proc. R. Soc. Lond.* A **452** 1369–1395
12. Moroz, I.M., Hide, R. & Soward, A.M. (1998) On self-exciting coupled Faraday disk homopolar dynamos driving series motors, *Physica* **117D** 128–144
13. Oprea, I., Chossat, P. & Armbruster, D. (1997) Simulating the kinematic dynamo forced by heteroclinic convective velocity fields, *Theor. Comput. Fluid Dyn.* **9** 293–309
14. Chossat, P., Guyard, F. & Lauterbach, R (1999) Generalized heteroclinic cycles in spherically invariant systems and their perturbations, *J. Nonlinear Sci.* **9** 479–524
15. Busse, F.H. (1970) Thermal instabilities in rapidly rotating systems, *J. Fluid Mech.* **44** 441–460
16. Jackson, A., Jonkers, A.R.T. & Walker, M.R. (2000) Four centuries of geomagnetic secular variation from historical records, *Phil. Trans. R. Soc. Lond. A* **358** 957–990
17. Kumar, S. & Roberts, P.H. (1975) A three-dimensional kinematic dynamo, *Proc. R. Soc. Lond.* A **344** 235–258
18. Gubbins, D., Barber, C.N., Gibbons, S. & Love, J.J. (2000) Kinematic dynamo action in a sphere. II. Symmetry selection, *Proc. R. Soc. Lond.* A **456** 1669–1683
19. Proctor, M.R.E. (1977) The role of mean circulation in parity selection by planetary magnetic fields, *Geophys. Astrophys. Fluid Dynamics* **8** 311–324
20. Krupa, M. & Melbourne, I. (1995) Asymptotic stability of heteroclinic cycles in systems with symmetry, *Ergod. Th. & Dynam. Sys.* **15** 121–147
21. Melbourne, I. (1989) Intermittency as a codimension three phenomenon, *J. Dyn. Diff. Eqns.* **1** 347–367
22. Melbourne, I., Proctor, M.R.E. & Rucklidge, A.M. (2001) In preparation

# ASPECTS OF THE DYNAMICS UNDERLYING SOLAR AND STELLAR DYNAMO MODELS

REZA TAVAKOL, EURICO COVAS
*Astronomy Unit*
*School of Mathematical Sciences*
*Queen Mary & Westfield College*
*Mile End Road*
*London E1 4NS, UK*

AND

DAVID MOSS
*Department of Mathematics*
*The University*
*Manchester M13 9PL, UK*

**Abstract.** Observations of the Sun and solar-type stars continue to reveal phenomena whose understanding is very likely to require a nonlinear framework. Here we shall concentrate on two such phenomena, namely the grand minima type behaviour observed in the Sun and solar-type stars and the recent dynamical variations of the differential rotation in the solar convection zone, deduced from the helioseismic observations, and discuss how their explanations have recently motivated the development/employment of novel ideas from nonlinear dynamics.

## 1. Introduction

An important characteristic of the Sun and solar-type stars is their variability over a very large range of time (and space) scales, including the intermediate time scales of $\sim 10^0 - 10^4$ years (Weiss 1990). Though ultimately spatiotemporal in character, these variabilities fall into two categories: (i) those whose main features can be explained in terms of temporal models and (ii) those which require a spatiotemporal approach for their explanations.



*P. Chossat et al. (eds.), Dynamo and Dynamics, a Mathematical Challenge,* 371–379.

Here we shall be focusing on two such forms of variability, namely the grand minima type episodes seen in in sunspot activity and the recently observed dynamical variations of the differential rotation in the solar convection zone.

Evidence for the presence of the grand–minima type behaviour comes from a variety of sources, including the studies of the historical records of the annual mean sunspot data since 1607 AD which show the occurrence of epochs of suppressed sunspot activity, such as the *Maunder minimum* (Eddy 1976, Foukal 1990, Wilson 1994, Ribes & Nesme-Ribes 1993, Hoyt & Schatten 1996). Further research, employing $^{14}C$ (Stuiver & Quey 1980; Stuiver & Braziunas 1988, 1989) and $^{10}B$ (Beer *et al.* 1990, 1994a,b, Weiss & Tobias 1997) as proxy indicators, has provided additional strong evidence that the occurrence of such epochs of reduced activity (referred to as *grand minima*) has persisted with irregular time intervals in the past. Further, they appear to be typical of solar-type stars (Baliunas *et al.* 1995).

Evidence for the variations in the dynamical behaviour of the differential rotation comes from recent analyses of the inversions of the helioseismic data from both the Michelson Doppler Imager (MDI) instrument on board the SOHO spacecraft and the Global Oscillation Network Group (GONG) project which provide evidence to show (i) that the previously observed surface torsional oscillations with periods of about 11 years (e.g. Howard & LaBonte 1980; Snodgrass, Howard & Webster 1985; Kosovichev & Schou 1997; Schou *et al.* 1998) extend significantly downwards into the solar convective zone to depths of about 10 percent in radius (Howe *et al.* 2000a and Antia & Basu 2000; see Covas *et al.* 2000a for a recent study), and (ii) that the dynamical regimes at the base of the convection zone may be different from those observed at the top, having either significantly shorter periods (Howe *et al.* 2000b) or non–periodic behaviour (Antia & Basu 2000).

Our aim here is to briefly describe some recent attempts at understanding these phenomena within a nonlinear theoretical framework.

## 2. Intermittency as a mechanism underlying grand minima type behaviour

The seemingly irregular variations of the grand minima type in the sunspot activity are of interest for at least two reasons. Firstly, they are theoretically challenging, especially in view of the absence of any naturally occurring mechanisms in the Sun and solar-type stars with appropriate time scales (Gough 1990). Secondly, given their time scales, such variations can be of potential consequence for the occurrence of climatic variability on similar time scales (e.g. Friis-Christensen & Lassen 1991, Beer *at al.* 1994b, Lean 1994, Stuiver, Grootes & Braziunas 1995, Baliunas & Soon 1995, Butler &

Johnston 1996, White *et al.* 1997).

These considerations have motivated considerable effort into trying to understand the mechanism(s) underlying such variations, employing a variety of approaches (e.g. Weiss *et al.* 1984; Sokoloff and Nesme–Ribes 1994; Tobias *et al.* 1995; Knobloch and Landsberg 1996; Schmitt *et al.* 1996, Knobloch *et al.* 1998; Tobias 1998a,b; Tobias *et al.* 1999).

In principle, there are essentially two frameworks within which such variabilities could be studied: stochastic and deterministic. In practice, however, it is difficult to distinguish between these two frameworks (Weiss 1990). Here we shall concentrate on the deterministic approach and recall that whatever the relevant framework may turn out to be, the deterministic components will still be present and are likely to play an important role.

Within this framework, a number of approaches have been adopted. Among these is the employment of low dimensional ODE models that are obtained using the Normal Form approach (Spiegel 1994, Tobias *et al.* 1995, Knobloch *et al.* 1996). These have led to robust modes of behaviour which are of potential importance in accounting for certain aspects of solar variability of the grand minima type, such as various forms of amplitude modulation of the magnetic field energy.

An alternative way to understand this type of variability in the sunspot record is to postulate that this type of behaviour is caused by some form of dynamical intermittency, an idea that in various forms goes back at least to the late 1970s (e.g. Tavakol 1978, Ruzmaikin 1981, Zeldovich *et al.* 1983, Weiss *et al.* 1984, Spiegel 1985, Feudel *et al.* 1993, Schmitt *et al.* 1996, Brooke 1997, Tworkowski *et al.* 1998).

Much effort has gone into producing evidence for the occurrence of various forms of dynamical intermittency in both truncated and PDE mean-field dynamo models. Given the large number of intermittency types that dynamical systems theory has produced and bearing in mind the complexity of the full dynamo equations, a natural approach would be to single out the main generic ingredients of stellar dynamos and to study their dynamical consequences. For axisymmetric dynamo models, these ingredients consist of the presence of invariant subspaces, non-normal parameters and the non-skew property (see e.g. Ashwin *et al.* 1999 for definitions). The dynamics underlying such systems has recently been studied (Covas *et al.* 1997c, 1999b; Ashwin *et al.* 1999), leading to a number of novel phenomena, including a new type of intermittency, referred to as *in–out intermittency* (see also Brooke 1997).

The easiest way to characterise in–out intermittency is by contrasting it with on–off intermittency. Briefly, let $M_I$ be the invariant subspace and $A$ the attractor which exhibits either on–off or in–out intermittency. One then has on–off or in–out intermittencies, depending respectively upon whether the intersection $A_0 = A \cap M_I$ is a minimal attractor or not. The name in–out is chosen to be indicative of the fact that in this case there can be different invariant sets in $A_0$ associated with the attraction and repulsion transverse to $A_0$. Another crucial difference between the two is that, as opposed to on–off intermittency, in the case of in–out intermittency the minimal attractors in the invariant subspaces do not necessarily need to be chaotic and hence the trajectories can (and often do) shadow a periodic orbit in the 'out' phases (see Ashwin *et al.* 1999 for details).

Given the genericity of the ingredients required for the presence of this type of intermittency, it is of interest to see if this type of behaviour can indeed occur in the context of dynamo models. Interestingly, this form of intermittency has been concretely shown to occur in a number of dynamo models, including both ODE (Covas et al. 1997c) and PDE (Covas et al. 2000c) mean field dynamo models. An example of such an occurrence is given in Fig. 1, where in–out intermittency occurs for an axisymmetric PDE mean–field dynamo model (See Covas *et al.* (2000d) for details).

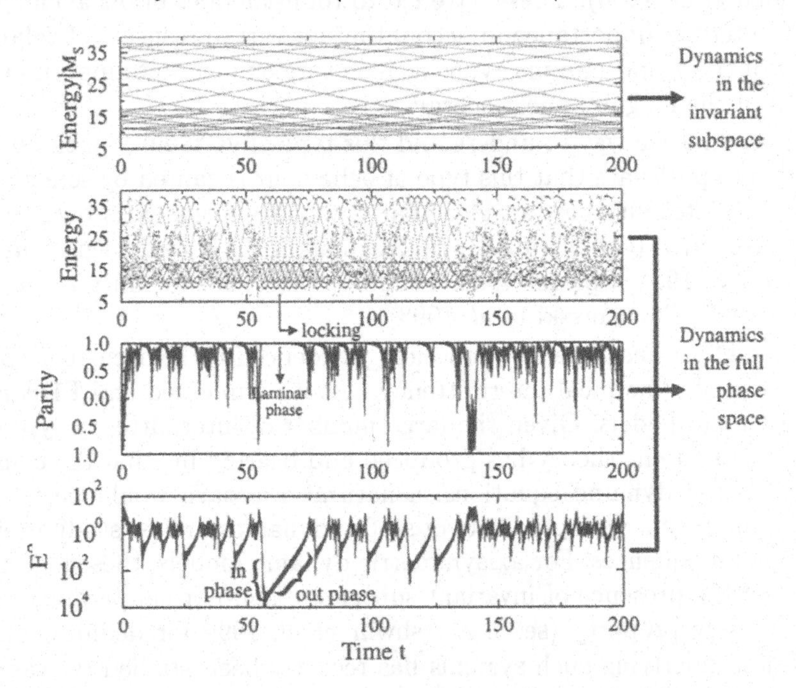


*Figure 1.* In–out intermittency in the axisymmetric PDE mean–field dynamo model. The parameters used were $r_0 = 0.4$, $C_\alpha = 1.505$, $C_\omega = -10^5$, $f = 0.7$, together with an algebraic form of $\alpha$. To enhance visually the periodic locking we have time sampled the series in the two upper panels. See Covas *et al.* (2000d) for details of the model.

We should add here that in addition to in–out intermittency, firm evidence has also been found for the occurrence of other forms of intermittency in mean field dynamo models. These include crisis (or attractor merging) intermittency, in both ODE (Covas & Tavakol 1997) and PDE (Covas & Tavakol 1999) dynamo models, and Type I Intermittency, in both ODE (Covas *et al.* 1997c) and PDE models (Covas & Tavakol 1999). We further note that on-off intermittency (Platt *et al.* 1993, see also Schmitt *et al.* 1996 and Ossendrijver 2000) is also likely to be relevant for such models, given their invariant subspaces.

The demonstration of the occurrence of various form of intermittency establishes an important element of the *multiple intermittency hypothesis* (Tavakol & Covas 1999), according to which the grand minima type variability in solar-type stars may be understood in terms of one or a number of types of dynamical intermittency. What remains to be shown is whether these types of intermittency still persist in more realistic models and ultimately, through detailed comparison with observational and proxy data, whether they actually occur in the Sun and stars, and whether they can in fact account for the grand minima type behaviour.

The important point to be emphasised here is that in addition to theoretical predictions for these types of intermittency, their occurrence has been conclusively demonstrated not only in truncated dynamo models, but also in non-truncated PDE dynamo models (in three spatial dimensions, although so far restricted to axisymmetry, and time).

## 3. Spatiotemporal bifurcations as a mechanism for different modes of behaviour in the solar convection zone

An important outcome of the recent inversion of the helioseismic data has been to provide increasing evidence to show that the Sun is likely to be very active dynamically throughout the convection zone. To establish firmly the precise nature of these variations, future observations are clearly required. However, despite the error bars that such inversions are bound to entail, the results so far seem to point to the very interesting possibility that the variations in the differential rotation can have different periodicities/behaviours at different depths in the solar convection zone, having either markedly reduced periods (Howe *et al.* 2000b) or non–periodic behaviour (Antia & Basu 2000) at the bottom of the convective zone.

There are in principle two ways to account for such behaviours. One could imagine different physical mechanisms at different physical locations in the convection zone, resulting in different dynamical behaviours in those locations. Alternatively, a dynamical mechanism could be sought that could produce such different behaviours at different spatial locations, without requiring different physical mechanisms at those locations.

As an example of the latter mechanism, spatiotemporal fragmentation/bifurcation has recently been proposed as a possible natural dynamical mechanism to account for such observed multi-mode behaviours in different parts of the solar convection zone (Covas *et al.* 2000b). In this way, dynamical regimes with different temporal behaviours can coexist at different spatial locations, for certain *given* values of the control parameters of the system. The important point here is that, in contrast to the usual temporal bifurcations, which result in identical temporal behaviour at each spatial point, and which require changes in parameters in order to be initiated, spatiotemporal bifurcations can result in different dynamical modes of behaviour at different locations without requiring changes in the control parameters. Also, importantly, the occurrence of such diverse modes of behaviour does not require different physical mechanisms at different locations.

Evidence for the occurrence of this mechanism was found in the context of a two dimensional axisymmetric mean field dynamo model operating in a spherical shell, with a semi–open outer boundary condition, in which the only nonlinearity is the action of the azimuthal component of the Lorentz force of the dynamo generated magnetic field on the solar angular velocity (Covas *et al.* 2000b). The zero order angular velocity was chosen to be consistent with the most recent helioseismological data (MDI). Subsequently it has been shown, through a detailed study (Covas *et al.* 2000c), that the occurrence of this type of behaviour does not depend upon the details of the model employed, thus providing strong evidence to support the idea that spatiotemporal fragmentation is likely to occur in general dynamo settings.

As an example of this type of spatiotemporal fragmentation/bifurcation, we have plotted in Fig. 2 the radial contours of the angular velocity residuals $\delta\Omega$, as a function of time for a cut at a fixed latitude. This demonstrates how, as a result of such spatiotemporal bifurcation, the period (as well ads the phase) of the angular velocity residuals vary going from the top to the bottom of the convection zone. The details of the model are given in Covas *et al.* (2000b,c).

To show the the presence of the spatiotemporal bifurcation more clearly, we also studied the bifurcation diagram by calculating the time between the minima of $\delta\Omega(r = 0.66R_{\odot}, \theta = 30°)$, the perturbation to the zero order rotation rate at a fixed latitude and radius, as a function of time.

Fig. 3 shows an example of such bifurcation, demonstrating clearly a sudden bifurcation from a fundamental 11 year cycle to a cycle with two periods (whose sum is 11 years). Thus this mechanism is capable of producing different modes of behaviour at different locations, for given values of the control parameters of the system.

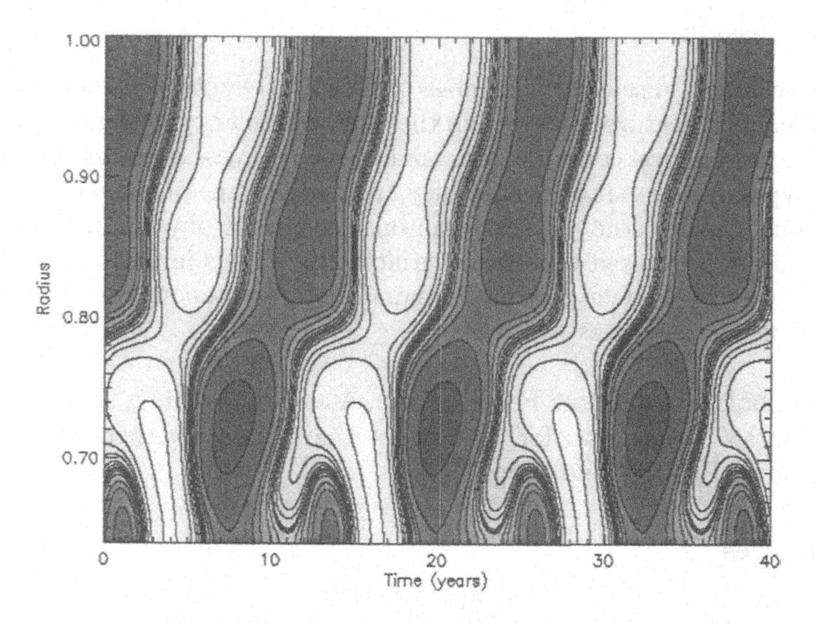


*Figure 2.* Radial contours of the angular velocity residuals $\delta\Omega$ as a function of time for a cut at latitude 30°. Parameter values are $R_\alpha = -11.0$, $P_r = 0.9$, $R_\omega = 44000$ for $\alpha(r,\theta) = \alpha_r \cos\theta \sin^2\theta$. See Covas *et al.* (2000b) for the details of the model.

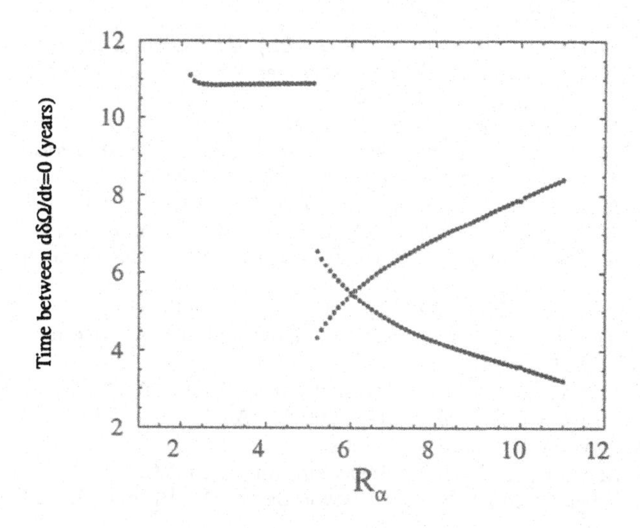


*Figure 3.* Bifurcation diagram showing the time between minima of the curve $\Omega(r,\theta,time)$. It shows clearly a sudden bifurcation from a fundamental 11 year cycle to a cycle with two timescales (with sum 11 years). Parameter values are $R_\alpha = -11.0$, $P_r = 1.0$, $R_\omega = 44000$ for a $\alpha(r,\theta) = \alpha_r \cos\theta \sin^2\theta$. See Covas *et al.* (2000b) for the details of the model.

## 4. Discussion

We have briefly discussed a number of recent attempts at understanding two dynamical features observed in the Sun (and solar type stars), in terms of a number of novel concepts from nonlinear dynamics. A crucial feature of these types of explanation is that they are based on the nonlinear nature of the regimes under study, rather than on any specific physical mechanisms possessing the required time scale variabilities. This is important, particularly in view of the fact that much remains unknown about the underlying physics in these regimes.

We would like to thank Peter Ashwin, Axel Brandenburg, John Brooke, and Andrew Tworkowski for the work we have done together. EC is supported by a PPARC fellowship.

## References

1. Antia H. M., and Basu S., 2000, ApJ, 541, 442
2. Ashwin, P., Covas, E. & Tavakol, R., 1999, *Nonlinearity*, **9**, 563.
3. Baliunas, S. L. & Soon, W., 1995, *Astrophy. J.*, **450**, 896.
4. Baliunas, S. L., *et al.*, 1995, *Astrophysical Journal*, **438**, 269.
5. Beer, J., *et al.*, 1994, in *The Sun as a Variable Star: Solar and Stellar Irradiance Variations*, J. M. Pap, C. Fröhlich, H. S. Hudson and S. K. Solaski, eds., Cambridge University Press, Cambridge, 291.
6. Beer, J. *et al.*, 1994b, in E. Nesme-Ribes (ed.), *The Solar Engine and its Influence on Terrestial Atmosphere and Climate*, Springer-Verlag, Berlin, p. 221.
7. Beer, J. *et al.*, 1990, *Nature*, **347**, 164.
8. Brooke, J. M., 1997, *Europhysics Letters*, **37**, 3.
9. Butler, C. J. & Johnston, D. J., 1996, *J. Atmospheric Terrest. Phys.*, **58**, 1657.
10. Covas, E. & Tavakol, R., 1997, *Phys. Rev. E*, **55**, 6641.
11. Covas, E., Ashwin, P. & Tavakol, R., 1997b, *Phys. Rev. E*, **56**, 6451.
12. Covas, E. & Tavakol, R., 1998, Proceedings of the 5th International Workshop "Planetary and Cosmic Dynamos", Trest, Czech Republic, Studia Geophysica et Geodaetica, 42.
13. Covas, E. & Tavakol, R., 1999, *Phys. Rev. E*, **60**, 5435.
14. Covas, E., Tavakol, R., Tworkowski, A., Brandenburg, A., Brooke, J. M. & Moss, D., 1999, *A&A*, **345**, 669.
15. Covas E., Tavakol R., Moss D., Tworkowski A., 2000a, A&A 360, L21.
16. Covas E., Tavakol R., Moss D., 2000b, A&A 363, L13, also available at http://www.eurico.web.com
17. Covas E., Tavakol R., Moss D., 2000c, submitted to A&A.
18. Covas, E., Tavakol, R., Ashwin, P., Tworkowski, A., Brooke, J. , 2000d, submitted to *Chaos*.
19. Eddy, J. A., 1976, *Science*, **192**, 1189.
20. Feudel, W. Jansen, & J. Kurths, 1993, *Int. J. of Bifurcation and Chaos*, **3**, 131.
21. Foukal, P. V., 1990, *Solar Astrophysics*, Wiley Interscience, New York.
22. Friis-Christensen, E. & Lassen, K., 1991, Science **254**, 698.
23. Gough, D., 1990, *Phil. Trans. R. Soc. Lond.*, **A330**, 627.
24. Howard R., and LaBonte B. J., 1980, ApJ, 239, L33.
25. Howe R., et al., 2000a, ApJ Lett., 533, L163.
26. Howe R., et al., 2000b, Science 287, 2456.

27. Hoyt, D. V. & Schatten, K. H., 1996, *Solar Phys.*, **165**, 181.
28. L. L. Kitchatinov, 1987, *Geophys. Astrophys. Fluid Dyn.*, **38**, 273.
29. Knobloch, E., and Landsberg, A. S., 1996, *Mon. Not. Royal Astron. Soc.*, **278**, 294.
30. Knobloch, E., Tobias, S. M. & Weiss, N. O., 1998, *MNRAS*, **297**, 1123.
31. Kosovichev A. G., and Schou J., 1997, ApJ, 482, L207.
32. Lean, J., 1994, in E. Nesme-Ribes (ed.), *The Solar Engine and its Influence on Terrestial Atmosphere and Climate*, Springer-Verlag, Berlin, p. 163.
33. Ossendrijver, M. A. J. H., 2000, *A&A*, **359**, 364.
34. Platt, N., Spiegel, E., and Tresser, C., 1993, *Phys. Rev. Lett.*, **70**, 279.
35. Ribes, J. C. & Nesme-Ribes, E., 1993, *A&A*, **276**, 549.
36. Ruzmaikin, A. A., 1981, *Comm. Astrophys.*, **9**, 88.
37. Schmitt, D., Schüssler, M., and Ferriz–Mas, A., 1996, *A&A*, **311**, L1.
38. Sokoloff, D., and Nesme–Ribes, E., 1994, *A&A*, **288**, 293.
39. Snodgrass H. B., Howard R. F., and Webster L., 1985, *Sol. Phys.*, **95**, 221.
40. Spiegel, E., Platt, N. & Tresser, C., 1993b, *Geophys. and Astrophys. Fluid Dyn.*, **73**, 146.
41. Spiegel, E.A. 1994, in Proctor M.R.E., Gilbert A.D., eds, *Lectures on Solar and Planetary Dynamos*, Cambridge Univ. Press, Cambridge.
42. Spiegel, in *Chaos in Astrophysics*, edited by J. R. Butcher, J. Perdang, & E. A. Spiegel (Reidel, Dordrecht, 1985).
43. Stuiver, M., Grootes, P. M. & Braziunas, T. F., 1995, *Quarternary Res.*, **44**, 341.
44. Stuiver, M. & Braziunas, T. F., 1988, in F. R. Stephenson & A. W. Wolfendale (eds.), *Secular Solar and Geomagnetic Variations in the Last 10 000 Years*, Kluwer Academic Publishers, Dordrecht, Holland, p. 245.
45. Stuiver, M. & Braziunas, T. F., 1989, *Nature*, **338**, 405.
46. Stuiver, M. & Quay, P. D., 1980, *Science*, **207**, 19.
47. Tavakol, R., 1978, *Nature*, **276**, 802.
48. Tavakol, R., and Covas, E., 1999, *Astron. Soc. of the Pacific Conference Series*, **178**, 173.
49. Tobias, S. M., Weiss, N. O., and Kirk, V., 1995, *Mon. Not. Royal Astron. Soc.*, **273**, 1150.
50. Tobias, S. M., 1998a, *Astron. Soc. of the Pacific Conference Series*, **154**, 1349.
51. Tobias, S. M., 1998b, *Mon. Not. Royal Astron. Soc.*, **296**, 653.
52. Tobias, S. M., Knobloch, E., and Weiss, N. O., 1999, *Astron. Soc. of the Pacific Conference Series*, **178**, 185.
53. Tworkowski, A., Tavakol, R., Brandenburg, A., Brooke, J. M., Moss, D. & Tuominen I., 1998, *MNRAS*, **296**, 287.
54. Weiss, N. O., Cattaneo, F., Jones, C. A., 1984, *Geophys. Astrophys. Fluid Dyn.*, **30**, 305.
55. Weiss, N. O., in *Lectures on Solar and Planetary Dynamos*, edited by Proctor, M.R.E. and Gilbert, A.D., Cambridge University Press, Cambridge (1994).
56. Weiss, N. O. & Tobias, S. M., 1997, in Solar and Heliospheric Plasma Physics, ed. G. M. Simnett, C. E. Alissandrakis & L. Vlahos, 25, Springer, Berlin.
57. Weiss, N. O., 1990, *Phil. Trans. R. Soc. Lond.*, **A330**, 617.
58. White, W. B., Lean, J., Cayan, D. & Dettinger, M. D., 1997, *J. Geophys. Res.*, **102**, 3255.
59. Wilson, P. R., 1994, *Solar and Stellar Activity Cycles*, Cambridge University Press, Cambridge.
60. Zeldovich, Ya. B., Ruzmaikin, A. A. & Sokoloff, D. D., 1983, *Magnetic Fields in Astrophysics*, Gordon and Breach, New York.

# MODULATION AND SYMMETRY-BREAKING IN LOW-ORDER MODELS OF THE SOLAR DYNAMO

N.O. WEISS
*Department of Applied Mathematics and Theoretical Physics, University of Cambridge, Cambridge CB3 9EW, UK*

AND

E. KNOBLOCH AND S.M. TOBIAS
*Department of Applied Mathematics, University of Leeds, Leeds LS2 9JT, UK*

**Abstract.** Modulation of cyclic magnetic activity associated with grand minima, as well as modulation associated with breaking of dipole or quadrupole symmetry, can be represented in low-order normal form equations. The behaviour found is robust and can be related to similar spatio-temporal patterns in mean field dynamo models.

## 1. Introduction

There are several possible approaches to astrophysical dynamos. Much work so far has relied on mean field dynamo theory: if one wants to go beyond that, there is a choice between massive computation (which has proved so successful for the geodynamo but would be premature for the Sun) at one extreme, and low-order models at the other. There is a long history of using low-order systems of ordinary differential equations (ODEs) to describe nonlinear dynamos, including some very early studies of chaotic behaviour [13]. Such systems can be constructed in three different ways: many authors have simply invented convenient evolution equations and studied their behaviour; others have obtained a low-order model by truncating the relevant partial differential equations (PDEs) – but there is then a risk that interesting behaviour will disappear as higher-order terms are included; so the safest route is to rely on normal form equations, which are structurally stable. Although they have no detailed predictive power, they yield patterns



*P. Chossat et al. (eds.), Dynamo and Dynamics, a Mathematical Challenge,* 381–390.

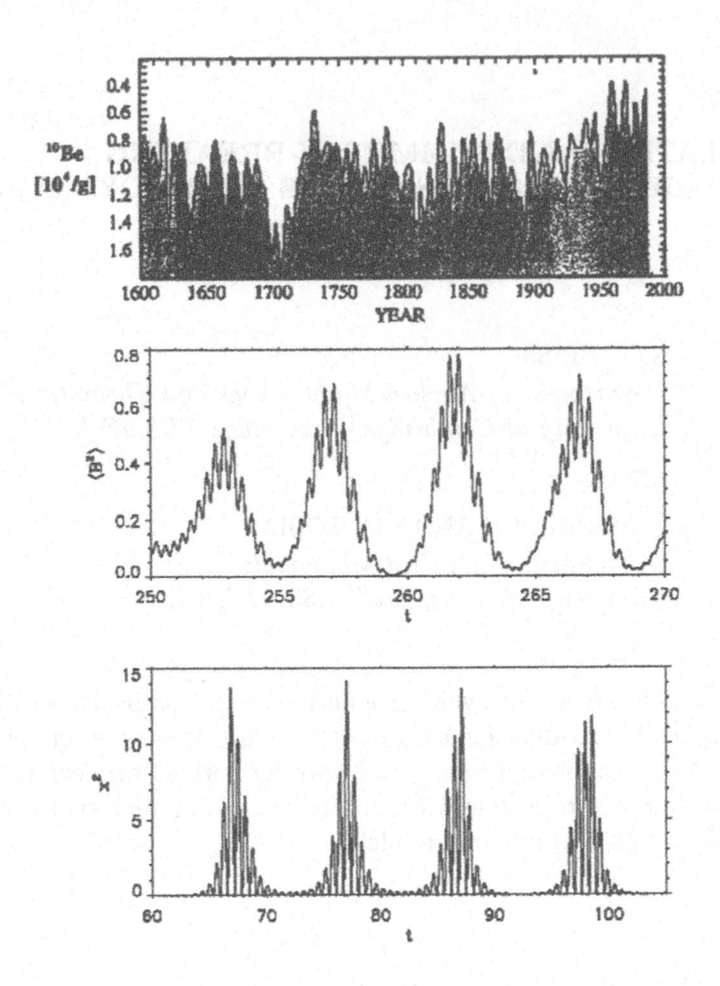


*Figure 1.* Modulation of solar magnetic activity. The top panel shows the $^{10}$Be record since 1600, which is anticorrelated with sunspots. The central panel shows the magnetic energy from a two-dimensional PDE model, while the lower panel shows the corresponding energy $x^2$ from the ODEs (1)–(3).

of behaviour which are robust. In this paper we show how systems based on normal form equations can be used to explain key features of the solar cycle.

The Sun is a typical middle-aged, middling-sized star; unlike the Earth, it has an inner core (about 70% by radius) in which energy is carried by radiation, enclosed by a turbulent outer convection zone. Magnetic fields at the solar surface are highly intermittent but nevertheless exhibit systematic patterns of cyclic activity (e.g. [14] [14]; [15] [15]). The incidence of sunspots, which are sites where strong toroidal magnetic fields emerge through the surface, varies aperiodically with an average period of 11 years; since the fields reverse, the magnetic cycle has a 22-yr period. This activity is modulated by grand minima, such as the Maunder Minimum in the 17th century (which coincided with the reign of the Roi Soleil in France). Proxy data, provided by cosmogenic isotopes such as $^{10}$Be and $^{14}$C whose abundance is modulated by magnetic fields in the solar wind, show that grand minima occur aperiodically but with a characteristic 210-yr periodicity (cf. Fig. 1*a*). Comparison with similar stars shows that their magnetic activity depends on the rate at which they rotate, which declines with age owing to magnetic braking. Slow rotators, like the Sun, exhibit cyclic activity, with evidence of grand minima.

In what follows, we first present a third-order model that describes modulation associated with grand minima, and then confirm that similar results can be obtained from PDEs describing a mean field dynamo. Then we turn to spatial structure and discuss modulation involving changes between dipolar and quadrupolar symmetries. Finally, we demonstrate that both types of modulation can be represented in a sixth-order system of nonlinear ODEs.

## 2. Grand Minima in a Third-Order Model

The observed patterns of activity in solar-type stars are consistent with the following bifurcation structure. As the relevant stability parameter, the dynamo number $D$ (which is a monotonically increasing function of the star's angular velocity $\Omega$), is increased, there is a transition from a trivial state with no large-scale magnetic field to one with periodic activity, at a Hopf bifurcation H1. Trajectories are then attracted to a limit cycle in the phase space of the system. After a second Hopf bifurcation H2, solutions are quasiperiodic, corresponding to periodically modulated cycles, and trajectories are attracted to a two-torus. Thereafter, a series of bifurcations lead to chaotically modulated cycles.

[12] [12] sought a minimal model that exhibits this bifurcation pattern. A third-order system is necessary for a two-torus and is also sufficient to allow chaotic behaviour. Generic properties can be guaranteed by choosing an appropriate normal form. Let the cartesian coordinates $x, y$ represent the toroidal and poloidal magnetic fields, respectively, while all the hydrodynamics is collapsed onto the invariant $v$-axis. We suppose that there are two fixed points on this axis, of which one is hydrodynamically stable but liable to magnetic instabilities, while the other is hydrodynamically unstable but

magnetically stable. These two points appear in a saddle-node bifurcation, followed by the Hopf bifurcation H1, which leads to cyclic dynamo action, so the system can be described by

$$\dot{z} = (\lambda + i\omega)\, z + avz + c|z|^2 v, \tag{1}$$
$$\dot{v} = \mu - v^2 - |z|^2 - bv^3 \ , \tag{2}$$

which are the normal form equations for a saddle-node/Hopf bifurcation [2] with the term $c|z|^2v$ added to allow for chaotic dynamics. Here the complex variable $z = x + iy$ represents the magnetic field and $v$ is real. The bifurcation parameters are $\lambda$ and $\mu$ and $a, b, c$ are constants; the other higher order terms are needed to eliminate degeneracies. Then there is a path in the $\lambda\mu$-parameter plane along which the primary Hopf bifurcation H1 is followed by a secondary Hopf bifurcation H2 (at $\lambda = 0$) which leads to periodically modulated cycles and then to chaotically modulated cycles as the path approaches a heteroclinic tangle [12]. Fig. 1*c* shows an example of activity in this parameter range. What this model demonstrates is that modulation associated with grand minima is likely to occur in any dynamo that generates oscillatory solutions. Indeed, similar behaviour can also be found for the simplified third-order system

$$\dot{z} = (\lambda + i\omega + v + cv^2)z + a|z|^2 z + b|z|^2 v \ , \tag{3}$$
$$\dot{v} = -v - |z|^2 \ , \tag{4}$$

in which the redundant hydrodynamically unstable fixed point is absent.

## 3. Partial Differential Equations

The third order model described above concentrates on the development of complicated temporal behaviour without reference to spatial structure of the magnetic fields. It is important to demonstrate that similar temporal behaviour can also be found in PDEs where spatial structure is included and to examine how the inclusion of this structure can lead to other types of modulation. Tobias (1996, 1997) considered a nonlinear mean field dynamo at the interface between the convective and radiative zones (cf. [8] [8]), where helioseismology has revealed strong radial gradients in $\Omega$ (the tachocline). This simplified model adopts two-dimensional cartesian geometry with the poles at $x = 0,\ 2L$ and the equator at $x = L$. The $\alpha$-effect is confined to the upper half-space ($0 < z < 1$: the convection zone) while differential rotation occurs in the lower half-space ($0 > z > -1$: the tachocline). Growth of the magnetic field is limited by the macrodynamic back-reaction of the Lorentz force on the differential rotation (the Malkus-Proctor effect). Then bifurcations from the trivial field-free solution give rise to two distinct families

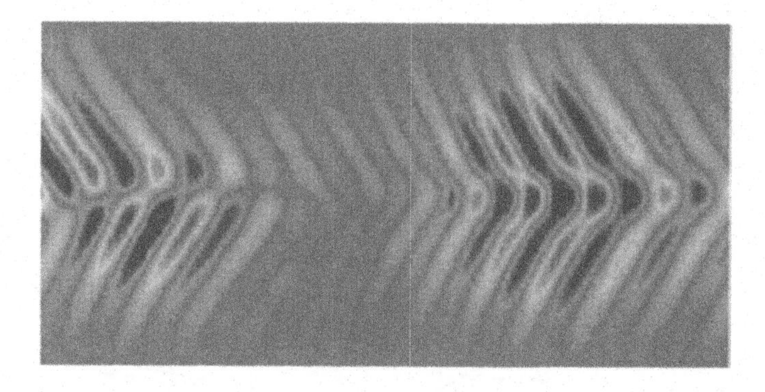

*Figure 2.* Butterfly diagram showing the toroidal field as a function of latitude and time for a PDE solution. The solution flips from dipole to quadrupole symmetry during the grand minimum.

of solutions with different symmetries: for *dipole* solutions the toroidal field $B$ is antisymmetric about the equator, while for *quadrupole* solutions $B$ is symmetric. The initial Hopf bifurcations are, as usual, very close. When the equations are restricted to the invariant dipole subspace the sequence of bifurcations described in Section 2 can indeed be found [10], and this is so for solutions of quadrupolar type as well. Moreover, for suitably chosen parameters there are aperiodically modulated solutions with grand minima, like that displayed in Fig. 1*b*.

Once the fixed parity constraint is relaxed, symmetry-breaking bifurcations lead to the appearance of branches of mixed-parity solutions, connecting the dipole and quadrupole branches [11], and a new type of modulation involving changes in the parity of the field without substantial changes in the level of activity (hereafter Type 1 modulation, to distinguish it from the Type 2 modulation discussed in Section 2) becomes possible. For example, one can obtain solutions with predominantly dipolar fields that nevertheless show hemispheric structure as they emerge from grand minima [1], much as occurred at the end of the Maunder Minimum, when sunspots were largely confined to one hemisphere of the Sun [9]. For similar parameter values more remarkable behaviour can be found. Fig. 2 shows a "butterfly" diagram of activity as a function of latitude and time, in which the solution switches from dipole to quadrupole symmetry as it emerges from a grand minimum. In Fig. 3*a* the corresponding attractor for this "flipping" behaviour is pro-

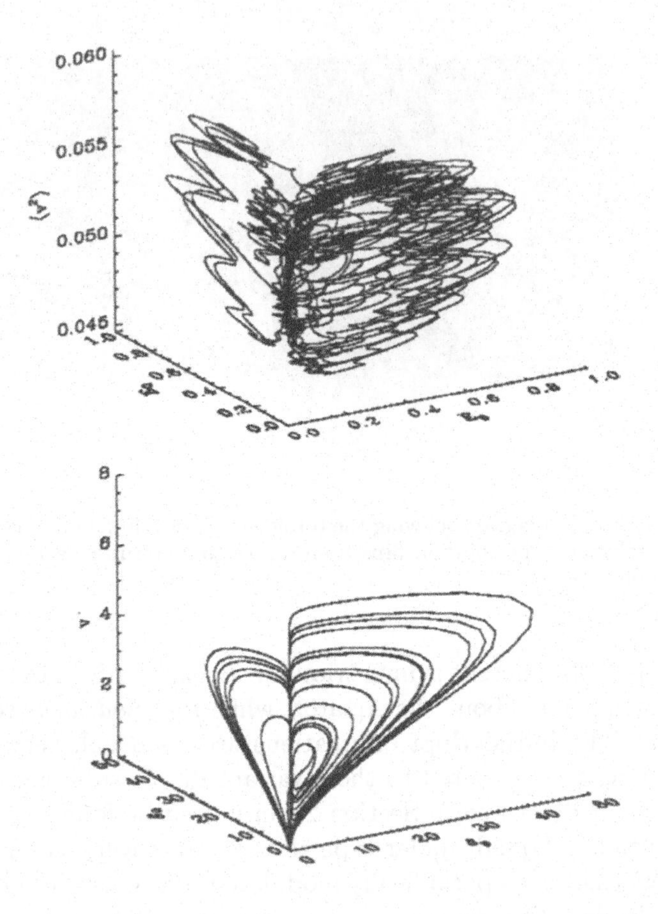


*Figure 3.* Three-dimensional phase portraits showing flipping for PDEs (upper panel) and for ODEs (lower panel). Trajectories are projected onto a space with co-ordinates corresponding to the dipole and quadrupole magnetic energies, and the kinetic energy of motion driven by the Lorentz force.

jected onto a three-dimensional phase space: the trajectory lies mainly in the dipole subspace but occasionally visits the quadrupole subspace [1]. These features can also be found in low-order models, as discussed below.

## 4. Interaction between dipole and quadrupole modes

As noted above the Sun's magnetic field remains largely dipolar. However, the parity of the solar field does not remain constant, and the field departs from dipole symmetry when the magnetic energy is small. The observations [9], together with results from PDE calculations, suggest that more detailed models of the sunspot cycle should represent the interaction between dipolar and quadrupolar fields. Here we describe a set of low-order models that include such an interaction and are capable of describing all the dynamics found in the PDEs, and so potentially the dynamics of magnetic fields in other solar-type stars.

Because both types of fields (dipole and quadrupole) appear in close succession and with similar frequencies it is appropriate to describe their interaction by the normal form for the Hopf bifurcation with 1:1 resonance. We let $z_{1,2}$ be the (complex) amplitudes of the dipolar and quadrupolar fields, respectively. The requirement that the model respect the symmetry of the Sun under reflection in the equator translates into the requirement that the equations for $z_{1,2}$ commute with the action $(z_1, z_2) \rightarrow (-z_1, z_2)$ (cf. Knobloch 1994). To third order the resulting normal form is

$$\dot{z}_1 = (\mu + \sigma + i\omega_1)\, z_1 + a\,|z_1|^2 z_1 + b\,|z_2|^2 z_1 + c\, z_2^2 \overline{z}_1, \tag{5}$$

$$\dot{z}_2 = (\mu + i\omega_2)\, z_2 + a'\,|z_2|^2 z_2 + b'\,|z_1|^2 z_2 + c'\, z_1^2 \overline{z}_2. \tag{6}$$

Here $\mu$ represents the dynamo number and $\sigma$ the splitting between the two modes. These equations were studied in detail by Knobloch and Landsberg (1996) and Landsberg and Knobloch (1996), and more recently by Moehlis and Knobloch (2000). Those authors identified solutions in the form of a pure dipole and a pure quadrupole, as well as periodic states of mixed parity. Of greater interest are states in which the amplitude of the basic dipole or quadrupole cycle also oscillates (i.e. the overall cycle is quasiperiodic), and the global bifurcations near which the system remains for a long time near a pure parity state before switching to a state of the opposite parity. This switching is an example of Type 1 modulation, and can be either periodic or aperiodic; it is often prominent near the 'extinction' limit, i.e., just before the dynamo number falls to values that would be insufficient to maintain a magnetic cycle as the star spins down.

The two types of modulation (Type 1 and Type 2) can be brought together in a single model in two apparently different ways. One could add additional equations to include the Malkus-Proctor dynamic nonlinearity in the dipole/quadrupole modulation model described above. Alternatively one could couple together (respecting the symmetries of the system) two versions of the third-order model of section 2. In fact both procedures lead to the same set of equations, viz.

$$\begin{aligned} \dot{z}_1 &= (\mu + \sigma + i\omega_1)\, z_1 + a\,|z_1|^2 z_1 + b\,|z_2|^2 z_1 + c\, z_2^2 \overline{z}_1, \\ &\quad +(\epsilon v + \delta v^2 + \kappa w^2)\, z_1 + (\beta + \gamma v)\, w z_2, \end{aligned} \tag{7}$$

$$\begin{aligned} \dot{z}_2 &= (\mu + i\omega_2)\, z_2 + a'\,|z_2|^2 z_2 + b'\,|z_1|^2 z_2 + c'\, z_1^2 \overline{z}_2 \\ &\quad +(\epsilon' v + \delta' v^2 + \kappa' w^2)\, z_2 + (\beta' + \gamma' v)\, w z_1, \end{aligned} \tag{8}$$

$$\dot{v} = -\tau_1 v + e_1\,(|z_1|^2 + |z_2|^2), \tag{9}$$

$$\dot{w} = -\tau_2 w + e_2\,(z_1 \overline{z}_2 + z_2 \overline{z}_1). \tag{10}$$

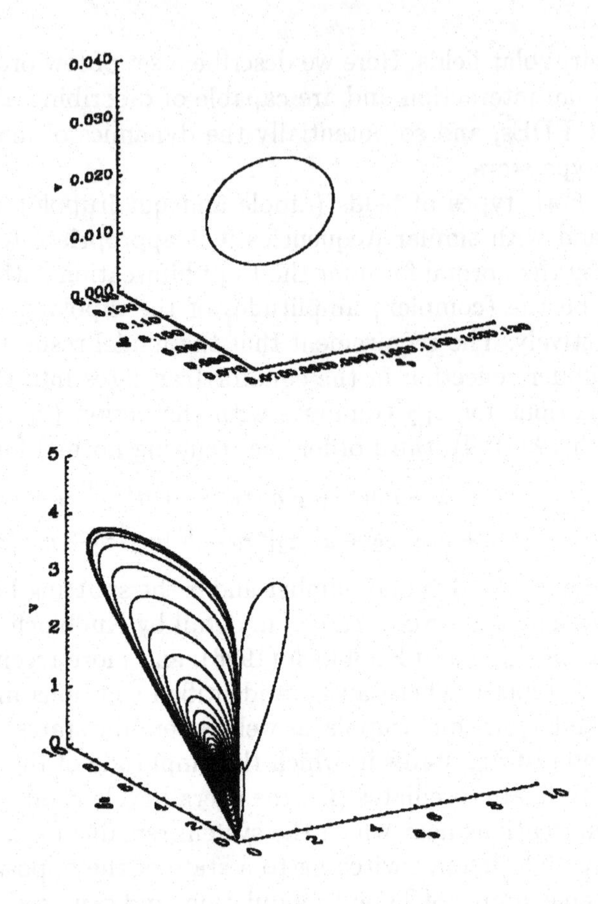


*Figure 4.* **Phase portraits showing different types of modulation for solutions of equations (7-10). Upper Panel: Type 1 modulation, with changes in parity but scarcely any variation of magnetic energy. Lower panel: Type 2 modulation with large changes in amplitude for a solution that is attracted to the invariant quadrupole subspace. Trajectories are projected onto a phase space with co-ordinates $\mathcal{E}_D$, $\mathcal{E}_Q$, $v$.**

Here $a$, $b$, $c$, $\beta$, $\gamma$, $\delta$, $\epsilon$, $\kappa$, $a'$, $b'$, $c'$, $\beta'$, $\gamma'$, $\delta'$, $\epsilon'$, $\kappa'$ are all complex coefficients whilst $\tau_1$, $\tau_2$, $e_1$ and $e_2$ are real. Full details of the derivation and the behaviour of the model can be found in Knobloch *et al.* (1998). The relationship of the above model to that of Knobloch & Landsberg (1996) is transparent, evolution equations for the symmetric and antisymmetric parts of the velocity perturbation driven by the Lorentz force ($v$ and $w$, respectively) having been added to those for the dipole and quadrupole fields. Moreover the connection with the third order system can be made by observing that if $z_2 = w = 0$ the equations reduce to (3,4), with $b = 0$.

The sixth-order system described by equations (7-10) thus has three important subsystems: the pure dipole system (3,4) studied by Tobias *et al.* (1995), together with a corresponding pure quadrupole system, and the system (5,6) describing the interaction of dipole and quadrupole fields in the absence of the Malkus-Proctor effect ($e_1 = e_2 = 0$), which was investigated by Knobloch and Landsberg (1996). Consequently, the sixth order system (7-10) is capable of describing nearly all the dynamics of both the earlier models. Solutions are best represented by projecting trajectories onto the three-dimensional phase space with coordinates $\mathcal{E}_{\rm D} = |z_1|^2$, $\mathcal{E}_{\rm Q} = |z_2|^2$, representing the energies in the dipole and quadrupole fields, respectively, and $v$. The phase portrait in Fig. 4*a* shows an example of Type 1 modulation: the original trajectory lies on a two-torus, which reduces to a limit cycle, since this projection has the effect of filtering out the basic cycle and revealing only the modulation. Type 2 modulation is illustrated in Fig. 4*b*, for a quasiperiodic solution that is attracted to the invariant quadrupole subspace. Note that the simplification mentioned above prevents the modulation from being chaotic within the dipole and quadrupole subspaces; we rely here on the interaction between the modes to produce chaotic behaviour.

This model is, moreover, capable of describing the interaction between the two types of modulation found in the PDEs, including the "flipping" between dipole and quadrupole solutions as the solution emerges from grand minima. Fig. 3(b) shows the projection of a trajectory obtained from the model (7-10) that exhibits the type of "flipping" behaviour shown in Figs 2 and 3(a) and obtained from the PDE model. The basic features of the PDE solution are reproduced extremely well.

## 5. Discussion

In this paper we have discussed how progress can be made in understanding the complicated temporal behaviour found in solar and stellar dynamos by utilising low-order normal-form equations that respect the underlying symmetries of the star and utilise the correct form of the nonlinearity. The equations derived are capable of reproducing even the most complicated qualitative behaviour of mean-field PDE models of solar and stellar activity. They suggest that this behaviour is not model-dependent and should therefore be found in most models of stellar activity and indeed in the stars themselves. Given the uncertainties in dynamo models that rely solely on mean-field theory, the approach summarized above appears to provide a promising alternative for understanding the variety of dynamo activity in the Sun and other stars.

## References

1. Beer, J., Tobias, S.M. and Weiss, N.O. (1998) An active Sun throughout the Maunder Minimum, *Solar Phys.*, **181**, pp. 237–249.
2. Guckenheimer, J. and Holmes, P. (1986) *Nonlinear oscillations, dynamical systems and bifurcations of vector fields*, 2nd printing, Springer, New York.
3. Knobloch, E. (1994) Bifurcations in rotating systems, in *Lectures on Solar and Planetary Dynamos*, M.R.E. Proctor and A.D. Gilbert (eds), Cambridge University Press, Cambridge, pp. 331–372.
4. Knobloch, E. and Landsberg, A.S. (1996) A new model of the solar cycle, *Mon. Not. R. Astron. Soc.*, **278**, pp. 294–302.
5. Knobloch, E., Tobias, S.M. and Weiss, N.O. (1998) Modulation and symmetry changes in stellar dynamos, *Mon. Not. R. Astron. Soc.*, **297**, pp. 1123–1138.
6. Landsberg, A.S and Knobloch, E. (1996) Oscillatory bifurcation with broken translation symmetry. *Phys. Rev. E* **53**, pp. 3579–3600.
7. Moehlis, J. and Knobloch, E. (2000) Bursts in oscillatory systems with broken $D_4$ symmetry, *Physica D*, **135**, pp. 263–304.
8. Parker, E.N. (1993) A solar dynamo surface-wave at the interface between convection and non-uniform rotation, *Astrophys. J.*, **408**, pp. 707–719.
9. Ribes, J.C. and Nesme-Ribes, E. (1993) The solar sunspot cycle in the Maunder minimum AD 1645 to AD 1715. *Astron. Astrophys.*, **276**, pp. 549–563.
10. Tobias, S.M. (1996) Grand Minima in nonlinear dynamos, *Astron. Astrophys.*, **307**, pp. L21–L24.
11. Tobias, S.M. (1997) The solar cycle: parity interactions and amplitude modulation, *Astron. Astrophys.*, **322**, pp. 1007–1017.
12. Tobias, S.M., Weiss, N.O. and Kirk, V. (1995) Chaotically modulated stellar dynamos, *Mon. Not. R. Astron. Soc.*, **273**, pp. 1150–1166.
13. Weiss, N.O. (1993) Bifurcations and symmetry-breaking in simple models of nonlinear dynamos, in *The Cosmic Dynamo*, F. Krause, K.-H. Rädler and G. Rüdiger (eds), Kluwer, Dordrecht, pp. 219–229.
14. Weiss, N.O. (1994) Solar and stellar dynamos, in *Lectures on Solar and Planetary Dynamos*, M.R.E. Proctor and A.D. Gilbert (eds), Cambridge University Press, Cambridge, pp. 59–95.
15. Weiss, N.O. and Tobias, S.M. (2000) Physical causes of solar activity, *Space Sci. Rev.*, in press.